THE ETHICS OF DIET

THE ETHICS OF DIET

A Catena

of Authorities Deprecatory

of the Practice of Flesh-Eating

HOWARD WILLIAMS

Introduction by Carol J. Adams

UNIVERSITY OF ILLINOIS PRESS

Urbana and Chicago

Library of Congress Cataloging-in-Publication Data
Williams, Howard, 1837–1931.
The ethics of diet : a catena of authorities deprecatory of the practice
of flesh-eating / Howard Williams ; introduction by Carol J. Adams.
p.   cm.
Originally published: London : F. Pitman, 1883.
Includes bibliographical references and index.
ISBN 0-252-02851-1 (cloth : alk. paper)
ISBN 0-252-07130–1 (pbk. : alk. paper)
1. Vegetarianism. 2. Diet. I. Adams, Carol J. II. Title.
TX392.w63   2003
613.2'62—dc21   2002073288

Man by Nature was never made to be a carnivorous animal,
nor is he armed at all for prey and rapine.
—*Ray*

Hommes, soyez *humains*! c'est votre premier devoir.
Quelle sagesse y-a-t-il pour vous hors de l'humanité?
—*Rousseau*

Der Mensch ist was er isst.
—*German proverb*

CONTENTS.

APPENDIX.

APPENDIX TO THE
ILLINOIS EDITION.

INTRODUCTION TO
THE ILLINOIS EDITION

Challenging Kreophagy and
"Meat-and-Potatoes" History

———

CAROL J. ADAMS

[I]f it is true that all thought begins with remembrance, it is also true that
no remembrance remains secure until it is condensed and distilled into a
framework of conceptual notions within which it can further exercise itself. . . .
What saves the affairs of mortal men from their inherent futility is nothing
but this incessant talk about them, which in its turn remains futile unless
certain concepts, certain guideposts for future remembrance,
and even for sheer reference, arise out of it.
—Hannah Arendt, *On Revolution*

The average kreophagist is by no means convinced that kreophagy is
the perfect way in diet.
—*The Saturday Review*, 1881

No matter how much cachet social history may have from time to time,
politics and the men who write about it are the "meat and potatoes"
of great history.
—Bonnie G. Smith, *The Gender of History*

IN 1877 or thereabouts, Howard Williams was perhaps sitting around in London with other like-minded humanitarians discussing the odd reception to vegetarianism that they encountered. Vegetarians were being assailed with questions such as "Do vegetarians eat meat *by night?*" and "What on earth do you
live on?" They were constantly being drawn into arguments with flesh eaters
who attempted to defend the practice of eating nonhumans. Vegetarians would
be told, "The animals were 'sent' to humans for food." Or it would be demanded of them, "What would otherwise become of the animals?"[1] Vegetarianism

———

1. For these specific questions, asked of Henry Salt around this time, see Henry S. Salt,
Seventy Years among Savages (London: George Allen and Unwin, 1921), pp. 63, 68, 71.

was accused of being (merely) sentimental, if not "womanish." To some, including Howard Williams, the most galling attack came when vegetarianism was called a fad.

These flesh eaters, "kreophagists" in his words, and their conversational dynamics infuriated Williams.[2] "All they offer are popular fallacies, specious objections, and subterfuges!" Williams bemoaned. "Their arguments are unthinking and ignorant." Williams would have been tired of them by now, after five years of a nonflesh diet. He continued: "Why, throughout history there are records of individuals who have provided answers to these puerile attempts at defending flesh eating." Then thundering with impatience he focused on flesh eaters' historical illiteracy: "And those blokes who call vegetarianism faddish just aren't aware of the history of ethical vegetarianism! People have always challenged the Regime of Blood!"

"Well, man," one associate responded, recognizing both the indignant reformer and the careful scholar in Williams's exclamation. "What are you—you with your Fourth in History from St. John's College, Cambridge—going to do about it?"

"We are in the midst of a Reformation in Dietetics!" Williams announced, leaping from his chair, "And I'll write its history!"

And so Howard Williams, M.A., began the work that reached fruition several years later in *The Ethics of Diet.*

It makes a good story, and Williams, in the following pages, reveals just how much this topic animated him, but I do not know if such a conversation actually happened in this precise way. What we do know is that articles by Williams that became *The Ethics of Diet* initially appeared in the *Dietetic Reformer* and then were published as a book in 1883. The volume was reissued in a second edition in 1896, and in 1907 an abridged popular version appeared.

We know that Gandhi read it.[3] Tolstoy read it, too, and determined to visit a slaughterhouse as a result of his encounter with it. In fact, Tolstoy provided

2. For the problem of flesh eaters' argumentativeness, see Williams, in this volume, pp. 338–44. For an analysis of twenty-first-century conversational dynamics between vegetarians and flesh eaters, and the concomitant evidence that some things never change, see the chapters entitled "Talking with Meat Eaters" and "Stopping the Conversation" in Carol J. Adams, *Living among Meat Eaters* (New York: Three Rivers Press, 2001).

3. Gandhi wrote, "My faith in vegetarianism grew on me from day to day. Salt's book whetted my appetite for dietetic studies. I went in for all books available on vegetarianism and read them. One of these, Howard Williams' *The Ethics of Diet,* was 'biographical history of the literature of humane dietetics from the earliest period to the present day.' It tried to make out, that all philosophers and prophets from Pythagoras and Jesus down to those of the present age were vegetarians." Mohandas K. Gandhi, *An Autobiography: The Story of My Experiments with Truth* (Boston: Beacon Press, 1957, 1972), p. 48.

a preface to *The Ethics of Diet*—"that excellent book" as he called it[4]—when it was translated in 1892 into Russian. His essay "The First Step" is that preface and describes in precise detail that slaughterhouse visit.

The Ethics of Diet had a profound impact on the reformer Henry Salt, who called it a book of "rare merit."[5] Indeed, one Salt biographer credits it with helping to change Salt's life.[6] Salt, then an Eton master, describes his changing consciousness about the ethics of eating nonhumans, of how gradually "the conviction had been forced on me that we Eton masters, however irreproachable our surroundings, were but cannibals in cap and gown—almost literally cannibals, as devouring the flesh and blood of the higher non-human animals so closely akin to us, and indirectly cannibals, as living by the sweat and toil of the classes who do the hard work of the world."[7] Williams was one of those influences that brought about this conviction. Williams hoped his book would have a similar effect, not just on a few literate people like Salt, but on many.

Over the past 120 years, countless others have read this book and been moved by it. But as copies became harder to acquire, many people made pilgrimages to libraries to read the book; others were happy to get a photocopy of the volume to pore over in their own homes. Now, a truly joyful publishing event has occurred: Williams is restored to print, and all readers have the opportunity to encounter his life-changing work.

In the intervening 120 years some new arguments on behalf of vegetarianism have appeared, prompted by the emergence of factory farming and its horrific treatment of nonhumans. The environmental consequences of flesh eating, alluded to in Plato's *Republic,* have become more apparent. In addition, we now have critical theory that proposes the interlocking nature of the oppression of women, people of color, and nonhumans. Yet many of the basic ethical arguments found in this book are as timely today as they were when they were first written. This is one reason to welcome this edition. It shows that thoughtful engagement with the issue of eating nonhuman animals is not a recent event.

Vegetarians of the twenty-first century well acquainted with the various arguments presented in defense of flesh eating will find it interesting that many of these arguments were the common defenses of flesh eaters in the nineteenth century as well. Williams answered these contentions in a way he knew best—

---

4. Leo Tolstoy, "The First Step," in *Leo Tolstoy: Selected Essays,* translated by Aylmer Maude, selected by Ernest J. Simmons (New York: Random House, 1964), p. 232.

5. Salt, *Seventy Years,* p. 63.

6. George Hendrick, *Henry Salt: Humanitarian Reformer and Man of Letters* (Urbana: University of Illinois Press, 1977), p. 56.

7. Salt, *Seventy Years,* p. 64.

by using the historical survey. One may experience a sense of sadness as one realizes that these flesh-eating defenses had been asked and answered so many years ago. But one may experience a thrill, too. Something changed in 1883. With Williams's book came proof of a tradition that countered flesh eating. Williams anticipated and answered every tedious flesh-eater defense. His book represented the type of work evoked by Hannah Arendt, when she observed, "If it is true that all thought begins with remembrance, it is also true that no remembrance remains secure until it is condensed and distilled into a framework of conceptual notions within which it can further exercise itself."[8] Indeed, Williams's effort not only put vegetarianism in historical perspective but also placed flesh-eating defenses in historical perspective, too.

Threading a Tradition

Before a theory of history or a critique of a historical approach can exist, there must be a sense that there is a history to critique or theorize. This has been the traditional problem for vegetarianism in the West—no sense of continuity. The opprobrium of the label "faddishness" suggests that the practice is greeted as though it were an adolescent eruption against authority. Within that label rests the assumption that once maturity is achieved it will fade with time, it has no lasting power. This prejudice is perhaps one reason why the vegetarians who appear in the following pages refer to how many years they have been vegetarian. They are testifying to the long-lasting nature of vegetarianism.

Faddishness as a label is, of course, ironic, since similar arguments against flesh-eating flower throughout the ages. They incorporate new information and changes in the treatment of nonhuman animals, but the basic argument is in Ovid's depiction of Pythagoras and in Plutarch's essays on flesh eating and carries forward from there. Is vegetarianism, something that recurs throughout time, actually a fad, that is, short-lived, or is it only imputed to be a fad so that its disruption of the ordered world is minimized?

Repeatedly, vegetarian writers cite their predecessors. This was an implicit way of demonstrating that vegetarians have a history—the citations prove it so. Howard Williams took this basic vegetarian methodology to the next logical step: his "catena of authorities deprecatory of the practice of flesh-eating."

By the time the second edition was published in 1896, Williams had changed the subtitle to a "biographical history of the literature of humane dietetics from the earliest period to the present day." Maybe he had been told that "catena" was not a helpful term. But let us consider that word. If we use

---

8. Hannah Arendt, *On Revolution* (New York: Viking Press, 1963, 1965), p. 222.

e-mail frequently, we probably understand the mechanism of a catena—it
is threading.

A catena: in one, precisely chosen word, Williams answered those who dismissed vegetarianism as a fad. According to the *Oxford English Dictionary*,
a catena is "a chronological series of extracts to prove the existence of a continuous tradition." The existence of a catena of authorities demonstrates that
vegetarianism is not a fad at all but its opposite, a continuous and continuing ethical position followed throughout centuries of writings and actions by
thoughtful people.

The Ethics of Diet is the benchmark against which everything that has come
after must be measured. For vegetarian historians this is the Ur-text, although
we may argue with Williams here and there on some points. I know I did when
I first encountered his table of contents and found no women in the first edition.

In initially calling his authorities "Deprecatory" Williams decides that he
doesn't have to determine definitively whether these writers were vegetarian. The question was simply, Were they writing against flesh eating?

In the second edition, he threaded in some additional authorities, for instance, Henry David Thoreau as well as Dr. Anna Kingsford. He saw Thoreau
differently from Henry Salt, and Salt (a biographer of Thoreau) might not have
agreed with Williams's analysis. If we are classicists, or early modern historians, or Shelley specialists, we might have points to quibble about with Williams when he discusses "authorities" from our own specialized field. Vegans
may be disappointed that Williams failed to challenge the use of female animals for dairy and egg production, though he endorsed the use of nonanimal
products instead of leather.

But with Williams, *catena* was and is the operative word. The whole is greater than the sum of its parts. In the century before Williams, there had been
John Oswald, and Joseph Ritson, and John Frank Newton and Percy Shelley,
who wrote specific anti-kreophagist tracts. With Williams's opus, we have
something new, a *tradition;* in Hannah Arendt's words, we have a "guidepost"
for "future remembrance, and even for sheer reference."

Biblio-Imperialism and Vegetarianism

The nineteenth century—particularly the last decades, when Williams was
writing, revising, expanding, and amending *The Ethics of Diet*—was an encyclopedic time, a time of gathering and classifying and defining. It was an era
when the British Empire found its confirmation in various elaborate and definitive publications, a sort of biblio-imperialism that mirrored its political impe-

rialism. The fruits of this imperialist enthusiasm, found in the publication of *The Dictionary of National Biography* (*DNB*) and the *Oxford English Dictionary*, are exhibited in Williams's project, too. As the vegetarian historian Rynn Berry comments about the British, "Williams epitomized the colonial mindset that they had exhaustive knowledge and they could write the definitive work on any subject."[9]

Yet Williams was using the encyclopedic methodology of his time to stand against, rather than with, his culture. Consider the state of vegetarianism at that time. The word "vegetarian" had been coined only some thirty years earlier. Before that you might be called any number of names, such as "Pythagorean." People laughed at your diet. Regarding the vegetarian Sir Richard Phillips, *The Dictionary of National Biography* tells us that "Tom Moore considered him a bore, and laughed at his 'Pythagorean Diet.'"

Indeed, *The Dictionary of National Biography,* begun in 1882—the year before the publication of *The Ethics of Diet*—often, like old Tom Moore, betrays its culture's attitude toward vegetarians. It demonstrates what Williams was up against. For instance, discussing Joseph Ritson's vegetarianism the *DNB* reports that "to this depressing diet he adhered, in the face of much ridicule until death, and it was doubtless in part responsible for the moroseness of his temper which characterized his latter years." Further, the *DNB* offers a pseudo-medical analysis of Joseph Ritson's *Essay on Abstinence from Animal Food.* We are told it bears marks of "incipient insanity."[10] Regarding another notable discussed in the *DNB*, William Lambe, we are told by the carnivore-centric author of his biography that "Lambe was accounted an eccentric by his contemporaries, mainly on the ground that he was a strict, though by no means fanatical, vegetarian." John Oswald, yet another vegetarian, fares little better. "Repugnance to flesh" is one of several "curious beliefs" he is credited with having. Each of these writers—Phillips, Ritson, Lambe, and Oswald—is correctly rehabilitated by Williams.[11]

---

9. Rynn Berry, conversation with the author at Vegetarian Summerfest, Johnstown, Pennsylvania, July 2001.

10. Keith Thomas refers to the author of this *DNB* entry as "very hostile" (Keith Thomas, *Man and the Natural World* [New York: Pantheon, 1983], p. 401 n. 49). I have pointed out the contradiction that Ritson's penultimate book, his vegetarian one, is derided for exhibiting signs of the author's insanity, yet his final work is regarded as the work of a mature scholar that "reveals the author as a scholar of no ordinary attainments." See my discussion in my book *The Sexual Politics of Meat: A Feminist-Vegetarian Critical Theory, Tenth Anniversary Edition* (New York: Continuum, 2000), p. 116.

11. The quotations from the *Dictionary of National Biography: From the Earliest Times to 1900* in this and the previous paragraph come from the original edition, edited by Sir Leslie Stephen and Sir Sidney Lee, reprinted by Oxford University Press, 1921–22, 1937–38, 1949–50, 1959–60, 1963–64, s.v. "Phillips, Sir Richard," "Ritson, Joseph," "Lambe, William," and "Oswald, John."

In several ways Williams undercut the hegemony of the imperialist discourse. The British "beefeater," who was proudly subduing the "vegetable eaters" of the world, was implicitly scrutinized in these pages.[12] In the introduction to the 1896 edition, Williams reminds his English-speaking audience of "the incontestable fact of resistance on the part of the *least carnivorous* people of the globe, during so many ages, to the onslaughts of one of the *most carnivorous* races, so often with success." When one challenged flesh eating, cultural divisions were challenged as well. Cultures could not be so neatly described as either "civilized" (flesh eating) or "primitive" (vegetable eating). Further, with a work such as Williams's, the biblical justification for flesh eating was challenged. If the biblical basis for flesh eating could be questioned, what other biblically claimed rights could be destabilized?

The hurdle ethical vegetarians faced then was the influence of a biblical view of history. Genesis 1:29 suggests a vegan diet in paradise; Isaiah's eschatological vision of the perfected society, the peaceable kingdom, reinstates vegetarianism. The problem is the in-between time—historical time. Samuel Dresner's discussion of Jewish dietary laws suggests how this positioning of vegetarianism influences one's understanding of history: "At the 'beginning' and at the 'end' man is, thus, in his ideal state, herbivorous. His life is not maintained at the expense of the life of the beast. In 'history' which takes place here and now, and in which man, with all his frailties and relativities, lives and works out his destiny, he may be carnivorous."[13] In other words, what is interposed between Genesis 1:29 and Isaiah's vision is human history. If vegetarianism is out of time, then we mortals need not concern ourselves with it. Idealizing or spiritualizing the practice of vegetarianism makes it ahistorical, before and after, but never "now," not only without a history but outside of history. This is the challenge Howard Williams faced: to reinstate vegetarianism as an ethical imperative *within* history by giving it a history. To establish a vegetarian history, Williams would confront not only kreophagy and the imperialist uplifting of the British beefeater but also the conventional, triumphant, historical method.

When the feminist historian Bonnie G. Smith refers to the "meat and potatoes" of history she means the topics that were deemed important by the emerging class of professional historians (political history, matters of the nation-state, and the triumph of great leaders) rather than the issues that interested "amateur historians."[14] Amateur history was concerned with the domes-

---

12. See my discussion of the racial politics of meat in the writings of nineteenth-century medical writers such as George Beard, in *The Sexual Politics of Meat,* pp. 40–41.

13. Samuel H. Dresner, *The Jewish Dietary Laws: Their Meaning for Our Time,* rev. and expanded ed. (New York: Rabbinical Assembly of America, 1982), p. 24.

14. Bonnie G. Smith, *The Gender of History: Men, Women, and Historical Practice* (Cambridge, Mass.: Harvard University Press, 1998), p. 71.

tic, the trivial, the social. This history had connotations of the feminine; for many different reasons, vegetarianism carried those connotations as well. During the latter part of the nineteenth century, surveys in England of the traditional foods eaten by men and women found that it was assumed that men "needed" meat and women did not. The first national food survey of British dietary habits, in 1863, revealed that the major differences in the diet of men and women in the same family lay in the amount of meat consumed.[15] Where poverty forced a conscious distribution of meat, men received it. A 1913 survey confirmed this practice among the poor: "In the household which spends 10s or even less on food, only one kind of diet is possible, and that is the man's diet." The husband consumed meat almost daily; the wife and her children, once a week.[16]

Challenging flesh eating required a distinctive approach to traditional male history as it was being codified at this time. Analyzing the traditions associated with men, who also happened to be the ones associated more closely with triumphant flesh eating, Howard Williams was able to restore a vibrant tradition of those who refused the privilege their sex and class provided them—the right to eat nonhumans. Williams challenged not just a meat-and-potatoes diet but also a meat-and-potatoes history.

The Embodied versus the Cartesian Historian

Good history animates the dead, bringing to life the actions and experiences of those whose lives are now inaccessible. Ethical vegetarian theory performs a similar function. For ethical vegetarians, dead flesh on a table is not devoid of specificity. It is the dead flesh of what was once a living, feeling being. When humans turn a nonhuman into "meat," someone who has a very particular, situated life—a unique being—is converted into something that has no distinctiveness, no uniqueness, no individuality. Thus, behind every meal of butchered flesh is an absence: the death of the nonhuman whose place the product takes. Once nonhumans have been transformed into food, they conceptually disappear from the act of eating flesh. Through flesh eating, therefore, nonhumans are deprived both of a biography and a history.[17]

---

15. Edward Smith, *Practical Dietary for Families, Schools and the Labouring Class* (London: Walton and Maberly, 1864), p. 199.

16. Maud Pember Reeves, *Round About a Pound a Week* (G. Bell and Sons, 1913; London: Virago Press, 1979), pp. 97, 144 (quote).

17. On nonhumans' having a biography, see Tom Regan's *The Case for Animal Rights* (Berkeley: University of California Press, 1983). Of interest as well is Barbara Noske's proposal for an anthropology of nonhumans. See her *Beyond Boundaries: Humans and Animals* (Montreal: Black Rose Books, 1997).

Plutarch advises that it is better to ask who first started eating flesh than who of late has left it off. In Williams's text there are many answers to Plutarch's prompting. One theoretical understanding that evolved in the late twentieth century places flesh eating, especially as practiced in the Western world, within a dualistic and hierarchical framework. It is within this same oppressive framework that the idea of what is historically valuable has evolved as well. Dualistic thinking that conceptualized humans both as "not animals" and as "better than animals" colored the ethical practices and historical practices of Williams's day. Dualistic thinking privileged the "mind" and "reason" and identified them as "male." It denigrated the "body" and "emotion" and identified them as "female."

As feminist philosophers have pointed out, the Western definition of the man of reason cast him as one who could overcome body, history, and social situations and thereby gain knowledge of others whom he examined as objects. Reason is defined as the antithesis of what is thought to be female. "[T]he feminine has been associated with what rational knowledge transcends, dominates or simply leaves behind"—emotions, feelings, and the body.[18] This Cartesian legacy prevailed throughout the nineteenth century: the mind, or "reason," was constructed to exclude nature, so nonhuman animals, a part of nature, were constructed as mindless. Such radical exclusiveness keeps nonhuman animals as oppositional and alien to humans. Like the "animal"—cut off from any association with the "human"—the animal-like, especially the body (and particularly the body of the professional historian), is disowned. The body, identified with nonhuman animals, is what must be transcended. Consequently, emotions, associated as they are with the body, have been denigrated as untrustworthy and unreliable, as invalid sources of knowledge.

The lower ontological status of the body influenced the idea of the professional historian. The traditional historical method ostensibly involved a rational, objective historian imposing an interpretation on thoughts or events. Complicating a historical approach that denies the body is an ethical approach that also does so; often with vegetarianism the mind justifies eating the bodies of other beings while disempowering any emotional response to the knowledge of butchering and loss of life.

Being concerned with the human practice of kreophagy was thought to be emotional; because the emotional was deemed unreliable, vegetarianism was rejected along gendered terms for several reasons: its concern for nonhumans instead of humans, its ostensibly emotional nature, and its valuation of the body

---

18. Genevieve Lloyd, *The Man of Reason: "Male" and "Female" in Western Philosophy* (Minneapolis: University of Minnesota Press, 1984), p. 2.

over the mind or spirit. So the response of the traditional or professional historian was that vegetarianism was trivial, too involved with the daily and the body, a diet for the conquered not the conquerer. Nonhuman animals and vegetarians were both unvalued.

Bonnie Smith describes how amateur history (the history associated with women) consists of "the writing of multiple traumas."[19] She identifies the traumas that women historians of the early nineteenth century would have experienced. These women were aware that their rights were eroding at a time when universal rights were being (supposedly) championed. They or family members had survived revolutions and wars, and at a personal level, they had experienced the threat or actuality of rape, poverty, violence, and abuse. Smith explains that "death, representing history's immanence, was always on hand."[20] The death of nonhumans, representing a violation of the nonhuman and a violation of the humane desire for the good and the just, is always on hand for vegetarians. The knowledge that other animals are being butchered to feed humans, even though other foods were available that required no such butchery, is also a form of traumatic knowledge.

Sir Richard Phillips traced his vegetarianism to his experience at twelve years of age, when he "was struck with such horror in accidentally seeing the barbarities of a London slaughter-house, that since that hour he has never eaten anything but vegetables" (see the text, p. 243). John Oswald would take long detours to avoid passing slaughterhouses and butcher shops. Even if one avoided passing the slaughterhouse, the existence of butchery was everywhere visible to the vegetarian, encountered at restaurants and over meals with flesh eaters.

The gratuitous nature of butchery heightens the trauma. Repeatedly in the following pages we encounter the perspective that killing animals is violent *and* unnecessary because of the abundance of vegetable food. We find this perspective in Ovid's description of the argument Pythagoras offered for vegetarianism: "The lavish Earth heaps up her riches and her gentle foods, and offers you dainties without blood and without slaughter" (see the text, p. 26).

Traumatic experiences, re-encountered regularly, were, and are, an inescapable factor in the vegetarian writer's life. Williams cannot leave this out of his text. In the following pages, we encounter not only arguments against flesh eating, but exasperation, anger, frustration. The boorish flesh eater (alluded to in the opening of this introduction, as well as in Williams's preface) reintroduces reminders of the trauma of butchering by his need to reiterate a defense of his right to eat flesh.

---

19. Smith, *The Gender of History*, p. 38.
20. Ibid., p. 21.

The vegetarian writer, at least when discussing his or her experience of vegetarianism, is not writing from an objectified, supposedly objective, disembodied, mind-privileging state, but from the experience of literally embodying vegetarianism. By testifying to its salutary effects, the writer addresses his or her body's experience of vegetarianism. Not only that, the writer examines "the structure of the human frame" (in Shelley's words [p. 226]) and concludes that humans do not possess a carnivorous body. Both the body, associated with the feminine or the female, and the diet, similarly associated with women, not men, are inserted into the text and so subvert the traditional approach to historical writing.

Even though *The Ethics of Diet* adheres somewhat to the tradition of focusing historical interest on elevated male worthies, in doing so it brings attention not only to the minds of these writers but to their embodiment of an ethical issue. Bonnie Smith proposes that "[t]hrough the physically present historian," for instance, an embodied writer like Germaine de Stael, "Cartesianism and the disembodied rights-bearing acquisitive individual as the fundament of either history or the nation" were brought into question.[21] Unlike the professional historian of his day, but like de Stael, Williams refuses to disappear from the text. He destabilizes the gender politics associated with the appearance of a class of professional historians by assuming this authorial position: he refuses to be invisible and he refuses to be calm. By publishing body-mediated writing that reclaimed animals' bodies, Williams undercuts Cartesianism on multiple levels.

Editions, Additions, and Subtractions: *The Evolving* Ethics of Diet

One cannot encounter this book without recognizing the energy and passion that Howard Williams put into his "catena." He collected, collated, interpreted, promulgated. Like his contemporaries, he was making history by cataloging history. At least the *Oxford English Dictionary* and the *Dictionary of National Biography* made history. But what happened to *The Ethics of Diet?* It was passed around from vegetarian to vegetarian; it was even translated into Russian and Swedish. As noted earlier, during Williams's lifetime *The Ethics of Diet* appeared in three different editions. The first edition—the one that had begun its life as a series of biographical articles—was published by F. Pitman and John Heywood in 1883. The second edition, in 1896, was published by Swan Sonnenschein and Company and the Ideal Publishing Company. For

---

21. Ibid., p. 29.

this revised and expanded 1896 edition, Williams corrected, added, and omitted. Because he limited his discussions to individuals who were dead, Anna
Kingsford, the lone woman included as an authority, makes her appearance
in the 1896 edition; in 1883, she was still alive. The final version, published
by Albert Broadbent and Richard J. James in 1907, shrank to a modest 242
pages from the original 332 at the recommendation, apparently, of the Vegetarian Society. Clearly, Williams's vegetarian friends believed a book of more
manageable size would help to get the word out.

Jon Gregerson, in *Vegetarianism: A History*, assesses the impact of *The
Ethics of Diet:* "Not unimportant in the momentum gathered by the Vegetarian Movement in late Victorian England was a book by one Howard Williams
entitled *The Ethics of Diet*."[22] Henry Salt, seemingly anticipating the neglect
the book would soon experience, claimed, "[I]f ethical societies were intended to be anything more than places of debate, they would long ago have included this work among their standard text-books."[23] But that is not what happened. As Gregerson observes, instead of becoming a standard textbook, "the
scarcity of this book today places it in the 'rare book' category."[24]

Out of three competing editions, how was it decided now to reprint the first
edition? I consulted vegetarian historians, vegetarian antiquarian book collectors, and a publisher who had overseen the republication of other earlier books
from the humane tradition. I told them I wanted to propose the reprinting of
The Ethics of Diet. They were all enthusiastic. And what edition should I recommend? Each of us had at least a copy (though perhaps only a photocopy of
a copy) of one edition, and a couple of us had copies of two editions. We began examining our copies, comparing the differences, and assessing the impact those differences had on the text.

It was easy to rule out the third edition. While the goal of having a popular-
ized presentation of *The Ethics of Diet* that might extend its circulation was a
laudable one, too much was lost in the editing. The 1907 version, while helpful, is inadequate to communicate the breadth of Williams's knowledge and
vision. At this point, we want to encounter Williams—footnotes and all—as
well as his "Catena of Authorities."

Eliminating the popular edition from consideration did not simplify the
task of selecting between the 1883 and 1896 editions. Both have their strengths,
and both have their weaknesses. The 1883 edition has 50 biographical entries, 17 appendixes, and a 4-page index; the 1896 edition has 58 biographical entries, fewer appendixes, and an expanded 37-page index. Williams

---

22. Jon Gregerson, *Vegetarianism: A History* (Fremont, Calif.: Jain, 1994), p. 78.
23. Salt, *Seventy Years,* p. 124.
24. Gregerson, *Vegetarianism,* p. 78.

pulled some of the people who appeared in the 1883 appendix into the 1896 text (specifically, Musonius, Lessio, Tryon, Hecquet, and Pressavin), while dropping others: Ray, Gay, Haller, Linne, Hawkesworth, and Graham. He added "M.D." where appropriate to the names in the table of contents, perhaps a sign of the increasing authority of medical professionals. He divided the 1896 table of contents into two parts, from Hesiod to Chrysostom and from Cornaro to Kingsford. (Only in the 1907 edition did he name these divisions: "Part I: Earlier Ethical Dietetics" and "Part II: Modern Ethical Dietetics.") Finally, to the two later editions Williams also added an epigraph from Tennyson: "Move upward, working out the beast, / And let the wolf and tiger die."

While the chance to remove some repetition, add names, and rearrange his entries must have made Williams happy with the 1896 revised and enlarged edition, his switch to a new publisher resulted in a product that was 597 pages long. That length might have seemed a little daunting to the average reader.

Still, an equally compelling case could be made for either edition. Tolstoy read the first edition; we should use that one. But Gandhi read the second edition; we should use this one. The second edition contains additional biographies—use that one. The second edition omits some biographies—don't use it. Finally, the question became not which edition would Williams want reprinted, but which edition has the most significance for us today?

I picked the first edition because this is the one that marks a new stage in vegetarian history—the stage that moves beyond futility, in Hannah Arendt's word, by offering "certain concepts, certain guideposts." This first edition was the initiating point of a "framework of conceptual notions." But besides reprinting the first edition to preserve its historic role as initiator, my advisers and I conceived the idea of adding a new appendix for this 2003 edition that would include the new material introduced in the 1896 edition. This would bring the path-breaking first edition forward to present-day readers while also respecting Williams's additional scholarly work.

So, we have the best of both worlds, the original text and Williams's additions.

Howard Williams, Narrator and Humanitarian

I wish I had met Williams, this preacher's son, this confident, seemingly tireless man. In these pages we encounter Williams's passion, his imagination, and his quarrel with the hegemonic flesh-eating culture, but we do not meet Williams the gardener, Williams the person who enjoyed canoeing. We can imagine this man producing the *Life and Letters of Swift and Pope* (1885),

Selected Dialogues of Lucian (1887, translated from the Greek), and *Pioneers for Humanity* (1907), the latter for the Humanitarian League. We can imagine Williams's suggesting to Henry Salt that they create an organization whose premise was "respect and kindness toward all sentient life" (this is how the Humanitarian League was founded in 1890). We can understand why Salt said of Williams that he is "one who by word and by pen, in private and in public, was a quiet but untiring champion of the humanitarian cause." Williams, he continues, "was in the truest sense a student and professor of *literae humaniores.*"[25]

No entry for this true professor of humane letters appears in *The Dictionary of National Biography* after his death in 1931. To those professional historians who compiled the *DNB,* Williams's landmark book must have been freighted with too many unacceptable ideas—the value of nonhumans, the body, and feelings; the repetitious traumatic figuration of the death of non-humans; the insinuation of the narrator. Yet, from our position in the twenty-first century, we can see how Williams's pioneering work anticipates the work of scholars and activist writers today.

Williams understood that those who have the power to name have the power to control the way in which reality is experienced. Joan Dunayer's recent innovative book *Animal Equality: Language and Liberation* provides a thesaurus for "Alternatives to Speciesist Terms."[26] Williams in his own way undermines speciesist language with correctives, too. In one place he inserts the word "other" in front of someone else's reference to animals, a reminder that nonhumans are not the only animals. Dunayer probably would disagree with Williams's choice of "lower" animals but agree with his use of "Fellow beings" or "great commonwealth of Living Beings," and "innocent sentient life, irrespective of nationality, creed, or species." Like Williams she would challenge a "'live stock' (as they complacently are termed)" (p. xxxii). Williams referred to the conveying of nonhuman animals across water for flesh-eating purposes as "the middle passage" (p. xxxii n.), anticipating the comparison Marjorie Spiegel made in 1988 in *The Dreaded Comparison: Animal and Human Slavery.*[27] In his reference to the vegetarianism with which Jesus was associated, including that of James, Peter, Matthew, the *Clementine Homilies,* and "the original, Ebionite, Church," he anticipated, too, important work such as Keith

---

25. Salt, *Seventy Years,* p. 124.

26. Joan Dunayer, *Animal Equality: Language and Liberation* (Derwood, Md.: Ryce, 2001), pp. 187–201.

27. Marjorie Spiegel, *The Dreaded Comparison: Human and Animal Slavery* (New York: Mirror Books, 1988, 1989); see especially "Transportation, or The Unbearable Journey," pp. 49–55.

Akers's *The Lost Religion of Jesus: Simple Living and Nonviolence in Early Christianity*.[28]

Williams could not shed completely some common assumptions of the nineteenth century. As readers, we will object to his references to Native Americans as "savage tribes" while realizing how strong the racism and imperialism of the time was when a humanitarian who was against flogging, the death penalty, and experimentation on animals, who wanted to expand rights for humans and nonhumans, can fall prey himself to mirroring the attitudes of the British beefeaters. And we prepare ourselves to recognize that some of his "authorities" in the following pages, as they meditate on the causes of human sacrifices and propose that cannibalism follows from flesh eating, drew upon prevailing colonialist views and inherited the legacy of racist misunderstandings of Africans and Aztecs.[29]

Williams was concerned not just with historical memory but with reforming the world by restoring that memory. By so doing, he would be creating a living memory with a purpose—to let nonhumans live.

The Challenge to Present-day Readers

Williams thought about us, his future readers, as he wrote. His preface gestures toward the future. He expected that "our more enlightened descendants (let us suppose, of the 2001st century of the Christian era)" might be surprised by how little flesh eating was critiqued before *The Ethics of Diet* appeared. Williams assumed that his enlightened descendants would no longer be the thoughtful few. He expected that the Dietetic Reformation would have truly happened. Instead, we descendants in the twenty-first century, still in the midst of a kreophagic world, are grateful that he gathered these ideas together.

Reformers are optimists; they believe in the humanity of their opponents and the ability of human beings to change. Would Williams's optimism have waned if he had known that in the century that intervened between his book and our new edition flesh production would evolve to a new stage of cruelty—farmed nonhumans in warehouses?

Even with the absence of women writers, Williams's book was invaluable to me as I wrote *The Sexual Politics of Meat: A Feminist-Vegetarian Critical*

---

28. See Keith Akers, *The Lost Religion of Jesus: Simple Living and Nonviolence in Early Christianity* (New York: Lantern Books, 2000). The phrase "the original, Ebionite, Church" is found on pages 128–29 of the 1896 edition of *The Ethics of Diet*.

29. For background on the colonialist influence on writers see *Critical Inquiry* 12 (1985), which includes Patrick Brantlinger's "Victorians and Africans: The Genealogy of the Myth of the Dark Continent" (pp. 166–203) and Mary Louise Pratt's "Scratches on the Face of the Country; or, What Mr. Barrow Saw in the Land of the Bushmen" (pp. 119–43).

Theory. It is a theoretician's goldmine, a historian's raw material, a vegetarian's ethical confirmation, and he hoped—and as Tolstoy and Salt and countless others demonstrated—the flesh eater's conscience-disturber. All during the time of the flowering of encyclopedic enthusiasms, Howard Williams directed his readers toward something both ethical and practical, something that countered the meat-and-potatoes diet of the average westerner and the average professional historian.

Now we can join Gandhi and Tolstoy and nameless others who encountered this vigorous and invigorating book. Welcome to a company of radicals who believed we could and should stop eating nonhuman animals. They brought vegetarianism out of history and into the here and now. Williams tried to ensure that this is where vegetarianism would stay: embodied in the present. That is his legacy and his challenge to future generations.

I deeply appreciate the support of the University of Illinois Press staff, especially Liz Dulany and Carol Betts, in ensuring that Williams's great work, *The Ethics of Diet,* is again available to the general public. In addition, a group of scholars and vegetarian writers has helped me think about Howard Williams, the various editions of *The Ethics of Diet,* and the ways of bringing it back into circulation. Thank you to Keith Akers, Rynn Berry, Andrew Rowan, Kim Stallwood, Berni Unti, and Jon Wynne-Tyson for their help in acquiring copies and for their comments and encouragement. Thanks especially to Dawn Letson and Glenda Kallman of the Woman's Collection at the library of Texas Women's University for providing a copy of the 1883 edition of *The Ethics of Diet* from which to produce the new edition; the staff of the University of Alabama Libraries for sharing a copy of the 1896 edition; and the staff of the Wellcome Library for the History and Understanding of Medicine, in London, for assistance with this project.

PREFACE.

AT the present day, in all parts of the civilised world, the once orthodox practices of cannibalism and human sacrifice universally are regarded with astonishment and horror. The history of human development in the past, and the slow but sure progressive movements in the present time, make it absolutely certain that, with the same astonishment and horror will the now prevailing habits of living by the slaughter and suffering of the inferior species—habits different in degree rather than in kind from the old-world barbarism—be regarded by an age more enlightened and more refined than ours. Of such certainty no one, whose *beau idéal* of civilisation is not a State crowded with jails, penitentiaries, reformatories, and asylums, and who does not measure Progress by the imposing but delusive standard of an ostentatious Materialism—by the statistics of commerce, by the amount of wealth accumulated in the hands of a small part of the community, by the increase of populations which are mainly recruited from the impoverished classes, by the number and popularity of churches and chapels, or even by the number of school buildings and lecture halls, or the number and variety of charitable institutions throughout the country—will pretend to have any reasonable doubt.

In searching the records of this nineteenth century—the minutes and proceedings of innumerable learned and scientific societies, especially those of Social and Sanitary Science Congresses—our more enlightened descendants (let us suppose, of the 2001st century of the Christian era), it is equally impossible to doubt, will observe with amazement that, amid all the immeasurable talking and writing upon social and moral science, there is discoverable little or no trace of serious inquiry in regard to a subject which the more thoughtful Few, in all times, have agreed in placing at the very foundation of all public or private well-being. Nor, probably, will the astonishment diminish when, further, it is found that, amid all the vast mass of theologico-religious

spite, that is to say, of the most cherished prejudices, traditions, and
sophisms of Education.

If we seek the historical origin of anti-kreophagist philosophy, it is to
the Pythagorean School, in the later development of the Platonic
philosophy especially, that the western world is indebted for the first
systematic enunciation of the principle, and inculcation of the practice,
of anti-materialistic living—the first historical protest against the *practical*
materialism of every-day eating and drinking.   How Christianity, which,
in its first origin, owes so much to, and was so deeply imbued with, on
the one hand, Essenian, and, on the other, Platonic principles, to the
incalculable loss of all the succeeding ages, has failed to propagate and
develope this true and vital spiritualism—in spite, too, of the convictions of some of its earliest and best exponents, an Origen or Clemens,
seems to be explained, in the first instance, by the hostility of the triumphant and orthodox Church to the "Gnostic" element which, in its various
shapes, long predominated in the Christian Faith, and which at one time
seemed destined to be the ruling sentiment in the Church ; and, secondly,
by the natural growth of materialistic principles and practice in proportion
to the growth of ecclesiastical wealth and power ; for, although the virtues
of "asceticism," derived from Essenism and Platonism, obtained a high
reputation in the orthodox Church, they were relegated and appropriated
to the ecclesiastical order (theoretically at least), or rather to certain
departments of it.

Such was what may be termed the sectarian cause of this fatal
abandonment of the more spiritual elements of the new Faith, operating in
conjunction with the corrupting influences of wealth and power. As regards
the *humanitarian* reason of anti-materialistic living, the failure and seeming
incapacity of Christianity to recognise this, the most significant of all the
underlying principles of reformation in Diet—the cause is not far to seek.
It lay, essentially, in the (theoretical) depreciation of, and contempt for,
present as compared with *future* existence. All the fatal consequence of
this theoretical teaching (which yet has had no extensive influence, even in
the way it might have been supposed to act beneficially), in regard to the
status and rights of the non-human species, has been well indicated by a
distinguished authority. "It should seem," writes Dr. Arnold, "as
if the primitive Christians, by laying so much stress upon a future
life, and placing the lower beings out of the pale of hope [of extended
existence], placed them at the same time out of the pale of sympathy,
and thus laid the foundation for this utter disregard of [other] animals

in the light of our fellow-beings. Their definition of *Virtue* was the same as that of Paley—that it was good performed for the sake of ensuring everlasting happiness ; which, of course, excluded all the [so-called] brute creatures."* Hence it comes about that Humanitarianism and, in particular, Humane Dietetics, finds no place whatever in the religionism or pseudo-philosophy of the whole of the ages distinguished as the *Mediæval*—that is to say, from about the fifth or sixth to the sixteenth century—and, in fact, there existed not only a negative indifferentism, but even a positive tendency towards the still further depreciation and debasement of the extra-human races, of which the great doctor of mediæval theology, St. Thomas Aquinas (in his famous *Summa Totius Theologiæ*—the standard text book of the orthodox church), is especially the exponent. After the revival of reason and learning in the sixteenth century, to Montaigne, who, following Plutarch and Porphyry, reasserted the rights of the non-human species in general ; and to Gassendi, who reasserted the right of innocent beings to life, in particular, among philosophers, belongs the supreme merit of being the first to dispel the long-dominant prejudices, ignorance, and selfishness of the common-place teachers of Morals and Religion. For orthodox Protestantism, in spite of its high-sounding name, so far at least as its theology is concerned, has done little in *protesting* against the infringement of the moral rights of the most helpless and the most harmless of all the members of the great commonwealth of Living Beings.

The principles of Dietary Reform are widely and deeply founded upon the teaching of (1) Comparative Anatomy and Physiology ; (2) Humaneness, in the two-fold meaning of Refinement of Living, and of what is commonly called "Humanity ;" (3) National Economy ; (4) Social Reform ; (5) Domestic and Individual Economy ; (6) Hygienic Philosophy, all of which are amply displayed in the following pages. Various minds are variously affected by the same arguments, and the force of each separate one will appear to be of different weight according to the special bias of the inquirer. The *accumulated* weight of all, for those who are able to form a calm and impartial judgment, cannot but cause the subject to appear one which demands and requires the most serious attention. To the present writer, the humanitarian argument appears to be of

---

* Quoted by Sir Arthur Helps in his *Animals and their Masters*. (Strahan, 1873.) The further just remark of Arnold upon this subject may here be quoted :—"Kind, loving, submissive, conscientious, much-enduring we know them to be ; but *because* we deprive them of all stake in the future—*because* they have no selfish, calculated aims—these are not virtues. Yet, if we say a 'vicious' Horse, why not say a 'virtuous' Horse?"

publications, periodical or other (supposing, indeed, any considerable proportion of them to survive to that age), no consciousness appeared to exist of the reality of such virtues as Humaneness and Universal Compassion, or of any obligation upon the writers to exhibit them to the serious consideration of the world : and this, notwithstanding the contemporary existence of a long-established association of humanitarian reformers who, though few in number, and not in the position of dignity and power which compels the attention of mankind, none the less by every means at their disposal—upon the platform and in the press, by pamphlets and treatises appealing at once to physical science, to reason, to conscience, to the authority of the most earnest thinkers, to the logic of facts—had been protesting against the cruel barbarisms, the criminal waste, and the demoralising influences of Butchery; and demonstrating by their own example, and by that of vast numbers of persons in the most different parts of the globe, the entire practicability of Humane Living.

When, further, it is revealed in the popular literature, as well as in the scientific books and journals of this nineteenth century, that the innocent victims of the luxurious gluttony of the richer classes in all communities, subjected as they were to every conceivable kind of brutal atrocity, were yet, by the science of the time, acknowledged, without controversy, to be beings essentially of the same physical and mental organisation with their human devourers ; to be as susceptible to physical suffering and pain as they; to be endowed—at all events, a very large proportion of them—with reasoning and mental faculties in very high degrees, and far from destitute of moral perceptions, the amazement may well be conjectured to give way to incredulity, that such knowledge and such practices could possibly co-exist. That the outward signs of all this gross barbarism—the entire or mangled bodies of the victims of the Table—were accustomed to be put up for public exhibition in every street and thoroughfare, without manifestations of disgust or abhorrence from the passers-by—even from those pretending to most culture or fashion—such outward proofs of extraordinary insensibility on the part of all classes to finer feeling may, nevertheless, scarcely provoke so much astonishment from an enlightened posterity as the fact that every public gathering of the governors or civil dignitaries of the country; every celebration of ecclesiastical or religious festivals appeared to be made the special occasion of the sacrifice and suffering of a greater number and variety than usual of their harmless fellow-beings; and all this often in the near neighbourhood of starving thousands, starving from want of the merest necessaries of life.

Happily, however, there will be visible to the philosopher of the Future signs of the dawn of the better day in this last quarter of the nineteenth century. He will find, in the midst of the general barbarism of life, and in spite of the prevailing indifferentism and infidelity to truth, that there was a gradually increasing number of dissenters and protesters; that already, at the beginning of that period, there were associations of dietary reformers—offshoots from the English parent society, founded in 1847—successively established in America, in Germany, in Switzerland, in France, and, finally, in Italy; small indeed in numbers, but strenuous in efforts to spread their principles and practice; that in some of the larger cities, both in this country and in other parts of Europe, there had also been set on foot *Reformed Restaurants*, which supplied to considerable numbers of persons at once better food and better knowledge.

If the truth or importance of any Principle or Feeling is to be measured, not by its popularity, indeed—not by the *quod ab omnibus*—but by the extent of its recognition by the most refined and the most earnest thinkers in all the most enlightened times—by the *quod a sapientibus*—the value of no principle has better been established than that which insists upon the vital importance of a radical reform in Diet. The number of the protesters against the barbarism of human living who, at various periods in the known history of our world, have more or less strongly denounced it, is a fact which cannot fail to arrest the attention of the most superficial inquirer. But a still more striking characteristic of this large body of protestation is the *variety* of the witnesses. Gautama Buddha and Pythagoras, Plato and Epikurus, Seneca and Ovid, Plutarch and Clement (of Alexandria), Porphyry and Chrysostom, Gassendi and Mandeville, Milton and Evelyn, Newton and Pope, Ray and Linné, Tryon and Hecquet, Cocchi and Cheyne, Thomson and Hartley, Chesterfield and Ritson, Voltaire and Swedenborg, Wesley and Rousseau, Franklin and Howard, Lambe and Pressavin, Shelley and Byron, Hufeland and Graham, Gleïzès and Phillips, Lamartine and Michelet, Daumer and Struve—such are some of the more or less famous, or meritorious, names in the Past to be found among the prophets of Reformed Dietetics, who, in various degrees of abhorrence, have shrunk from the *régime* of blood. Of many of those who have revolted from it, it may almost be said that they revolted *in spite of themselves*—in double weight; for it is founded upon the irrefragable principles of Justice and Compassion—universal Justice and universal Compassion— the two principles most essential in any system of ethics worthy of the name. That this argument seems to have so limited an influence—even with persons otherwise humanely disposed, and of finer feeling in respect to their own, and, also, in a general way, to other species—can be attributed only to the deadening power of custom and habit, of traditional prejudice, and educational bias. If they could be brought to reflect upon the simple ethics of the question, divesting their minds of these distorting media, it must appear in a light very different from that in which they accustom themselves to consider it. This sub- ject, however, has been abundantly insisted upon with eloquence and ability much greater than the present writer has any pretensions to. It is necessary to add here, upon this particular branch of the subject, only one or two observations. The popular objections to the disuse of the flesh-diet may be classified under the two heads of fallacies and subter- fuges. Not a few candid inquirers, doubtless, there are who sincerely allege certain *specious* objections to the humanitarian argument, which have a considerable amount of *apparent* force; and these fallacies seem alone to deserve a serious examination.

In the general constitution of life on our globe, suffering and slaughter, it is objected, are the normal and constant condition of things—the strong relentlessly and cruelly preying upon the weak in endless succession—and, it is asked, why, then, should the human species form an exception to the general rule, and hopelessly fight against Nature? To this it is to be replied, first : *that*, although, too certainly, an unceasing and cruel internecine warfare has been waged upon this atomic globe of ours from the first origin of Life until now, yet, apparently, there has been going on a slow, but not uncertain, progress towards the ultimate elimination of the crueller phenomena of Life; *that*, if the *carnivora* form a very large proportion of Living Beings, yet the *non-carnivora* are in the majority; and, lastly, what is still more to the purpose, *that* Man, most evidently, by his origin and physical organisation, belongs not to the former but to the latter; besides and beyond which, *that* in proportion as he boasts himself—and as he is seen *at his best* (and only so far) he boasts himself with justness—to be the highest of all the gradually ascending and co-ordinated series of Living Beings, so is he, in that proportion, bound to prove his right to the supreme place and power, and his asserted claims to moral as well as mental superiority, by his conduct.

In brief, in so far only as he proves himself to be the *beneficent ruler and pacificator*—and not the selfish Tyrant—of the world, can he have any just title to the moral pre-eminence.

If the philosophical fallacy (the *eidolon specûs*) thus vanishes under a near examination; the next considerable objection, upon a superficial view, not wholly unnatural, that, if slaughtering for food were to be abolished, there would be a failure of manufacturing material for the ordinary uses of social life, is, in reality, based upon a contracted apprehension of facts and phenomena. For it is a reasonable and sufficient reply, that the whole history of civilisation, as it has been a history of the slow but, upon the whole, continuous advance of the human race in the arts of Refinement, so, also, has it proved that *demand creates supply*—that it is the absence of the former alone which permits the various substances, no less than the various forces, yet latent in Nature to remain uninvestigated and unused. Nor can any thoughtful person, who knows anything of the history of Science and Discovery, doubt that the resources of Nature and the mechanical ingenuity of man are all but boundless. Already, notwithstanding the absence of any demand for them, excepting within the ranks of anti-kreophagists, various non-animal substances have been proposed, in some cases used, as substitutes for the prepared skins of the victims of the Slaughter-house; and that, in the event of a general demand for such substitutes, there would spring up an active competition among inventors and manufacturers in this direction there is not the least reason for doubt. Besides, it must be taken into account that the process of conversion of the flesh-eating (that is to say, of the richer) sections of communities to the bloodless diet will, only too certainly, be very slow and gradual.

As for the popular—perhaps the most popular—fallacy (the *eidolon fori*), which exhibits little of philosophical accuracy, or, indeed, of common reason, involved in the questions: "What is to become of *the animals?*" and, "Why were they created, if they are not intended for Slaughter and for human food?"—it is scarcely possible to return a grave reply. The brief answer, of course, is—that those variously-tortured beings have been brought into existence, and their numbers maintained, by selfish human invention only. Cease to breed for the butcher, and they will cease to exist beyond the numbers necessary for lawful and innocent use; they were "created" indeed, but they have been created by man, since he has vastly modified and, by no means, for the benefit of his helpless dependants, the natural form and organisation of the original types, the parent stocks of the domesticated Ox, Sheep, and Swine, now very remote from the native grandeur and vigour of the Bison, the Mouflon, and the wild Boar.

There remains one fallacy of quite recent origin. An association has been formed—somewhat late in the day, it must be allowed—consisting of a few sanitary reformers, who put forward, also, humane reasons, for "Reform of the Slaughter-Houses," one of the secondary propositions of which is, that the savagery and brutality of the Butchers' trade could be obviated by the partial or general use of less lingering and revolting modes of killing than those of the universal knife and axe. No humanitarian will refuse to welcome any sign, however feeble, of the awakening of the conscience of the Community, or rather of the more thoughtful part of it, to the paramount obligations of common Humanity, and of the recognition of the claims of the subject species to *some* consideration and to *some* compassion, if not of the recognition of the claims of Justice ; or will refuse to welcome any sort of proposition to lessen the enormous sum total of atrocities to which the lower animals are constantly subjected by human avarice, gluttony, and brutality. But, at the same time, no earnest humanitarian can accept the sophism, that an attempt at a mitigation of cruelty and suffering which, fundamentally, are *unnecessary*, ought to satisfy the educated conscience or reason. Vainly do the more feeling persons, who happen to have some scruples of conscience in respect to the sanction of the barbarous practice of Butchering, think to abolish the cruelties, while still indulging the appetite for the flesh·luxuries, of the Table. The vastness of the demands upon the butchers—demands constantly increasing with the pecuniary resources of the nation, and stimulated by the pernicious example of the wealthy classes ; the immensity of the traffic in "live stock" (as they complacently are termed) by rail and by ship,* the frightful horrors of which it has often been attempted, though inadequately, to describe ; the utter impossibility of efficiently supervising and regulating such traffic and such slaughter—even supposing the desire to do so to exist to any considerable extent—and the inveterate

---

* That the indescribable atrocities inflicted in the final scene of the slaughter-house, are far from being the only sufferings to which the victims of the Table are liable, is a fact upon which, at this day, it ought to be superfluous to insist. The frightful sufferings during "the middle passage," in rough weather, and especially in severe storms, have over and over again been recounted even by spectators the least likely to be easily affected by the spectacles of lower animal suffering. Thousands of Oxen and Sheep, year by year, are thrown *living* into the sea during the passage from the United States alone. In the year 1879, according to the official report, 14,000 thus perished, while 1,240 were landed dead, and 450 were slaughtered on the quay upon landing to prevent death from wounds.—See, among other recent works on humane Dietetics, the *Perfect Way in Diet* of Dr. Anna Kingsford for some most instructive details upon this subject. The reader is also referred to the Lecture recently addressed to the Students of Girton College, Cambridge, by the same able and eloquent writer, for other aspects of the humanitarian argument.

indifferentism of the Legislature and of the influential classes, sufficiently declare the futility of such expectation and of the indulgence of such comfortable hope.   It is, in brief, as with other attempts at patching and mending, or at applying salves to a hopelessly festered and gangrened wound, merely to put the "flattering unction" of compromise to the conscience.   "Diseases, desperate grown, by desperate appliances are relieved, or not at all;" the foul stream of cruelty must be stopped at its source; the fountain and origin of the evil—the Slaughter-House itself—must be abolished.   *Delendum est Macellum.*

It has been well said by one of the most eloquent of the prophets of Humane Living, that there are steps on the way to the summit of Dietetic Reform, and, if only one step be taken, yet that that single step will be not without importance and without influence in the world.   The step, which leaves for ever behind it the barbarism of slaughtering our fellow-beings, the Mammals and Birds, is, it is superfluous to add, the most important and most influential of all.

As for the plan of the present work, living writers and authorities— numerous and important as they are—necessarily have been excluded. Its bulk, already extended beyond the original conception of its limits, otherwise would have been swollen to a considerably larger size.   For its entire execution, as well as for the collection and arrangement of the matter, the compiler alone is responsible; and, conscious that it must fall short of the completeness at which he aimed, he can pretend only to the merits of careful research and an eclectic impartiality.   To the fact that the work already has appeared in the pages of the *Dietetic Reformer*, to which it has been contributed periodically during a space of time extending over five years, is owing some repetition of matter, which also, necessarily, is due to the nature of the subject.   Errors of inadvertence, it is hoped, will be found to be few and inconsiderable.   For the rest, he leaves the *Ethics of Diet* to the candour of the critics and of the public.

THE ETHICS OF DIET

THE ETHICS OF DIET.

I.

HESIOD. EIGHTH CENTURY B.C.

HESIOD—the poet *par excellence* of peace and of agriculture, as Homer is of war and of the "heroic" virtues—was born at Ascra, a village in Bœotia, a part of Hellas, which, in spite of its proverbial fame for beef-eating and stupidity, gave birth to three other eminent persons—Pindar, the lyric poet, Epameinondas, the great military genius and statesman, and Plutarch, the most amiable moralist of antiquity.

The little that is known of the life of Hesiod is derived from his *Works and Days*. From this celebrated poem we learn that his father was an emigrant from Æolia, the Greek portion of the north-west corner of the Lesser Asia; that his elder brother, Perses, had, by collusion with the judges, deprived him of his just inheritance; that after this he settled at Orchomenos, a neighbouring town—in the pre-historical ages a powerful and renowned city. This is all that is certainly known of the author of the *Works and Days*, and *The Theogony*. Of the genuineness of the former there has been little or no doubt; that of the latter—at least in part—has been called in question. Besides these two chief works, there is extant a piece entitled *The Shield of Herakles*, in imitation of the Homeric Shield (*Iliad* xviii.) The *Catalogues of Women*—a poem commemorating the heroines beloved by the gods, and who were thus the ancestresses of the long line of heroes, the reputed founders of the ruling families in Hellas—is lost.

The charm of the *Works and Days*—the first didactic poem extant—is its apparent earnestness of purpose and simplicity of style. The author's frequent references to, and rebuke of, legal injustices—his sense of which had been quickened by the iniquitous decisions of the judges already referred to—are as *naïve* as they are pathetic.

Of the *Theogony*, the subject, as the title implies, is the history of the generation and successive dynasties of the Olympian divinities—the objects of Greek worship.   It may, indeed, be styled the Hellenic Bible, and, with the Homeric Epics, it formed the principal theology of the old Greeks, and of the later Romans or Latins.   The "Procemium," or introductory verses—in which the Muses are represented as appearing to their votary at the foot of the sacred Helicon, and consecrating him to the work of revealing the divine mysteries by the gift of a laurel-branch—and the following verses, describing their return to the celestial mansions, where they hymn the omnipotent Father, are very charming. To the long description of the tremendous struggle of the warring gods and Titans, fighting for the possession of heaven, Milton was indebted for his famous delineation of a similar conflict.

The *Works and Days,* in striking contrast with the military spirit of the Homeric epic, deals in plain and simple verse with questions ethical, political, and economic.   The ethical portion exhibits much true feeling, and a conviction of the evils brought upon the earth by the triumph of injustice and of violence.   The well-known passages in which the poet figures the gradual declension and degeneracy of men from the golden to the present iron race, are the remote original of all the later pleasing poetic fictions of golden ages and times of innocence.

According to Hesiod, there are two everlastingly antagonistic agents at work on the Earth ; the spirit of war and fighting, and the peaceful spirit of agriculture and mechanical industry.   And in the apostrophe in which he bitterly reproaches his unrighteous judges—

> " O fools ! they know not, in their selfish soul,
> How far the half is better than the whole :
> The good which Asphodel and Mallows yield,
> The feast of herbs, the dainties of the field "—

he seems to have a profound conviction of the truth taught by Vegetarianism—that luxurious living is the fruitful parent of selfishness in its manifold forms.*

That Hesiod regarded that diet which depends mainly or entirely upon agriculture and upon fruits as the highest and best mode of life is sufficiently evident in the following verses descriptive of the "Golden Age " life :—

---

* Cf. Horace (whom, however, we do not quote as an authority)—
> " Let olives, endives, mallows light
> Be all my fare ; "

and Virgil thus indicates the charm of a rural existence for him who realises it :—
> " Whatever fruit the branches and the mead
> Spontaneous bring, he gathers for his need."

"Like gods, they lived with calm, untroubled mind,
   Free from the toil and anguish of our kind,
   Nor did decrepid age mis-shape their frame.

Pleased with earth's unbought feasts : all ills removed,
   Wealthy in flocks,* and of the Blest beloved,
   Death, as a slumber, pressed their eyelids down :
   All Nature's common blessings were their own.
   The life-bestowing tilth its fruitage bore,
   A full, spontaneous, and ungrudging store.
   They with abundant goods, 'midst quiet lands,
   All willing, shared the gatherings of their hands.
   When Earth's dark breast had closed this race around,
   Great Zeus, as demons,† raised them from the ground ;
   Earth-hovering spirits, they their charge began—
   The ministers of good, and guards of men.
   Mantled with mist of darkling air they glide,
   And compass Earth, and pass on every side ;
   And mark, with earnest vigilance of eyes,
   Where just deeds live, or crooked ways arise,
   And shower the wealth of seasons from above." ‡

The second race—the " Silver Age "—inferior to the first and wholly innocent people, were, nevertheless, guiltless of bloodshed in the preparation of their food ; nor did they offer sacrifices—in the poet's judgment, it appears, a damnable error. For the third—the "Brazen Age"—it was reserved to inaugurate the feast of blood :—

"Strong with the ashen spear, and fierce and bold,
   Their thoughts were bent on violence alone,
   The deed of battle, and the dying groan.
   Bloody their feasts, with wheaten food unblessed."

According to Hesiod, who is followed by the later poets, the "immortals inhabiting the Olympian mansions" feast ever on the pure and bloodless food of *Ambrosia*, and their drink is *Nectar*, which may be taken to be a sort of refined dew. He represents the divine Muses of Helicon, who inspire his song, as reproaching the shepherds, his neighbours, "that tend the flocks," with the possession of "mere fleshly appetites."

Ovid, amongst the Latins, is the most charming painter of the innocence of the " Golden Age." Amongst our own poets, Pope, Thomson, and Shelley—the last as a prophet of the future and actual rather than the poet of a past and fictitious age of innocence—have contributed to embellish the fable of the Past and the hope of the Future.

---

* The same apparent contradiction—the co-existence of "flocks and herds " with the prevalence of the non-flesh diet—appears in the Jewish theology, in *Genesis*. It is obvious, however, that in both cases the "flocks and herds" might be existing for other purposes than for slaughter.

† *Daimones.* The *dæmon* in Greek theology was simply a lesser divinity—an *angel*.

‡ Compare Spenser's charming verses ("Faery Queen," Book ii., canto 8): "And is there care in heaven," &c.

II.

PYTHAGORAS.   570—470 B.C.

" A GREATER good never came, nor ever will come, to mankind, than that which was imparted by the gods through Pythagoras." Such is the expression of enthusiastic admiration of one of his biographers. To those who are unacquainted with the historical development of Greek thought and Greek philosophy it may seem to be merely the utterance of the partiality of hero-worship. Those, on the other hand, who know anything of that most important history, and of the influence, direct or indirect, of Pythagoras upon the most intellectual and earnest minds of his countrymen—in particular upon Plato and his followers, and through them upon the later Jewish and upon very early Christian ideas— will acknowledge, at least, that the name of the prophet of Samos is that of one of the most important and influential factors in the production and progress of higher human thought.

There is a true and there is a false hero-worship. The latter, whatever it may have done to preserve the blind and unreasoning subservience of mankind, has not tended to accelerate the progress of the world towards the attainment of truth. The old-world occupants of the popular Pantheon—" the patrons of mankind, gods and sons of gods, destroyers rightlier called and plagues of men "—are indeed fast losing, if they have not entirely lost, their ancient credit, but their vacant places have yet to be filled by the representatives of the most exalted ideals of humanity. Whenever, in the place of the representatives of mere physical and mental force, the *true* heroes shall be enthroned, amongst the moral luminaries and pioneers who have contributed to lessen the thick darkness of ignorance, barbarism, and selfishness, the name of the first western apostle of humanitarianism and of spiritualism must assume a prominent position.

It is a natural and legitimate curiosity which leads us to wish to know, with something of certainty and fulness, the outer and inner life of the master spirits of our race. Unfortunately, the *personality* of many of the most interesting and illustrious of them is of a vague and shadowy kind. But when we reflect that little more is known of the personal life of Shakspere than of that of Pythagoras or Plato—not to mention other eminent names—our surprise is lessened that, in an age long preceding the discovery of printing, the records of a life even so important and influential as that of the founder of Pythagoreanism are meagre and scanty.

The earliest account of his teaching is given by Philolaus ("Lover of the People," an auspicious name) of Tarentum, who, born about forty or fifty years after the death of his master – was thus contemporary with Sokrates and Plato. His *Pythagorean System*, in three books, was so highly esteemed by Plato that he is said to have given £400 or £500 for a copy, and to have incorporated the principal part of it in his *Timœus*. Sharing the fate of so many other valuable products of the Greek genius, it has long since perished. Our remaining authorities for the Life are Diogenes of Laerte, Porphyry, one of the most erudite writers of any age, and Iamblichus. Of these, the biography of the last is the fullest, if not the most critical; that of Porphyry wants the beginning and the end; whilst of the ten books of Iamblichus *On the Pythagorean Sect* (Περὶ Πυθαγόρου Αἱρέσεως), of which only five remain, the first was devoted to the life of the founder. Diogenes, who seems to have been of the school of Epikurus, belongs to the second, while Porphyry and Iamblichus, the well-known exponents of Neo-Platonism, wrote in the third and fourth centuries of our era.

Pythagoras was born in the Island of Samos, somewhere about the year 570 B.C. At some period in his youth, Polykrates—celebrated by the fine story of Herodotus—had acquired the *tyranny* of Samos, and his rule, like that of most of his compeers, has deserved the stigma of the modern meaning of the Greek equivalent for princely and monarchical government. The future philosopher, we are told, unable to descend to the ordinary arts of sycophancy and dissimulation, left his country, and entered, like the Sirian philosopher of Voltaire, upon an extensive course of travels—extensive for the age in which he lived. How far he actually travelled is uncertain. He visited Egypt, the great nurse of the old-world science, and Syria, and it is not impossible that he may have penetrated eastwards as far as Babylon, perhaps as the captive of the recent conqueror of Egypt—the Persian Kambyses. It was in the East, and particularly in Egypt, that he probably imbibed the dogma of the immortality of the soul, or, as he chose to represent it to the public, that of the *metempsychosis*—a fancy widely spread in the eastern theologies.

It has been asserted that he had already abandoned the orthodox diet at the age of nineteen or twenty. If this was actually the fact, he has the additional merit of having adopted the higher life by his own original force of mind and refinement of feeling. If not, he may have derived the most characteristic as well as the most important of his teachings from the Egyptians or Persians, or, through them, even from the Hindus—the most religiously strict abstainers from the flesh of animals. It is remarkable that the two great apostles of abstinence—Pythagoras and Sakya-Muni, or Buddha—were almost contemporaries; nor is it impossible that the Greek may, in whatever way, have become acquainted with the sublime tenets of the Hindu prophet, who had lately seceded from Brahminism, the established sacerdotal and exclusive religion of the Peninsula, and promulgated his great revelation—until then new to the world—that religion, at least his religion, was to be " a religion of mercy to all beings," human and non-human.*

As the natural and necessary result of his pure living, we are told by Iamblichus that " his sleep was brief, his soul vigilant and pure, and his body confirmed in a state of perfect and invariable health." He appears to have passed the period of middle life when he returned to Samos, where his reputation had preceded him. Either, however, finding his countrymen hopelessly debased by the corrupting influence of despotism, or believing that he would find a better field for the propagandism of his new revelation, he not long afterwards set out for Southern Italy, then known as " Great Greece," by reason of its numerous Greek colonies, or, rather, autonomous communities. At Krotona his fame and eloquence soon attracted, it seems, a select if not numerous auditory; and there he founded his famous society—the first historical anti-flesh-eating association in the western world—the prototype, in some respects, of the ascetic establishments of Greek and Catholic Christendom. It consisted of about three hundred young men belonging to the most influential families of the city and neighbourhood.

It was the practice of the Egyptian priestly caste and of other exclusive institutions to reserve their better ideas (of a more satisfactory sort, at all events, than the system of theology that was promulgated to the mass of the community), into which only privileged persons were initiated. This esoteric method, which under the name of the *mysteries* has exercised the learned ingenuity of modern writers—who have, for the most part, vainly laboured to penetrate the obscurity enveloping the most remarkable institution of the Hellenic theology—was accompanied with the strictest vows and circumstances of silence and secrecy. As for the priestly order, it was their evident policy to maintain the superstitious ignorance of the people and to overawe their minds, while in regard to the philosophic sects, it was perhaps to shield themselves from the priestly or popular suspicion that they shrouded their scepticism in this dark and convenient disguise. The parabolic or esoteric method was, perhaps, almost a necessity of the earlier ages. It is to be lamented that it should be still in favour in this safer age, and that the old exclusiveness

* His moral principles are reduced to these:—"1. Mercy established on an immovable basis. 2. Aversion to all cruelty. 3. A boundless compassion for all creatures." Quoted from Klaproth by Huc, *Chinese Empire.* xv Buddhism was to Brahminism, sacerdotally, what early Christianity was to Mosaism.

of the *mysteries* is in esteem with many modern authorities, who seem to hold that to unveil the spotless Truth to the multitude is " to cast pearls before swine."

It was probably from the philosophic motive that the founder of the new society instituted his grades of catechumens and probationary course, as well as vows of the strictest secrecy. The exact nature of all his interior instruction is necessarily very much matter of conjecture, inasmuch as, whether he committed his system to writing or not, nothing from his own hand has come down to us. However this may be, it is evident that the general spirit and characteristic of his teaching was self-denial or self-control, founded upon the great principles of justice and temperance; and that communism and asceticism were the principal aim of his sociology. He was the founder of communism in the West—his communistic ideas, however, being of an aristocratic and exclusive rather than of a democratic and cosmopolitan kind. "He first taught," says Diogenes, "that the property of friends was to be held in common—that friendship is equality—and his disciples laid down their money and goods at his feet, and had all things common."

The moral precepts of the great master were much in advance of the conventional morality of the day. He enjoined upon his disciples, the same biographer informs us, each time they entered their houses to interrogate themselves—" How have I transgressed? What have I done? What have I left undone that I ought to have done?" He exhorted them to live in perfect harmony, to do good to their enemies and by kindness to convert them into friends. "He forbade them either to pray for themselves, seeing that they were ignorant of what was best for them; or to offer slain victims (σφαγια) as sacrifices; and taught them to respect a *bloodless* altar only." Cakes and fruits, and other innocent offerings were the only sacrifices he would allow. This, and the sublime commandment " Not to kill or injure any innocent animal," are the grand distinguishing doctrines of his moral religion. So far did he carry his respect for the beautiful and beneficent in Nature, that he specially prohibited wanton injury to cultivated and useful trees and plants.

By confining themselves to the innocent, pure, and spiritual dietary he promised his followers the enjoyment of health and equanimity, undisturbed and invigorating sleep, as well as a superiority of mental and moral perceptions. As for his own diet, " he was satisfied," says Porphyry, " with honey or the honeycomb, or with bread only, and he did not taste wine from morning to night (μεθ'ἡμεραν); or his principal dish was often kitchen herbs, cooked or uncooked. Fish he ate rarely."

Humanitarianism—the extension of the sublime principles of justice and of compassion to all innocent sentient life, irrespective of nationality, creed, or species—is a very modern and even now very inadequately recognised creed ; and, although there have been here and there a few, like Plutarch and Seneca, who were "splendidly false," to the spirit of their age, the recognition of the obligation (the *practice* has always been a very different thing) of benevolence and beneficence, so far from being extended to the non-human races, until a comparatively recent time has been limited to the narrow bounds of country and citizenship ; and patriotism and internationalism are, apparently, two very opposite principles.

The obligation to abstain from the flesh of animals was founded by Pythagoras on mental and spiritual rather than on humanitarian grounds. Yet that the latter were not ignored by the prophet of *akreophagy* is evident equally by his prohibition of the infliction of pain, no less than of death, upon the lower animals, and by his injunction to abstain from the bloody sacrifices of the altar.   Such was his abhorrence of the Slaughter-House, Porphyry tells us, that not only did he carefully abstain from the flesh of its victims, but that he could never bring himself to endure contact with, or even the sight of, butchers and cooks.

While thus careful of the lives and feelings of the innocent non-human races, he recognised the necessity of making war upon the ferocious *carnivora.*   Yet to such a degree had he become familiar with the habits and dispositions of the lower animals that he is said, by the exclusive use of vegetable food, not only to have tamed a formidable bear, which by its devastations on their crops had become the terror of the country people, but even to have accustomed it to eat that food only for the remainder of its life.   The story may be true or fictitious, but it is not incredible ; for there are well-authenticated instances, even in our own times, of true *carnivora* that have been fed, for longer or shorter periods, upon the non-flesh diet.*

"Amongst other reasons, Pythagoras," says Iamblichus, "enjoined abstinence from the flesh of animals because it is conducive to peace. For those who are accustomed to abominate the slaughter of other animals, as iniquitous and unnatural, will think it still more unjust and unlawful to kill a man or to engage in war."   Specially, he "exhorted those politicians who are legislators to abstain.   For if they were willing to act justly in the highest degree, it was indubitably incumbent upon them not to injure any of the lower animals.   Since how could they persuade others to act justly, if they themselves were proved to be indulging an insatiable avidity by devouring these animals that are allied to us.

---

* All the varieties of the bear tribe, it is perhaps scarcely necessary to observe, are by organisation, and therefore by preference, frugivorous.   It is from necessity only, for the most part, that they seek for flesh.

For through the communion of life and the same elements, and the sympathy thus existing, they are, as it were, conjoined to us by a fraternal alliance."* Maxims how different from those in favour in the present "year of grace," 1877! If the refined thinker of the sixth century B.C. were now living, what would be his indignation at the enormous slaughter of innocent life for the public banquets at which our statesmen and others are constantly *fêted*, and which are recorded in our journals with so much magniloquence and minuteness? His hopes for the regeneration of his fellow-men would surely be terribly shattered. We may apply the words of the great Latin satirist, Juvenal, who so frequently denounces in burning language the luxurious gluttony of his countrymen under the Empire—" What would not Pythagoras denounce, or whither would he not flee, could he see these monstrous sights— he who abstained from the flesh of all other animals as though they were human?" (*Satire* xv.)

How long the communistic society of Krotona remained undisturbed is uncertain. Inasmuch as its reputation and influence were widely spread, it may be supposed that the outbreak of the populace (the origin of which is obscure), by which the society was broken up and his disciples massacred, did not happen until many years after its establishment. At all events, it is commonly believed that Pythagoras lived to an advanced age, variously computed at eighty, ninety, or one hundred years.

It is not within our purpose to discuss minutely the scientific or theological theories of Pythagoras. In accordance with the abstruse speculative character of the Ionic school of science, which inclined to refer the origin of the universe to some one primordial principle, he was led by his mathematical predilections to discover the cosmic element in numbers, or proportion—a theory which savours of John Dalton's philosophy, now accepted in chemistry, and a virtual enunciation of what we now call *quantitative* science. Pythagoras taught the Kopernican theory prematurely. He regarded the sun as more *divine* than the earth, and therefore set it in the *centre* of the earth and planets. The argument was surely a mark of genius, but it was too transcendental for his contemporaries, even for Plato and Aristotle. His elder contemporary, the celebrated Thales of Miletus, with whom in his early youth he may have been acquainted, may claim, indeed, to be the remote originator of the famous nebular hypothesis of Laplace and modern astronomy. Another cardinal doctrine of the Pythagorean school was the musical, from whence the idea, so popular with the poets, of the "music of the spheres." To

---

* Compare Montaigne (*Essais*, Book II., chap. 12), who, to the shame of the popular opinion of the present day, ably maintains the same thesis.

music was attributed the greatest influence in the control of the passions. In its larger sense, by the Greeks generally, the term "Music" (*Musice*—pertaining to the Muses) denoted, it is to be remembered, not alone the "concord of sweet sounds," but also an artistic and æsthetic education in general—all humanising and refining instruction.

The famous doctrine of the Metempsychosis or Transmigration or Souls also was, doubtless, a prominent feature in the Pythagorean system; but it is probable that we may presume that by it Pythagoras intended merely to convey to the "uninstructed," by parable, the sublime idea that the soul is gradually purified by a severe course of discipline until finally it becomes fitted for a fleshless life of immortality.* We are chiefly concerned with his attitude in regard to flesh eating. There can be no question that abstinence was a fundamental part of his system, yet certain modern critics—little in sympathy with so practical a manifestation of the higher life, or, indeed, with self-denial of any kind—have sometimes affected either to doubt the fact or to pass it by in contemptuous silence, thus ignoring what for the after ages stands out as by far the most important residuum of Pythagoreanism. In support of this scepticism the fact of the celebrated athlete Milo, whose prodigies of strength have become proverbial, has been quoted. Yet if these critics had been at the pains of inquiring somewhat further, they would have learned, on the contrary, that the non-flesh diet is exactly that which is most conducive to physical vigour; that in the East there are at this day non-flesh eaters, who in feats of strength might put even our strongest men to the blush. The extraordinary powers of the porters and boatmen of Constantinople have been remarked by many travellers; and the Chinese coolies and others are almost equally notorious for their marvellous powers of endurance. Yet their food is not only of the simplest—rice, dhourra (*i.e.*, millet), onions, &c.—but of the scantiest possible. Moreover, the elder Greek athletes themselves, for the most part, trained on vegetarian diet. Not to multiply details, the fact that, upon a moderate calculation, two-thirds at least of the population of our globe—including the mass of the inhabitants of these islands—live, *nolentes, volentes,* on a dietary from which flesh is almost altogether necessarily excluded, is on the face of it sufficient proof in itself of the non-necessity of the diet of the rich.

While the general consent of antiquity and of later times has received as undoubted the obligation of strict abstinence on the part of the immediate followers of Pythagoras, it seems that as regards the un-

---

* The allegory of the trials and final purification of the soul was a favourite one with the Greeks, in the charming story of the loves and sorrows of Psyche and Eros. Apuleius inserted it in his fiction of *The Golden Ass,* and it constantly occurs in Greek and modern art.

initiated, or (to use the ecclesiastical term) *catechumens*, the obligation was not so strict. Indeed relaxation of the rules of the higher life was simply a *sine quâ non* of securing the attention of the mass of the community at all ; and, like one still more eminent than himself in an after age, he found it a matter of necessity to present a teaching and a mode of living not too exalted and unattainable by the grossness and " hardness of heart " of the multitude. Hence, in all probability, the seeming contradictions in his teaching on this point found in the narratives of his followers.

If his critics had been more intent on discovering the excellence of his rules of abstinence than on discussing, with frivolous diligence, the probable or possible reasons of his alleged prohibition of beans, it would have redounded more to their credit for wisdom and love of truth. Assuming the fact of the prohibition, in place of collecting all the most absurd gossip of antiquity, they might perhaps have found a more rational and more solid reason in the hypothesis that the bean being, as used in the ballot, a symbol and outward and visible sign of political life, was employed by Pythagoras parabolically to dissuade his followers from participating in the idle strife of party faction, and to exhort them to concentrate their efforts upon an attempt to achieve the solid and lasting reformation of mankind.* But to be much concerned in a patient inquiry after truth unhappily has been not always the characteristic of professional commentators.

Blind hero-worship or idolatry of genius or intellect, even when directed to high moral aims, is no part of our creed ; and it is sufficient to be assured that he was human, to be free to confess that the historical founder of *akreophagy* was not exempt from human infirmity, and that he could not wholly rise above the wonder-loving spirit of an uncritical age. Deducting all that has been imputed to him of the fanciful or fantastic, enough still remains to force us to recognise in the philosopher-prophet of Samos one of the master-spirits of the world. †

---

* Beans, like lean flesh, are very nitrogenous, and it is possible that Pythagoras may have deemed them too invigorating a diet for the more aspiring ascetics. This may seem at least a more solid reason than the absurd conjectures to which we have referred.

† " As regards the fruits of this system of training or belief (the Pythagorean), it is interesting to remark," says the author of the article Pythagoras in Dr. Smith's *Dictionary of Greek and Roman Biography*, that, wherever we have notices of distinguished Pythagoreans, we usually hear of them as men of great uprightness, conscientiousness, and self-restraint, and as capable of devoted and enduring friendship." Amongst them the names of Archytas, and Damon, and Phintias are particularly eminent. Archytas was one of the very greatest geniuses of antiquity : he was distinguished alike as a philosopher, mathematician, statesman, and general. In mechanics he was the inventor of the wooden flying dove—one of the wonders of the older world. Empedokles (the Apollonius of the 5th century B.C.), who devoted his marvellous attainments to the service of humanity, may be claimed as, at least in part, a follower of Pythagoras.

III.

PLATO.   428—347 B.C.

THE most renowned of all the prose writers of antiquity may be said to have been almost the lineal descendant, in philosophy, of the teacher of Samos.   He belonged to the aristocratic families of Athens—"the eye of Greece"—then and for long afterwards the centre of art and science.   His original name was Aristokles, which he might well have retained.   Like another equally famous leader in literature, François Marie Arouet, he abandoned his birth-name, and he assumed or acquired the name by which he is immortalised, to characterise, as it is said, either the breadth of his brow or the extensiveness of his mental powers.   In very early youth he seems to have displayed his literary aptitude and tastes in the various kinds of poetry—epic, tragic, and lyric—as well as to have distinguished himself as an athlete in the great national contests or "games," as they were called, the grand object of ambition of every Greek.   He was instructed in the chief and necessary parts of a liberal Greek education by the most able professors of the time. He devoted himself with ardour to the pursuit of knowledge, and sedulously studied the systems of philosophy which then divided the literary world.

In his twentieth year he attached himself to Sokrates, who was then at the height of his reputation as a moralist and dialectician.   After the judicial murder of his master, 399, he withdrew from his native city, which, with a theological intolerance extremely rare in pagan antiquity, had already been disgraced by the previous persecution of another eminent teacher—Anaxagoras—the instructor of Euripides and of Perikles.   Plato then resided for some time at Megara, at a very short distance from Athens, and afterwards set out, according to the custom of the eager searchers after knowledge of that age, on a course of travels.

He traversed the countries which had been visited by Pythagoras, but his alleged visit to the further East is as traditional as that of his predecessor.   The most interesting fact or tradition in his first travels is his alleged intimacy with the Greek prince of Syracuse, the elder Dionysius, and his invitation to the western capital of the Hellenic world.   The story that he was given up by his perfidious host to the Spartan envoy, and by him sold into slavery, though not disprovable, may be merely an exaggerated account of the ill-treatment which he actually received.

His grand purpose in going to Italy was, without doubt, the desire to become personally known to the eminent Pythagoreans whose head-quarters were in the southern part of the Peninsula, and to secure the best opportunities of making himself thoroughly acquainted with their philosophic tenets. At that time the most eminent representative of the school was the celebrated Archytas, one of the most extraordinary mathematical geniuses and mechanicians of any age. Upon his return to Athens, at about the age of forty, he established his ever-memorable school in the suburban groves or "gardens" known as Ἀκαδημία—whence the well-known *Academy* by which the Platonic philosophy is distinguished, and which, in modern days, has been so much vulgarised. All the most eminent Athenians, present and future, attended his lectures, and among them was Aristotle, who was destined to rival the fame of his master. From about 388 to 347, the date of his death, he continued to lecture in the Academy and to compose his Dialogues.

In the intervals of his literary and didactic labours he twice visited Sicily; the first time at the invitation of his friend Dion, the relative and minister of the two Dionysii, the younger of whom had succeeded to his father's throne, and whom Dion hoped to win to justice and moderation by the eloquent wisdom of the Athenian sage. Such hopes were doomed to bitter disappointment. His second visit to Syracuse was undertaken at the urgent entreaties of his Pythagorean friends, of whose tenets and dietetic principles he always remained- an ardent admirer. For whatever reason, it proved unsuccessful. Dion was driven into exile, and Plato himself escaped only by the interposition of Archytas. Thus the only chance of attempting the realisation of his ideal of a communistic commonwealth—if he ever actually entertained the hope of realising it—was frustrated. Almost the only source of the biographies of Plato are the *Letters* ascribed to him, commonly held to be fictitious, but maintained to be genuine by Grote. The narrative of the first visit to Sicily is found in the seventh Letter.

We can refer but briefly to the nature of the philosophy and writings of Plato. In the notice of Pythagoras it has been stated that Plato valued very highly that teacher's methods and principles. Pythagoreanism, in fact, enters very largely into the principal writings of the great disciple and exponent (and, it may safely be added, improver) of Sokrates, especially in the *Republic* and the *Timæus*. The four cardinal virtues inculcated in the *Republic*—justice or righteousness (Δικαιοσύνη), temperance or self-control (Ἐγκράτεια or Σωφροσύνη), prudence or wisdom (Φρονήσις), fortitude (Ἀνδρεία)—are eminently pythagorean.

The characteristic of the purely speculative portion of Platonism is the theory of *ideas* (used by the author in the new sense of *unities*, the original meaning being *forms* and *figures*), of which it may be said that its merit depends upon its poetic fancy rather than upon its scientific value.

Divesting it of the verbiage of the commentators, who have not succeeded in making it more intelligible, all that need be said of this abstruse and fantastic notion is, that by it he intended to convey that all sensible objects which, according to him, are but the shadows and phantoms of things unseen, are ultimately referable to certain abstract conceptions or ideas, which he termed *unities*, that can only be reached by pure thinking. Hence he asserted that "not being in a condition to grasp the idea of the Good with full distinctness, we are able to approximate to it only so far as we elevate the power of thinking to its proper purity." Whatever may be thought of the premiss, the truth and utility of the deduction may be allowed to be as unquestionable as they are unheeded. This characteristic theory may be traced to the belief of Plato not only in the immortality, but also in the past eternity of the soul. In the *Phœdrus*, under the form of allegory, he describes the soul in its former state of existence as traversing the circuit of the universe where, if reason duly control the appetite, it is initiated, as it were, into the essences of things which are there disclosed to its gaze. And it is this ante-natal experience, which supplies the fleshly mind or soul with its ideas of the beautiful and the true.

The subtlety of the Greek intellect and language was, apparently, an irresistible temptation to their greatest ornaments to indulge in the nicest and most mystic speculation, which, to the possessors of less subtle intellects and of a far less flexible language, seems often strangely unpractical and hyperbolic. Thus while it is impossible not to be lost in admiration of the marvellous powers of the Greek *dialectics*, one cannot but at the same time regret that faculties so extraordinary should have been expended (we will not say altogether wasted) in so many instances on unsubstantial phantoms. If, however, the transcendentalism of the Platonic and other schools of Greek thought is matter for regret, how must we not deplore the enormous waste of time and labour apparent in the theological controversies of the first three or four centuries of Christendom—at least of Greek Christendom—when the omission or insertion of a single letter could profoundly agitate the whole ecclesiastical world and originate volumes upon volumes of refined, indeed, but useless verbiage. Yet even the ecclesiastical Greek writers of the early centuries may lay claim to a certain originality and merit of style which cannot be conceded to the "schoolmen" of the mediæval ages, and of still later times, whose solemn trifling—under the proud titles of Platonists and Aristotelians, or Nominalists and Realists, and the numerous other appellations assumed by them—for centuries was received with patience and even applause. Nor, unfortunately, is this war of Phantoms by any means unknown or extinct in our day. It was the lament f Seneca, often echoed by the most earnest minds, that all, or at least the greater part of, our learning is expended upon words rather than upon the acquisition of wisdom.*

Plato deserves his high place among the Immortals not so much on account of any very definite results from his philosophy as on account of its general *tendency* to elevate and direct human thought and aspirations to sublime speculations and aims. Of all his *Dialogues*, the most valuable and interesting, without doubt, is the *Republic*—the one of his writings upon which he seems to have bestowed the most pains, and in which he has recorded the outcome of his most mature reflections. Next may be ranked the *Phædo* and the *Phædrus*—the former, it is well known, being a disquisition on the immortality of the soul. In spite of certain fantastic conceptions, it must always retain its interest, as well by reason of its speculations on a subject which is (or rather which ought to be) the most interesting that can engage the mind, as because it purports to be the last discourse of Sokrates, who was expecting in his prison the approaching sentence of death. The *Phædrus* derives its unusual merit from the beauty of the language and style, and from the fact of its being one of the few writings of antiquity in which the charms of rural nature are described with enthusiasm.

The *Republic*, with which we are here chiefly concerned, since it is in that important work that the author reproduces the dietetic principles of Pythagoras, may have been first published amongst his earlier writings, about the year 395; but that it was published in a larger and revised edition at a later period is sufficiently evident. It consists of ten Books. The question of Dietetics is touched upon in the second and third, in which Plato takes care to point out the essential importance to the well-being of his ideal state, that both the mass of the community and, in a special degree, the *guardians* or rulers, should be educated and trained in proper dietetic principles, which, if not so definitely insisted upon as we could wish them to have been, sufficiently reveal the bias of his mind towards Vegetarianism. In the second Book the discussion turns principally upon the nature of Justice; and there is one passage which, still more significant for the age in which it was written, is not without instruction for the present. While Sokrates is discussing the subject with his interlocutors, one of them is represented as objecting:

"With much respect be it spoken, you who profess to be admirers of justice, beginning with the heroes of old, have every one of you, without exception, made the praise of Justice and the condemnation of Injustice turn solely upon the reputation and honour and gifts resulting from them. But what each is in itself, by its own peculiar force

---

* "Quæ Philosophia fuit, facta Philologia est." (Ep. cviii.) Compare Montaigne, *Essais,* i., 24, on Pedantry, where he admirably distinguishes between *wisdom* and *learning*.

as it resides in the soul of its possessor, unseen either by gods or men, has never, in poetry or prose, been adequately discussed, so as to show that Injustice is the greatest bane that a soul can receive into itself, and Justice the greatest blessing. Had this been the language held by you all from the first, and had you tried to persuade us of this from our childhood, we should not be on the watch to check one another in the commission of injustice, because everyone would be his own watchman, fearful lest by committing injustice he might attach to himself the greatest of evils."

Very useful and necessary for those times, and not wholly inapplicable to less remote ages, is the incidental remark in the same book, that " there are quacks and soothsayers who flock to the rich man's doors, and try to persuade him that they have a power at command which they procure from heaven, and which enables them, by sacrifices and incantations, performed amid feasting and indulgence, to make amends for any crime committed either by the individual himself or by his ancestors. . . . . And in support of all these assertions they produce the evidence of poets—some, to exhibit the facilities of vice, quoting the words :—

" Whoso wickedness seeks, may even in masses obtain it
    Easily.   Smooth is the way, and short, for nigh is her dwelling.
    Virtue, heaven has ordained, shall be reached by the sweat of the forehead."
                   —*Hesiod, Works and Days,* 287.*

It is the fifth Book, however, which has always excited the greatest interest and controversy, for therein he introduces his Communistic views. Our interest in it is increased by the fact that it is the original of the ideal Communisms of modern writers—the prototype of the *Utopia* of More, of the *New Atlantis* of Francis Bacon, the *Oceanica* of Harrington, and the *Gaudentio* of Berkeley, &c.

In maintaining the perfect natural equality of women to men,† and insisting upon an identity of education and training, he advances propositions which perhaps only the more advanced of the assertors of women's rights might be prepared to entertain. Whatever may have been said by the various admirers of Plato, who have been anxious to present his political or social views in a light which might render them less in conflict with modern Conservatism, there can be no doubt for any candid reader of the *Republic* that the author published to the world his *bonâ fide* convictions. One of the *dramatis personæ* of the dialogue, while expressing his concurrence in the Communistic legislation of Sokrates, at the same time objects to the difficulty of realising it in actual life, and desires Sokrates to point out whether, and how, it could be really practicable. Whereupon Sokrates (who it is scarcely

---

* *The Republic of Plato.* By Davies and Vaughan.

† In support of this thesis Plato adduces arguments derived from analogy. Amongst the non-human species the sexes, he points out, are nearly equal in strength and intelligence. In human savage life the difference is far less marked than in artificial conditions of life.

necessary to remark, is the convenient mouthpiece of Plato) replies: "Do you think any the worse of an artist who has painted the *beau idéal* of human beauty, and has left nothing wanting in the picture, because he cannot prove that such a one as he has painted might possibly exist? Were not we, likewise, proposing to construct, in theory, the pattern of a perfect State? Will our theory suffer at all in your good opinion if we cannot prove that it is *possible* for a city to be organised in the manner proposed?"

As has been well paraphrased by the interpreters to whom we are indebted for the English version: "The possibilities of realising such a commonwealth in actual practice is quite a secondary consideration, which does not in the least affect the soundness of the method or the truth of the results. All that can fairly be demanded of him is to show how the imperfect politics at present existing may be brought most nearly into harmony with the perfect State which has just been described. To bring about this great result one fundamental change is necessary, and only one: the highest political power must, by some means or other, be vested in philosophers." The next point to be determined is, What is, or ought to be, implied by the term *philosopher*, and what are the characteristics of the true philosophic disposition? "They are— (1) an eager desire for the knowledge of all real existence; (2) hatred of falsehood, and devoted love of truth; (3) contempt for the pleasures of the body; (4) indifference to money; (5) high-mindedness and liberality; (6) justice and gentleness; (7) a quick apprehension and a good memory; (8) a musical, regular, and harmonious disposition." But how is this disposition to be secured? Under the present condition of things, and the corrupting influences of various kinds, where temptations abound to compromise truth and substitute expediency and self-interest, it would seem wellnigh impossible and Utopian to expect it.

"How is this evil to be remedied? The State itself must regulate the study of philosophy, and must take care that the students pursue it on right principles, and at a right age. And now, surely, we may expect to be believed when we assert that if a State is to prosper it must be governed by philosophers. If such a contingency should ever take place (and why should it not?), our ideal State will undoubtedly be realised. So that, upon the whole, we come to this conclusion: The constitution just described is the best, if it can be realised; and to realise it is difficult, but not impossible." At this moment, when the question of compulsory education, under the immediate superintendence of the State, is being fought with so much fierceness—on one side, at least—to recur to Plato might not be without advantage.

In the most famous dialogue of Plato—the *Republic*, or, as it might

B

be termed *On Justice*—the principal interlocutors, besides Sokrates, are Glaukon, Polymachus, and Adeimantus; and the whole piece originates in the chance question which rose between them, "What is Justice?" In the second Book, from which the following passage is taken, the discussion turns upon the origin of society, which gives opportunity to Sokrates to develop his opinions upon the diet best adapted for the community—at all events, for the great majority :—

" 'They [the artisans and work-people generally] will live, I suppose, on barley and wheat, baking cakes of the meal, and kneading loaves of the flour. And spreading these excellent cakes and loaves upon mats of straw or on clean leaves, and themselves reclining on rude beds of yew or myrtle-boughs, they will make merry, themselves and their children, drinking their wine, weaving garlands, and singing the praises of the gods, enjoying one another's society, and not begetting children beyond their means, through a prudent fear of poverty or war.'

"Glaukon here interrupted me, remarking, 'Apparently you describe your men as feasting, without anything to relish their bread.' *

" 'True,' I said, 'I had forgotten. Of course they will have something to relish their food. Salt, no doubt, and olives, and cheese, together with the country fare of boiled onions and cabbage. We shall also set before them a dessert, I imagine, of figs, pease, and beans : they may roast myrtle-berries and beech-nuts at the fire, taking wine with their fruit in moderation. And thus, passing their days in tranquillity and sound health, they will, in all probability, live to an advanced age, and dying, bequeath to their children a life in which their own will be reproduced.'

"Upon this Glaukon exclaimed, 'Why, Sokrates, if you were founding a community of swine, this is just the style in which you would feed them up !'

" 'How, then,' said I, 'would you have them live, Glaukon ?'

" 'In a civilised manner,' he replied. 'They ought to recline on couches, I should think, if they are not to have a hard life of it, and dine off tables, and have the usual dishes and dessert of a modern dinner.'

" 'Very good : I understand. Apparently we are considering the growth, not of a city merely, but of a *luxurious* city. I dare say it is not a bad plan, for by this extension of our inquiry we shall perhaps discover how it is that justice and injustice take root in cities. Now, it appears to me that the city which we have described is the *genuine* and, so to speak, *healthy* city. But if you wish us also to contemplate a city that is suffering from inflammation, there is nothing to hinder us. Some people will not be satisfied, it seems, with the fare or the mode of life which we have described, but must have, in addition, couches and tables and every other article of furniture, as well as viands . . . . . . Swineherds again are among the additions we shall require—a class of persons not to be found, because not wanted, in our former city, but needed among the rest in this. We shall also need great quantities of all kinds of cattle for those who may wish to eat them, shall we not ?'

" 'Of course we shall.'

" 'Then shall we not experience the need of medical men also to a much greater extent under this than under the former *régime* ?'

" 'Yes, indeed.'

---

* Ὄψον—the name given by the Greeks generally to everything which they considered rather as a "relish" than a necessary. Bread was held to be—not only in name but in fact—the veritable "staff of life." Olives, figs, cheese, and, at Athens especially, fish were the ordinary Ὄψον.

" 'The country, too, I presume, which was formerly adequate to the support of its then inhabitants, will be now too small, and adequate no longer. Shall we say so ?'

" 'Certainly.'

" 'Then must we not cut ourselves a slice of our neighbours' territory, if we are to have land enough both for pasture and tillage ? While they will do the same to ours if they, like us, permit themselves to overstep the limit of necessaries, and plunge into the unbounded acquisition of wealth.'

" 'It must inevitably be so, Sokrates.'

" 'Will our next step be to go to war, Glaukon, or how will it be ?'

" 'As you say.'

" At this stage of our inquiry let us avoid asserting either that war does good or that it does harm, confining ourselves to this statement—that we have further traced the origin of war to causes which are the most fruitful sources of whatever evils befall a State, either in its corporate capacity or in its individual members." (Book II.) *

Justly holding that the best laws will be of little avail unless the administrators of them shall be just and virtuous, Sokrates, in the Third Book, proceeds to lay down rules for the education and diet of the magistrates or executive, whom he calls—in conformity with the Communistic system—*guardians :*—

" 'We have already said,' proceeds Sokrates, 'that the persons in question must refrain from drunkenness ; for a guardian is the last person in the world, I should think, to be allowed to get drunk, and not know where he is.'

" 'Truly it would be ridiculous for a guardian to require a guard.'

" 'But about eating : our men are combatants in a most important arena, are they not ?'

" 'They are.'

" 'Then will the habit of body which is cultivated by the trained fighters of the Palæstra be suitable to such persons ?'

" 'Perhaps it will.'

" 'Well, but this is a sleepy kind of regimen, and produces a precarious state of health ; for do you not observe that men in the regular training sleep their life away, and, if they depart only slightly from the prescribed diet, are attacked by serious maladies in their worst form ?'

" 'I do.'

*   *   *   *   *   *

" 'In fact, it would not be amiss, I imagine, to compare this whole system of feeding and living to that kind of music and singing which is adapted to the panharmonicum, and composed in every variety of rhythm.'

" 'Undoubtedly it would be a just comparison.'

" 'Is it not true, then, that as in music variety begat dissoluteness in the soul, so here it begets disease in the body, while simplicity in gymnastic [diet] is as productive of health as in music it was productive of temperance ?'

" 'Most true.'

" 'But when dissoluteness and diseases abound in a city, are not law courts and

---

* Translated by Davies and Vaughan. 1874.

surgeries opened in abundance, and do not Law and Physic begin to hold their heads high, when numbers even of well-born persons devote themselves with eagerness to these professions ?'

"'What else can we expect ?'

*   *   *   *   *   *

"'And do you not hold it disgraceful to require medical aid, unless it be for a wound, or an attack of illness incidental to the time of the year—to require it, I mean, owing to our laziness and the life we lead, and to get ourselves so stuffed with humours and wind, like quagmires, as to compel the clever sons of Asklepios to call diseases by such names as *flatulence* and *catarrh ?*'

"'To be sure, these are very strange and new-fangled names for disorders.'" (Book III.)

Elsewhere, in a well-known passage (in *The Laws*,), Plato pronounces that the springs of human conduct and moral worth depend principally on diet. "I observe," says he, "that men's thoughts and actions are intimately connected with the threefold need and desire (accordingly as they are properly used or abused, virtue or its opposite is the result) of eating, drinking, and sexual love." He himself was remarkable for the extreme frugality of his living. Like most of his countrymen, he was a great eater of figs; and so much did he affect that frugal repast that he was called, *par excellence*, the "lover of figs" (φιλόσυκος).

The Greeks, in general, were noted among the Europeans for their abstemiousness; and Antiphanes, the comic poet (in Athenæus), terms them "leaf-eaters" (φυλλοτρῶγες). Amongst the Greeks, the Athenians and Spartans were specially noted for frugal living. That of the latter is proverbial. The comic poets frequently refer, in terms of ridicule, to what seemed to them so unaccountable an indifferentism to the "good things" of life on the part of the witty and refined people of Attica. See the *Deipnosophists* (dinner-philosophers) of Athenæus (the great repertory of the *bon-vivantism* of the time), and Plutarch's *Symposiacs*.

It has been pointed out by Professor Mahaffy, in his recent work on old Greek life, that slaughter-houses and butchers are seldom, or never, mentioned in Greek literature. "The eating of [flesh] meat," he observes, "must have been almost confined to sacrificial feasts; for, in ordinary language, butchers' meat was called *victim* (ἱερεῖον). The most esteemed, or popular, dishes were *madsa*, a sort of porridge of wheat or barley; various kinds of bread (see *Deipn.* iii.); honey, beans, lupines, lettuce and salad, onions and leeks. Olives, dates, and figs formed the usual fruit portion of their meals. In regard to non-vegetable food, fish was the most sought after and preferred to anything else; and the well-known term *opson*, which so frequently recurs in Greek literature, was specially appropriated to it.

Contemporary with the great master of language was the great master
of medicine, Hippokrates, (460—357) who is to his science what Homer
is to poetry and Herodotus to history—the first historical founder of
the art of healing. He was a native of Kōs, a small island of the S.W.
coast of Lesser Asia, the traditional cradle and home of the disciples of
Asklepios, or Æsculapius (as he was termed by the Latins), the semi-
divine author and patron of medicine. And it may be remarked, in
passing, that the College of Asklepiads of Kōs were careful to exercise a
despotism as severe and exclusive as that which obtains, for the most
part, with the modern orthodox schools.

Amongst a large number of writings of various kinds attributed to
Hippokrates is the treatise *On Regimen in Acute Diseases* (περὶ Διαίτης
Ὀξέων), which is generally received as genuine; and *On the Healthful
Regimen* (περὶ Διαίτης Ὑγιεινῆς), which belongs to the same age, though
not to the *canonical* writings of the founder of the school himself. He
was the author, real or reputed, of some of the most valuable
apophthegms of Greek antiquity. *Ars longa—Vita brevis* (education is
slow; life is short) is the best known, and most often quoted. What is
still more to our purpose is his maxim—"Over-drinking is *almost as bad*
as over-eating." Of all the productions of this most voluminous of
writers, his *Aphorisms* (Ἀφορισμοί), in which these specimens of laconic
wisdom are collected, and which consists of some four hundred short
practical sentences, are the most popular.

About a century after the death of Plato appeared a popular exposition
of the Pythagorean teaching, in hexameters, which is known by the title
given to it by Iamblichus—the *Golden Verses*. "More than half of
them," says Professor Clifford, "consist of a sort of versified 'Duty to
God and my Neighbour,' except that it is not designed by the rich to be
obeyed by the poor; that it lays stress on the laws of health; and that it
is just such sensible counsel for the good and right conduct of life as an
Englishman might now-a-days give to his son."

Hierokles, an eminent Neo-Platonist of the fifth century, A.D., gave
a course of lectures upon them at Alexandria—which since the time of
the Ptolemies had been one of the chief centres of Greek learning and
science—and his commentary is sufficiently interesting. Suïdas, the
lexicographer, speaks of his matter and style in the highest terms of
praise. "He astonished his hearers everywhere," he tells us, "by the
calm, the magnificence, the width of his superlative intellect, and by
the sweetness of his speech, full of the most beautiful words and things."
The Alexandrian lecturer quotes the old Pythagorean maxims:

"You shall honour God best by becoming godlike] in your thoughts. Whoso
giveth God honour as to one that needeth it, that man in his folly hath made himself

greater than God. The wise man only is a priest, is a lover of God, is skilful to pray ; ... for that man only knows how to worship, who begins by offering himself as the victim, fashions his own soul into a divine image, and furnishes his mind as a temple for the reception of the divine light."

The following extracts will serve as a specimen of the religious or moral character of the *Golden Verses :—*

"Let not sleep come upon thine eyelids till thou hast pondered thy deeds of the day.

" Wherein have I sinned? What work have I done, what left undone that I ought to have done ?

" Beginning at the first, go through even unto the last, and then let thy heart smite thee for the evil deeds, but rejoice in the good work.

"Work at these commandments and think upon them : these commandments shalt thou love.

"They shall surely set thee in the way of divine righteousness : yea, by Him who gave into our soul the *Tetrad,** well-spring of life everlasting.

*   *   *   *   *   *   *

"Know so far as is permitted thee, that Nature in all things is like unto herself :

"That thou mayest not hope that of which there is no hope, nor be ignorant of that which may be.

"Know thou also, that *the woes of men are the work of their own hands.*

"Miserable are they, *because they see not and hear not the good that is very nigh them :* and the way of escape from evil few there be that understand it.

*   *   *   *   *   *   *

"Verily, Father Zeus, thou wouldst free all men from much evil, if thou wouldst teach all men what manner of spirit they are of.

*   *   *   *   *   *   *

"Keep from the meats aforesaid, using judgment both in cleansing and setting free the soul.

"Give heed to every matter, and set reason on high, who best holdeth the reins of guidance.†

"Then when thou leavest the body, and comest into the free æther, thou shalt be a god undying, everlasting, neither shall death have any more dominion over thee."

Referring to these verses, which inculcate that the human race is itself responsible for the evils which men, for the most part, prefer to regret than to remedy, Professor Clifford, to whom we are indebted for the above version of the *Golden Verses,* remarks on the merits of this teaching, that it reminds us that " men suffer from *preventible* evils, that the people perish for lack of knowledge." ‡   Thus we find that the

---

* The *four* sacred Pythagorean virtues--justice, temperance, wisdom, fortitude. See notice of Plato above.

† Upon which excellent maxim Hierokles justly remarks : " The judge here appointed is the most just of all, and the one which is [ought to be] most at home with us, viz.: conscience and right reason."

‡ *Nineteenth Century,* October, 1877. The Greek original of the *Golden Verses* is found in the text of Mullach, in *Fragmenta Philosophorum Græcorum.* Paris, 1860.

principal obstructions, in all ages, to human progress and perfectibility may be ever found in IGNORANCE and SELFISHNESS.

———◆•◦•◆———

IV.

OVID. 43 B.C.—18 A.D.

THE school of Pythagoras and of Plato, although it was not the fashionable or popular religion of Rome, counted amongst its disciples some distinguished Italians, and the name of Cicero, who belonged to the "New Academy," is sufficiently illustrious. The Italians, however, who borrowed their religion as well as their literature from the Greeks, were never distinguished, like their masters, for that refinement of thought which might have led them to attach themselves to the Pythagorean teaching. Under the bloody despotism of the Empire, the philosophy which was most affected by the *literati* and those who were driven to the consolations of philosophy was the *stoical*, which taught its disciples to consider *apathy* as the *summum bonum* of existence. This school of philosophy, whatever its other merits, was too much centred in self—paradoxical as the assertion may seem—to have much regard for the rest of mankind, much less for the non-human species. Nor, while they professed supreme contempt for the luxuries and even comforts of life, did the disciples of the "Porch," in general, practice abstinence from any exalted motive, humanitarian or spiritual. They preached indifference for the "good things" of this life, not so much to elevate the spiritual and moral side of human nature as to show their contempt for human life altogether.

That the Italian was essentially of a more barbarous nature than the Greek is apparent in the national spectacles and amusements. The savage scenes of gladiatorial and non-human combat and internecine slaughter of the Latin amphitheatres, of which the famous Colosseum in the capital was the model of many others in the provinces, were abhorrent to the more refined Greek mind.* In view of scenes so sanguinary—the "Roman holiday"—it is scarcely necessary to observe that humanitarianism was a creed unknown to the Italians ; and it was not likely that a people, addicted throughout their career as a dominant race to the most bloody wars, not only foreign but also internecine, with whom fighting and slaughter of their own kind was an almost daily occupation, should entertain any feeling of pity (to say nothing of justice)

---

* The Romans, we may remark, imported the gladiatorial fights from Spain.

towards their non-human dependants. Nevertheless, even they were not wholly inaccessible, on occasion, to the prompting of pity. Referring to a grand spectacle given by Pompeius at the dedication of his theatre (B.C. 55), in which a large number of elephants, amongst others, were forced to fight, the elder Pliny tells us :—

"When they lost the hope of escape, they sought the compassion of the crowd with an appearance that is indescribable, bewailing themselves with a sort of lamentation so much to the pain of the populace that, forgetful of the imperator and the elaborate munificence displayed for their honour, they all rose up in tears and bestowed imprecations on Pompeius, of which he soon after experienced the effect."*

Cicero, who was himself present at the spectacle of the Circus, in a letter to a friend, Marcus Marius, writes :—

"What followed, for five days, was successive combats between a man and a wild beast. (*Venationes binæ.*) It was magnificent. No one disputes it. But what pleasure can it be to a person of refinement, when either a weak man is torn to pieces by a very powerful beast, or a noble animal is struck through by a hunting spear ? . .
The last day was that of the elephants, in which there was great astonishment on the part of the populace and crowd, but no enjoyment. Indeed there followed a degree of compassion, and a certain idea that there is a sort of fellowship between that huge animal and the human race." (Cicero, *Ep. ad Diversos* vii., 1.)

Testimonies which might induce one almost to think that, had not they been systematically and industriously accustomed to these horrible and gigantic butcheries by their rulers, even the Roman populace might have been susceptible of better feelings and desires than those inspired by their amphitheatres, though these savage exhibitions were perhaps hardly worse than the combats and slaughter in the bull-rings of Seville or Madrid, or at the courts of the Mohammedan princes of India recently sanctioned by the presence of English royalty. It is worth noting, in passing, that while the *gladiatorial* slaughters were discontinued some years after the triumph of Christianity, the other part of the entertainment—the indiscriminate combats and slaughter of the *non-human* victims—continued to be exhibited to a much later period.

If we reflect that the rise of the humanitarian spirit in Christian Europe, or rather in the better section of it, is of very recent origin, it might appear unreasonable to look for any distinct exhibition of so exalted a feeling in the younger age of the world. Yet, to the shame of more advanced civilisations, we find manifestations of it in the writings of a few of the more refined minds of Greece and Italy ; and Plutarch

---

* *Hist. Naturalis VIII. 7.* His nephew says of these huge slaughter-houses that "there is no novelty, no variety, or anything that could not be seen once for all." On one occasion, in the year A.D. 284, we are credibly informed that 1,000 ostriches, 1,000 stags, 1,000 fallow-deer, besides numerous wild sheep and goats, were mingled together for indiscriminate slaughter by the wild beasts of the forest or the equally wild beasts of the city. (See *Decline and Fall.*)

and Seneca—the former particularly—occupy a distinguished place amongst the first preachers of that sacred truth.*

Publius Ovidius Naso, the Latin versifier of the Pythagorean philosophy, was born B.C. 43. He belonged to the equestrian order, a position in the social scale which corresponds with the "higher middle class" of modern days. Like so many other names eminent in literature, he was in the first instance educated for the law, for which, also like many other literary celebrities, he soon showed his genius to be unfitted and uncongenial. He studied at the great University of that age—Athens—where he acquired a knowledge of the Greek language, and probably of its rich literature. The most memorable event in his life—which, in accordance with the fashion of his contemporaries of the same rank, was for the most part devoted to "gallantry" and the accustomed amatory licence—is his mysterious banishment from Rome to the inhospitable and savage shores of the Euxine, where he passed the last seven years of his existence, dying there in the sixtieth year of his age. The cause of his sudden exile from the Court of Augustus, where he had been in high favour, is one of those secrets of history which have exercised the ingenuity of his successive biographers. According to the terms of the imperial edict, the freedom of the poet's *Ars Amatoria* was the offence. That this was a mere pretext is plain, as well from the long interval of time which had passed since the publication of the poem as from the character of the fashionable society of the capital. Ovid himself attributes his misfortune to the fact of his having become the involuntary witness of some secret of the palace, the nature of which is not divulged.

His most important poems are (1) *The Metamorphoses*, in fifteen books, so called from its being a collection of the numerous transformations of the popular theology. It is, perhaps, the most *charming* of Latin poems that have come down to us. Particular passages have a special beauty. (2) *The Fasti*, in twelve books, of which only six are extant, is the Roman Calendar in verse. Its interest, apart from the poetic genius of the author, is great, as being the grand repertory of the Latin feasts and their popular origin. Besides these two principal poems he was the author of the famous *Loves*, in three books; the *Letters of the Heroines*, *The Remedies of Love*, and *The Tristia, or Sad Thoughts*. He also wrote a tragedy—*Medea*—which, unfortunately has not come down to us. All his poems are characterised by elegance and a remark-

---

* Some traces of it may be found, *e.g.*, in Lucretius *(De Rerum Nat. II.*, where see his touching picture of the bereaved mother-cow, whose young is ravished from her for the horrid sacrificial altar); Virgil *(Æneis VII.)*, in his story of Silvia's deer—the most touching passage in the poem; Pliny, *Hist. Nat.* In earlier Greek literature, Euripides seems most in sympathy with suffering—at least as regards his own species.

able smoothness and regularity of versification, and in much of his productions there is an unusual beauty and picturesqueness of poetic ideas.

The following passage from the fifteenth book of the *Metamorphoses*
has been justly said by Dryden, his translator, to be the finest part of
the whole poem. It is almost impossible to believe but that, in spite of
his misspent life, he must have felt, in his better moments at least, something of the truth and beauty of the Pythagorean principles which he so
exquisitely versifies. In the touching words which he puts into the
mouth of the jealous Medea—the murderess of her children—he might
have exclaimed in his own case—

> "Video meliora proboque
> Deteriora sequor." *

"He [Pythagoras], too, was the first to forbid animals to be served up at the table,
and he was first to open his lips, indeed full of wisdom yet all unheeded, in the
following words : 'Forbear, O mortals! to pollute your bodies with such abominable
food. There are the *jarinacea (fruges)*, there are the fruits which bear down the
branches with their weight, and there are the grapes swelling on the vines ; there are
the sweet herbs ; there are those that may be softened by the flame and become
tender. Nor is the milky juice denied you ; nor honey, redolent of the flower of
thyme. The lavish Earth heaps up her riches and her gentle foods, and offers you
dainties without blood and without slaughter. The lower animals satisfy their
ravenous hunger with flesh. And yet not all of them ; for the horse, the sheep, the
cows and oxen subsist on grass ; while those whose disposition is cruel and fierce, the
tigers of Armenia and the raging lions, and the wolves and bears, revel in their
bloody diet.

"'Alas! what a monstrous crime it is *(scelus)* that entrails should be entombed in
entrails ; that one ravening body should grow fat on others which it crams into it ;
that one living creature should live by the death of another living creature ! Amid
so great an abundance which the Earth—that best of mothers—produces does, indeed,
nothing delight you but to gnaw with savage teeth the sad produce of the wounds you
inflict and to imitate the habits of the Cyclops ? Can you not appease the hunger of
a voracious and ill-regulated stomach unless you first destroy another being ? Yet
that age of old, to which we have given the name of *golden*, was blest in the produce
of the trees and in the herbs which the earth brings forth, and the human mouth was
not polluted with blood.

"'Then the birds moved their wings secure in the air, and the hare, without fear,
wandered in the open fields. Then the fish did not fall a victim to the hook and its
own credulity. Every place was void of treachery ; there was no dread of injury—all
things were full of peace. In later ages some one—a mischievous innovator *(non
utilis auctor)*, whoever he was—set at naught and scorned this pure and simple food,
and engulfed in his greedy paunch victuals made from a carcase. It was he that
opened the road to wickedness. I can believe that the steel, since stained with blood,
was first dipped in the gore of savage wild beasts ; and that was lawful enough. We
hold that the bodies of animals that seek our destruction are put to death without any
breach of the sacred laws of morality. But although they might be put to death

---

* I see and approve the better way ; I pursue the worse.—*Metam.* vii., 20.

they were not to be eaten as well. From this time the abomination advanced rapidly. The swine is believed to have been the first victim destined to slaughter, because it grubbed up the seeds with its broad snout, and so cut short the hopes of the year. For gnawing and injuring the vine the goat was led to slaughter at the altars of the avenging Bacchus. Its own fault was the ruin of each of these victims.

" 'But how have you deserved to die, ye sheep, you harmless breed that have come into existence for the service of men—who carry nectar in your full udders—who give your wool as soft coverings for us—who assist us more by your life than by your death ? Why have the oxen deserved this—beings without guile and without deceit—innocent, mild, born for the endurance of labour ? Ungrateful, indeed, is man, and unworthy of the bounteous gifts of the harvest who, after unyoking him from the plough, can slaughter the tiller of his fields—who can strike with the axe that neck worn bare with labour, through which he had so often turned up the hard ground, and which had afforded so many a harvest.

" 'And it is not enough that such wickedness is committed by men. They have involved the gods themselves in this abomination, and they believe that a Deity in the heavens can rejoice in the slaughter of the laborious and useful ox. The spotless victim, excelling in the beauty of its form (for its very beauty is the cause of its destruction), decked out with garlands and with gold is placed before their altars, and, ignorant of the purport of the proceedings, it hears the prayers of the priest. It sees the fruits which it cultivated placed on its head between its horns, and, struck down, with its life-blood it dyes the sacrificial knife which it had perhaps already seen in the clear water. Immediately they inspect the nerves and fibres torn from the yet living being, and scrutinise the will of the gods in them.

" 'From whence such a hunger in man after unnatural and unlawful food ? Do you dare, O mortal race, to continue to feed on flesh ? Do it not, I beseech you, and give heed to my admonitions. And when you present to your palates the limbs of slaughtered oxen, know and feel that you are feeding on the tillers of the ground.' "— *Metam.* xv., 73—142.

V.

SENECA. DIED 65 A.D.

LUCIUS ANNÆUS SENECA, the greatest name in the stoic school of philosophy, and the first of Latin moralists, was born at Corduba (Cordova) almost contemporaneously with the beginning of the Christian era. His family, like that of Ovid, was of the equestrian order. He was of a weakly constitution ; and bodily feebleness, as with many other great intellects, served to intensify if not originate, the activity of the mind. At Rome, with which he early made acquaintance, he soon gained great distinction at the bar ; and the eloquence and fervour he displayed in the Senate before the Emperor Caligula excited the jealous hatred of that insane tyrant. Later in life he obtained a prætorship, and he was also appointed to the tutorship of the young Domitius, afterwards the Emperor Nero. On the accession of that prince, at the age of seventeen, to the imperial throne, Seneca became one of his chief advisers.

Unfortunately for his credit as a philosopher, while exerting his influence to restrain the vicious propensities of his old pupil, he seems to have been too anxious to acquire, not only a fair proportion of wealth, but even an enormous fortune, and his villas and gardens were of so splendid a kind as to provoke the jealousy and covetousness of Nero. This, added to his alleged disparagement of the prince's talents, especially in singing and driving, for which Nero particularly desired to be famous, was the cause of his subsequent disgrace and death. The philosopher prudently attempted to anticipate the will of Nero by a voluntary surrender of all his accumulated possessions, and he sought to disarm the jealous suspicions of the tyrant by a retired and unostentatious life. These precautions were of no avail ; his death was already decided. He was accused of complicity in the conspiracy of Piso, and the only grace allowed him was to be his own executioner. The despair of his wife, Pompeia Paulina, he attempted to mitigate by the reflection that his life had been always directed by the standard of a higher morality. Nothing, however, could dissuade her from sharing her husband's fate, and the two faithful friends laid open their veins by the same blow.

Advanced age and his extremely meagre diet had left little blood in Seneca's veins, and it flowed with painful slowness. His tortures were excessive and, to avoid the intolerable grief of being witnesses of each other's suffering, they shut themselves up in separate apartments. With that marvellous intrepid tranquillity which characterised some of the old sages, Seneca calmly dictated his last thoughts to his surrounding friends. These were afterwards published. His agonies being still prolonged, he took hemlock ; and this also failing, he was carried into a vapour-stove, where he was suffocated, and thus at length ceased to suffer.

In estimating the character of Seneca, it is just that we should consider all the circumstances of the exceptional times in which his life was cast. Perhaps there has never been an age or people more utterly corrupt and abandoned than that of the period of the earlier Roman Cæsars and that of Rome and the large cities of the empire. Allowing the utmost that his detractors have brought against him, the moral character of the author of the *Consolations* and *Letters* stands out in bright relief as compared with that of the immense majority of his contemporaries of equal rank and position, who were sunk in the depths of licentiousness and of selfish indifference to the miseries of the surrounding world. That his public career was not of so exalted a character altogether as are his moral precepts, is only too patent to be denied and, in this shortcoming of a loftier *ideal*, he must share reproach with some of the most esteemed of the world's luminaries. If, for instance, we compare him with Cicero or with Francis Bacon, the comparison would certainly be not unfavourable to Seneca. The darkest stigma on the reputation of the great Latin moralist is his connivance at the death of the infamous Agrippina, the mother of his pupil Nero. Although not to be excused, we may fairly attribute this act to conscientious, if mistaken, motives. His best apology is to be found in the fact that, so long as he assisted to direct the counsels of Nero, he contrived to restrain that prince's depraved disposition from those outbreaks which, after the death of the philosopher, have stigmatised the name of Nero with undying infamy.

The principal writings of Seneca are :—

1. *On Anger.* His earliest, and perhaps his best known, work.

2. *On Consolation.* Addressed to his mother, Helvia. An admirable philosophical exhortation.

3. *On Providence ; or, Why evils happen to good men though a divine Providence may exist."*

4. *On Tranquillity of Mind.*

5. *On Clemency.* Addressed to Nero Cæsar. One of the most 'meritorious writings of all antiquity. It is not unworthy of being classed with the humanitarian protests of Beccaria and Voltaire. The stoical distinction between clemency and pity *(misericordia)*, in book ii., is, as Seneca admits, merely a dispute about words.

6. *On the Shortness of Life.* In which the proper employment of time and the acquisition of wisdom are eloquently enforced as the best employment of a fleeting life.

7. *On a Happy Life.* In which he inculcates that there is no happiness without virtue. An excellent treatise.

8. *On Kindnesses.*

9. *Epistles to Lucilius.* 124 in number. They abound in lessons and precepts in morality and philosophy, and, excepting the *De Irâ*, have been the most read, perhaps, of all Seneca's productions.

10. *Questions on Natural History.* In seven books.

Besides these moral and philosophic works, he composed several tragedies. They were not intended for the stage, but rather as moral lessons. As in all his works, there is much of earnest thought and feeling, although expressed in rhetorical and declamatory language.

What especially characterises Seneca's writings is their remarkably *humanitarian* spirit. Altogether he is imbued with this, for the most part, very modern feeling in a greater degree than any other writer, Greek or Latin. Plutarch indeed, in his noble *Essay on Flesh Eating*, is more expressly denunciatory of the barbarism of the Slaughter House, and of the horrible cruelties inseparably connected with it, and evidently felt more deeply the importance of exposing its evils. The Latin moralist, however, deals with a wider range of ethical questions, and on such subjects, as, *e.g.*, the relations of master and slave, is far ahead of his contemporaries. His treatment of *Dietetics*, in common with that of most of the old-world moralists, is rather from the spiritual and ascetic than from the purely humanitarian point of view. "The judgments on Seneca's writings," says the author of the article on Seneca in Dr. Smith's *Dictionary of Greek and Latin Biography*, "have been as various as the opinions about his character, and both in extremes. It has been said of him that he looks best in quotations; but this is an admission that there is something worth quoting, which cannot be said of all writers. That Seneca possessed great mental powers cannot be doubted. He had seen much of human life, and he knew well what man is. His philosophy, so far as he adopted a system, was the stoical; but it was rather an eclecticisim of stoicism than pure stoicism. His style is antithetical, and apparently laboured; and where there is much labour there is generally affectation. Yet his language is clear and forcible—it is not mere words—there is thought always. It would not be easy to name any modern writer, who has treated on morality and has said so much that is practically good and true, or has treated the matter in so attractive a way."

Jerome, in his *Ecclesiastical Writers*, hesitates to include him in the catalogue of his saints only because he is not certain of the genuineness of the alleged literary correspondence between Seneca and St. Paul. We may observe, in passing, on the remarkable coincidence of the presence of the two greatest teachers of the old and the new faiths in the capital of the Roman Empire at the same time; and it is possible, or rather highly probable, that St. Paul was acquainted with the writings of Seneca; while, from the total silence of the pagan philosopher, it seems that he knew nothing of the Pauline epistles or teaching. Amongst many testimonies to the superiority of Seneca, Tacitus, the great historian of the empire, speaks of the "splendour and celebrity of his philosophic writings," as well as of his "amiable genius"—*ingenium amœnum. (Annals,* xii., xiii.) The elder Pliny writes of him as "at the very head of all the learned men of that time." (xiv. 4.) Petrarch quotes the testimony of Plutarch, "that great man who, Greek though he was freely confesses 'that there is no Greek writer who could be brought into comparison with him in the department of *morals.*'"

The following passage is to be found in a letter to Lucilius, in which, after expatiating on the sublimity of the teaching of the philosopher Attalus in inculcating moderation and self-control in corporeal pleasures, Seneca thus enunciates his *dietetic* opinions:—

"Since I have begun to confide to you with what exceeding ardour I approached the study of philosophy in my youth, I shall not be ashamed to confess the affection with which Sotion [his preceptor] inspired me for the teaching of Pythagoras. He was wont to instruct me on what grounds he himself, and, after him, Sextius, had determined to abstain from the flesh of animals. Each had a different reason, but the reason in both instances was a grand one *(magnifica.)* Sotion held that man can find a sufficiency of nourishment without blood shedding, and that cruelty became habitual when once the practice of butchering was applied to the gratification of the appetite. He was wont to add that ' It is our bounden duty to limit the materials of luxury. That, moreover, variety of foods is injurious to health, and not natural to our bodies. If these maxims [of the Pythagorean school] are true, then to abstain from the flesh of animals is to encourage and foster *innocence;* if ill-founded, at least they teach us frugality and simplicity of living. And what loss have you in losing your cruelty ? (Quod istic crudelitatis tuæ damnum est ?) I merely deprive you of the food of lions and vultures.'

" Moved by these and similar arguments, I resolved to abstain from flesh meat, and at the end of a year the habit of abstinence was not only easy but delightful. I firmly believed that the faculties of my mind were more active,* and at this day I will not take pains to assure you whether they were so or not. You ask, then, ' Why did you go back and relinquish this mode of life ? ' I reply that the lot of my early days was cast in the reign of the emperor Tiberius. Certain foreign religions became the object of the imperial suspicion, and amongst the proofs of adherence to the foreign cultus or superstition was that of abstinence from the flesh of animals. At the entreaties of my father, therefore, who had no real fear of the practice being made a ground of accusation, but who had a hatred of philosophy,† I was induced to return to my former dietetic habits, nor had he much difficulty in persuading me to recur to more sumptuous repasts. . . .

" This I tell," he proceeds, " to prove to you how powerful are the early impetuses of youth to what is truest and best under the exhortations and incentives of virtuous teachers. We err partly through the fault of our guides, who teach us *how to dispute,* not *how to live;* partly by our own fault in expecting our teachers to cultivate not so much the *disposition of the mind* as the faculties of the intellect. Hence it is that in place of a love of wisdom there is only a love of words (Itaque quæ *philosophia* fuit, facta *philologia* est)."—*Epistola* cviii. ‡

Seneca here cautiously reveals the jealous suspicion with which the first Cæsars viewed all foreign, and especially quasi-religious, innovations, and his own *public* compliance, to some extent, with the orthodox dietetic practices. Yet that in private life he continued to practise, as well as to preach, a radical dietary reformation is sufficiently evident to all who are conversant with his various writings. The refinement and gentleness of his ethics are everywhere apparent, and exhibit him as a man of extraordinary sensibility and feeling.

As for *dietetics,* he makes it a matter of the first importance, on which he is never weary of insisting. " *We must so live, not as if we ought*

---

* In a note on this passage Lipsius, the famous Dutch commentator, remarks : "I am quite in accord with this feeling. The constant use of flesh meat (*assidua* κρεοφαγία) by Europeans makes them stupid and irrational (*brutos*).

† Lipsius suggests, with much reason, that Seneca actually wrote the opposite respecting his father, " who had no dislike for this philosophy, but who feared calumny," &c

‡ On this melancholy truth compare Montaigne's *Essais.*

to live for, but as though we could not do without, the body."   He quotes
Epikurus : " *If you live according to nature, you will never be poor ; if
according to conventionalism, you will never be rich.   Nature demands
little ; fashion* (opinio) *superfluity."*   In one of his letters he eloquently
describes the riotous feasting of the period which corresponds to our
festival of Christmas—another illustration of the proverb, " History
repeats itself " :—

"December is the month," he begins his letter, "when the city [Rome] most
especially gives itself up to riotous living *(desudat).* Free licence is allowed to the
public luxury. Every place resounds with the gigantic preparations for eating and
gorging, just as if," he adds, " the whole year were not a sort of *Saturnalia.*"

He contrasts with all this waste and gluttony the simplicity and
frugality of Epikurus, who, in a letter to his friend Polyænus, declares
that his own food does not cost him sixpence a day; while his friend
Metrodorus, who had not advanced so far in frugality, expended the
whole of that small sum :—

"Do you ask if that can supply due nourishment? Yes ; and pleasure too. Not,
indeed, that fleeting and superficial pleasure which needs to be perpetually recruited,
but a solid and substantial one. Bread and pearl-barley *(polenta)* certainly is not
luxurious feeding, but it is no little advantage to be able to receive pleasure from a
simple diet of which no change of fortune can deprive one. . . . Nature demands
bread and water only : no one is poor in regard to those necessaries." *

Again, Seneca writes :—

"How long shall we weary heaven with petitions for superfluous luxuries, as though
we had not at hand wherewithal to feed ourselves ? How long shall we fill our plains

---

* Ep. xxv.   Lipsius here quotes Lucan "still more a philosopher than a poet " :—

> "*Discite quam parvo liceat producere vitam,
> Et quantum natura petat.
>   . . Satis est populis fluviusque Ceresque."*

"Learn by how little life may be sustained, and how much nature requires. The gifts of Ceres
and water are sufficient nourishment for all peoples."—(*Pharsalia.*)

Also Euripides :—

> " 'Επεὶ τί δεῖ βροτοῖσι . . . .
>   . . . πλὴν δύοιν μόνον,
> Δημητρὸς ἀκτῆς, πώματος θ'ὑδρηχόου,
> ῞Απερ πάρεστι καὶ πέφυχ' ἡμᾶς τρέφειν ;
> ῟Ων οὐκ ἀπαρκεῖ πλησμονή · τρυφῇ γέ τοι
> 'Αλλων ἐδεστῶν μηχανὰς θηρεύομεν."

Which may be translated :—

> "*Since what need mortals, save twain things alone,
> Crush'd grain (heaven's gift), and streaming water-draught ?
> Food nigh at hand, and nature's aliment—
> Of which no glut contents us. Pampered taste
> Hunts out device of other eatables."*

(Fragment of lost drama of Euripides, preserved in *Athenæus* iv. and in *Gellius* vii.)

See, too, the elder Pliny, who professes his conviction that "the plainest food is also the most
beneficial " *(cibus simplex utilissimus),* and asserts that it is from his eating that man derives
most of his diseases, and from thence that all the drugs and all the arts of physicians abound.
(Hist. Nat. xxvi., 28.)

with huge cities ? How long shall the people slave for us unnecessarily ? How long shall countless numbers of ships from every sea bring us provisions for the consumption of a single month ? An Ox is satisfied with the pasture of an acre or two : one wood suffices for several Elephants. Man alone supports himself by the pillage of the whole earth and sea. What ! Has Nature indeed given us so insatiable a stomach, while she has given us so insignificant bodies ? No : it is not the hunger of our stomachs, but insatiable covetousness (*ambitio*) which costs so much. The slaves of the belly (as says Sallust) are to be counted in the number of the lower animals, not of men. Nay, not of them, but rather of the dead. . . . You might inscribe on their doors, ' These have anticipated death.' "—(*Ep.* lx.)

The extreme difficulty of abstinence is oftentimes alleged :—

"It is disagreeable, you say, to abstain from the pleasures of the customary diet. Such abstinence is, I grant, difficult at first. But in course of time the desire for that diet will begin to languish ; the incentives to our unnatural wants failing, the stomach, at first rebellious, will after a time feel an aversion for what formerly it eagerly coveted. The desire dies of itself, and it is no severe loss to be without those things that you have ceased to long for. Add to this that there is no disease, no pain, which is not certainly intermitted or relieved, or cured altogether. Moreover it is possible for you to be on your guard against a threatened return of the disease, and to oppose remedies if it comes upon you."—(*Ep.* lxxviii.)

On the occasion of a shipwreck, when his fellow-passengers found themselves forced to live upon the scantiest fare, he takes the opportunity to point out how extravagantly superfluous must be the ordinary living of the richer part of the community :—

" How easily we can dispense with these superfluities, which, when necessity takes them from us, we do not feel the want of. . . . Whenever I happen to be in the company of richly-living people I cannot prevent a blush of shame, because I see evident proof that the principles which I approve and commend have as yet no sure and firm faith placed in them. . . . A warning voice needs to be published abroad in opposition to the prevailing opinion of the human race : ' You are out of your senses *(insanitis)* ; you are wandering from the path of right ; you are lost in stupid admiration for superfluous luxuries ; you value no one thing for its proper worth.' "— (*Ep.* lxxxvii.)

Again :—

" I now turn to you, whose insatiable and unfathomable gluttony *(profunda et insatiabilis gula)* searches every land and every sea. Some animals it persecutes with snares and traps, with hunting-nets [the customary method of the *battue* of that period], with hooks, sparing no sort of toil to obtain them. Excepting from mere caprice or daintiness, there is no peace allowed to any species of beings. Yet how much of all these feasts which you obtain by the agency of innumerable hands do you even so much as touch with your lips, satiated as they are with luxuries ? How much of that animal, which has been caught with so much expense or peril, does the dyspeptic and bilious owner taste ? Unhappy even in this ! that you perceive not that you hunger more than your belly. Study," he concludes his exhortation to his friend, "not to know *more*, but to know *better*."

Again :—

"If the human race would but listen to the voice of reason, it would recognise that [fashionable] cooks are as superfluous as soldiers. . . Wisdom engages in all

C

useful things, is favourable to peace, and summons the whole human species to concord."—(*Ep.* xc.)

" In the simpler times there was no need of so large a supernumerary force of medical men, nor of so many surgical instruments or of so many boxes of drugs. Health was simple for a simple reason. Many dishes have induced many diseases. Note how *vast a quantity of lives one stomach absorbs*—devastator of land and sea.* No wonder that with so discordant diet disease is ever varying. . . . Count the cooks : you will no longer wonder at the innumerable number of human maladies."—(*Ep.* xcv.)

We must be content with giving our readers only one more of Seneca's exhortations to a reform in diet :—

" You think it a great matter that you can bring yourself to live without all the apparatus of fashionable dishes ; that you do not desire wild boars of a thousand pounds weight or the tongues of rare birds, and other portents of a luxury which now despises whole carcases,† and chooses only certain parts of each victim. I shall admire you then only when you scorn not plain bread, when you have persuaded yourself that herbs exist not for other animals only, but for man also—if you shall recognise that vegetables are sufficient food for the stomach into which we now stuff valuable lives, as though it were to keep them for ever. For what matters it what it receives, since it will soon lose all that it has devoured ? The apparatus of dishes, containing the spoils of sea and land, gives you pleasure, you say. . . . The splendour of all this, heightened by art, gives you pleasure. Ah ! those very things so solicitously sought for and served up so variously—no sooner have they entered the belly than one and the same foulness shall take possession of them all. Would you contemn the pleasures of the table ? Consider their final destination" *(exitum specta)*.‡

If Seneca makes *dietetics* of the first importance, he at the same time by no means neglects the other departments of *ethics*, which, for the most part, ultimately depend upon that fundamental reformation ; and he is equally excellent on them all. Space will not allow us to present our readers with all the admirable *dicta* of this great moralist. We cannot resist, however, the temptation to quote some of his unique teaching on certain branches of humanitarianism and philosophy little regarded either in his own time or in later ages. Slaves, both in pagan and Christian Europe, were regarded very much as the domesticated non-human species are at the present day, as born merely for the will and pleasure of their masters. Such seems to have been the universal estimate of their *status.* While often superior to their lords, nationally and individually, by birth, by mind, and by education, they were at the arbitrary disposal of too often cruel and capricious owners :—

"Are they slaves? " eloquently demands Seneca. " Nay, they are men. Are they slaves ? Nay, they live under the same roof *(contubernales).* Are they slaves? Nay, they are humble friends. Are they slaves ? Nay, they are fellow-servants *(conservi),*

---

* Cf. Pope's accusation of the gluttony of his species :—
       " Of half that live, the butcher and the tomb."
                                              —*Essay on Man.*
   † Compare Juvenal *passim*, Martial, Athenæus, Plutarch, and Clement of Alexandria.
   ‡ *Ep.* *x.* Cf. St. Chrysostom (*Hom.* i. on *Coloss.* i.) who seems to have borrowed his equally forcible admonition on the same subject from Seneca.

if you will consider that both master and servant are equally the creatures of chance. I smile, then, at the prevalent opinion which thinks it a disgrace for one to sit down to a meal with his servant. Why is it thought a disgrace, but because arrogant *Custom* allows a master a crowd of servants to stand round him while he is feasting ? "

He expressly denounces their cruel and contemptuous treatment, and demands in noble language (afterwards used by Epictetus, himself a slave):—

" Would you suppose that he whom you call a slave has the same origin and birth as yourself ? has the same free air of heaven with yourself ? that he breathes, lives, and dies like yourself ? "

He denounces the haughty and insulting attitude of masters towards their helpless dependants, and lays down the precept : " So live with your dependant as you would wish your superior to live with you." He laments the use of the term " slaves," or " servants " *(servi)*, in place of the old " domestics " *(familiares)*. He declaims against the common prejudice which judges by the *outward* appearance :—

" That man," he asserts, " is of the stupidest sort who values another either by his dress or by his condition. Is he a slave ? He is, it may be, *free in mind.* He is the *true* slave who is a slave to cruelty, to ambition, to avarice, to pleasure. Love," he declares, insisting upon humanity, " cannot co-exist with fear."—(*Ep.* xlviii.)

He is equally clear upon the ferocity and barbarity of the gladiatorial and other shows of the *Circus,* which were looked upon by his contemporaries as not only interesting spectacles, but as a useful school for war and endurance—much for the same reason as that on which the "sports" of the present day are defended. Cicero uses this argument, and only expresses the general sentiment. Not so Seneca. He speaks of a chance visit to the Circus (the gigantic Colosseum was not yet built), for the sake of mental relaxation, expecting to see, at the period of the day he had chosen, only innocent exercises. He indignantly narrates the horrid and bloody scenes of suffering, and demands, with only too much reason, whether it is not evident that such evil examples receive their righteous retribution in the deterioration of character of those who encourage them :—

" Ah ! what dense mists of darkness do power and prosperity cast over the human mind. He [the magistrate] believes himself to be raised above the common lot of mortality, and to be at the pinnacle of glory, when he has offered so many crowds of wretched human beings to the assaults of wild beasts ; when he forces animals of the most different species to engage in conflict ; when in the full presence of the Roman populace he causes torrents of blood to flow, a fitting school for the future scenes of still greater bloodshed." *

---

* *Epistola* vii. and *De Brevitate Vitæ* xiv. As to the effect of the gross diet of the later *athletes,* Ariston (as quoted by Lipsius) compared them to columns in the *gymnasium,* at once "sleek and stony"—λιπαροὺς καὶ λιθίνους. Diogenes of Sinope, being asked why the athletes seemed always so void of sense and intelligence, replied, " Because they are made up of ox and swine flesh." Galen, the great Greek medical writer of the second century of our æra, makes the same remark upon the proverbial stupidity of this class, and adds : " And this is the universal experience of mankind—that a gross stomach does not make a refined mind." The Greek proverb, " παχεῖα γαστὴρ λεπτὸν οὐ τίκτει νόον," exactly expresses the same experience.

In his treatise *On Clemency*, dedicated to his youthful pupil Nero, he anticipates the very modern theory—*theory*, for the prevalent *practice* is a very different thing—that *prevention* is better than *punishment*, and he denounces the cruel and selfish policy of princes and magistrates, who are, for the most part, concerned only to punish the criminals produced by unjust and unequal laws:—

"Will not that man," he asks, "appear to be a very bad father who punishes his children, even for the slightest causes, with constant blows? Which preceptor is the worthier to teach—the one who scarifies his pupils' backs if their memory happens to fail them, or if their eyes make a slight blunder in reading, or he who chooses rather to correct and instruct by admonition and the influence of shame? . . You will find that those crimes are most often committed which are most often punished. . . . Many capital punishments are no less disgraceful to a ruler than are many deaths to a physician. Men are more easily governed by mild laws. The human mind is naturally stubborn and inclined to be perverse, and it more readily follows than is forced. The disposition to cruelty which takes delight in blood and wounds is the characteristic of wild beasts; it is to throw away the human character and to pass into that of a denizen of the woods."

Speaking of giving assistance to the needy, he says that the genuine philanthropist will give his money—

"Not in that insulting way in which the great majority of those who wish to seem merciful disdain and despise those whom they help, and shrink from contact with them, but as one mortal to a fellow-mortal he will give as though out of a treasury that should be common to all." *

Next to the *De Clementiâ* and the *De Irâ* ("On Anger"), his treatise *On the Happy Life* is most admirable. In the abundance of what is unusually good and useful it is difficult to choose. His warning (so unheeded) against implicit confidence in authority and tradition cannot be too often repeated:—

"There is nothing against which we ought to be more on our guard than, like a flock of sheep, following the crowd of those who have preceded us—going, as we do, not where we ought to go, but where men have walked before. And yet there is nothing which involves us in greater evils than following and settling our faith upon authority—considering those dogmas or practices best which have been received heretofore with the greatest applause, and which have a multitude of great names. We live not according to reason, but according to mere fashion and tradition, from whence that enormous heap of bodies, which fall one over the other. It happens as in a great slaughter of men, when the crowd presses upon itself. Not one falls without dragging with him another. The first to fall are the cause of destruction to the succeeding ranks. It runs through the whole of human life. No-one's error is limited to himself alone, but he is the author and cause of

---

* *De Clementiâ* i. and ii. The author has been accused of flattering a notorious tyrant. The charge is, however, unjust, since Nero, at the period of the dedication of the treatise to him, had not yet discovered his latent viciousness and cruelty. Like Voltaire, in recent times, Seneca bestowed perhaps unmerited praise, in the hope of flattering the powerful into the practice of justice and virtue.

another's error. . . . We shall recover our sound health if only we shall separate ourselves from the herd, for the crowd of mankind stands opposed to right reason—the defender of its own evils and miseries.* . . . Human history is not so well conducted, that the better way is pleasing to the mass. The very fact of the approbation of the multitude is a proof of the badness of the opinion or practice. Let us ask what is *best*, not what is *most customary ;* what may place us firmly in the possession of an everlasting felicity, not what has received the approbation of the vulgar—the worst interpreter of the truth. Now I call "the vulgar" *the common herd of all ranks and conditions" (Tam chlamydatos quam coronatos).*—(*De Vitâ Beatâ* i. and ii.)

Again :—

" I will do nothing for the sake of opinion ; everything for the sake of conscience."

He repudiates the doctrines of Egoism for those of Altruism :—

" I will so live, as knowing myself to have come into the world for others. . . I shall recognise the *world* as my proper country. Whenever nature or reason shall demand my last breath I shall depart with the testimony that I have loved a good conscience, useful pursuits—that I have encroached upon the liberty of no one, least of all my own."

Very admirable are his rebukes of unjust and insensate anger in regard to the non-human species :—

" As it is the characteristic of a madman to be in a rage with lifeless objects, so also is it to be angry with dumb animals,† inasmuch as there can be no injury unless *intentional.* Hurt us they can—as a stone or iron—*injure* us they cannot. Nevertheless, there are persons who consider themselves insulted when horses that will readily obey one rider are obstinate in the case of another ; just as if they are more tractable to some individuals than to others of *set purpose,* not from custom or *owing to treatment."*—(*De Irâ* ii., xxvi.)

Again, of anger, as between human beings :—

"The faults of others we keep constantly before us ; our own we hide behind us. . . . A large proportion of mankind are angry, not with the *sins,* but with the *sinners.* In regard to reported offences, *many speak falsely to deceive, many because they are themselves deceived."*

Of the use of self-examination, he quotes the example of his excellent preceptor, Sextius, who strictly followed the Pythagorean precept to examine oneself each night before sleep :—

"Of what bad practice have you cured yourself to-day ? What vice have you resisted ? In what respect are you the better ? Rash anger will be moderated and finally cease when it finds itself daily confronted with its judge. What, then, is more useful than this custom of thoroughly weighing the actions of the entire day ?"

He adduces the feebleness and shortness of human life as one of the most forcible arguments against the indulgence of malevolence :—

---

* Cf. the sad experiences of the great Jewish prophet. " The prophets prophesy falsely," &c.

† In the original, " dumb animals" *(mutis animalibus)*—a term which, it deserves special note, Seneca usually employs, rather than the traditional expressions " beasts " and " brutes.' The term " dumb animals " is not strictly accurate, seeing that almost all *terrestrials* have the use of voice though it may not be intelligible to human ears. Yet it is, at all events, preferable to the old traditional terms still in general use.

"Nothing will be of more avail than reflections on the nature of mortality. Let each one say to himself, as to another, 'What good is it to declare enmity against such and such persons, as though we were born to live for ever, and to thus waste our very brief existence? What profit is it to employ time which might be spent in honourable pleasures in inflicting pain and torture upon any of our fellow-beings?' . . . Why rush we to battle? Why do we provoke quarrels? Why, forgetful of our mortal weakness, do we engage in huge hatreds? Fragile beings as we are, why will we rise up to crush others? . . . Why do we tumultuously and seditiously set life in an uproar? Death stands staring us in the face, and approaches ever nearer and nearer. That moment which you destine for another's destruction perchance may be for your own. . . . Behold! death comes, which makes us all equal. Whilst we are in this mortal life, let us cultivate humanity; let us not be a cause of fear or of danger to any of our fellow-mortals. Let us contemn losses, injuries, insults. Let us bear with magnanimity the brief inconveniences of life."

Again, in dealing with the weak and defenceless :—

" Let each one say to himself, whenever he is provoked, 'What right have I to punish with whips or fetters a slave who has offended me by voice or manner? Who am I, whose ears it is such a monstrous crime to offend? Many grant pardon to their enemies; shall I not pardon simply idle, negligent, or garrulous slaves?' Tender years should shield childhood—their sex, women—individual liberty, a stranger— the common roof, a domestic. Does he offend now for the first time? Let us think how often he may have pleased us."—(*De Irâ* iii., passim.)

As to the conduct of life :—

"We ought so to live, as though in the sight of all men. We ought so to employ our thoughts, as though someone were able to inspect our inmost soul—and there is one able. For what advantages it that a thing is hidden from men; nothing is hidden from God. (*Ep.* 83.) . . . Would you propitiate heaven? Be good. He worships the gods, who imitates [the higher ideal of] them. How do we act? What principles do we lay down? That we are to refrain from human bloodshed? Is it a great matter to refrain from injuring him to whom you are bound to do good? The whole of human and divine teaching is summed up in this one principle—we are all members of one mighty body. Nature has made us of one kin *(cognatos)*, since she has produced us from the same elements and will resolve us into the same elements. She has implanted in us love one for another, and made us for living together in society. She has laid down the laws of right and justice, by which ordinance it is more wretched to injure than to be injured; and by her ordering, our hands are given us to help each the other. . . . Let us ask what things *are*, not what they *are called*. Let us value each thing on its own merits, without thought of the world's opinion. Let us love temperance; let us, before all things, cherish justice. . . Our actions will not be right unless the will is first right, for from that proceeds the act."

Again :—

" The will will not be right unless the *habits* of mind are right, for from these results the will. The habits of thought, however, will not be at the best unless they shall have been based upon *the laws of the whole of life;* unless they shall have tried all things by the test of truth."—(*Ep.* xcv.)

Excellent is his advice on the choice of books and of reading :—

"Be careful that the reading of many authors, and of every sort of books, does not induce a certain vagueness and uncertainty of mind. We ought to linger over and nourish our minds with, writers of assured genius and worth, if we wish to extract something which may usefully remain fixed in the mind. A multitude of books distracts the mind. Read always, then, books of approved merit. If ever you have a wish to go for a time to other kinds of books, yet always return to the former."*—(Ep. ii.)

In his 88th Letter Seneca well exposes the folly of a learning which begins and ends in *mere words,* which has no real bearing on the conduct of life and the instruction of the *moral* faculties :—

"In testing the value of books and writers, let us see whether or no they teach *virtue.* . . You inquire minutely about the wanderings of Ulysses rather than work for the prevention of error in your own case. We have no leisure to hear exactly how and where he was tossed about between Italy and Sicily. . . The tempests of the soul are ever tossing us, and evildoing urges us into all the miseries of Ulysses. . . Oh marvellously excellent education ! By it you can measure circles and squares, and all the distances of the stars. There is nothing that is not within the reach of your geometry. Since you are so able a mechanician, measure the human mind. Tell me how great it is, how small it is *(pusillus).* You know what a straight line is. What does it profit you, if you know not what is straight *(rectum)* in life."† What then ? Are liberal studies of no avail ? For other things much ; for virtue nothing. . . . They do not lead the mind to virtue—they only clear the way.

"Humanity forbids us to be arrogant towards our fellows ; forbids us to be grasping ; shows itself kind and courteous to all, in word, deed, and thought ; thinks no evil of another, but rather loves its own highest good, chiefly because it will be of good to another. Do liberal studies [always] inculcate these maxims ? No more than they do simplicity of character and moderation ; no more than they do frugality and economy of living ; no more than they do mercy, which is as sparing of another's blood as it is of its own, and recognises that man is not to use the services of his fellows unnecessarily or prodigally.

"Wisdom is a great, a vast subject. It needs all the spare time that can be given to it. . . . Whatever amount of natural and moral questions you may have mastered, you will still be wearied with the vast abundance of questions to be asked and solved. So many, so great, are these questions, all superfluous things must be removed from the mind, that it may have free scope for exercise. Shall I waste my life in mere words *(syllabis)* ? Thus does it come about that the learned are more anxious to talk than to live. Mark what mischief *excessive* subtlety of mind produces, and how dangerous it may be to truth."—(*Ep.* lxxxviii.)

Elsewhere he indignantly demands :—

"What is more vile or disgraceful than a learning which catches at popular applause *(clamores)* ?"—(*Ep.* lii.)

Anticipating the ultimate triumph of Truth, he well says :—

"No virtue is really lost—that it has to remain hidden for a time is no loss to itself. A day will come which will publish the truth at present neglected and oppressed by the malignity *(malignitas)* of its age. He who thinks the world to be of his own age only, is born for the few. Many thousands of years, many millions of people, will

---

* Compare the advice of the younger Pliny—" Read much rather than many books." (*Letters* vii., 9 in the excellent revision of Mr. Bosanquet, Bell and Daldy, 1877) and Gibbon's just remarks (*Miscellaneous Works*).

† See this finely and wittily illustrated in *Micromégas* (one of the most exquisite satires ever written), where the philosopher of the star Sirius proposes the same questions to the contending metaphysicians and *savans* of our planet.

supervene. Look forward to that time. Though the envy of your own day shall have condemned you to obscurity, there will come those who will judge you without fear or favour. If there is any reward for virtue from fame, that is imperishable. The talk of posterity, indeed, will be nothing to us. Yet it will revere us, even though we are insensible to its praise ; and it will frequently consult us. . . . What now deceives has not the elements of duration. Falsehood is thinly disguised ; it is ransparent, if only you look close enough.''—(*Ep.* lxix.)

In his *Questions on Nature,* in which he often shows himself to have been much in advance of his contemporaries, and, indeed, of the whole mediæval ages, in scientific acumen, he takes occasion to reprobate the common practice of glorifying the lives and deeds of worthless princes and others, and exclaims in the modern spirit :—

" How much better to try to extinguish the evils of our own age than to glorify the bad deeds of others to posterity ! How much better to celebrate the works of Nature [*deorum*] than the piracies of a Philip or Alexander and of the rest who, become illustrious by the calamities of nations, have been no less the pests of mankind than an inundation which devastates a whole country, or a conflagration in which a large proportion of living creatures is consumed."—(*Quæst. Nat.* iii.)

It will be sufficiently apparent, from what we have presented to our readers, that Seneca, though nominally of the Stoic school, belonged in reality to no special sect or party. *Nullius addictus jurare in verba magistri.* Bound to the words of no one master, he sought for truth everywhere. The authority whom he most frequently quotes with approval is Epicurus, the arch-enemy of Stoicism. Wiser and more candid than the great mass of sectaries, he scorns the tactics of partisanship. He justly recognises the fact that the " luxurious egoists have not derived their impulse or sanction from Epicurus ; but, abandoned to their vices, they disguise their selfishness in the name of his philosophy." He professes his own conviction to be " against the common prejudice of the popular writers of my own school, that the teaching of Epicurus was just and holy, and, on a close examination, essentially grave and sober. . . . . I affirm this, that he is ill-understood, defamed, and depreciated." (*De Vitâ Beatâ,* xii., xiii.)

It will also be sufficiently clear that the ethics of Seneca consist of no mere trials of skill in logomachy ; in finely-drawn distinctions between words and names, as do so large a proportion both of modern and ancient dialectics. If so daring a heresy may possibly be forgiven us, we would venture to suggest that the authorities of our schools and universities might, with no inconsiderable advantage, substitute judicious excerpts from the *Morals* of Seneca for the *Ethics* of Aristotle ; or, as Latin literature is now in question, even for the *De Officiis* of Cicero. This, however, is perhaps to indulge Utopian speculation too greatly. The mediæval spirit of scholasticism is not yet sufficiently out of favour at the ancient schools of Aquinas and Scotus.

VI

PLUTARCH. 40—120 A.D. (?)

THE years of the birth and death of the first of biographers and the most amiable of moralists are unknown. We learn from himself that he was studying philosophy at Athens under Ammonius, the Peripatetic, at the time when Nero was making his ridiculous progress through Greece. This was in 66 A.D., and the date of his birth may therefore be approximately placed somewhere about the year 40. He was thus a younger contemporary of Seneca. Chæronea, in Bœotia, claims the honour of giving him birth.

He lived several years at Rome and in other parts of Italy, where, according to the fashion of the age and the custom of the philosophic rhetoricians (of whom, probably, he was one of the very few whose *prælections* were of any real value), he gave public lectures, attended by the most eminent literary as well as social personages of the time, among whom were Tacitus, the younger Pliny, Quintilian, and perhaps Juvenal. These lectures may have formed the basis, if not the entire matter, of the miscellaneous essays which he afterwards published. When in Italy he neglected altogether the Latin language and literature, and the reason he gives proves the estimation in which he was held : "I had so many public commissions, and so many people came to me to receive instruction in philosophy. . . . . it was, therefore, not till a late period in life that I began to read the Latin writers." In fact, the very general indifference, or at least silence, of the Greek masters in regard to Latin literature is not a little remarkable.

It is asserted, on doubtful authority (Suidas), that he was preceptor of Trajan, in the beginning of whose reign he held the high post of Procurator of Greece ; and he also filled the honourable office of *Archon*, or Chief Magistrate of his native city, as well as of priest of the Delphic Apollo. He passed the later and larger portion of his life in quiet retirement at Chæronea. The reason he assigns for clinging to that dull and decaying provincial town, although residence there was not a little inconvenient for him, is creditable to his citizen-feeling, since he believed that by quitting it he, as a person of influence, might contribute to its ruin. In all the relations of social life Plutarch appears to have been exemplary, and he was evidently held in high esteem by his fellow-citizens. As husband and father he was particularly admirable. The death of a young daughter, one of a numerous progeny, was the occasion of one of his most affecting productions—the *Consolation*—addressed to his wife Timoxena. He himself died at an advanced age, in the reign of Hadrian.

Plutarch's writings are sufficiently numerous. The *Parallel Lives,* forty-six in number, in which he brings together a Greek and a Roman celebrity by way of comparison, is perhaps the book of Greek and Latin literature which has been the most widely read in all languages. "The reason of its popularity," justly observes a writer in Dr. Smith's *Dictionary,* "is that Plutarch has rightly conceived the business of a biographer—his biography is true portraiture. Other biography is often a dull, tedious enumeration of facts in the order of time, with perhaps a summing up of character at the end. The reflections of Plutarch are neither impertinent nor trifling; his sound good sense is always there; his honest purpose is transparent; his love of humanity warms the whole. His work is and will remain, in spite of all the fault that can be found with it by plodding collectors of facts and small critics, the book of those who can nobly think and dare and do."

His miscellaneous writings—indiscriminately classed under the title *Moralia,* or *Morals,* but including historical, antiquarian, literary, political, and religious disquisitions—are about eighty in number. As might be expected of so miscellaneous a collection, these essays are of various merit, and some of them are, doubtless, the product of other minds than Plutarch's. Next to the *Essay on Flesh Eating** may be distinguished as amongst the most important or interesting, *That the Lower Animals Reason,*† *On the Sagacity of the Lower Animals*—highly meritorious treatises, far beyond the ethical or intellectual standard of the mass of "educated" people even of our day—*Rules for the Preservation of Health, A Discourse on the Training of Children, Marriage Precepts, or Advice to the Newly Married, On Justice, On the Soul, Symposiacs*—in which he deals with a variety of interesting or curious questions—*Isis and Osiris,* a theological disquisition; *On the Opinions of the Philosophers, On the Face that Appears in the Moon,*‡ *Political Precepts, Platonic Questions,* and last, not least, his *Consolation,* addressed to Timoxena. Plutarch also wrote his autobiography. If it had come down to us it would have been one of the most interesting remains

---

* This essay ranks among the most valuable productions that have come down to us from antiquity. Its sagacious anticipation of the modern argument from comparative physiology and anatomy, as well as the earnestness and true feeling of its eloquent appeal to the higher instincts of human nature, gives it a special interest and importance. We have therefore placed it separately at the end of this article.

† Περὶ τοῦ Τὰ Ἄλογα Λογῷ Χρῆσθαι—"An Essay to prove that the Lower Animals reason."

‡ This essay is remarkable as being, perhaps, the first speculation as to the existence of other *worlds* than ours.

of Antiquity, dealing, as we may well imagine it did deal, with some of the most important phenomena of the age. Possibly we might have had the expression of his feeling and attitude in regard to the new religion (established some 200 years later), which, strangely enough, is altogether overlooked or ignored as well by himself as by the other eminent writers of Greece and Italy.*

Plutarch was an especial admirer of Plato and] his school, but he attached himself exclusively to no sect or system. He was essentially eclectic: he chose what his reason and conscience informed him to be the most good and useful from the various philosophies. As to the influence of his literary labours in instructing the world, it has been truly remarked by the author of the article in the *Penny Cyclopædia* that, "a kind, humane disposition, and a love of everything that is ennobling and excellent, pervades his writings, and gives the reader the same kind of pleasure that he has in the company of an esteemed friend, whose singleness of heart appears in everything that he says or does." His personal character is, in fact, exactly reflected in his publications. That he was somewhat superstitious and of a conservative bias is sufficiently apparent; † but it is also equally clear, in his case, that the moral perceptions were not obscured by a selfishness which is too often the product of optimism, or self-complacent contentment with things as they are. In metaphysics, with all earnest minds oppressed by the terrible fact of the dominance of evil and error in the world, he vainly attempted to find a solution of the enigma in that prevalent Western Asiatic prejudice of a dualism of contending powers. He found consolation in the persuasion that the two antagonistic principles are not of *equal* power, and that the Good must eventually prevail over the Evil.

The *Lives* has gone through numerous editions in all languages. Of the *Morals*, the first translation in this country was made by Philemon Holland, M.D., London, 1603 and 1657. The next English version was published in 1684—1694, "by several hands." The fifth edition, "revised and corrected from the many errors of the former edition," appeared in 1718. The latest English version is that of Professor Goodwin, of Harvard University (1870), with an introduction by R. W. Emerson. It is, for the most part, a reprint of the revision of 1718, and consists of five octavo volumes. It is a matter equally for

---

* As regards this complete silence of Plutarch, it may be attributed to his eminently *conservative* temperament, which shrank from an exclusive system that so completely broke with the sacred traditions of "the venerable Past." Besides, Christianity had not assumed the imposing proportions of the age of Lucian, whose indifference is therefore more surprising than that of Plutarch.

† See, for example, the *Isis and Osiris*, 49. And yet, with Francis Bacon, and Bayle, and Addison, he prefers Atheism to fanatical Superstition.

surprise and regret that, in an age of so much literary, or at least publishing, enterprise, a judicious selection from the productions of so estimable a mind has never yet been attempted in a form accessible to ordinary readers.*

In his *Symposiacs*, discussing (*Quest.* ii.), "whether the sea or land affords the better food," and summing up the arguments, he proceeds :—

" We can claim no great right over land animals which are nourished with the same food, inspire the same air, wash in and drink the same water that we do ourselves ; and when they are slaughtered they make us ashamed of our work by their terrible cries ; and then, again, by living amongst us they arrive at some degree of familiarity and intimacy with us. But sea creatures are altogether strangers to us, and are brought up, as it were, in another world. Neither does their voice, look, or any service they have done us plead for their life. This kind of animals are of no use at all to us, nor is there any obligation upon us that we should love them. The element we inhabit is a hell to them, and as soon as ever they enter upon it they die."

We may infer that Plutarch advanced gradually to the perfect knowledge of the truth, and it is probable that his essay on *Flesh-eating* was published at a comparatively late period in his life, since in some of his miscellaneous writings, in alluding to the subject, he speaks in less decided and emphatic terms of its barbarism and inhumanity : *e.g.*, in his *Rules for the Preservation of Health,* while recommending moderation in eating, and professing abstinence from flesh, he does not so expressly denounce the prevalent practice. Yet he is sufficiently pronounced even here in favour of the reformed diet on the score of health :—

"Ill-digestion," says he, "is most to be feared after flesh-eating, for it very soon clogs us and leaves ill consequences behind it. It would be best to accustom oneself *to eat no flesh at all,* for the earth affords plenty enough of things fit not only for nourishment but for delight and enjoyment ; some of which you may eat without much preparation, and others you may make pleasant by adding various other things."

That the non-Christian humanitarian of the first century was far ahead—we will not say of his contemporaries, but of the common crowd of writers and speakers of the present age in his estimate of the just rights and position of the innocent non-human races—will be sufficiently apparent from the following extract from his remarkable essay entitled, *That the Lower Animals Reason,* to which Montaigne seems to have been indebted. The essay is in the form of a dialogue between Odysseus (Ulysses) and Gryllus, who is one of the transformed captives of the sorceress Circe (see *Odyssey* ix.) Gryllus maintains the superiority

---

* Of the many eminent persons who have been indebted to, or who have professed the greatest admiration for, the writings of Plutarch are Eusebius, who places him at the head of all Greek philosophers, Origen, Theodoret, Aulus Gellius, Photius, Suidas, Lipsius. Theodore of Gaza, when asked what writer he would first save from a general conflagration of libraries, answered, " Plutarch ; for he considered his philosophical writings the most beneficial to society, and the best substitute for all other books." Amongst moderns, Montaigne, Montesquieu, Voltaire, and especially Rousseau, recognise him as one of the first of moralists.

of the non-human races generally in very many qualities and in regard to many of their habits—*e.g.*, in eating and drinking :—

" Being thus wicked and incontinent in inordinate desires, it is no less easy to be proved that men are more intemperate than other animals even in those things which are necessary—*e.g.*, in eating and drinking—the pleasures of which we [the non-human races] always enjoy with some benefit to ourselves. But you, pursuing the pleasures of eating and drinking beyond the satisfaction of nature, are punished with many and lingering diseases* which, arising from the single fountain of superfluous gormandising, fill your bodies with all manner of wind and vapours not easy for purgation to expel. In the first place, all species of the lower animals, according to their kind, feed upon one sort of food which is proper to their natures— some upon grass, some upon roots, and others upon fruits. Neither do they rob the weaker of their nourishment. But man, such is his voracity, *falls upon all* to satisfy the pleasures of his appetite, tries all things, tastes all things ; and, as if he were yet to seek what was the most proper diet and most agreeable to his nature, among all animals is the only *all-devourer*.† He makes use of flesh *not out of want and necessity*, seeing that he has the liberty to make his choice of herbs and fruits, the plenty of which is inexhaustible ; but out of luxury and being cloyed with necessaries, he seeks after impure and inconvenient diet, purchased by the slaughter of living beings ; by this showing himself more cruel than the most savage of wild beasts. For blood, murder, and flesh are proper to nourish the kite, the wolf, and the serpent : *to men they are superfluous viands.* The lower animals abstain from most of other kinds and are at enmity with only a few, and that only compelled by necessities of hunger ; but neither fish, nor fowl, nor anything that lives upon the land escapes your tables, though they bear the name of humane and *hospitable.*"

Reprobating the harshness and inhumanity of Cato the Censor, who is usually regarded as the type of old Roman virtue, Plutarch, with his accustomed good feeling, declares :—

" For my part, I cannot but charge his using his servants like so many horses and oxen, or turning them off or selling them when grown old, to the account of a mean and ungenerous spirit, which thinks that the sole tie between man and man is interest or necessity. But goodness moves in a larger sphere than [so-called] justice. The obligations of law and equity reach only to mankind, but kindness and beneficence should be extended to beings of every species. And these always flow from the breast of a well-natured man, as streams that flow from the living fountain.

A good man will take care of his horses and dogs, not only while they are young, but when old and past service. Thus the people of Athens, when they had finished the temple of *Hecatompedon*, set at liberty the lower animals that had been chiefly employed in that work, suffering them to pasture at large, free from any further service. . . . We certainly ought not to treat living beings like shoes or household goods, which, when worn out with use, we throw away ; and *were it only to learn benevolence to human kind,* we should be compassionate to other beings. For my own part, I would not sell even an old ox that had laboured for me ; much less would I remove, for the sake of a little money, a man, grown old in my service, from his accustomed place—for to him, poor man, it would be as bad as banishment, since he could be of no more use to the buyer than he was to the seller. But Cato, as if he took a pride in these things, tells us that, when Consul, he left his war-horse in

---

* See Milton (*Paradise Lost*, xi.), and Shelley (*Queen Mab*).
† Cf. Pope :—" Of half that live, the butcher and the tomb."—*Moral Essays.*

Spain, to save the public the charge of his freight.   Whether such things as these are instances of greatness or of littleness of soul, let the reader judge for himself."*

If we shall compare these sentiments of the pagan humanitarian with the every-day practices of modern christian society in the matter, *e.g.*, of "knackers' yards," and other similar methods of getting rid of dumb dependants after a life-time of continuous hard labour—perhaps of bad usage, and even semi-starvation—the comparison scarcely will be in favour of christian ethics.   From the essay *On Flesh-Eating* we extract the principal and most significant passages :—

PLUTARCH—ESSAY ON FLESH-EATING.

"You ask me upon what grounds Pythagoras abstained from feeding on the flesh of animals.   I, for my part, marvel of what sort of feeling, mind, or reason, that man was possessed who was the first to pollute his mouth with gore, and to allow his lips to touch the flesh of a murdered being : who spread his table with the mangled forms of dead bodies, and claimed as his daily food what were but now beings endowed with movement, with perception, and with voice.

"How could his eyes endure the spectacle of the flayed and dismembered limbs ?   How could his sense of smell endure the horrid *effluvium* ?   How, I ask, was his taste not sickened by contact with festering wounds, with the pollution of corrupted blood and juices ?   'The very hides began to creep, and the flesh, both roast and raw, groaned on the spits, and the slaughtered oxen were endowed, as it might seem, with human voice.'†   This is poetic fiction ; but the actual feast of ordinary life is, of a truth, a veritable portent—that a human being should hunger after the flesh of oxen actually bellowing before him, and teach upon what parts one should feast, and lay down elaborate rules about joints and roastings and dishes.   The first man who set the example of this savagery is the person to arraign ; not, assuredly, that great mind which, in a later age, determined to have nothing to do with such horrors.

"For the wretches who first applied to flesh-eating may justly be alleged in excuse their utter resourcelessness and destitution, inasmuch as it was not to indulge in lawless desires, or amidst the superfluities of necessaries, for the pleasure of wanton indulgence in unnatural luxuries that they [the primeval peoples] betook themselves to carnivorous habits.

"If *they* could now assume consciousness and speech they might exclaim, ' O blest and God-loved men who live at this day !   What a happy age in the world's history has fallen to *your* lot, you who plant and reap an inheritance of all good things which grow for you in ungrudging abundance !   What rich harvests do you not gather in ?   What wealth from the plains, what innocent pleasures is it not in your power to reap from the rich vegetation surrounding you on all sides !   *You* may indulge in luxurious food without staining your hands with innocent blood.   While as for us wretches, *our* lot was cast in an age of the world the most savage and frightful conceivable.   *We* were plunged into the midst of an all-prevailing and fatal want of the commonest necessaries of life from the period of the earth's first genesis, while yet the gross atmosphere of the globe hid the cheerful heavens from view, while the stars were yet

---

* *Parallel Lives: Cato the Censor*.   Translated by John and William Langhorne, 1826.
† See *Odyssey*, xii., 395, of the oxen of the sun impiously slaughtered by the companions of Ulysses.

wrapped in a dense and gloomy mist of fiery vapours, and the sun [earth] itself had no firm and regular course. Our globe was then a savage and uncultivated wilderness, perpetually overwhelmed with the floods of the disorderly rivers, abounding in shape-less and impenetrable morasses and forests. Not for us the gathering in of domesti-cated fruits ; no mechanical instrument of any kind wherewith to fight against nature. Famines gave us no time, nor could there be any periods of seed-time and harvest.

" ' What wonder, then, if, contrary to nature, we had recourse to the flesh of living beings, when all our other means of subsistence consisted in wild corn [or a sort of grass—*ἄγρωστιν*], and the bark of trees, and even slimy mud, and when we deemed ourselves fortunate to find some chance wild root or herb ? When we tasted an acorn or beech-nut we danced with grateful joy around the tree, hailing it as our bounteous mother and nurse. Such was the gala-feast of those primeval days, when the whole earth was one universal scene of passion and violence, engendered by the struggle for the very means of existence.

" ' But what struggle for existence, or what goading madness has incited *you* to imbrue your hands in blood—you who have, we repeat, a superabundance of all the necessaries and comforts of existence ? Why do you belie the Earth [*τί καταψεύδεσθε τῆς Γῆς*] as though it were unable to feed and nourish you ? Why do you do despite to the bounteous [goddess] Ceres, and blaspheme the sweet and mellow gifts of Bacchus, as though you received not a sufficiency from them ?

" ' Does it not shame you to mingle murder and blood with their beneficent fruits ? Other *carnivora* you call savage and ferocious—lions and tigers and serpents—while yourselves come behind them in no species of barbarity. And yet for them murder is the only means of sustenance ; whereas to you it is a superfluous luxury and crime.'

" For, in point of fact, we do not kill and eat lions and wolves, as we might do in self-defence—on the contrary, we leave them unmolested ; and yet the innocent and the domesticated and helpless and unprovided with weapons of offence—these we hunt and kill, whom Nature seems to have brought into existence for their beauty and gracefulness. . . . .

" Nothing puts us out of countenance [*δυσωπεῖ*], not the charming beauty of their form, not the plaintive sweetness of their voice or cry, not their mental intelligence [*πανουργία ψυχῆς*], not the purity of their diet, not superiority of understanding. For the sake of a part of their flesh only, we deprive them of the glorious light of the sun—of the life for which they were born. The plaintive cries they utter we affect to take to be meaningless ; whereas, in fact, they are entreaties and supplications and prayers addressed to us by each which say, ' It is not the satisfaction of your real necessities we deprecate, but the wanton indulgence [*ὕβριν*] of your appetites. Kill to eat, if you must or will, but do not slay me that you may feed *luxuriously.*'

" Alas for our savage inhumanity ! It is a terrible thing to see the table of rich men decked out by those layers out of corpses [*νεκρόκοσμους*], the butchers and cooks : a still more terrible sight is the same table *after* the feast—for the wasted relics are even more than the consumption. These victims, then, have given up their lives uselessly. At other times, from mere niggardliness, the host will grudge to distribute his dishes, and yet he grudged not to deprive innocent beings of their existence !

" Well, I have taken away the excuse of those who allege that they have the authority and sanction of Nature. For that man is not, by nature, carnivorous is proved, in the first place, by the external frame of his body—seeing that to none of the animals designed for living on flesh has the human body any resemblance. He has no curved beak, no sharp talons and claws, no pointed teeth, no intense power of stomach [*κοιλίας εὐτονία*] or heat of blood which might help him to masticate and digest the gross and tough flesh-substance. On the contrary, by the smoothness of his teeth, the small capacity of his mouth, the softness of his tongue, and the sluggishness of his digestive apparatus, Nature sternly forbids him [ἐξομνύται] to feed on flesh.

"If, in spite of all this, you still affirm that you were intended by nature for such a diet, then, to begin with, kill *yourself* what you wish to eat—but do it yourself with your own *natural* weapons, without the use of butcher's knife, or axe, or club. No ; as the wolves and lions and bears themselves slay all they feed on, so, in like manner, do you kill the cow or ox with a gripe of your jaws, or the pig with your teeth, or a hare or a lamb by falling upon and rending them there and then. Having gone through all these preliminaries, *then* sit down to your repast. If, however, you wait until the living and intelligent existence be deprived of life, and if it would disgust you to have to rend out the heart and shed the life-blood of your victim, why, I ask, in the very face of Nature, and in despite of her. do you feed on beings endowed with sentient life ? But more than this—not even, after your victims have been killed, will you eat them just as they are from the slaughter-house. You boil, roast, and altogether metamorphose them by fire and condiments. You entirely alter and disguise the murdered animal by the use of ten thousand sweet herbs and spices, that your natural taste may be deceived and be prepared to take the unnatural food. A proper and witty rebuke was that of the Spartan who bought a fish and gave it to his cook to dress. When the latter asked for butter, and olive oil, and vinegar, he replied, ' Why, if I had all these things, I should not have bought the fish !'

"To such a degree do we make luxuries of bloodshed, that we call flesh 'a delicacy,' and forthwith require delicate sauces [ὄψων] for this same flesh-meat, and mix together oil and wine and honey and pickle and vinegar with all the spices of Syria and Arabia—for all the world as though we were embalming a human corpse. After all these heterogeneous matters have been mixed and dissolved and, in a manner, corrupted, it is for the stomach, forsooth, to masticate and assimilate them—if it can. And though this may be, for the time, accomplished, the natural sequence is a variety of diseases, produced by imperfect digestion and repletion.*

"Diogenes (the Cynic) had the courage, on one occasion, to swallow a *polypus* without any cooking preparation, to dispense with the time and trouble expended in the kitchen. In the presence of a numerous concourse of priests and others, unwrapping the morsel from his tattered cloak, and putting it to his lips, ' For your sakes,' cried he, ' I perform this extravagant action and incur this danger.' A self-sacrifice truly meritorious ! Not like Pelopidas, for the freedom of Thebes, or like Harmodius and Aristogeiton, on behalf of the citizens of Athens, did the philosopher submit to this hazardous experiments; for *he* acted thus that he might *unbarbarise*, if possible, the life of human kind.

"Flesh-eating is not unnatural to our physical constitution only. The mind and intellect are made gross by gorging and repletion ; for flesh-meat and wine may possibly tend to give robustness to the body, but it gives only feebleness to the mind. Not to incur the resentment of the prize-fighters [the *athletes*], I will avail myself of examples nearer home. The wits of Athens, it is well known, bestow on us Bœotians the epithets ' gross,' ' dull-brained,' and ' stupid,' chiefly on account of our gross feeding. We are even called ' hogs.' Menander nicknames us the ' jaw-people [οἱ γνάϑους ἔχοντες]. Pindar has it that ' mind is a very secondary consideration with

---

* " Hinc subitæ mortes , atque intestata Senectus."—" Hence sudden deaths, and age without a will." Juvenal, *Sat.* I.

them.' 'A fine understanding of clouded brilliancy' is the ironical phrase of Herakleitus. . . . .

"Besides and beyond all these reasons, does it not seem admirable to foster habits of philanthropy ? Who that is so kindly and gently disposed towards beings of another species would ever be inclined to do injury to his own kind ? I remember in conversation hearing, as a saying of Xenokrates, that the Athenians imposed a penalty upon a man for flaying a sheep alive, and he who tortures a living being is little worse (it seems to me) than he who needlessly deprives of life and murders outright. We have, it appears, clearer perceptions of what is contrary to propriety and custom than of what is contrary to nature. . . . .

"Reason proves both by our thoughts and our desires that we are (comparatively) new to the reeking feasts [ἕωλα] of kreophagy. Yet it is hard, as says Cato, to argue with stomachs since they have no ears ; and the inebriating potion of Custom* has been drunk, like Circe's, with all its deceptions and witcheries. Now that men are saturated and penetrated, as it were, with love of pleasure, it is not an easy task to attempt to pluck out from their bodies the flesh-baited hook. Well would it be if, as the people of Egypt turning their back to the pure light of day disembowelled their dead and cast away the offal, as the very source and origin of their sins, we, too, in like manner, were to eradicate bloodshed and gluttony from ourselves and purify the remainder of our lives: If the irreproachable diet be impossible to any by reason of inveterate habit, at least let them devour their flesh as driven to it by hunger, not in luxurious wantonness, but with feelings of shame. Slay your victim, but at least do so with feelings of pity and pain, not with callous heedlessness and with torture. And yet that is what is done in a variety of ways.

"In slaughtering swine, for example, they thrust red-hot irons into their living bodies, so that, by sucking up or diffusing the blood, they may render the flesh soft and tender. Some butchers jump upon or kick the udders of pregnant sows, that by mingling the blood and milk and matter of the *embryos* that have been murdered together in the very pangs of parturition, they may enjoy the pleasure of feeding upon unnaturally and highly inflamed flesh ! † Again, it is a common practice to stitch up the eyes of cranes and swans, and shut them up in dark places to fatten. In this and other similar ways are manufactured their dainty dishes, with all the varieties of sauces and spices [καρυκείαις—Lydian sauces, composed of blood and spices]—from all which it is sufficiently evident that men have indulged their lawless appetites in the pleasures of luxury, not for necessary food, and from no necessity, but only out of the merest wantonness, and gluttony, and display. . . . . . ." ‡

---

* "The anarch Custom's reign."
Shelley : *Revolt of Islam.*

† Such it seems, were some of the popular methods of torture in the Slaughter Houses in the first century of our æra. Whether the "calf-bleeding," and the preliminary operations which produce the *pâté de foie gras*, &c., or the older methods, bear away the palm for ingenuity in culinary torture, may be a question.

‡ See Περὶ Σαρκοφαγίας Λόγος—in the Latin title, *De Esu Carnium*—"On Flesh-Eating," Parts 1 and 2. We shall here add the authority of Pliny, who professes his conviction that "the plainest food is the most beneficial." (*Hist. Nat.* xi., 117); and asserts that it is from his eating that man derives most of his diseases. (xxv., 28.) Compare the feeling of Ovid, whom we have already quoted—*Metamorphoses* xv. We may here refer our readers also to the celebration, by the same poet, of the innocent and peaceful gifts of *Ceres*, and of the superiority of her pure table and altar—*Fasti* iv., 395-416.

> Pace, Ceres, lœta est. At vos optate, Coloni,
> Perpetuam pacem, perpetuumque ducem.
> Farra Deæ, micæque licet salientis honorem
> Detis : et in veteres thure grana focos.

D

Among the illustrious earlier contemporaries of Plutarch who practised no less than preached rigid abstinence, Apollonius of Tyana, the Pythagorean, one of the most extraordinary men of any age, deserves particular notice. He came into the world in the same year with the founder of Christianity, B.C. 4. The facts and fictions of his life we owe to Philostratus, who wrote his memoirs at the express desire of the Empress Julia Domna, the wife of Severus.

Apollonius, according to his biographer, came of noble ancestry. He early applied himself to severe study at the ever memorable Tarsus, where he may have known the great persecutor, and afterwards second founder, of Christianity. Disgusted with the luxury of the people, he soon exiled himself to a more congenial atmosphere, and applied himself to the examination of the various schools of philosophy—the Epicurean, the Stoic, the Peripatetic, &c.—finally giving the preference to the Pythagorean. He embraced the strictest ascetic life, and travelled extensively, visiting, in the first instance, Nineveh, Babylon, and, it is said, India, and afterwards Greece, Italy, Spain, and Roman Africa and Ethiopia. At the accession of Domitian, he narrowly escaped from the hands of that tyrant, after having voluntarily given himself up to his tribunal, by an exertion of his reputed supernatural power. He passed the last years of his life at Ephesus, where, according to the well-known story, he is said to have announced the death of Domitian at the very moment of the event at Rome. His alleged miracles were so celebrated, and so curiously resemble the Christian miracles, that they have excited an unusual amount of attention.*

Et, si thura aberant, unctas accendite tædas.
Parva bonæ Cereri, *sint modo casta*, placent.
A Bove succincti cultros removete ministri :
Bos aret   .   .   .   .   .
Apta jugo cervix non est ferienda securi :
Vivat, et in durâ sæpe laboret humo.

And the fine picture of Virgil of the agricultural life in the ideal "Golden Age," in which slaughter for food and war was unknown :—

Ante
Impia quam cœsis gens est epulata juvenca.
" Before
An impious world the labouring oxen slew."—*Georgics II.*

---

* "The proclamation of the birth of Apollonius to his mother by Proteus, and the incarnation of Proteus himself—the chorus of swans which sang for joy on the occasion—the casting out of devils, raising the dead, and healing the sick—the sudden disappearances and reappearances of Apollonius--his adventures in the Cave of Trophonius, and the sacred Voice which called him at his death, to which may be added his claim as a teacher to reform the world – cannot fail to suggest the parallel passages in the Gospel history. .   Still, it must be allowed that the resemblances are very general, and on the whole it seems probable that the life of Apollonius was not written with a *controversial* aim, as the resemblances, though real, only indicate that a few things were borrowed, and exhibit no trace of a systematic parallel."—*Dictionary of Greek and Roman Biography.* Edited by Wm. Smith, LL. D. So great was the estimation in which he was held, that the emperor Alxeander Severus (one of the very few good Roman princes) placed his statue or bust in the imperial *Larium* or private Chapel, together with those of Orpheus and of Christ

Unfortunately, the life by Philostratus, in accordance with the taste of a necessarily uncritical age, is so full of the preternatural and marvellous that the real fact that the pythagorean philosopher had acquired and possessed extraordinary mental as well as moral faculties, which might well be deemed supernatural at that period, is too apt to be discredited. The Life was composed long after the death of the hero, and thus a considerable amount of inventive license was possible to the biographer; but that it rested upon an undoubted substratum of actual occurrences will scarcely be disputed. There is one passage which deserves to be transcribed as of wider application. The people of a town in Pamphylia (in the Lesser Asia), where the great Thaumaturgist chanced to be staying, were starving in the midst of plenty by the selfish policy of the monopolists of grain, and, driven to desperation, were on the point of attacking the responsible authorities. Apollonius, at this crisis, wrote the following address, and gave it to the magistrates to read aloud :—

"Apollonius to the Monopolists of Corn in Aspendos, greeting : The Earth is the common mother of all, for she is just.* You are unjust, for you have made her the mother of *yourselves only*. If you will not cease from acting thus, I will not suffer you to remain upon her."

Philostratus assures us that "intimidated by these indignant words they filled the market with grain, and the city recovered from its distress."

———◆◆———

VII.

TERTULLIAN. 160—240 (?) A.D.

THE earliest of the Latin Fathers extant is, also, one of the most esteemed by the Church,† notwithstanding the well-known heterodoxy of his later life, as the first Apologist of Christianity in the Western and Latin world. He was a native of Carthage, the son of an officer holding an important post under the imperial government. The facts of his life known to us are very few, nor is it ascertained at what period he became a convert to the new religion, or when he was ordained as *presbyter*. The ill-treatment to which he was subjected by his clerical brethren at Rome induced him, it seems, to throw in his lot with the Montanist sect, in whose defence he wrote several books. He lived to an advanced age.

---

* Cf. Virgil, *Georgics* II. : "Fundit humo *facilem* victum *justissima* Tellus."

† So greatly was he esteemed by the later and leading Fathers of the Church that Cyprian, the celebrated Bishop of Carthage, and "the doctor and guide of all the Western Churches," was accustomed to say, whenever he applied himself to the study of his writings, "*Da mihi magistrum*" ("Give me my master").—Jerome, *De Viris Illustribus* I., 284.

Of his numerous works the best known (by name at least) is his *Apologeticus* ("An Apology for Christianity"). Amongst his other treatises we may enumerate *De Spectaculis* ("On Shows"), *On Idolatry*, *On the Soldier's Crown* (in which Tertullian raises the question of the lawfulness of the "violent and sanguinary occupation" of the soldier, but rather, however, for the reason of the circumstances of the pagan ceremonial), *On Monogamy*, *On the Dress of Women* (upon the extravagance of which the "Old Fathers" were eloquently denunciative), *Address to his Wife*. The treatise which here concerns us is his *De Jejuniis Adversus Psychicos*.*

Tertullian sets himself to expose the subterfuge of a large proportion of the professing Christians in his day who appealed to the pretended authority of Christ and his Apostles for the lawfulness of flesh-eating. Especially does he refute the (supposed) defence of kreophagy in I. *Tim*. iv., 3.† As to the celebrated verse in *Genesis* which solemnly enjoins the vegetable diet, the opponents of abstinence allege the permission afterwards given to the "post-diluvians."

"To this we reply," says Tertullian, "that it was not proper that man should be burdened with an express command to abstain, who had not been able in fact, to support even so slight a prohibition as that of not to eat one single species of fruit ; and, therefore, he was released from that stringency that, by the very enjoyment of freedom, he might learn to acquire strength of mind ; and after the 'flood,' in the reformation of the human species, the simple command to abstain from blood sufficed, and the use of other things was freely left to his choice. Inasmuch as God had displayed his judgment through the 'flood,' and had threatened, moreover, exquisition of blood, whether at the hand of man or of beast, giving evident proof beforehand of the justice of his sentence, he left them liberty of choice and responsibility, supplying the material for discipline by the freedom of will, intending to enjoin abstinence by the very indulgence granted, in order, as we have said, that the primordial offence might be the better expiated by greater abstinence under the opportunity of greater license." (*Quo magis, ut diximus, primordiale delictum expiaretur majoris abstinentiæ operatione in majoris licentiæ occasione.*)

---

* *On Fasting or Abstinence Against the Carnal-Minded.* The style of Tertullian, we may remark, is, for the most part, obscure and abrupt.

† It is worth noting that neither the original ($\beta\rho\omega\mu\acute{\alpha}\tau\omega\nu$) of the "Authorised Version," nor the *meats* of the "A. V." itself, says anything about *flesh-eating* in this favourite resort of its apologists. Both expressions merely signify foods of *any kind;* so that the passage in question of this Pastoral Letter—which is apparently post-Pauline—can be made to condemn *absolute* fasting only : nor does the context warrant any other interpretation. As to St. Paul, the great opponent of the earlier Christian belief and practice, it must be conceded that he seems not to have shared the abhorrence of the immediately accredited disciples of Jesus for the sanguinary diet, especially of St. Matthew, of St. James, and of St. Peter, who, as we are expressly assured by Clement of Alexandria, St. Augustine, and others, lived entirely on *non-flesh* meats. The apparent indifferentism of St. Paul upon the question of abstinence is best and most briefly explained by his avowed principle of action—from the missionary point of view useful, doubtless, but from the point of view of abstract ethics not always satisfactory—the being "all things to all men."

He quotes the various passages in the Jewish Scriptures, in which the causes of the, idolatrous proclivities and the crimes of the earlier Jews are connected by Jehovah and his prophets with flesh-eating and gross living :—

" Whether or no," he proceeds, " I have unreasonably explained the cause of the condemnation of the ordinary food by God, and of the obligation upon us, through the divine will, to denounce it, let us consult the common conscience of men. Nature herself will inform us whether, before gross eating and drinking, we were not of much more powerful intellect, of much more sensitive feeling, than when the entire domicile of men's interior has been stuffed with meats, inundated with wines, and, fermenting with filth in course of digestion, turned into a mere preparatory place for the draught (*Præmeditatorium latrinarum.*)*

" I greatly mistake *(mentior)* if God himself, upbraiding the forgetfulness of himself by Israel, does not attribute it to fulness of stomach. In fine,. in the book of Deuteronomy, bidding them to be on their guard against the same cause, he says, ‹ Lest when thou hast eaten and art full—when thy flocks and thy herds multiply,' &c. He makes the enormity of gluttony an evil superior to any other corrupting result of riches. . . . So great is the privilege (prerogative) of a circumscribed diet that it makes God a dweller with men (*contubernalem*—literally, 'a fellow-guest'), and, indeed, to live (as it were) on equal terms with them. For if the eternal God—as he testifies through Isaiah—feels no hunger, man, too, may become equal to the Deity when he subsists without gross nourishment."

He instances Daniel and his countrymen, "who preferred vegetable food and water to the royal dishes and goblets, and so became more comely than the rest, in order that no one might fear for his personal appearance ; while, at the same time, they were still more improved in understanding." As to the priesthood :—

" God said to Aaron, 'Wine and strong liquor shall ye not drink, you and your sons after you,' &c. So, also, he upbraids Israel : ' And ye gave the Nazarites wine to drink.' (Amos ii., 3.) Now this prohibition of drink is essentially connected with the vegetable diet. Thus, where abstinence from wine is required. by God, or is vowed by man, there, too, may be understood suppression of gross feeding, *for as is the eating, so is the drinking (qualis enim esus, talis et potus).* It is not consistent with truth that a man should sacrifice *half* of his stomach (*gulam*) only to God—that he should be sober in drinking, but intemperate in eating.†

" You reply, finally, that this [abstinence] is to be observed according to the will of each individual, not by imperious obligation. But what sort of thing is this, that you should allow to your arbitrary inclinations what you will not allow to the will of God? Shall more licence be conceded to the human inclinations than to the divine power? I, for my part, hold that, free from obligation to follow the fashions of the world, I am not free from obligation to God."

---

* Compare Seneca, *Epistles,* cx., and Chrysostom, *Homilies.*

† *Aquis sobrius, et cibis ebrius.* This important truth we venture to commend to the earnest attention of those philanthropists, or hygeists, who are adherents of what may be termed the *semi*-temperance Clause—who abstain from alcoholic drinks but not from flesh.

In regard to St. Paul's well-known sentences (*Rom.* xiv., **1**, &c.),
Tertullian maintains that he refers to certain teachers of abstinence who
acted from pride, not from a sense of right :—

"And even if he has handed over to you the keys of the slaughter-house or
butcher's shop (*Macelli*) in permitting you to eat all things, excepting sacrifices to
idols, at least he has not made the kingdom of heaven to consist in *butchery ;* 'for,'
says he, 'eating and drinking is not the kingdom of God, and food commends us not
to God.' You are not to suppose it said of vegetable, but of gross and luxurious,
food, since he adds, 'Neither if we eat have we anything the more, nor if we eat not
have we anything the less.'* How unworthily, too, do you press the example of
Christ as having come 'eating and drinking' into the service of your lusts. I think
that He who pronounced not the full but the hungry and thirsty 'blessed,' who professed His work to be (not as His disciples understood it) the completion of His
Father's will, I think that He was wont to abstain—instructing them to labour for
that 'meat' which lasts to eternal life, and enjoining in their common prayers petition,
not for rich and gross food, but for bread only.

"And if there be One who prefers the works of justice, not, however, without
sacrifice—that is to say, a spirit exercised by abstinence—it is surely that God to
whom neither a gluttonous people nor priest was acceptable—monuments of whose
concupiscence remain to this day, where was buried [a large proportion of] a people
greedy and clamorous for flesh-meats, gorging quails even to the point of inducing
jaundice.†

* A more accurate version of the original than that of the *A. V.* (1 *Cor.* viii., 8-13). We may here
quote the conclusion of the argument of the Greek-Jew Apostle—"Wherefore, if [the kind of] meat
is a cause of offence to my brother, I will eat no flesh while the world stands, that I may not be
a cause of offence to my brother"—and press it, more particularly, upon the attention of English
residents, and especially of Christian *missionaries,* amongst the sensitive and refined Hindus who
form so overwhelming a proportion of the population of the British Empire. According to the
evidence of the missionaries of the various Christian churches themselves, their habits of flesh-
eating have not infrequently been found to prejudice all but the lowest caste of Hindus against
the reception of other ideas of Christian and Western "civilisation."

† *Usque ad choleram ortygometras cruditando.* In the present case it seems that the wanderers
in the Arabian deserts were not so much clamorous for flesh as for *some* kind of sustenance, or
rather for something more than the *manna* with which they were supplied ; since the late Egyptian
slaves are reported to have said, "We remember the fish that we did eat in Egypt freely—the
cucumbers, the melons, and the leeks, and the onions, and the garlic ; but now our soul
is dried away: *there is nothing at all* besides this manna before our eyes."
We may here take occasion to observe that the fact of the existence of *sacrifice* throughout their
history necessarily involves the practice of flesh-eating—indeed, the two practices are, historically,
clearly connected. What, however, we may fairly deduce from their simple and frugal living in the
Egyptian slavery, lasting, as it did, through several centuries, during which period they must have
been weaned from the gross living of their previous barbarous *pastoral* life, is this—that but for the
sacrificial rites (and, perhaps, the necessities of the desert) the Jews would have, like other Eastern
peoples, probably adopted this *frugal* living—of cucumbers, melons, onions &c.—in their new
homes. Such, at least, seems to be a legitimate inference from the highly-significant fact that,
throughout their sacred scriptures, not flesh-meats but corn, and oil, and honey, and pomegranates,
and figs, and other vegetable products (in which their land originally abounded), are their highest
dietary *ideal—e.g.,* "O that my people would have hearkened to me ; for if Israel had walked in
my ways. . . . He should have fed them with the finest wheat flour: and with honey out of
the stony rock should I have satisfied thee." (Ps. lxxxi., 17 ; cf. also Ps. civ., 14, 15.) It is equally
significant of the latent and secret consciousness of the *unspiritual* nature of the products of the
Slaughter-House, even in the Western world, that in the *liturgies* or "public services" of the
Christian churches, wherever food is prayed for or whenever thanks are returned for it, there is
(as it seems) a natural shrinking from mention of that which is obtained only by cruelty and
bloodshed, and it is "the kindly fruits of the earth" which represent the legitimate dietary wants
of the petitioners.

"'Your belly is your god,'" [thus he indignantly reproaches the apologists of kreophagy,] "your liver is your temple, your paunch is your altar, the cook is your priest, and the fat steam is your Holy Spirit; the seasonings and the sauces are your chrisms, and your eructations are your prophesyings. I ever," continues Tertullian with bitter irony, "recognise Esau the hunter as a man of taste (*sapere*), and as his were so are your whole skill and interest given to hunting and trapping—just like him you come in 'from the field' of your licentious chase. Were I to offer you 'a mess of pottage,' you would, doubtless, straightway sell all your 'birthright.' It is in the cooking-pots that your love is inflamed—it is in the kitchen that your faith grows fervid—it is in the flesh dishes that all your hope lies hid. . . . Who is held in so much esteem with you as the frequent giver of dinners, as the sumptuous entertainer, as the practised toaster of healths?

"Consistently do you men of flesh reject the things of the spirit. But if your prophets are complacent towards such persons, they are not *my* prophets. Why preach *you* not constantly, 'Let us eat and drink, *for* to-morrow we die,' just as *we* preach, 'Let us abstain, brothers and sisters, *lest* to-morrow, perchance, we die'?

"Let us openly and boldly vindicate our teaching. We are sure that they 'who are in the flesh cannot please God.'* Not, surely, meaning 'in the covering or substance of the flesh,' but in the care, the affection, the desire for it. As for us, less grossness (*macies*) of the body is no cause of regret, for neither does God give *flesh by weight* any more than he gives *spirit by measure*. . . . Let prize-fighters and pugilists fatten themselves up (*saginentur*)—for them a mere corporeal ambition suffices. And yet even they become stronger by living on vegetable food (*xerophagia*—literally, 'eating of dry foods'). But other strength and vigour is our aim, as other contests are ours, who fight not against flesh and blood. Against our antagonists we must fight—not by means of flesh and blood, but with faith and a strong mind. For the rest, a grossly-feeding Christian is akin (*necessarius*) to lions and bears rather than to God, although even as against wild beasts it should be our interest to practice abstinence.†

---

* "For they that are after the Flesh do mind the things of the Flesh ; but they that are after the Spirit the things of the Spirit. For to be *carnally minded is death* ; but to be *spiritually minded is life and peace.* . . . So then they that are in the flesh cannot please God. . . . Therefore, brethren, we are debtors not to the flesh, to live after the flesh. For if ye live after the flesh, ye shall die ; but if ye, through the spirit, do mortify the deeds of the body, ye shall live." (*Rom.* viii., 5, &c.) A more spiritual apprehension of 'divine verities,' if we may so say, than the apparently more equivocal utterance of the same great reformer elsewhere. Here it is well to observe, once for all that the whole significance of the utterances of St. Paul upon flesh-eating depends upon the bitter controversies between the older Jew and the newer Greek or Roman sections of the rising Church. It is, in fact, a question of the lawfulness of eating the flesh of the victims of the Pagan and Jewish sacrificial altars—not of the question of flesh-eating in the *abstract* at all. In fine, it is a question not of *ethics*, but of theological ritual. It is greatly to be lamented that the confused and obscure translation of the *A. V.* has for so many centuries hopelessly mystified the whole subject—as far, at least, as the mass of the community is concerned.

† See *De Jejuniis Adversus Psychicos.* (Quinti. Sept. Flor. Tertulliani Opera. Edited by Gersdorf, Tauchnitz.)

VIII.

CLEMENT OF ALEXANDRIA; DIED 220 (?) A.D

THE attitude of the first great Christian writers and apologists in regard to total abstinence was somewhat peculiar. Trained in the school of Plato, in the later development of neo-platonism, their strongest convictions and their personal sympathies were, naturally, anti-kreophagistic. The traditions, too, of the earliest period in the history of Christianity coincided with their pre-Christian convictions, since the immediate and accredited representatives of the Founder of the new religion, who pre sided over the first Christian society, were commonly held to have been, equally with their predecessors and contemporaries the Essenes, strict abstinents from flesh-eating.*

Moreover, the very numerous party in the Church—the most diametrically opposed in other respects to the Jewish or Ebionite Christians— the Gnostics or philosophical Christians, "the most polite, the most learned, and the most wealthy of the Christian name," for the most part agreed with their rivals for orthodox supremacy in aversion from flesh, and, as it seems, for nearly the same reason—a belief in the essential and inherent evil of matter, a persuasion, it may be said, however unscientific, not unnatural, perhaps, in any age, and certainly not surprising in an age especially characterised by the grossest materialism,

---

* In the *Clementine Homilies*, which had a great authority and reputation in the earlier times of Christianity, St. Peter is represented, in describing his way of living to Clement of Rome, a⁸ professing the *strictest* Vegetarianism. "I live," he declares, "upon bread and olives only, with the addition, rarely, of kitchen herbs" (ἄρτῳ μόνῳ καὶ ἐλαίαις χρῶμαι καὶ σπανίως λαχάνοις XII. 6.) Clement of Alexandria (*Pædagogus* ii. 1) assures us that "Matthew the apostle lived upon seeds, and hard-shelled fruits, and other vegetables, without touching flesh;" while Hegesippus, the historian of the Church (as quoted by Eusebius, *Ecclesiastical Hist.* ii. 2, 3) asserts of St. James that "he never ate any animal food"—οὐδε εμψυχον ἔφαγε : an assertion repeated by St. Augustine (*Ad. Faust*, xxii. 3) who states that James, the brother of the Lord, lived upon seeds and vegetables, never tasting flesh or wine " (*Jacobus, frater Domini, seminibus et oleribus usus est, non carne nec vino*). The connexion of the beginnings of Christianity with the sublime and simple tenets of the Essenes, whose communistic and abstinent principles were strikingly coincident with those of the earliest Christians, is at once one of the most interesting and one of the most obscure phenomena in its nascent history. The Essenes, "the sober thinkers," as their assumed name implies, seem to have been to the more noisy and ostentatious Jewish sects, what the Pythagoreans were to the other Greek schools of philosophy—*practical moralists* rather than mere talkers and theorisers. They first appear in Jewish history in the first century B.C. Their communities were settled in the recesses of the Jordan valley, yet their members were sometimes found in the towns and villages. Like the Pythagoreans, they extorted respect even from the worldly and self-seeking religionists and politicians of the capital. See Josephus (*Antiquities* xiii. and xviii.), and Philo, who speak in the highest terms of admiration of the simplicity of their life and the purity of their morality. Dean Stanley (*Lectures on the Jewish Church*, vol. iii.) regards St. John the Baptist as Essenian in his substitution of "reformation of life" for "the sanguinary, costly gifts of the sacrificial slaughter-house."

selfishness, and cruelty. But the creed of the Christian church, which eventually became the prevailing and ruling dogma, like that of the English Church at the Revolution of the sixteenth century, was a compromise—a compromise between the two opposite parties of those who received and those who rejected the old Jewish revelation.

On the one hand Christianity, in its later and more developed form, had insensibly cast off the rigid formalism and exclusiveness of Mosaism, and, on the other, had stamped with the brand of heresy the Greek infusion of philosophy and liberalism. Unfortunately, unable clearly to distinguish between the true and the false—between the accidental and fanciful and the permanent and real—timidly cautious of approving anything which seemed connected with heresy—the leaders of the dominant body were prone to seek refuge in a middle course, in regard to the question of flesh-eating, scarcely consistent with strict logic or strict reason. While advocating abstinence as the highest spiritual exercise or aspiration, they seem to have been unduly anxious to disclaim any motives other than *ascetic*—to disclaim, in fine, humanitarian or "secular" reason, such as that of the Pythagoreans.

Such was the feeling, apparently, of the later orthodox church, at least in the West. While, however, we thus find, occasionally, a certain constraint and even contradiction in the *theory* of the first great teachers of the Church, the *practice* was much more consistent. That, in fact, during the first three or four centuries the most esteemed of the Christian heroes and saints were not only non-flesh-eaters but Vegetarians of the extremest kind (far surpassing, if we give any credit to the accounts we have of them, the *most frugal* of modern abstainers) is well known to everyone at all acquainted with ecclesiastical and, especially, eremitical history—and it is unnecessary to further insist upon a notorious fact.*

Titus Flavius Clemens, the founder of the famous Alexandrian school of Christian theology, and at once the most learned and most philosophic of all the Christian Fathers, is generally supposed to have been a native of Athens. His Latin name suggests some connexion with the family of Clemens, cousin of the emperor Domitian, who is said to have been put to death for the crime of *atheism,* as the new religion was commonly termed by the orthodox pagans.

---

* It is a curious and remarkable inconsistency, we may here observe, that the modern ardent admirers of the Fathers and Saints of the Church, while professing unbounded respect for their *doctrines,* for the most part ignore the one of their *practices* at once the most ancient, the most highly reputed, and the most universal. *Quod semper, quod ubique,* &c., the favourite maxim of St. Augustine and the orthodox church, is, in this case, "more honoured in the breach than in the observance." Partial and periodical Abstinence, it is scarcely necessary to add, however consecrated by later ecclesiasticism, is sufficiently remote from the daily *frugal* living of a St. James, a St. Anthony, or a St. Chrysostom.

He travelled and studied the various philosophies in the East and West. On accepting the Christian faith he sought information in the schools of its most reputed teachers, of whom the name of Pantænus is the only one known to us. At the death of Pantænus, in 190, Clement succeeded to the chair of theology in Alexandria, and at the same time, perhaps, he became a presbyter. He continued to lecture with great reputation till the year 202, when the persecution under Severus forced him to retire from the Egyptian capital. He then took refuge in Palestine, and appears not to have returned to Alexandria. The time and manner of his death are alike unknown. He is supposed to have died in the year 220. Amongst his pupils by far the most famous, hardly second to himself in learning and ability, was Origen, his successor in the Alexandrian professorship.

His three great works are: *A Hortatory Discourse Addressed to the Greeks* (Λόγος Προτρεπτικὸς πρὸς Ἕλληνας), *The Instructor* (*Paidagogos*—strictly, *Tutor*, or Conductor to school), and the *Miscellanies* (*Stromateis*, or *Stromata*—lit. "Patch-work").† The three works were intended to form a graduated and complete initiation and instruction in Christian theology and ethics. The first is addressed to the pagan Greek world, the second to the recent convert, and in the last he conducts the initiated to the higher *gnosis*, or knowledge. The *Miscellanies* originally consisted of eight books, the last of which is lost. The whole series is of unusual value, not only as the record of the opinions of the ablest and most philosophical of the mediators between Greek philosophy and the Christian creed, but also as containing an immense amount of information on Greek life and literature. Eloquence, earnestness, and erudition equally characterise the writings of Clement.

He assumes the name and character of a *Gnostic*,* or philosophic Christian, not in the historical but in his own sense of the word, and professes himself an eclectic—as far as a liberal interpretation of his religion admitted. " By philosophy," he says, " I do not mean the Stoic, the Platonic, the Epicurean, or the Aristotelian, but all that has been well said in each of those sects teaching righteousness with religious

---

† The full title of the treatise is—*The Miscellaneous Collection of T. F. Clemens of Gnostic (or Speculative) Memoirs upon the true Philosophy.*

* This celebrated term distinguished the superority of *knowledge* (*gnosis*) of "the most polite, the most learned, and the most wealthy of the Christian name." During the first three or four centuries the Gnostics formed an extremely numerous as well as influential section of the Church. They sub-divided themselves into more than fifty particular sects, of whom the followers of Marcion and the Manicheans are the most celebrated. Holding opinions regarding the Jewish sacred scriptures and their authority the opposite to those of the Ebionites or Jewish Christians, they agreed, at least a large proportion of them, with the latter on the question of kreophagy.

science—all this selected truth (τοῦτο σύμπαν τὸ ἐκλεκτικὸν) I call philosophy." Again, he echoes the sentiments of Seneca in lamenting that "we incline more to beliefs that are in repute (τὰ ἔνδοξα), even when they are contradictory, than to the truth" (*Miscellanies*, i. and vii.). "It would have been well for Christianity if the principles, which he set forth with such an array of profound scholarship and ingenious reasoning, had been adopted more generally by those who came after him . . . If anyone, even in a Protestant community, were to assert the liberal and comprehensive principles of the great Father of Alexandria, he would be told that he wished to compromise the distinctive claims of theology, and that he was little better than a heathen and a publican."*

It is in his second treatise, the *Instructor* or *Tutor*, that Clement displays his opinions on the subject of flesh-eating :—

"Some men live that they may eat, as the irrational beings 'whose life is their belly and nothing else.' But the Instructor enjoins us to eat that we may live. For neither is food our business, nor is pleasure our aim. Therefore discrimination is to be used in reference to food : it must be plain, truly simple, suiting precisely simple and artless children—as ministering to life not to luxury. And the life to which it conduces consists of two things, health and strength : to which plainness of fare is most suitable, being conducive both to digestion and lightness of body, from which come growth, and health, and right strength : not strength that is violent or dangerous, and wretched, as is that of the *athletes* which is produced by artificial feeding."

Referring to the injunction of Jesus, " When thou makest an entertainment, call the poor," for " whose sake chiefly a supper ought to be made," Clement says of the rich :—

"They have not yet learned that God has provided for his creature (man, I mean) food and drink for *sustenance* not for pleasure : since the body derives no advantage from extravagance in viands. On the contrary, those *who use the most frugal fare are the strongest and the healthiest, and the noblest :* as domestics are healthier and stronger than their masters, and agricultural labourers than proprietors, and not only more vigorous but wiser than rich men. For they have not buried the mind beneath food. Wholly unnatural and inhuman is it for those who are of the earth, fattening themselves like cattle, to *feed themselves up for death.*† Looking downwards on

---

* *History of the Literature of Ancient Greece*, by K. O. Müller, continued by J. W. Donaldson, D.D., vol. iii., 58.

† The argument here suggested, although rarely, if ever, adduced, may well be deemed worthy of the most serious consideration. It is, to our mind, one of the most forcible of all the many reasons for abstinence. That the life even of a really useful member of the human community should be supported by the slaughter of hundreds of innocent and intelligent beings is surely enough to "give us pause." What, then, shall be said of the appalling fact, that every day thousands of worthless, and too often worse than useless, human lives go down to the grave (to be thenceforth altogether forgotten) after having been the cause of the slaughter and suffering of countless beings, surely far superior to themselves in all real worth? To object the privilege of an " immortal soul" is, in this case, merely a miserable subterfuge. Sidney Smith calculated that *forty-four* wagon-loads of flesh had been consumed by himself during a life of seventy years ! (See his letter to Lord Murray.)

the earth, bending ever over tables, leading a life of gluttony, burying all the good of existence here in a life that by and by will end for ever: so that cooks are held in higher esteem than the tillers of the ground. We do not abolish social intercourse, but we look with suspicion on the snares of Custom and regard them as a fatal mischief. Therefore daintiness must be spurned, and we are to partake of few and necessary things......Nor is it suitable to eat and drink simultaneously. For it is the very extreme of intemperance to confound the times whose uses are discordant. And 'whether ye eat or drink, do all to the glory of God,' aiming after true frugality, which Christ also seems to me to have hinted at when he blessed the loaves and the cooked fishes with which he feasted the disciples, introducing a beautiful example of simple diet. And the fish which, at the command of the Lord, Peter caught, points to digestible and God-given and moderate food. . . .

We must guard against those sorts of food which persuade us to eat when we are not hungry, bewitching the appetite. For is there not, within a temperate simplicity, a wholesome variety of eatables—vegetables, roots, olives, herbs, milk, cheese, fruits, and all kinds of dry food? 'Have you anything here to eat?' said the Lord to the disciples after the resurrection : and they, as taught by Him to practice frugality, 'gave him a piece of broiled fish,' and besides this, it is not to be overlooked that those who feed according to the Word are not debarred from dainties—such as honey combs. For of sorts of food those are the most proper which are fit for immediate use without fire, since they are readiest : and second to these *are those which are the simplest*, as we said before. But those who bend around inflammatory tables, nourishing their own diseases, are ruled by a most licentious disease which I shall venture to call the demon of the belly : and the worst and most vile of demons. It is far better to be happy than to have a devil dwelling in us : and happiness is found only in the practice of virtue. Accordingly the Apostle Matthew lived upon seeds and nuts, (Ακρόδρυα—hard-shelled fruits) and vegetables without the use of flesh. And John, who carried temperance to the extreme, 'ate locusts and wild honey.' "

As to the Jewish laws : " The Jews," says Clement, " had frugality enjoined on them by the Law in the most systematic manner. For the Instructor, by Moses, deprived them of the use of innumerable things, adding reasons—the spiritual ones hidden, the carnal ones apparent—to which latter, indeed, they have trusted " :—

" So that, altogether, but a few [animals] were left proper for their food. And of those which he permitted them to touch, he prohibited such as had died, or were offered to idols, or had been strangled : inasmuch as to touch these was unlawful. . . . Pleasure has often produced in men harm and pain, and full feeding begets in the soul uneasiness, and forgetfulness, and foolishness. It is said, moreover, that the bodies of children, when shooting up to their height, are made to grow right by abstinence in diet ; for then the spirit which pervades the body, in order to its growth, is not checked by abundance of food obstructing the freedom of its course. Whence that truth-seeking philosopher, Plato, fanning the spark of the Hebrew philosophy, when condemning a life of luxury, says : 'On my coming hither [to Syracuse] the life which is here called happy pleased me not by any means. For not one man under heaven, if brought up from his youth in such practices, will ever turn out a *wise* man, with however admirable genius he may be endowed.' For Plato was not unacquainted with David,* who placed the sacred ark in his city in the midst of the tabernacle, and bidding all his subjects rejoice 'before the Lord, divided to the whole host of Israel, men and women, to each a loaf of bread, and baked bread, and a cake from the frying-pan.'† This was the *sufficient* sustenance of the Israelites. But that of the Gentiles was over-abundant, and no one who uses it will ever study to become temperate, burying, as he does, his mind in his belly, very like the fish called *onos* which, Aristotle says, alone of all creatures has 'its heart in its stomach. This fish Epicharmus, the comic poet, calls 'monster-paunch.' Such are the men who believe in their stomach, 'whose God is their belly, whose glory is in their shame, who mind earthly things.' To them the apostle predicted no good when he said 'whose end is destruction.' " ‡

In treating of the subject of sacrifices, upon which he uses a good deal of sarcasm (in regard to the *pagan* sacrifices at least), Clement incidentally allows us to see, still further, his opinion respecting gross feeding. He quotes several of the Greek poets who ridicule the practice and pretence of sacrificial propitiation, *e.g.*, Menander :—

> " the end of the loin,
> The gall, the bones uneatable, they give
> Alone to Heaven : the rest *themselves* consume."

" If, in fact," remarks Clement, " the savour is the special desire of the Gods of the Greeks, should they not first deify the *cooks*, and worship the Chimney itself which is still closer to the much-prized savour ? "

" If," he justly adds, " the deity need nothing, what need has he of food ? Now, if nourishing matters taken in by the nostrils are diviner than those taken in by the mouth, yet they imply respiration. What then do they say of God ? Does He *exhale*, like the oaks, or does he only *inhale*, like the aquatic animals by the dilatation of the gills, or does he breathe all around like the insects ? "

The only innocent altar he asserts to be the one allowed by Pythagoras :—

" The very ancient altar in Delos was celebrated for its purity, to which alone, as being undefiled by slaughter and death, they say that Pythagoras would permit approach. And will they not believe us when we say that the righteous soul is the truly sacred altar ? But I believe that sacrifices were invented by men *to be a pretext for eating flesh*, and yet, without such idolatry, they might have partaken of it."

---

* It was the fond belief of the *mediating* Christian writers that the best parts of Greek philosophy were derived, in whole or in part, from the Jewish Sacred Scriptures. For this belief, which has prevailed so widely, which, perhaps, still lingers amongst us, and which has engaged the useless speculation of so many minds, an Alexandrian Jew of the age of the later Ptolemies is responsible.   It is now well known that he deliberately forged passages in the (so-called) Orphic poems and " Sybilline " predictions, in order to gain the respect of the Greek rulers of his country for the Jewish Scriptures. This patriotic but unscrupulous Jew is known by his Greek name of Aristobulus. He was preceptor or counsellor of Ptolemy VI.

† 2 *Sam.* vi., 19. Clement, in common with all the first Christian writers, quotes from the *Septuagint* version, which differs considerably from the Hebrew. The English translators of the latter, presuming that " flesh " must have formed part of the royal bounty, gratuitously insert that word in the context.

‡ *Pædagogus* ii. 1, " On Eating."

He next glances at the *popular* reason for the Pythagorean abstinence, and declares :—

"If any righteous man does not burden his soul by the eating of flesh, he has the advantage of a rational motive, not, as Pythagoras and his followers dream, of the transmigration of the soul. Now Xenokrates, treating of 'Food derived from Animals,'* and Polemon in his work 'On Life according to Nature,'* seem clearly to affirm that animal food is unwholesome. If it be said that the lower animals were assigned to man—and we partly admit it—yet it was not entirely for food ; nor were all animals, but *such as do not work.* And so the comic poet, Plato, says not badly in the drama of *The Feasts :*—

> 'For of the quadrupeds we should not slay
> In future aught but swine. For they have flesh
> Most delicate : and about the swine is nought
> For us : excepting bristles, dirt, and noise.'

Some eat them as being useless, others as destructive of fruits, and others do not eat them because they are said to have strong propensity to coition. It is alleged that the greatest amount of fatty substance is produced by swine's flesh : it may, then, be appropriate for those whose ambition is for the body ; it is not so for those who cultivate the soul, by reason of the dulling of the faculties resulting from eating of flesh. The Gnostic, perhaps, too, will abstain for the sake of training, and that the body may not grow wanton in amorousness. 'For wine,' says Andokides, 'and gluttonous feeds of flesh make the body strong, but the soul more sluggish.' Accordingly such food, in order to a clear understanding, is to be rejected." †

In a chapter in his *Miscellanies*, discussing the comparative merits of the Pagan and of the Jewish code of ethics, he displays much eloquence in attempting to prove the superiority of the latter. In the course of his argument he is led to make some acknowledgment of the claims of the lower animals which, however incomplete, is remarkable as being almost unique in Christian theology. He quotes certain of the "Proverbs," *e.g.*, 'The merciful man is long-suffering, and in every one who shows solicitude there is wisdom,' and proceeds (assuming the indebtedness of the Greeks to the Jews) :—

"Pythagoras seems to me to have derived his mildness towards irrational animals from the Law. For instance, he interdicted the employment of the young of sheep and goats and cows for some time after their birth ; not even on the pretext of sacrifice allowing it, on account both of the young ones and of the mother ; training men to gentleness by their conduct towards those beneath them. 'Resign,' he says, 'the young one to the mother for the proper time.' For if nothing takes place without a cause, and milk is produced in large quantity in parturition for the sustenance of the progeny, he who tears away the young one from the supply of the milk and the breast of the mother, dishonours Nature."

Reverting to the Jewish religion, he asserts :—

"The Law, too, expressly prohibits the slaying of such animals as are pregnant till they have brought forth, remotely restraining the proneness of men to do wrong to

---

* These works, which would have been highly interesting, have, with so many other valuable productions of Greek genius, long since perished.

†*Miscellanies*. vii. "On Sacrifices."

men ; and thus also it has extended its clemency to the irrational animals, that by the exercise of humanity to beings of different races we may practise amongst those of the same species a larger abundance of it. Those too that kick the bellies of certain animals before parturition, in order to feast on flesh mixed with milk, make the womb created for the birth of the fœtus its grave, though the Law expressly commands 'but neither shalt thou seethe a lamb in his mother's milk.* For the nourishment of the living animal, it is meant, may not be converted into sauce for that which has been deprived of life ; and that which is the cause of life may not co-operate in the consumption of its flesh.†

IX.

PORPHYRY. 233—306 (?) A.D.

ONE of the most erudite, as well as one of the most spiritual, of the *literati* of any age or people, and certainly the most estimable of all the extant Greek philosophers after the days of Plutarch, was born either at Tyre or at some neighbouring town. His original name, Malchus, the Greek form of the Syrian Melech (king), and the name by which he is known to us, Porphyrius (purple-robed), we may well take deservedly to mark his philosophic superiority. He was exceptionally fortunate in his preceptors—Longinus, the most eloquent and elegant of the later Greek critics, under whom he studied at Athens ; Origen, the most independent and learned of the Christian Fathers, from whom, probably, he derived his vast knowledge of theological literature ; and, finally, Plotinus, the famous founder of New-Platonism, who had established his school at Rome in the year 244.

Upon first joining the school of Plotinus, he had ventured to contest some of the characteristic doctrines of his new teacher, and he even wrote a book to refute them. Amerius, his fellow-disciple,

---

* See Plutarch's denunciation of the very same practice of the butchers of his day, *Essay on Flesh Eating*. Unfortunately for the credit of Jewish humanity, it must be added that the method of butchering (enjoined, it is alleged, by their religious laws) entails a greater amount of suffering and torture to the victim than even the Christian. This fact has been abundantly proved by the evidence of many competent witnesses. The cruelty of the Jewish method of slaughter was especially exposed at one of the recent International Congresses of representatives of European Societies for Prevention of Cruelty.

† *Miscellanies* ii., 18. We have used for the most part the translation of the writings of Clement, published in the Ante-Nicene Library, by Messrs. Clarke, Edinburgh, 1869. The Greek text is corrupt.

was chosen to reply to this attack. After a second trial of strength by each antagonist, Amerius, by weight of argument induced Porphyry to confess his errors, and to read his recantation before the assembled Platonists. Porphyry ever after remained an attached and enthusiastic follower of the beloved master, with the final revision and edition of whose voluminous works he was entrusted. He had lived with him six years when, becoming so far unsettled in his mind as even to contemplate suicide in order to free himself from the shackles of the flesh, by the persuasion of his preceptor he made a voyage to Sicily for the restoration of his health and serenity of mind. This was in 270, in the thirty-seventh year of his age. Returning to the capital upon the death of his master, he continued the amiable but vain work of attempting the reform of the established religion, which had then sunk to its lowest degradation, and to this labour of love he may be said to have devoted his whole life. At an advanced age he married Marcella, the widow of one of his friends, who was a Christian and the mother of a rather numerous progeny, with the view, as he tells us, of superintending the education of her children.

About sixty separate works of Porphyry are enumerated by Fabricius, published, unpublished, or lost; the last numbering some forty-three distinct productions. The most important of his writings are—

(1) *On Abstinence from the Flesh of Living Beings,** in four books, addressed to a certain Firmus Castricius, a Pythagorean, who for some reason or other had become a renegade to the principles, or at least to the practice, of his old faith. Next to the inculcation of abstinence as a spiritual or moral obligation, Porphyry's "chief object seems to have been to recommend a more spiritual worship in the place of the sacrificial system of the pagan world, with all its false notions and practical abuses. This work," adds Dr. Donaldson, "is valuable on many accounts, and full of information."

(2) His criticism on Christianity, which he entitled a *Treatise against the Christians*—his most celebrated production. It was divided into fifteen books. All our knowledge of it is derived from Eusebius, Jerome, and other ecclesiastical writers. Several years after its appearance the courtly Bishop of Cæsarea, the well-known historian of the· first ages of Christianity, replied to it in a work extending to twenty-five books. More than a century later, Theodosius II. caused the obnoxious volume to be publicly burned, and Porphyry's criticism shared the fate of those "many elaborate treatises which have since

---

* Περὶ ᾿Αποχῆς Τῶν ᾿Εμψύχων.

been committed to the flames" by the theological or political zeal of orthodox emperors and princes.*

(3) *The Life of Pythagoras*—a fragment, but, as far as it goes, the most interesting of the Pythagorean biographies.

(4) *On the Life of Plotinus and the Arrangement of his Works.* It is to this biography we are indebted for our knowledge of the estimable elaborator of New-Platonism. We learn that he was the pupil of Ammonius, who disputes with Numenius the fame of having originated the principles of the new school of thought of which Plotinus, however, was the St. Paul—the actual founder. Of a naturally feeble constitution, he had early betaken himself to the consolations of divine philosophy. After vainly seeking rest for his truth-loving and aspiring spirit in other systems, he at last found in Ammonius the teacher and teaching which his intellectual and spiritual sympathies demanded. His great ambition was to visit the country of Buddha and of Zerdusht or Zoroaster, and, for that purpose, he joined the expedition of the Emperor Gordian against the Persians. The defeat and death of that prince frustrated his plans. He then settled at Rome, where he established his school, and he remained in Italy until his death in 270. By the earnest solicitations of his disciples, Porphyry and Amerius, he was induced with much reluctance to publish his oral discourses, and eventually they appeared in fifty-four books, edited by Porphyry, who gave them the name of the *Enneads*, as being arranged in six groups of *nine* treatises. Perhaps no teacher ever engaged to so unbounded an extent the admiration and affection of his followers.

"During the long period of his residence at Rome, Plotinus enjoyed an estimation almost approaching to a belief in his superhuman wisdom and sanctity. His ascetic virtue, and the mysterious transcendentalism of his conversation, which made him the Coleridge of the day, seems to have carried away the minds of his associates, and raised them to a state of imaginative exaltation. He was regarded as a sort of prophet, divine himself, and capable of elevating his disciples to a participation in his divinity. . . . These coincidences or collusions [his alleged miracles] show how sacred a character had attached to Plotinus. And we see the same evidenced in his social influence. Men and women of the highest rank crowded round him, and his house was filled with young persons of both sexes whom their parents when dying had committed to his care. Rogatian, a senator and prætor-elect, gave up his wealth and dignities, and lived as the humble bedesman of his friends, devoting himself to

---

* "The first book discussed alleged contradictions and other marks of human fallibility in the Scriptures; the third treated of Scriptural interpretation, and, strangely enough, repudiated the allegories of Origen; the fourth examined the ancient history of the Jews; and, the twelfth and thirteenth maintained the point now generally admitted by scholars—that *Daniel* is not a prophecy, but a retrospective history of the age of Antiochus Epiphanes."—*Donaldson* (*Hist. of Gr. Lit.*)

F

ascetic and contemplative philosophy. His self-denial obtained for him the approbation of Plotinus, who held him up as a pattern of philosophy; and he gained the more solid advantage of a perfect cure from the worst kind of rheumatic gout. The influence of Plotinus extended to the imperial throne itself. The weak-minded Gallienus, and his Empress Salonina, were so completely guided by the philosopher, that he had actually obtained permission to convert a ruined City of Campania into a *Platonopolis*, in which the laws of Plato's *Republic* were to be tested by a practical experiment; and the philosopher had promised to retire thither accompanied by his chief friends."

The "practical common sense" (which usually may be interpreted to mean cynical indifferentism), of the statesmen and politicians of the day interposed to prevent this attempt at a realisation of Plato's great ideal; and, considering the prematurity of such ideas in the then condition of the world—and, it must be added, the extravagance of some of them— we can, perhaps, hardly regret that his "Republic" was never instituted. As to the essence and spirit of the teaching of Plotinus,

"He cannot be termed, strictly or exclusively, a Neo-Platonist: he is equally a Neo-Aristotelian and a *Neo-Philosopher* in general. He has himself one pervading idea, to which he is always recurring, and to which he accommodates, as far as he can, the reasonings of all his predecessors. It is his object to proclaim and exalt the immanent divinity of man, and to raise the soul to a contemplation of the good and the true, and to vindicate its independence of all that is sensuous, transitory, and special. With an enthusiasm bordering on fanaticism, he proclaims his philosophical faith in an unseen world: and, rejecting with indignation the humiliating attempt to make out that the spiritual world is no better than an essence or elixir drained off from the material—that thoughts are 'merely the shadows and ghosts of sensations,' he tells his disciples that the inward eyes of consciousness and conscience were to be purged and unsealed at the fountain of heavenly radiance, before they can discern the true form and colours and value of spiritual objects."

The personal humility of this sublime teacher, we may add, seems to have equalled the loftiness of his inspiration.

Of the other writings of Porphyry, space allows us to refer only to his *Epistle to Anebo*—a critical refutation of some of the popular prejudices of Pagan theology, such as the grosser dæmonism, necromancy, and incantation,* and, above all, animal sacrifice, to which his keen spiritual sense was essentially antagonistic. It is known only by fragments preserved in Eusebius. As to the theological or metaphysical opinions of Porphyry, "it is clear," remarks Dr. Donaldson, "that he

---

* In justice to the old Greek Theology which, as it really was, has enough to answer for, it must be remarked that its Demonology, or belief in the powers of subordinate divinities—in the first instance merely the internunciaries, or mediators, or *angels* between Heaven and Earth— was a very different thing from the *Diabolism* of Christian theology, a fact which, perhaps, can be adequately recognised by those only who happen to be acquainted with the history of that most widely-spread and most fearful of all superstitions. Necessarily, from the vague and, for the most part, merely secular character of the earlier theologies, the *infernal* horrors, with the frightful creed, tortures, burnings, &c., which characterised the faith of Christendom, were wholly unknown to the religion of Apollo and of Jupiter.

had but little faith in the old polytheism of the Greeks. He expressly tells his wife (Letter to Marcella) that outward worship does neither good nor harm." In truth, as regards the better parts of Christianity, he was nearer to the religion of Jesus than of Jupiter, although he found himself in opposition to what he considered the evils or errors of dogmatic Christian theology. In common with most of the principal expounders of Neo-Platonism,* his sympathies were with much that was contained in the Christian Scriptures, and, in particular, with the fourth Gospel, the sublime beginning of which, we are assured, the disciples of Plato regarded as " an exact transcript of their own opinions," and which, as St. Augustin informs us (De Civ. Dei x., 29), they declared to be worthy to be written in letters of gold, and inscribed in the most conspicuous place in every Christian church.

As for the learning, as well as lofty ideas, of the author of the treatise On Abstinence, there has been a general consensus of opinion even from his theological opponents. Augustin, himself among the most learned of the Latin Fathers, styles him doctissimus philosophorum ("the most learned of the philosophers"), and, again, philosophus nobilis ("a noble philosopher"), "a man of no common mind" (De Civit. Dei); and elsewhere he calls him "the great philosopher of the heathen." Even Eusebius, his immediate antagonist, concedes to him the titles of "the noble philosopher," "the wonderful theologian," "the great prophet of ineffable doctrines" (ὁ τῶν ἀπορρήτων μύστης). Donaldson, endorsing the common admiration of the moderns, describes his learning and erudition as "stupendous."

Amongst modern testimonies to the merits of Porphyry's treatise, On Abstinence, the sympathising remarks of Voltaire are worth transcribing :—

"It is well known that Pythagoras embraced this humane doctrine [of abstinence from flesh-eating] and carried it into Italy. His disciples followed it through a long period of time. The celebrated philosophers, Plotinus, Iamblichus, and Porphyry, recommended and practised it, although it is sufficiently rare to practice what one preaches. The work of Porphyry, written in the middle of our third century, and very well translated into our language by M. de Burigni, is much esteemed by the learned—but he has made no more converts amongst us than has the book of the

---

* Neo or New-Platonism may be briefly defined as a *spiritual* development of the Socratic or Platonic teaching. In the hands of some of its less judicious and rational advocates it tended to degenerate into puerile, though harmless, superstition. With the superior intellects of a Plotinus, Porphyry, Longinus, Hypatia, or Proclus, on the other hand, it was, in the main at least, a sublime attempt at the purification and spiritualisation of the established orthodox creed. It occupied a position midway between the old and the new religion, which was so soon to celebrate its triumph over its effete rival. That Christianity, on its spiritual side (whatever the ingratitude of its later authorities), owes far more than is generally acknowledged to both the old and newer Platonism, is sufficiently apparent to the attentive student of theological history.

† Author of a *Treatise on the Abandonment of the Flesh Diet*, 1709. He died in the year 1737.

physician Hecquet.† It is in vain that Porphyry alleges the example of the Buddhists and Persian Magi of the first class, who held in abhorrence the practice of engulfing the entrails of other beings in their own—he is followed at present only by the Fathers of La Trappe.* The treatise of Porphyry is addressed to one of his old disciples, named Firmus, who became a Christian, it is said, to recover his liberty to eat flesh and drink wine.

"He remonstrates with Firmus, that in abstaining from flesh and from strong liquors the health of the soul and of the body is preserved; that one lives longer and with more innocence. All his reflections are those of a scrupulous theologian, of a rigid philosopher, and of a gentle and sensitive spirit. One might believe, in reading him, that this great enemy of the Church is a Father of the Church. He does not speak of the *Metempsychosis,* but he regards other animals as our brothers—*because* they are endowed with life as we, *because* they have the same *principles* of life, the same feelings, ideas, memory, industry, as we. Speech alone is wanting to them. If they had it, should we dare to kill and eat them? Should we dare to commit those fratricides? What barbarian is there who would cause a lamb to be slaughtered and roasted, if that lamb conjured him, by an affecting appeal, not to be at once assassin and cannibal?

"This book, at least, proves that there were, among the 'Gentiles,' philosophers of the strictest and purest virtue. Yet they could not prevail against the butchers and the *gourmands.* It is to be remarked that Porphyry makes a very beautiful eulogy on the Essenians. At that time the rivalship was who could be the most virtuous— Essenians, Pythagoreans, Stoics, Christians. When churches form but a small flock their manners are pure; they degenerate as soon as they get powerful."†

Of this famous treatise there is, it appears, only one English translation, that of Taylor (1851), long out of print; and there is a German version by Herr Ed. Baltzer, President of the Vegetarian Society of Germany; thus we have to lament for Porphyry, no less than for Plutarch, the indifferentism of the publishers, or rather of the public, which allows a production, of an inspiration far above that of the common herd of writers, to continue to be a sealed book for the community in general.

It has been already stated that it consists of four Divisions. The first treats of Abstinence from the point of view of Temperance and Reason. In the second is considered the lawfulness or otherwise of animal sacrifice. In the third Porphyry treats the subject from the side of Justice. In the fourth he reviews the practice of some of the nations of antiquity and of the East—of the Egyptians, Hindus, and others. This last Book, by its abrupt termination, is evidently unfinished.

---

* Voltaire might have added the examples of the Greek *Coenobites.* There is at least one celebrated and long-established religious community, in the Sinaitic peninsula, which has always rigidly excluded all flesh from their diet. Like the community of La Trappe, these religious Vegetarians are notoriously the most free from disease and most long-lived of their countrymen.

† Article *Viande (Dict. Phil.)* In other passages in his writings the philosopher of Ferney, we may here remark, expresses his sympathy with the humane diet. See especially his *Essai sur les Mœurs et l'Esprit des Nations* (introduction), and his Romance of *La Princesse de Babylone.*

Porphyry begins with an expression of surprise and regret at the apostasy of the Pythagorean renegade :—

"For when I reflect with myself upon the cause of your change of mind [so he addresses his former associate], I cannot believe, as the vulgar herd will suppose, that it has anything to do with reasons of health or strength, inasmuch as you yourself were used to assert that the fleshless diet is more consonant to healthfulness and to an even and proportionate endurance of philosophic toils (σύμμετρον ὑπομονὴν τῶν περὶ φιλοσοφίαν πόνων), and experience fully proved the truth of your conviction. Whether then it was through some other fallacy or delusion, or through a later notion that this or that diet makes no difference to the intellectual powers, or whether it was from the fear of incurring odium by opposition to orthodox customs, or what the reason may have been, I am unable to conjecture."

He expresses his hope, or rather his belief, that, at least, the lapse was not due in this case to natural intemperance, or regret for the gluttonous habits (λαιμαργίας) of flesh-eating.

He then proceeds to quote and refute the fallacies of the ordinary systems and sects, and, in particular, the objections of one Clodius, a Neapolitan, who had published a treatise against Pythagoreanism. He professes that he does not hope to influence those who are engaged in sordid and selfish, or in sanguinary, pursuits. Rather he addresses himself to the man

"Who considers what he is, whence he came, and whither he ought to tend ; and who, in what pertains to the nourishment of the body and other necessary concerns, is of really thoughtful and earnest mind—who resolves that he shall not be led astray and governed by his passions. And let such a man tell me whether a rich flesh diet is more easily procured, or incites less to the indulgence of irregular passions and appetites, than a light vegetable dietary. But if neither he, nor a physician, nor, indeed, any reasonable man whosoever, dares to affirm this, why do we persist in oppressing ourselves with gross feeding ? And why do we not, together with that luxurious indulgence, throw off the encumbrances and snares which attend it ?

It is not from those who have lived on innocent foods that murderers, tyrants, robbers, and sycophants have come, but from eaters of flesh. The necessaries of life are few and easily procured, without violation of justice, liberty, or peace of mind ; whereas luxury obliges those ordinary souls who take delight in it to covet riches, to give up their liberty, to sell justice, to misspend their time, to ruin their health, and to renounce the satisfaction of an upright conscience."

In condemning animal sacrifice, he declares that "it is by means of an exalted and purified intellect alone that we can approximate to the Supreme Being, to whom nothing material should be offered." He distinguishes four degrees of virtue, the lowest being that of the man who attempts to moderate his passions; the highest, the life of pure reason, by which man becomes one with the Supreme Existence.

In the third book, maintaining that other animals are endowed with high degrees of reasoning and of mental faculties, and, in some measure, even with moral perception, Porphyry proceeds logically to insist that they are, *therefore*, the proper objects of Justice :—

"By these arguments, and others which I shall afterwards adduce in recording the opinions of the old peoples, it is demonstrated that [many species of] the lower animals are rational. In very many, reason is imperfect indeed—of which, nevertheless, they are by no means destitute. Since then justice is due to rational beings, as our opponents allow, how is it possible to evade the admission also that we are bound to act justly towards the races of beings below us ? We do not extend the obligations of justice to plants, because there appears in them no indication of reason ; although, even in the case of these, while we eat the fruits we do not, with the fruits, cut away the trunks. We use corn and leguminous vegetables when they have fallen on the earth and are dead. But no one uses for food the flesh of dead animals, unless they have been killed by violence, so that there is in these things a radical injustice. As Plutarch says, it does not follow, because we are in need of many things, that we should therefore act unjustly towards *all beings*. Inanimate things we are allowed to injure to a certain extent, to procure the necessary means of existence—if to take anything from plants while they are growing can be said to be an injury—but to destroy living and conscious beings merely for luxury and pleasure is truly barbarous and unjust. And to refrain from killing them neither diminishes our sustenance nor hinders our living happily. If indeed the destruction of other animals and the eating of flesh were as requisite as air and water, plants and fruits, then there could be no injustice, as they would be necessary to our nature."

Porphyry, it is scarcely necessary to remark, by these arguments proves himself to have been, in moral as well as mental perception, as far ahead of the average thinkers of the present day as he was of his own times. He justly maintains that

"Sensation and perception are the principle of the kinship of all living beings. And [he reminds his opponents] Zeno and his followers [the Stoics] admit that alliance or *kinship* (οἰκείωσις)* is the foundation of justice. Now, to the lower animals pertain perception and the sensations of pain and fear and injury. Is it not absurd, then, whereas we see that many of our own species live by brute sense alone, and exhibit neither reason nor intellect, and that very many of them surpass the most terrible wild beasts in cruelty, rage, rapine ; that they murder even their own relatives ; that they are tyrants and the tools of tyrants—seeing all this, is it not absurd, I say, to hold that we are obliged by nature to act leniently towards them, while no kindness is due from us to the Ox that ploughs, the Dog that is brought up with us, and those who nourish us with their milk and cover our bodies with their wool ? Is not such a prejudice most irrational and absurd ?"

To the objection of Chrysippus (the second founder of the school of the Porch) that the gods made us for themselves and for the sake of each other, and that they made the non-human species for us—a convenient subterfuge by no means unknown to writers and talkers of our own times —Porphyry unanswerably replies :—

"Let him to whom this sophism may appear to have weight or probability, consider

---

* Οἰκείωσις strictly means adoption, admission to intimacy and family life, or "domestication."

how he would meet the dictum of Karneades* that 'everything in nature is benefited, when it obtains the ends to which it is adapted and for which it was generated.' Now, *benefit* is to be understood in a more general way as meaning what the Stoics call *useful.* 'The hog, however,' says Chrysippus, 'was produced by nature for the purpose of being slaughtered and used for food, and when it undergoes this, it obtains the end for which it is adapted, and it is therefore benefited!' But if God brought other animals into existence for the use of men, what use do we make of flies, beetles, lice, vipers, and scorpions? Some of these are hateful to the sight, defile the touch, are intolerable to the smell, while others are actually destructive to human beings who fall in their way.† With respect to the *cetacea*, in particular, which Homer tells us live by myriads in the seas, does not the Demiurgus‡ teach us that they have come into being for the good of things in general? And unless they affirm that all things were indeed made for us and on our sole account, how can they escape the imputation of wrong-doing in treating injuriously beings that came into existence according to the *general arrangement* of Nature?

"I omit to insist on the fact that, if we depend on the argument of necessity or utility, we cannot avoid admitting by implication that we ourselves were created only for the sake of certain destructive animals, such as crocodiles and snakes and other monsters, for we are not in the least *benefited* by them. On the contrary, they seize and destroy and devour men whom they meet—in so doing acting not at all more cruelly than we. Nay, *they* act thus savagely through want and hunger; *we* from

---

* The founder of the new Academy at Athens, and the vigorous opponent of the Stoics.

† That unreasoning arrogance of human selfishness, which pretends that all other living beings have come into existence for the sole pleasure and benefit of man, has often been exposed by the wiser, and therefore more humble, thinkers of our race. Pope has well rebuked this sort of monstrous arrogance :—
" Has God, thou fool, worked solely for thy good,
Thy joy, thy pastime, thy attire, thy food?

.   .   .   .

Know, Nature's children all divide her care,
The fur that warms a monarch, warmed a bear.
While man exclaims : ' See, all things for my use !'
' See, man for mine,' replies a pampered goose.
And just as short of reason he must fall,
Who thinks *all made for one, not one for all.*"
Essay on Man, III.

And, as a commentary upon these truly philosophic verses, we may quote the words of a recent able writer, answering the objection, "Why were sheep and oxen created, if not for the use of man?" He replies to the same effect as Porphyry 1600 years ago: "It is only pride and imbecility in man to imagine all things made for his sole use. There exist millions of suns and their revolving orbs which the eye of man has never perceived. Myriads of animals enjoy their pastime unheeded and unseen by him—many are injurious and destructive to him. All exist for purposes but partially known. Yet we must believe, in general, that all were created for their own enjoyment, for mutual advantage, and for the preservation of universal harmony in Nature. If, mere y because we can eat sheep pleasantly, we are to believe that they exist only to supply us with food, we may as well say that man was created solely for various parasitical animals to feed on, *because* they do feed on him."--(*Fruits and Farinacea : the Proper Food of Man.* By J. Smith. Edited by Professor Newman. Heywood, Manchester; Pitman, London.) See, also, amongst other philosophic writers, the remarks of Joseph Ritson in his "Essay on Abstinence from Animal Food a Moral Duty"—(Phillips, London, 1802). As to Oxen and Sheep, it must be further remarked that they have been made what they are by the intervention of man alone. The original and wild stocks (especially that of sheep) are very different from the metamorphosed and almost helpless domesticated varieties. Naturam violant, pacem appellant.

‡ T e Artificer or *Creator, par excellence.* In the Platonic language, the usual distinguishing name of the subordinate creator of our imperfect world.

insolent wantonness and luxurious pleasure*, amusing ourselves as we do also in the Circus and in the murderous sports of the chase. By thus acting, a barbarous and brutal nature becomes strengthened in us, which renders men insensible to the feeling of pity and compassion. Those who first perpetrated these iniquities fatally blunted the most important part of the civilised mind. Therefore it is that Pythagoreans consider kindness and gentleness to the lower animals to be an exercise of philanthropy and gentleness."

Porphyry unanswerably and eloquently concludes this division of his subject with the *à fortiori* argument :—

" By admitting that [selfish] pleasure is the legitimate end of our action, justice is evidently destroyed. For to whom must it not be clear that the feeling of justice is fostered by abstinence ? He who abstains from injuring other species will be so much the more careful not to injure his own kind. For he who loves all animated Nature will not hate any one tribe of innocent beings, and by how much greater his love for the whole, by so much the more will he cultivate justice towards a part of them, and to that part to which he is most allied."

In fine, according to Porphyry, he who extends his sympathies to *all* innocent life is nearest to the Divine nature. Well would it have been for all the after-ages had this, the only sure foundation of any code of ethics worthy of the name, found favour with the constituted instructors and rulers of the western world. The fourth and final Book reviews the dietetic habits of some of the leading peoples of antiquity, and of certain of the philosophic societies which practised abstinence more or less rigidly. As for the Essenes, Porphyry describes their code of morals and manner of living in terms of high praise. We can here give only an abstract of his eloquent eulogium :—

" They are despisers of mere riches, and the communistic principle with them is admirably carried out. Nor is it possible to find amongst them a single person distinguished by the possession of wealth, for all who enter the society are obliged by their laws to divide property for the common good. There is neither the humiliation of poverty nor the arrogance of wealth. Their managers or guardians are elected by vote, and each of them is chosen with a view to the welfare and needs of all. They have no city or town, but dwell together in separate communities. . . . They do not discard their dress for a new one, before the first is really worn out by length of time. There is no buying and selling amongst them. Each gives to each according to his or her wants, and there is a free interchange between them. . . . They come to their dining-hall as to some pure and undefiled temple, and when they have taken their seats quietly, the baker sets their loaves before them in order, and the cook gives them one dish each of one sort, while their priest first recites a form of thanksgiving for their pure and refined food (τροφῆς ἁγνῆς οὔσης καὶ καθαρᾶς)."

The testimony of the national historian of the Jews, it is interesting to observe, is equally favourable to those pioneers of the modern communisms. " The Essenes, as we call a sect of ours," writes Josephus,

---

* Cf. Ovid's *Metam*, xv. ; Plutarch's *Essay on Flesh-Eating* ; Thomson's *Seasons*.

"pursue the same kind of life as those whom the Greeks call Pythagoreans. They are long-lived also, insomuch that many of them exist above a hundred years by means of their simplicity of diet and the regular course of their lives" (*Antiquities of the Jews.*). Upon entering the society and partaking of the common meal (which, with baptism, was the outward and visible sign of initiation) three solemn oaths were administered to each aspirant :—

"First, that he would reverence the divine ideal (τὸ θεῖον); second, that he would carefully practise justice towards his fellow-beings and refrain from injury, whether by his own or another's will; that he would always hate 'the Unjust and fight earnestly on the side of (συναγωνίζεσθαι) the Just and lovers of justice; keep faith with all men; if in power, never use authority insolently or violently; nor surpass his subordinates in dress and ornaments; above all things always to love Truth."

As for their food, while they seem not to have been bound to total abstinence from every kind of flesh, they may be considered to have been almost Vegetarian in practice. To kill any innocent individual of the non-human species that had sought refuge or an asylum amongst them was a breach of the most sacred laws : to spare the domesticated races, or fellow-workers with man, even in an enemy's country, was a solemn duty. For, says Porphyry, their founder had no groundless fear that there could be any overabundance of life productive of famine to ourselves, inasmuch as he knew, first, that those animals who bring forth many young at a time are short lived, and, secondly, that their too rapid increase is kept down by other hostile animals. "A proof of which is," he continues, "that though we abstain from eating very many, such as dogs, wild beasts, rats, lizards, and others, there is yet no fear that we should ever suffer from famine in consequence of their excessive multiplication; and, again, it is one thing to have to kill, and another to eat, since we have to kill many ferocious animals whom we do not also eat."

He quotes the historians of Syria who allege that, in the earlier period, the inhabitants of that part of the world abstained from all flesh, and, therefore, from sacrifice; and that when, afterwards, to avert some impending misfortune they were induced to offer up propitiatory victims, the practice of flesh-eating was by no means general. And Asklepiades says, in his History of Cyprus and Phœnicia, that "no living being was sacrificed to heaven, nor was there even any express law on the subject, *since it was forbidden by the law of Nature* (νομῷ φυσικῷ):" that, in course of time, they took to occasional propitiatory sacrifice : and that, at one of these times, the sacrificing priest happened to place his blood-smeared finger on his mouth, was tempted to repeat the action, and thus introduced the habit of flesh-eating, whence the general practice. As for the Persian *Magi* (the successors of Zerdusht), we are informed that the principal and most esteemed of their order neither eat nor kill any living being, while those of the second class eat the flesh of some, but not of domesticated, animals ; nor do even the third order eat indiscriminately. Instances are adduced of certain peoples who, being compelled by necessity to live upon flesh, have evidently deteriorated and been rendered savage and ferocious, " from which examples it is clearly unbecoming men of good disposition to belie their human nature (τῆς ἀνθρωπίνης καταψεύδεσθαι φύσεως)."

Amongst individuals he instances the example of the traditionary Athenian legislator Triptolemus—

"Of whom Hermippus, in his second book on the legislators, writes : Of his laws, according to Xenokrates the philosopher, the three following remain in force at Eleusis—'to gratify Heaven with the offering of fruits,' 'to harass or harm no [innocent] living being.' . . . As to the third, he is in doubt for what particular reason Triptolemus charged them to abstain—whether from believing it to be criminal to kill those that have an identical origin with ourselves (ὁμογενὲς), or from a consciousness that the slaughter of all the most useful animals would be the inevitable consequence of addiction to it, and wishing to render human life mild and innocent, and to preserve those species that are tame and gentle and domesticated with man.*

Somewhat later than Porphyry, the name of Julian (331—363), the Roman emperor, may here be fitly introduced. During his brief reign of sixteen months he proved himself, if not always a judicious, yet a sincere and earnest reformer of abuses of various kinds, and he may claim to be one of the very few virtuous princes, pagan or christian. Unfortunately the just blame attaching to his ill-judged attempt to suppress the religion of Constantine, from whose family his relatives and himself had suffered the greatest injury and insult, has enabled the lovers of party rather than of truth successfully to conceal from view his undoubted merits.

In his manner of living, with which alone we are now concerned, he seems to have almost rivalled the most ascetic of the Platonists or of the Christian anchorets. One of his most intimate friends, the celebrated

---

* Περὶ ʼΕποχῆς κ. τ. λ. In the number of the traditionary reformers and civilisers of the earlier nations, the name of Orpheus has always held a foremost place. In early Christian times Orpheus and the literature with which his name is connected occupy a very prominent and important position, and some celebrated forged prophecies passed current as the utterances of that half-legendary hero. Horace adopts the popular belief as to his radical dietetic reform in the following verses:—

Silvestres homines sacer, interpresque Deorum,
Cœdibus et fœdo victu deterruit Orpheus.
—*Ars Poetica.*

Virgil assigns him a place in the first rank of the Just in the Elysian paradise.—*Æn.* vi.

orator, Libanius, who had often shared the frugal simplicity of his table, has remarked that his "light and sparing diet, which was usually of the vegetable kind, left his mind and body always free and active for the various and important business of an author, a pontiff, a magistrate, a general, and a prince." That his *frugal* diet had not impaired his powers, either physica. or mental, may sufficiently appear from the fact that—

"In one and the same day he gave audience to several ambassadors, and wrote or ictated a great number of letters to his generals, his civil magistrates, his private friends, and the different cities of his dominions. He listened to the memorials which had been received, considered the subject of the petitions, and signified his intentions more rapidly than they could be taken in shorthand by the diligence of his secretaries. He possessed such flexibility of thought, and such firmness of attention, that he could employ his hand to write, his ear to listen, and his voice to dictate, and pursue at once three several trains of ideas without hesitation and without error. While his ministers reposed, the prince flew with agility from one labour to another, and, after a hasty dinner, retired into his library till the public business, which he had appointed for the evening, summonec him to interrupt the prosecution of his studies. The supper of che emperor was still less substantial than the former meal; his sleep was never clouded by the fumes of indigestion. . . . He was soon awakened by the entrance of fresh secretaries who had slept the preceding day, and his servants were obliged to wait alternately, while their indefatigable master allowed himself scarcely any other refreshment than the change of occupation. The predecessors of Julian, his uncle, his brother, and his cousin, indulged their puerile taste for the games of the circus under the specious pretence of complying with the inclination of the people, and they frequently remain d the greater part of the day as idle spectators. . . On solemn festivals Julian, who felt and professed an unfashionable dislike to these frivolous amusements, condescended to appear in the Circus, and, after bestowing a careless glance on five or six of the races, he hastily withdrew with the impatience of a philosopher who considered every moment as lost that was not devoted to the advantage of the public, or the improvement of his own mind. By this avarice of time he s emed to protract the short duration of his reign, and, if the dates were less securely ascertained, we should refuse to believe that only *sixteen months* elapsed between the de th of Constantius and the departure of his successor for the Persian war in which he perished."

Following tne principles of Platonism, "he justly concluded that the man who presumes to reign should aspire to the perfection of the divine nature—that he should purify his soul from her mortal and terrestrial part—that he should extinguish his appetites, enlighten his understanding, regulate his passions, and subdue the wild beast which, according to the lively metaphor of Aristotle, seldom fails to ascend the

* In his witty satire, the *Misopogon* or *Beard-Hater* —"a sort of inoffensive retaliation, which it would be in the power of few princes to employ"—directed against the luxurious people of Antioch, who had ridiculed his frugal meals and simple mode of living, "he himself mentions his vegetable diet, and upbraids the gross and sensual appetite" of that orthodox but corrupt Christian city. When they complained of the high prices of flesh-meats, "Julian publicly leclared that a *frugal city ought to be satisfied with a regular supply of wine, oil, and bread.*"— *Decline and Fall of the Roman Empire,* xxiv.

throne of a despot." With all these virtues, unfortunately for his credit as a philosopher and humanitarian, the imperial Stoic allowed his natural goodness of heart to be corrupted by superstition and fanaticism. Conceiving himself to be the special and chosen instrument of the Deity for the restoration of the fallen religion, which he regarded as the true faith, he made it the foremost object of his pious but misdirected ambition to re-establish its sumptuous temples, priesthoods, and sacrificial altars with all their imposing ritual, and "he was heard to declare, with the enthusiasm of a missionary, that if he could render each individual richer than Midas and every city greater than Babylon, he should not esteem himself the benefactor of mankind, unless at the same time he could reclaim his subjects from their impious revolt against the immortal gods."* Inspired by this religious zeal, he forgot the maxims of his master, Plato, so far as to rival, if not surpass, the ancient Jewish or Pagan ritual in the number of the sacrificial victims offered up in the name of religion and of the Deity. Happily for the future of the world, the fanatical piety of this youthful champion of the religion of Homer proved ineffectual to turn back the slow onward march of the Western mind, through fearful mazes of evil and error indeed, towards that "diviner day" which is yet to dawn for the Earth.

X.

CHRYSOSTOM.   347—407 A.D.

THE most eloquent, and one of the most estimable, of the "Fathers" was born at Antioch, the Christian city *par excellence*. His family held a distinguished position, and his father was in high command in the Syrian division of the imperial army. He studied for the law, and was instructed in oratory by the famous rhetorician Libanius (the intimate friend and counsellor of the young Emperor Julian), who pronounced his pupil worthy to succeed to his chair, if he had not adopted the Christian faith. He soon gave up the law for theology, and retired to a monastery, near Antioch, where he passed four years, rigidly abstaining from flesh-meat and, like the Essenes, abandoning the rights of private property and living a life of the strictest asceticism.

---

* Gibbon, *Decline and Fall of the Roman Empire*, xxii. The philosophical fable of Julian—*The Cæsars*—has been pronounced by the same historian to be "one of the most agreeable and instructive productions of ancient wit." Its purpose is to estimate the merits or demerits of the various Emperors from Augustus to Constantine. As for the *Enemy of the Beard*, it may be ranked, for sarcastic wit, almost with the *Jupiter in Tragedy* of Lucian.

Having submitted himself in solitude to the severest auste ities during a considerable length of time, he entered the Church, and soon gained the highest reputation for his extraordinary eloquence and zeal. On the death of the Archbishop of Constantinople, he was unanimously elected to fill the vacant Primacy. The *nolo me episcopari* seems, in his case, to have been no unmeaning formula. His beneficence and charity in the new position attracted general admiration. From the revenues of his See he founded a hospital for the sick—one of the very first of those rather modern institutions. The fame of the " Golden-mouthed " drew to his cathedral immense crowds of people, who before had frequented the theatre and the circus rather than the churches, and the building constantly resounded with their enthusiastic plaudits. He was, however, no mere popular preacher ; he fearlessly exposed the corrupt and selfish life of the large body of the clergy. At one time he deposed, it is said, no less than thirteen bishops, in Lesser Asia, from their Sees ; and in one of his *Homilies* he does not hesitate to charge " the whole ecclesiastical body with avarice and licentiousness, asserting that the number of bishops who could be saved bore a very small proportion to those who would be damned." *

At length, his repeated denunciations of the too notorious scandals of the Court and the Church excited the bitter enmity of his brother-prelates, and, by their intrigues at the Imperial Court of Constantinople, he was deposed from his See and exiled to the wildest parts of the Euxine coasts, where, exposed to every sort of privation, he caught a violent fever and died. So far did the hostility of the Episcopacy extend, that one of his rivals, a bishop, named Theophilus, in a book expressly written against him, amongst other vituperative epithets had proceeded to the length of styling him " a filthy demon," and of solemnly consigning his soul to Satan. With the poor, however, Chrysostom enjoyed unbounded popularity and esteem. His greatest fault was his theological intolerance—a fault, it is just to add, of the age rather than of the man.

The writings of Chrysostom are exceedingly voluminous—700 homilies, orations, doctrinal treatises, and 242 epistles. Their " chief value consists in the illustrations they furnish of the manners of the fourth and fifth centuries—of the moral and social state of the period. The circus, spectacles, theatres, baths, houses, domestic economy, banquets, dresses, fashions, pictures, processions, tight-rope dancing, funerals—in fine, everything has a place in the picture of licentious luxury which it is the object of Chrysostom to denounce." Next to his profession of

---

* Article, " Chrysostom," in the *Penny Cyclopædia*.

faith in the efficacy and virtues of a non-flesh diet, amongst the most interesting of his productions is his *Golden Book* on the education of the young. He recommends that children should be inured to habits of temperance, by abstaining, at least, twice a week from the ordinary grosser food with which they are supplied. As might be expected from the age, and from his order, the practice of Chrysostom, and of the numerous other ecclesiastical abstinents from the gross diet of the richer part of the community, reposed upon ascetic and traditionary principles, rather than on the more secular and modern motives of justice, humanity, and general social improvement. So, in fact, Origen, one of the most learned of the Fathers, expressly says (*Contra Celsum*, v.) : "We [the Christian leaders] practise abstinence from the flesh of animals to buffet our bodies and treat them as slaves (ὑπωπιάζομεν καί δουλαγωγοῦμεν), and we wish to mortify our members upon earth," &c.

Accordingly, the *Apostolical Canons* distinguished, as Bingham (*Antiquities of the Christian Church*) reports them, between abstinents, διὰ τὴν ἄσκησιν and διὰ τὴν βδελυρίαν, *i.e.*, between those who abstained to exercise self-control, and those who did so from disgust and abhorrence of what, in ordinary and orthodox language, are too complacently and confidently termed "the good creatures of God." This distinction, it must be added, holds only of the prevailing sentiment of the Orthodox Church as finally established. During several centuries—even so late as the Paulicians in the seventh, or even as the Albigeois of the thirteenth, century—*Manicheism*, as it is called, or a belief in the inherent evil of all matter, was widely spread in large and influential sections of the Christian Church—nor, indeed, were some of its most famous Fathers without suspicion of this heretical taint. According to the *Clementine Homilies*, "the unnatural eating of flesh-meat is of demoniacal origin, and was introduced by those giants who, from their bastard nature, took no pleasure in pure nourishment, and only lusted after blood. Therefore the eating of flesh is as polluting as the heathen worship of demons, with its sacrifices and its impure feasts; through participation in which, a man becomes a fellow-dietist (ὁμοδίαιτος) with demons."* That superstition was often, in the minds of the followers both of Plato and of St. Paul, mixed up with, and, indeed, usually dominated over, the reasonable motives of the more philosophic advocates of the higher life, there can be no sort of doubt; nor can we claim a monopoly of rational motives for the mass of the adherents of either Christian or Pythagorean abstinence. Yet an impartial judgment must allow almost equal credit to the earnestness of mind and purity of motive

---

* Baur's *Life and Work of St. Paul.* Part ii., chap. 3.

which, mingled though they undoubtedly were with (in the pre-scientific ages) a necessary infusion of superstition, urged the followers of the better way—Christian and non-Christian—to discard the "social lies" of the dead world around them.   At all events, it is not for the selfish egoists to sneer at the sublime—if error-infected—efforts of the earlier pioneers of moral progress for their own and the world's redemption from the bonds of the prevailing vile materialism in life and dietary habits.

We have already shown that the earliest Jewish-Christian communities, both in Palestine and elsewhere—the immediate disciples of the original Twelve—enjoined abstinence as one of the primary obligations of the New Faith; and that the earliest traditions represent the foremost of them as the strictest sort of Vegetarians.*   If then we impartially review the history of the practice, the teaching, and the traditions of the first Christian authorities, it cannot but appear surprising that the Orthodox Church, ignoring the practice and highest ideal of the most sacred period of its annals, has, even within its own Order, deemed it consistent with its claim of being representative of the Apostolic period to substitute partial and periodic for total and constant abstinence.

The following passages in the *Homilies*, or Congregational Discourses, of Chrysostom will serve as specimens of his feeling on the propriety of dietary reform.   The eloquent but diffusive style of the Greek Bossuet, it must be noted, is necessarily but feebly represented in the literal English version :—

"No streams of blood are among them [the ascetics] ; no butchering and cutting up of flesh ; no dainty cookery ; no heaviness of head.   Nor are there horrible smells

---

* We here take occasion to observe that, while final appeals to our sacred Scriptures to determine any sociological question—whether of slavery, polygamy, war, or of dietetics—cannot be too strongly deprecated, a candid and impartial inquirer, nevertheless, will gladly recognise traces of a consciousness of the unspiritual nature of the sacrificial altar and shambles.   He will gladly recognise that if—as might be expected in so various a collection of sacred writings produced by different minds in different ages—frequent sanction of the materialist mode of living may be urged on the one side; on the other hand, the inspiration of the more exalted minds is in accord with the practice of the true spiritual life.   Cf. *Gen.* i., 29, 30 ; *Isaiah* i., 11-17, and xi., 9 *Ps.* l., 9-14 ; *Ps.* lxxxi., 14-17 ; *Ps.* civ., 14, 15 ; *Prov.* xxiii., 2, 3, 20, 21 ; *Prov.* xxvii., 25-27 : *Prov.* xxx., 8, 22 ; *Prov.* xxxi., 4 ; *Eccl.* vi., 7 ; *Matt.* vi. 31 ; 1 *Cor.* viii., 13, and ix., 25 ; *Rom.* viii., 5-8, 12, 13 ; *Phil.* iii., 19, and iv., 8 ; *James* ii., 13, 4, and iv., 1-3 ; 1 *Pet.* ii., 11.   Perhaps, next to the alleged authority of *Gen.* ix. (noticed and refuted by Tertullian, as already quoted), the trance vision of St. Peter is most often urged by the *bibliolaters* (or those who revere the *letter* rather than the *true inspiration* of the Sacred Books) as a triumphant proof of biblical sanction of materialism.   Yet, unless, indeed, *literalism* is to over-ride the most ordinary rules of common sense, as well as of criticism, all that can be extracted from the "Vision" (in which were presented to the sleeper "all manner of four-footed beasts of the earth, and *wild beasts* and *creeping things*," which it will hardly be contented he was expected to eat) is the fact of a mental illumination, by which the Jewish Apostle recognises the folly of his countrymen in arrogating to themselves the exclusive privileges of the "Chosen People."   Besides, as has already been pointed out, the earliest traditions concur in representing St. Peter as always a strict abstinent, insomuch that he is stated to have celebrated the "Eucharist" with nothing but bread and salt.—*Clement Hom.*, xiv., 1.

of flesh-meats among them, or disagreeable fumes from the kitchen. No tumult and disturbance and wearisome clamours, but bread and water—the latter from a pure fountain, the former from honest labour. If, at any time, however, they may wish to feast more sumptuously, the sumptuousness consists in fruits, and their pleasure in these is greater than at royal tables. With this repast [of fruits and vegetables], even angels from Heaven, as they behold it, are delighted and pleased. For if over one sinner who repents they rejoice, over so many just men imitating them what will they not do ? No master and servant are there. All are servants—all free men. And think not this a mere form of speech, for they are servants one of another and masters one of another. Wherein, therefore, are we different from, or superior to, Ants, if we compare ourselves with them ? For as they care for the things of the body only, so also do we. And would it were for these alone ! But, alas ! it is for things far worse. For not for necessary things only do we care, but also for things superfluous. Those animals pursue an innocent life, while we follow after all covetousness. Nay, we do not so much as imitate the ways of Ants. *We follow the ways of Wolves, the habits of Tigers ; or, rather, we are worse even than they. To them Nature has assigned that they should be thus* [carnivorously] *fed, while God has honoured us with rational speech and a sense of equity. And yet we are become worse than the wild beasts."**

Again he protests :—

"Neither am I leading you to the lofty peak of total renunciation of possessions [ἀκτημοσύνη] ; but for the present I require you to cut off superfluities, and to desire a sufficiency alone. Now, the boundary of sufficiency is the using those things which it is impossible to live without. No one debars you from these, nor forbids you your daily food. I say 'food,' not 'luxury' [τροφὴν οὐ τρυφὴν λέγω]—'raiment,' not 'ornament.' Rather, this frugality—to speak correctly—is, in the best sense, luxury. For consider who should we say more truly feasted—he whose diet is herbs, and who is in sound health and suffered no uneasiness, or he who has the table of a Sybarite and is full of a thousand disorders ? Clearly, the former. Therefore let us seek nothing more than these, if we would at once live luxuriously and healthfully. And let him who can be satisfied with pulse, and can keep in good health, seek for nothing more. But let him who is weaker, and needs to be [more richly] dieted with other vegetables and fruits, not be debarred from them. . . . We do not advise this for the harm and injury of men, but to lop off what is superfluous—and that is superfluous which is more than we need. When we are able to live without a thing, healthfully and respectably, certainly the addition of that thing is a superfluity."—*Hom.* xix. 2 *Cor.*

Denouncing the grossness of the ordinary mode of living, he eloquently descants on the evil results, physical as well as mental :—

"A man who lives in pleasure [*i.e*, in selfish luxury] is dead while he lives, for he lives only to his belly. In his other senses he lives not. He sees not what he ought to see ; he hears not what he ought to hear ; he speaks not what he ought to speak. . . . Look not at the superficial countenance, but examine the interior, and you will see it full of deep dejection. If it were possible to bring the soul into view, and to behold it with our bodily eyes, that of the luxurious would seem depressed, mournful, miserable, and wasted with leanness, for the more the body grows sleek and gross, the more lean and weakly is the soul. The more the one is pampered, the more is the other hampered [θάλπεται—θάπτεται : the latter meaning, literally, buried]. As when the pupil of the eye has the external envelope too thick, it

---

* *Homily,* lxix. on *Mat.* xxii., 1-14.

cannot put forth the power of vision and look out, because the light is excluded by the dense covering, and darkness ensues ; so when the body is constantly full fed, the soul must be invested with grossness. The dead, say you, corrupt and rot, and a foul pestilential humour distils from them. So in her who lives in pleasure may be seen rheums, and phlegm, and catarrh, hiccough, vomiting, eructations, and the like, which, as too unseemly, I forbear to name. For such is the despotism of luxury, it makes us endure things which we do not think proper even to mention. . . . .

"'She that lives in pleasure is dead while she lives.' Hear this, ye women* who pass your time in revels and intemperance, and who neglect the poor, pining and perishing with hunger, whilst you are destroying yourselves with continual luxury. Thus you are the cause of two deaths—of those who are dying of want and of your own, both through ill-measure. If, out of your fulness, you tempered their want, you would save two lives. Why do you thus gorge your own body with excess, and waste that of the poor with want ? Consider what comes of food—into what it is changed. Are you not disgusted at its being named ? Why, then, be eager for such accumulations ? The increase of luxury is but the multiplication of filth.† For Nature has her limits, and what is beyond these is not nourishment, but injury and the increase of ordure.

"Nourish the body, but do not destroy it. Food is called nourishment, to show that its purpose is not to hurt, but to support us. For this reason, perhaps, food passes into excrement that we may not be lovers of luxury. If it were not so—if it were not useless and injurious to the body, we should hardly abstain from devouring one another. If the belly received as much as it pleased, digested it, and conveyed it to the body, we should see battles and wars innumerable. Even as it is, when part of our food passes into ordure, part into blood, part into spurious and useless phlegm, we are, nevertheless, so addicted to luxury that we spend, perhaps, whole estates on a meal. The more richly we live, the more noisome are the odours with which we are filled."—*Hom.* xiii. *Tim.* v.‡

From this period—the fifth century A.D. down to the sixteenth—Christian and Western literature contains little or nothing which comes within the purpose of this work. The merits of monastic asceticism were more or less preached during all those ages, although constant abstinence from flesh was by no means the general practice even with

---

* The *male* sex, according to our ideas, might have been more properly apostrophised ; and St. Chrysostom may seem, in this passage and elsewhere, to be somewhat partial in his invective. Candour, indeed, forces us to remark that the "Golden-mouthed," in common with many others of the Fathers, and with the Greek and Eastern world in general, depreciated the qualities, both moral and mental, of the feminine sex. That the weaker are what the stronger choose to make them, is an obvious truth generally ignored in all ages and countries—by modern satirists and other writers, as well as by a Simonides or Solomon. The *partial* severity of the Archbishop of Constantinople, it is proper to add, may be justified, in some measure, by the contemporary history of the Court of Byzantium, where the beautiful and licentious empress Eudoxia ruled supreme.

† St. Chrysostom seems to have derived this forcible appeal from Seneca. Compare the remarks of the latter, Ep. cx.: "At, mehercule, ista solicite scrutata varieque condita, cum subierint ventrem, una atque cadem fæditas occupabit. Vis ciborum voluptatem contemnere? *Exitum specta.'*

‡ The *Homilies* of St. John Chrysostom, Archbishop of Constantinople, Translated by Members of the English Church. Parker, Oxford. See *Hom.* vii. on *Phil.* ii. for a forcible representation of the inferiority, in many points, of our own to other species.

the inmates of the stricter monastic or conventual establishments—at all events in the Latin Church. But we look in vain for traces of anything like the humanitarian feeling of Plutarch or Porphyry. The mental intelligence as well as capacities for physical suffering of the non-human races—necessarily resulting from an organisation in all essential points like to our own—was apparently wholly ignored; their just rights and claims upon human justice were disregarded and trampled under foot. Consistently with the universal estimate, they were treated as beings destitute of all feeling—as if, in fine, they are the "automatic machines" they are alleged to be by the Cartesians of the present day. In those terrible ages of gross ignorance, of superstition, of violence, and cf injustice—in which human rights were seldom regarded—it would have been surprising indeed if any sort of regard had been displayed for the *non-human* slaves. And yet an underlying and latent consciousness of the falseness of the general estimate sometimes made itself apparent in certain extraordinary and perverse fancies.* To Montaigne, the first to revive the humanitarianism of Plutarch, belongs the great merit of reasserting the natural rights of the helpless slaves of human tyranny.

While Chrysostom seems to have been one of the last of Christian writers who manifested any sort of consciousness of the inhuman, as well as unspiritual nature of the ordinary gross foods, Platonism continued to bear aloft the flickering torch of a truer spiritualism; and "the golden chain" of the prophets of the dietary reformation reached down even so late as to the end of the sixth century. Hierokles, author of the commentary on the *Golden Verses* of Pythagoras, to which reference has already been made, and who lectured upon them with great success at Alexandria; Hypatia, the beautiful and accomplished daughter of Theon the great mathematician, who publicly taught the philosophy of Plato at the same great centre of Greek science and learning, and was barbarously murdered by the jealousy of her Christian rival Cyril, Archbishop of Alexandria; Proklus, surnamed the Successor, as having been considered the most illustrious disciple of Plato in the latter times, who left several treatises upon the Pythagorean system, and "whose sagacious mind explored the deepest questions of morals and metaphysics";† Olympiodorus, who wrote a life of Plato and commentaries on several of his dialogues, still extant, and lived in the reign of Justinian,

---

* For example, we may refer to the fact of trials of "criminal" dogs, and other non-human beings, with all the formalities of ordinary courts of justice, and in the gravest manner recorded by credible witnesses. The convicted "felons" were actually hanged with all the circumstances of human executions Instances of such trials are recorded even so late as the sixteenth century.

† His biographer, Marinus, writes in terms of the highest admiration of his virtues as well as of his genius, and of the perfection to which he had attained by his unmaterialistic diet and manner of living. He seems to have had a remarkably cosmopolitan mind, since he regarded with

by whose edict the illustrious school of Athens was finally closed, and with it the last vestiges of a sublime, if imperfect, attempt at the purification of human life—such are some of the most illustrious names which adorned the days of expiring Greek philosophy. Olympiodorus and six other Pythagoreans determined, if possible, to maintain their doctrines elsewhere ; and they sought refuge with the Persian Magi, with whose tenets, or, at least, manner of living, they believed themselves to be most in accord. The Persian customs were distasteful to the purer ideal of the Platonists, and, disappointed in other respects, they reluctantly relinquished their fond hopes of transplanting the doctrines of Plato into a foreign soil, and returned home. The Persian prince, Chosroes, we may add, acquired honour by his stipulation with the bigoted Justinian, that the seven sages should be allowed to live un- molested during the rest of their days. " Simplicius and his companions ended their lives in peace and obscurity ; and, as they left no disciples, they terminated the long list of Grecian philosophers who may be justly praised, notwithstanding their defects, as the wisest and most virtuous of their contemporaries. The writings of Simplicius are now extant. His physical and metaphysical commentaries on Aristotle have passed away with the fashion of the times, but his moral interpretation of Epiktetus is preserved in the library of nations as a classical book excel- lently adapted to direct the will, to purify the heart, and to confirm the understanding, by a just confidence in the nature both of God and Man." *

———◆•●•◆———

XI.
CORNARO. 1465—1566.

AFTER the extinction of Greek and Latin philosophy in the fifth century, a mental torpor seized upon and, during some thousand years, with rare exceptions, dominated the whole Western world. When this torpor was dispelled by the influence of returning knowledge and

---

equal respect the best parts of all the then existing religious systems ; and he is said even to have paid solemn honours to all the most illustrious, or rather most meritorious, of his philosophic predecessors. That his intellect, sublime and exalted as it was, had contracted the taint of superstition must excite our regret, though scarcely our wonder, in the absence of the light of modern science ; nor can there be any difficulty in perceiving how the miracles and celestial apparitions—which form a sort of halo around the great teachers—originated, viz., in the natural enthusiasm of his zealous but uncritical disciples. One of his principal works is *On the Theology of Plato*, in six books. Another of his productions was a Commentary on the *Works and Days of Hesiod*. Both are extant. He died at an advanced age in 485, having hastened his end by exces- sive asceticism.

* *Decline and Fall of the Roman Empire*, xl. This testimony of the great historian to the merits of the last of the New-Platonists is all the more weighty as coming from an authority notoriously the most unimpassioned and unenthusiastic, perhaps, of all writers. Compare his remarkable expression of personal feeling—guardedly stated as it is—upon the question of kreophagy in his chapter on the history and manners of the Tartar nations (chap. xxvi).

reason evoked by the various simultaneous discoveries in science and literature—in particular by the achievements of Gutenberg, Vasco da Gama, Christopher Colon, and, above all, Copernik—the moral sense then first, too, began to show signs of life. The renascence of the sixteenth century, however, with all the vigour of thought and action which accompanied it, proved to be rather a revival of mere verbal learning than of the higher moral feeling of the best minds of old Greece and Italy. Men, fettered as they were in the trammels of theological controversy and metaphysical subtleties, for the most part expended their energies and their intellect in the vain pursuit of phantoms. With the very few splendid exceptions of the more enlightened and earnest thinkers, *Ethics*, in the real and comprehensive meaning of the word, was an unknown science ; and a long period of time was yet to pass away before a perception of the universal obligations of Justice and of Right dawned upon the minds of men. In truth, it could not have been otherwise. Before the moral instincts can be developed, reason and knowledge must have sufficiently prepared the way. When attention to the importance of the neglected science of *Dietetics* had been in some degree aroused, the interest evoked was little connected with the higher sentiments of humanity.

Of all dietary reformers who have treated the subject from an exclusively sanitarian point of view, the most widely known and most popular name, perhaps, has been that of Luigi Cornaro ; and it is as a vehement protester against the follies, rather than against the barbarism, of the prevailing dietetic habits that he claims a place in this work. He belonged to one of the leading families of Venice, then at the height of its political power. Even in an age and in a city noted for luxuriousness and grossness of living of the rich and dominant classes, he had in his youth distinguished himself by his licentious habits in eating and drinking, as well as by other excesses. His constitution had been so impaired, and he had brought upon himself so many disorders by this course of living, that existence became a burden to him. He informs us that from his thirty-fifth to his fortieth year he passed his nights and days in continuous suffering. Every sort of known remedy was exhausted before his new medical adviser, superior to the prejudices of his profession and of the public, had the courage and the good sense to prescribe a total change of diet. At first Cornaro found his enforced regimen almost intolerable, and, as he tells us, he occasionally relapsed.

These relapses brought back his old sufferings, and, to save his life, he was driven at length to practise entire and uniform abstinence, the yolk of an egg often furnishing him the whole of his meal. In this way he assures us that he came to relish dry bread more than formerly he had enjoyed the most exquisite dishes of the ordinary table. At the end of the first year he found himself entirely freed from all his multiform maladies. In his eighty-third year he wrote and published his first exhortation to a radical change of diet under the title of *A Treatise on a Sober Life*,* in which he eloquently narrates his own case, and exhorts all who value health and immunity from physical or mental sufferings to follow his example. And his *exordium*, in which he takes occasion to denounce the waste and gluttony of the dinners of the rich, might be applied with little, or without any, modification of its language to the public and private tables of the present day :—

"It is very certain," he begins, "that Custom, with time, becomes a second nature, forcing men to use that, whether good or bad, to which they have been habituated ; and we see custom or habit get the better of reason in many things. . . Though all are agreed that intemperance (*la crapula*) is the offspring of gluttony, and sober living of abstemiousness, the former nevertheless is considered a virtue and a mark of distinction, and the latter as dishonourable and the badge of avarice. Such mistaken notions are entirely owing to the power of Custom, established by our senses and irregular appetites. These have blinded and besotted men to such a degree that, leaving the paths of virtue, they have followed those of vice, which lead them imperceptibly to an old age burdened with strange and mortal diseases. . . . . .

"O wretched and unhappy Italy ! [thus he apostrophises his own country] can you not see that gluttony murders every year more of your inhabitants than you could lose by the most cruel plague or by fire and sword in many battles ? Those truly shameful feasts (*i tuoi veramente disonesti banchetti*), now so much in fashion and so intolerably profuse that no tables are large enough to hold the infinite number of the dishes—those feasts, I say, are so many battles.† And how is it possible *to live* amongst such a multitude of jarring foods and disorders ? Put an end to this abuse, in heaven's name, for there is not—I am certain of it—a vice more abominable than this in the eyes of the divine Majesty. Drive away this plague, the worst you were ever afflicted with—this new [?] kind of death—as you have banished that disease which, though it formerly used to make such havoc, now does little or no mischief, owing to the laudable practice of attending more to the goodness of the provisions brought to our markets. Consider that there are means still left to banish intemperance, and such means, too, that every man may have recourse to them without any external assistance.

"Nothing more is requisite for this purpose than to live up to the simplicity dictated by nature, which teaches us to be content with little, to pursue the practice of holy abstemiousness and divine reason, and *accustom ourselves to eat no more than is absolutely necessary to support life ;* considering that what exceeds this is disease and death, and done merely to give the palate a satisfaction which, though but momentary, brings on the body a long and lasting train of disagreeable diseases, and at length kills it along with the soul. How many friends of mine—men of the finest

---

* *Trattato della Vita Sobria*, 1548.

† *Sævior armis Luxuria.* We may be tempted to ask ourselves whether we are reading denunciations of the gluttony and profusion of the sixteenth century or contemporary reports of public dinners in our own country, *e.g.*, of the Lord Mayor's annual dinner. The vast amount of slaughter of all kinds of victims to supply the various dishes of *one* of these exhibitions of national gluttony can be adequately described only by the use of the Homeric word *hecatomb*—slaughter of hundreds.

understanding and most amiable disposition—have I seen carried off by this plague in the flower of their youth! who, were they now living, would be an ornament to the public, and whose company I should enjoy with as much pleasure as I am now deprived of it with concern."

He tells us that he had undertaken his arduous task of proselytising with the more anxiety and zeal that he had been encouraged to it by many of his friends, men of "the finest intellect" (*di bellissimo intelletto*), who lamented the premature deaths of parents and relatives, and who observed so manifest a proof of the advantages of abstinence in the robust and vigorous frame of the dietetic missionary at the age of eighty. Cornaro was a thorough-going hygeist, and he followed a reformed *diet* in the widest meaning of the term, attending to the various requirements of a healthy condition of mind and body :—

"I likewise," he says with much candour, "did all that lay in my power to avoid those evils which we do not find it so easy to remove—melancholy, hatred, and other violent passions which appear to have the greatest influence over our bodies. However, I have not been able to guard so well against either one or the other kind of these disorders [passions] as not to suffer myself now and then to be hurried away by many, not to say all, of them ; but I reaped one great benefit from my weakness—that of knowing by experience that these passions have, in the main, no great influence over bodies governed by the two foregoing rules of eating and drinking, and therefore can do them but very little harm, so that it may, with great truth, be affirmed that whoever observes these two capital rules is liable to very little inconvenience from any other excess. This Galen, who was an eminent physician, observed before me. He affirms that so long as he followed these two rules relative to eating and drinking (*perchè si guardava da quelli due della bocca*) he suffered but little from other disorders—so little that they never gave him above a day's uneasiness. That what he says is true I am a living witness ; and so are many others who know me, and have seen how often I have been exposed to heats and colds and such other disagreeable changes of weather, and have likewise seen me (owing to various misfortunes which have more than once befallen me) greatly disturbed in mind. For not only can they say of me that such mental disturbance has affected me little, but they can aver of many others who did not lead a frugal and regular life that such failure proved very prejudicial to them, among whom was a brother of my own and others of my family who, trusting to the goodness of their constitution, did not follow my way of living."

At the age of seventy a serious accident befel him, which to the vast majority of men so far advanced in life would probably have been fatal. His coach was overturned, and he was dragged a considerable distance along the road before the horses could be stopped. He was taken up insensible, covered with severe wounds and bruises and with an arm and leg dislocated, and altogether he was in so dangerous a state that his physicians gave him only three days to live. As a matter of course they prescribed bleeding and purging as the only proper and effectual remedies :—

"But I, on the contrary, who knew that the sober life I had led for many years past had so well united, harmonised, and dispersed my humours as not to leave it in their power to ferment to such a degree [as to induce the expected high fever], refused to be either bled or purged. I simply caused my leg and arm to be set, and suffered myself to be rubbed with some oils, which they said were proper on the occasion. Thus, without using any other kind of remedy, I recovered, as I thought I should, without feeling the least alteration in myself or any other bad effects from the accident, a thing which appeared no less than miraculous in the eyes of the physicians."

It is, perhaps, hardly to be expected that "The Faculty" will endorse the opinions of Cornaro, that any person by attending strictly to his regimen "could never be sick again, as it removes every cause of illness; and so, for the future, would never want either physician or physic" :—

"Nay, by attending duly to what I have said he would become his own physician, and, indeed, the best he could have, since, in fact, no man can be a perfect physician to anyone but himself. The reason of which is that any man may, by repeated trials, acquire a perfect knowledge of his own constitution and the most hidden qualities of his body, and what food best agrees with his stomach. Now, it is so far from being an easy matter to know these things perfectly of another that we cannot, without much trouble, discover them in ourselves, since a great deal of time and repeated trials are required for that purpose."

Cornaro's second publication appeared three years later than his first, under the title of *A Compendium of a Sober Life* and the third, *An Earnest Exhortation to a Sober and Regular Life,** in the ninety-third year of his age. In these little treatises he repeats and enforces in the most earnest manner his previous exhortations and warnings. He also takes the opportunity of exposing some of the plausible sophisms employed in defence of luxurious living :—

"Some allege that many, without leading such a life, have lived to a hundred, and that in constant health, although they ate a great deal and used indiscriminately every kind of viands and wine, and therefore flatter themselves that they shall be equally fortunate. But in this they are guilty of two mistakes. The first is, that it is not one in one hundred thousand that ever attains that happiness ; the other mistake is, that such persons, in the end, most assuredly contract some illness which carries them off, nor can they ever be sure of ending their days otherwise, so that the safest way to obtain a long and healthy life is, at least after forty, to embrace abstinence. This is no difficult matter, since history informs us of very many who, in former times, lived with the greatest temperance, and I know that the present Age furnishes us with many such instances, reckoning myself one of the number. Now let us remember that we are human beings, and that man, being a rational animal, is himself master of his actions."

Amongst others :—

"There are old gluttons *(attempati)* who say that it is necessary they should eat and drink a great deal to keep up their natural heat, which is constantly diminishing as they advance in years, and that it is therefore necessary for them to eat heartily and of such things as please their palates, and that were they to lead a frugal life it would be a short one. To this I answer that our kind mother, Nature, in order that old men may live to a still greater age, has contrived matters so that they should be able to subsist on little, as I do, for large quantities of food cannot be digested by old

* *Amorevole Esortazione a Seguire La Vita Ordinata e Sobria.*

and feeble stomachs. Nor should such persons be afraid of shortening their lives by eating too little, *since when they are indisposed they recover by eating the smallest quantities.* Now, if by reducing themselves to a very small quantity of food they recover from the jaws of death, how can they doubt but that, with an increase of diet, still consistent, however, with sobriety, they will be able to support nature when in perfect health ?

" Others say that it is better for a man to suffer every year three or four returns of his usual disorders, such as gout, sciatica, and the like than to be tormented the whole year by not indulging his appetite, and eating everything his palate likes best, since by a good regimen alone he is sure to get the better of such attacks. To this I answer that, our natural heat growing less and less as we advance in years, no regimen can retain virtue enough to conquer the malignity with which disorders of repletion are ever attended, so that he must die at last of these periodical disorders, because they abridge life as health prolongs it. Others pretend that it is much better to live ten years less than not indulge one's appetite. My reply is that longevity ought to be highly valued by men of genius and intellect ; as to others it is of no great matter if it is not duly prized by them, since it is they who brutalise the world *(perchè questi fanno brutto il mondo),* so that *their* death is rather of service to mankind."

Cornaro frequently interrupts his discourse with apostrophes to the genius of Temperance, in which he seems to be at a loss for words to express his feeling of gratitude and thankfulness for the marvellous change effected in his constitution, by which he had been delivered from the terrible load of sufferings of his earlier life, and by which moreover he could fully appreciate, as he had never dreamed before, the beauties and charms of nature of the external world, as well as develope the mental faculties with which he had been endowed :—

" O thrice holy Sobriety, so useful to man by the services thou renderest him ! Thou prolongest his days, by which means he may greatly improve his understanding. Thou moreover freest him from the dreadful thoughts of death. How greatly is thy faithful disciple indebted to thee, since by thy assistance he enjoys this beautiful expanse of the visible world, which is really beautiful to such as know how to view it with a philosophic eye, as thou hast enabled me to do ! . . . O truly happy life which, besides these favours conferred on an old man, hast so improved and perfected him that he has now a better relish for his dry bread than he had formerly for the most exquisite dainties. And all this thou hast effected by acting rationally, knowing that bread is, above all things, man's proper food when seasoned by a good appetite. . . . It is for this reason that dry bread has so much relish for me ; and I know from experience, and can with truth affirm, that I find such sweetness in it that I should be afraid of sinning against temperance were it not for my being convinced of the absolute necessity of eating of it, and that we cannot make use of a more natural food."

The fourth and last of his appearances in print was a " Letter to Barbaro, Patriarch of Aquileia," written at the age of ninety-five. It describes in a very lively manner the health, vigour, and use of all his faculties of mind and body, of which he had the perfect enjoyment. He was far advanced in life when his daughter, his only child, was born, and he lived to see her an old woman. He informs us, at the age of ninety-one, with much eloquence and enthusiasm of the active interest and pleasure he experienced in all that concerned the prosperity of his native city : of his plans for improving its port ; for draining, recovering, and fertilizing the extensive marshes and barren sands in its neighbourhood. He died, having passed his one hundredth year, calmly and easily in his arm-chair at Padua in the year 1566.* His treatises, forming a small volume, have been "very frequently published in Italy, both in the vernacular Italian and in Latin. It has been translated into all the civilised languages of Europe, and was once a most popular book. There are several English translations of it, the best being one that bears the date 1779. Cornaro's system," says the writer in the *English Cyclopædia* whom we are quoting, "has had many followers." Recounting his many dignities and honours, and the distinguished part he took in the improvement of his native city, by which he acquired a great reputation amongst his fellow-citizens, the Italian editor of his writings justly adds :—

"But all these fine prerogatives of Luigi Cornaro would not have been sufficient to render his name famous in Europe if he had not left behind him the short treatises upon Temperance, composed at various times at the advanced ages of 85, 86, 91, and 95. The candour which breathes through their simplicity, the importance of the argument, and the fervour with which he urges upon all to study the means of prolonging our life, have obtained for them so great good fortune as to be praised to the skies by men of the best understanding. The many editions which have been published in Italy, and the translations which, together with an array of physiological and philological notes, have appeared out of Italy, at one time in Latin, at another in French, again in German, and again in English, prove their importance. These discourses, in fact, enjoyed all the reputation of a classical book, and, although occasionally somewhat unpolished, as "*Poca favilla gran fiamma seconda*," they have sufficed to inspire (*riscaldare*) a Lessio, a Bartolini, a Ramazzini, a Cheyne, a Hufeland, and so many others who have written works of greater weight upon the same subject."

Addison (*Spectator* 195) thus refers to him :—

"The most remarkable instance of the efficacy of temperance towards the procuring long life is what we meet with in a little book published by Lewis Cornaro, the Venetian, which I the rather mention because it is of undoubted credit, as the late Venetian Ambassador, who was of the same family, attested more than once in conversation when he resided in England. . . . After having passed his one hundredth year he died without pain or agony, and like one who falls asleep. The treatise I mention has been taken notice of by several eminent authors, and is written

---

* Cornaro's heterodoxy in dietetics was not allowed, as may well be supposed, to pass unchallenged by his contemporaries. One of his countrymen, a person of some note, Sperone Speroni, published a reply under the title of "Contra la Sobrietà;" but soon afterwards recanting his errors (*rimettendosi spontaneamente nel buon sentiero*) he wrote a Discourse in favour of Temperance. About the same time there appeared in Paris an "Anti-Cornaro," written "against all the rules of good taste," and which the editors of the *Biographie Universelle* characterise as full of remarks "*tout à fait oiseuses.*"

with such a spirit of cheerfulness, religion, and good sense as are the natural con-comitants of temperance and sobriety. The mixture of the old man in it is rather a recommendation than a discredit to it."

In fact he has exposed himself, it must be confessed, to the taunts of the "devotees of the Table" often cast at the *abstinents*, that they are too much given to parading their health and vigour, and certainly if any one can be justly obnoxious to them it is Luigi Cornaro.

<hr />

XII.

SIR THOMAS MORE.   1480—1535.

DURING part of the period covered by the long life of Cornaro there is one distinguished man, all reference to whose opinions—intimately though indirectly connected as they are with dietary reform—it would be improper to omit—Sir Thomas More. His eloquent denunciation of the grasping avarice and the ruinous policy which were rapidly con-verting the best part of the country into grazing lands, as well as his condemnation of the slaughter of innocent life, commonly euphemised by the name of "sport," are as instructive and almost as necessary for the present age as for the beginning of the sixteenth century.

Son of Sir John More, a judge of the King's Bench, he was brought up in the palace of the Cardinal Lord Chancellor Morton, an ecclesiastic who stands out in favourable contrast with the great majority of his order, and, indeed, of his contemporaries in general. In his twenty-first year he was returned to the House of Commons, where he distinguished himself by opposing a grant of a subsidy to the king (Henry VII.). In 1516 he published (in Latin) his world-famed *Utopia*—the most meri-torious production in sociological literature since the days of Plutarch. In 1523 he was elected Speaker of the House of Commons, and again he displayed his courage and integrity in resisting an illegal and oppressive subsidy bill, by which he was not in the way to advance his interests with Henry VIII. and his principal minister, Wolsey. Seven years later, however, upon the disgrace of the latter personage, Sir Thomas More succeeded to the vacant Chancellorship, in which office he maintained his reputation for integrity and laborious diligence. When the amorous and despotic king had determined upon the momentous divorce from Catherine, he resigned the Seals rather than sanction that equivocal proceeding; and soon afterwards he was sent to the Tower for refusing the Oath of Supremacy. After the interval of a year he was brought to trial before the King's Bench, and sentenced to the block (1535). In private life and in his domestic relations he exhibits a pleasing contrast to the ordinary harsh severity of his contemporaries. In learning and ability he occupies a foremost place in the annals of the period.

Unfortunately for his reputation with after ages, as Lord Chancellor he seems to have forgotten the maxims of toleration (political and theological) of his earlier career, so well set forth in his *Utopia ;* and he supplies a notable instance, not too rare, of retrogression with advancing years and dignities, and of " a head grown grey in vain." In fact, he belonged, ecclesiastically, to the school of conservative sceptics, of whom his intimate friend Erasmus was the most conspicuous representative, rather than to the party of practical reform. Yet, in spite of so lamentable a failure in practical philosophy, More may claim a high degree of merit both for his courage and for his sagacity in propounding views far in advance of his time.

In the *Utopia* his ideas in regard to labour and to crime exhibit him, indeed, as in advance of the received dogmas even of the present day. As to the former he held that the labourer, as the actual basis and support of the whole social system, was justly entitled to some consideration, and to a more rational existence than usually allowed him by the policy of the ruling classes ; and, in limiting the daily period of labour to nine hours, eh anticipated by 350 years the tardy legislation on that important matter. In exposing the equal absurdity and iniquity of the criminal code he preached the despised doctrine of *prevention* rather than punishment, and denounced the monstrous inequality of penalties by which thieving was placed in the same category with murder and crimes of violence :—

" For great and horrible punishments be awarded to thieves, whereas much rather provision should have been made that there were some means whereby they might get their living, so that no man should be driven to this extreme necessity—first to steal and then to die. . . . By suffering your youth to be wantonly and viciously brought up and to be infected, even from their tender age, by little and little with vice—then, in God's name, to be punished when they commit the same faults after being come to man's state, which from their youth they were ever like to do—in this point, I pray you, *what other thing do you than make thieves and then punish them.*"*

What we are immediately concerned with here is his feeling in regard to slaughter. The Utopians condemn—

---

* More points out very forcibly that to hang for theft is tantamount to offering a premium for *murder.* Two hundred and fifty years later Beccaria and other humanitarians vainly advanced similar objections to the criminal code of christian Europe. It is hardly necessary to remark that this Draconian bloodthirstiness of English criminal law remained to belie the name of " civilisation " so recently as fifty years ago.

"Hunters also and hawkers (falconers), for what delight can there be, and not rather displeasure, in hearing the barking and howling of dogs? Or what greater pleasure is there to be felt when a dog follows a hare than when a dog follows a dog? For one thing is done by both—that is to say, running, if you have pleasure in that. But if the hope of slaughter and the expectation of tearing the victim in pieces pleases you, you should rather be moved with pity to see an innocent hare murdered by a dog—the weak by the strong, the fearful by the fierce, the innocent by the cruel and pitiless.* Therefore this exercise of hunting, as a thing unworthy to be used of free men, the Utopians have rejected to their butchers, to the which craft (as we said before) they appoint their bondsmen. For they count hunting the lowest, the vilest, and most abject part of butchery; and the other parts of it more profitable and more honest as bringing much more commodity, in that they (the butchers) kill their victims from necessity, whereas the hunter seeks nothing but pleasure of the seely [simple, innocent] and woful animal's slaughter and murder. The which pleasure in beholding death, they say, doth rise in wild beasts, either of a cruel affection of mind or else by being changed, in continuance of time, into cruelty by long use of so cruel a pleasure. These, therefore, and all such like, which be innumerable, though the common sort of people do take them for pleasures, yet they, seeing that there is no natural pleasantness in them, plainly determine them to have no affinity with true and right feeling."

In telling us that his model people "permit not their free citizens to accustom themselves to the killing of 'beasts' through the use whereof they think clemency, gentlest affection of our nature, by little and little to decay and perish,"† More for ever condemns the immorality of the Slaughter-House, whether he intended to do so *in toto* or no. In relegating the business of slaughter to their bondsmen (criminals who had been degraded from the rights of citizenship), the Utopians, we may observe, exhibit less of justice than of refinement. To devolve the trade of slaughter upon a pariah-class is not the least immoral of the necessary concomitants of the shambles. That the author of *Utopia* should feel an instinctive aversion from the coarseness and cruelty of the shambles is not surprising; that he should have failed to banish it entirely from his ideal commonwealth is less to be wondered at

---

* Erasmus (who, to lash satirically and more effectively the various follies and crimes of men places the genius of Folly itself in the pulpit) seems to have shared the feeling of his friend in regard to the character of "sport." "When they (the 'sportsmen') have run down their victims, what strange pleasure they have in cutting them up! Cows and sheep may be slaughtered by common butchers, but those animals that are killed in hunting must be mangled by none under a gentleman, who will fall down on his knees, and drawing out a slashing dagger (for a common knife is not good enough) after several ceremonies shall dissect all the joints as artistically as the best skilled anatomist, while all who stand round shall look very intently and seem to be mightily surprised with the novelty, though they have seen the same thing a hundred times before; and he that can but dip his finger and taste of the blood shall think his own bettered by it. And yet the constant feeding on such diet does but assimilate them to the nature (?) of those animals they eat," &c.—*Encomium Moriæ,* or *Praise of Folly.* If we recall to mind that three centuries and a half have passed away since More and Erasmus raised their voices against the sanguinary pursuits of hunting, and that it is still necessary to reiterate the denunciation, we shall justly deplore the slow progress of the human mind in all that constitutes true morality and refinement of feeling.

† *Utopia* II.

than to be lamented. That he had at least a *latent* consciousness of the indefensibility of slaughter for food appears sufficiently clear from his remark upon the Utopian religion that "they kill no living animal in sacrifice, nor do they think that God has delight in blood and slaughter, *Who has given life to animals to the intent they should live.*"

Wiser than ourselves, the ideal people do not waste their corn in the manufacture of alcoholic drinks :—

"They sow corn only for bread. For their drink is either wine made of grapes, or else of apples or pears, or else it is clear water—and many times mead made of honey or liquorice sodden in water, for of that they have great store."

The selfish policy of converting arable into grazing land is emphatically denounced by More :—

"They (the oxen and sheep) consume, destroy, and devour whole fields, houses, and cities. For look in what parts of the realm doth grow the finest and therefore the dearest wool. There noblemen and gentlemen, yea, and certain abbots, holy men no doubt, not contenting themselves with the yearly revenues and profits that were wont to grow to their forefathers and predecessors of their lands, nor being content that they live in rest and pleasure nothing profiting, yea, much annoying, the public weal, leave no land for tillage—they enclose all into pasture, they throw down houses, they pluck down towns and leave nothing standing, but only the church to be made a sheep house ; and, as though you lost no small quantity of ground by forests, chases, lands, and parks, those good holy men turn dwelling-places and all glebe land into wilderness and desolation. . . . For one shepherd or herdsman is enough to eat up that ground with cattle, to the occupying whereof about husbandry many hands would be requisite. And this is also the cause why victuals be now in many places dearer ; yea, besides this, the price of wool is so risen that poor folks, which were wont to work it and make cloth thereof, be now able to buy none at all, and by this means very many be forced to forsake work and to give themselves to idleness. For after that so much land was enclosed for pasture, an infinite multitude of sheep died of the rot, such vengeance God took of their inordinate and insatiable covetousness, sending among the sheep that pestiferous murrain which much more justly should have fallen on the sheep-masters' own heads ; and though the number of sheep increase never so fast, yet the price falleth not one mite, because there be so few sellers," &c.

These sagacious and just reflections upon the evil social consequences of carnivorousness may be fitly commended to the earnest attention of our public writers and speakers of to-day. The periodical cattle plagues and foot-and-mouth diseases, which, in theological language, are vaguely assigned to national sins, might be more ingenuously and truthfully attributed to the one sufficient cause—to the general indulgence of selfish instincts, which closes the ear to all the promptings at once of humanity and of reason, and is, in truth, a national sin of the most serious character.*

---

* For a full and eloquent exposition of the social evils which threaten the country from the natural but mischievous greed of landowners and farmers, our readers are referred, in particular, to Professor Newman's admirable Lectures upon this aspect of the Vegetarian creed, delivered before the Society at various times. (Heywood : Manchester.)

The " wisdom of our ancestors," which has been so often invoked, both before and since the days of More, and which Bentham has so mercilessly exposed, apparently did not subdue the reason of the author of *Utopia ;* yet, with no little amount of applause it has been made to serve as a very conclusive argument against dietetic reformation, as against many other changes :—

" 'These things,' say they, 'pleased our forefathers and ancestors—would to God we could be so wise as they were !' And, as though they had wittily concluded the matter, and with this answer stopped every man's mouth, they sit down again as who should say, 'It were a very dangerous matter if a man in any point should be found wiser than his forefathers were.' And yet be we content to suffer the best and wittiest [wisest] of their decrees to be unexecuted ; but if in anything a better order might have been taken than by them was, there we take fast hold, finding therein many virtues."*

XIII.

MONTAIGNE. 1533—1592.

THE modern Plutarch and the first of essayists deserves his place in this work, if not so much for express and explicit denunciation, *totidem verbis,* of the barbarism of the Slaughter-House, at least for a sort of argument which logically and necessarily arrives at the same conclusion. In truth, if he had not "seen and approved the better way" (even though, with too many others, he may not have had the courage of his convictions), he would be no true disciple of the great humanitarian. It is necessary to remember that the " perfect day " was not yet come ; that a few rays only here and there enlightened the thick darkness of barbarism ; that, in fine, not even yet, with the light of truth shining full upon us, have reason and conscience triumphed, as regards the mass of the community, either in this country or elsewhere.

Michel de Montaigne descended from an old and influential house in Périgord (modern Périgeaux, in the department of the Dordogne). His youth was carefully trained, and his early inclination to learning fostered under his father's diligent superintendence. He became a member of the provincial parliament, and, by the universal suffrage of his fellow-citizens, was elected chief magistrate of Bordeaux, from the official routine of whose duties he soon retired to the more congenial atmosphere

---

* *Utopia.* Translated into English by Ralph Robinson, Fellow of Corpus Christi College. London : 1556 ; reprinted by Edward Arber, 1869. We have used this English edition as more nearly representing the style of Sir Thomas More than a modern version. It is a curious fact that no edition of the *Utopia* was published in England during the author's lifetime—or, indeed, before that of Robinson, in 1551. It was first printed at Louvain ; and, after revision by the author, it was reprinted at Basle, under the auspices of Erasmus, still in the original Latin.

of study and philosophic reflection.   In his château, at Montaigne, his studious tranquillity was violently interrupted by the savage contests then raging between the opposing factions of Catholics and Huguenots, from both of whom he received ill-treatment and loss.   To add to his troubles, the plague, which appeared in Guienne in 1586, broke up his household and compelled him, with his family, to abandon his home. Together they wandered through the country, exposed to the various dangers of a civil war; and he afterwards for some time settled in Paris. He had also travelled in Italy.   Montaigne returned to his home when the disturbances and atrocities had somewhat subsided, and there he died with the philosophic calmness with which he had lived.

The *Essais*—that book of "good faith," "without study and artifice," as its author justly calls it—appeared in the year 1580.   It is a book unique in modern literature, and the only other production to which it may be compared is the *Moralia* of Plutarch.   "It is not a book we are reading, but a conversation to which we are listening."   "It is," as another French critic observes, "less a book than a journal divided into chapters, which follow one another without connexion, which bear each a title without much regard to the fulfilment of their promise."

Montaigne treats of almost every phase of human thought and action ; and upon every subject he has something original and worth saying. Living in a savagely sectarian and persecuting age, he kept himself aloof and independent of either of the two contending theological sections, and contents himself with the *rôle* of a sceptical spectator.   It must be admitted that he is not always satisfactory in this character, since he sometimes seems to give forth an "uncertain sound."   Considering the age, however, his assertion of the proper authority of Reason deserves our respectful admiration, and is in pleasing contrast with the attitude of most of his contemporaries.   A few, like his friend De Thou, or the Italian Giordano Bruno—the latter of whom, indeed, had more of the martyr-spirit than Montaigne—contributed to keep alight the torch of Truth and Reason.   But we have only to recollect that it was the age *par excellence* of Diabolism in Catholic and Protestant theology alike, and of all the horrible superstitions and frightful tortures, both bodily and mental, of which the universal belief in the Devil's actual reign on earth was the fruitful cause.   About the very time of the appearance of the *Essais*, one of the most learned men of the period, the lawyer Jean Bodin published a work which he called the *Démonomanie des Sorciers* (the "Diabolic Inspiration of Witches"), in which he protested his unwavering faith in the most monstrous beliefs of the creed, and vehemently called upon the judges, ecclesiastical and civil, to punish the

reputed criminals (accused of an *impossible* crime) with the severest tortures. We have only to recognise this fact alone (the most astounding of
all the astounding facts and phases in the history of Superstition) to do
full justice to the reason and courage of this small band of protestors.

As for the influence of Montaigne on the modes of thought of after
times, and especially of his countrymen, it can scarcely be over estimated.
He is the literary progenitor of the most famous French writers of
the humanitarian eighteenth century. The most eminent of them,
Voltaire, perhaps, most resembles him, but naturally the style of
the eighteenth century philosopher is more concise and incisive, and
his opinions are more pronounced. "Both," says a French critic,
"laugh at the human species; but the laughter of Voltaire is more
bitter; his railleries are more terrible. Both, nevertheless, breathe
the love of humanity. That of Voltaire is more ardent, more
courageous, more unwearied. The hatred of both of them for char-
latanism and hypocrisy is well known. Their morality has for its
first principle benevolence towards others, without distinction of country,
of manners, or of religious beliefs; warning us not to think that we alone
hold the deposit of justice and of truth. It transports our soul, by
contempt of mortal things and by enthusiasm for great truths." It is to
be lamented that the countrymen of Montaigne and of Voltaire have
not profited to a larger extent by their humanitarian teaching and
tendencies. In reference to the almost incredible atrocities of war, and
especially of civil war, Montaigne protests :—

"Scarcely could I persuade myself, before I had seen it with my own eyes, that there
could be souls so ferocious as for the simple pleasure of murder to be ready to
perpetrate it; to hack and dismember the limbs of others; to ransack their invention
to discover unheard-of tortures and new kinds of deaths—and that without the incen-
tive of enmity or of profit—with the mere view of enjoying the pleasant spectacle of
pitiable actions and movements, of groans and lamentations, of a man dying in agony.
For this is the climax to which cruelty can attain—'for a man without anger, without
fear, to kill another merely to witness his sufferings.'

"For my part I have never been able to see, without displeasure, an innocent
and defenceless animal, from whom we receive no offence or harm, pursued
and slaughtered. And when a deer, as commonly happens, finding herself without
breath and strength, without other resource, throws herself down and surrenders, as it
were, to her pursuers, begging for mercy by her tears,

'Questuque cruentus
Atque imploranti similis.' *

---

* "With plaintive cries, all covered with blood, and in the attitude of a suppliant." See the
story of the death of Silvia's deer (*Æneis*, viii.)—the most touching episode in the whole epic of
Virgil. The affection of the Tuscan girl for her favourite, her anxious care of her, and the deep
indignation excited amongst her people by the murder of the deer by the son of Æneas and his
intruding followers—the cause of the war that ensued—are depicted with rare grace and feeling.

this has always appeared to me a very displeasing spectacle. I seldom, or never, take an animal alive whom I do not restore to the fields. Pythagoras was in the habit of buying their victims from the fowlers and fishermen for the same purpose.

> ' Primâque a cæde ferarum
> Incaluisse puto maculatum sanguine ferrum.' *

"Dispositions sanguinary in regard to other animals testify a natural inclination to cruelty towards their own kind. After they had accustomed themselves at Rome to the spectacle of the murders [*meurtres*] of other animals, they proceeded to those of men and gladiators. Nature has, I fear, herself attached some instinct of inhumanity to man's disposition. No one derives any amusement from seeing other animals enjoy themselves and caressing one another ; and no one fails to take pleasure in seeing them torn in pieces and dismembered. That I may not [he is cautious enough to add] be ridiculed for this sympathy which I have for them, even theology enjoins some respect for them,† and considering that one and the same Master has lodged us in this palatial world for his service, and that they are, as we, members of His family, it is right that it should enjoin some respect and affection towards them."

Quoting instances of the extreme respect in which some of the non-human races were held by people in Antiquity,‡ and Plutarch's interpretation of the meaning of the divine honours sometimes paid to them—that they adored certain qualities in them as types of divine faculties—Montaigne declares for himself that :—

"When I meet, amongst the more moderate opinions, arguments which go to prove our close resemblance to other animals, and how much they share in our greatest privileges, and with how much of probability they are compared to us, of a truth I abate much from our common presumption, and willingly abdicate that *imaginary* royalty which they assign us over other beings."

Wiser than the majority in later times, Montaigne well rebukes the arrogant presumption of the human animal who affects to hold all other life to be brought into being for his sole use and pleasure :—

"Let him shew me, by the most skilful argument, upon what foundations he has built these excessive prerogatives which he supposes himself to have over other existences. Who has persuaded him that that admirable impulse of the celestial vault, the eternal brightness of those Lights rolling so majestically over our heads, the tremendous motions of that infinite sea of Globes, were established and have continued so many ages for his advantage and for his service. Is it possible to imagine anything so ridiculous as that this pitiful [*chétive*], miserable creature, who is not even master of

---

* " It was in the slaughter, in the primæval times, of wild beasts (I suppose) the knife first was stained with the warm life-blood."—See *Ovid Metam.* xv.

† *Christian* theology, to which doubtless Montaigne here refers, the force of truth compels us to note, has always uttered a very "uncertain sound" in regard to the rights and even to the frightful sufferings of the non-human species. Excepting, indeed,[two or three isolated passages in the Jewish and Christian sacred Scriptures which, according to the theologians, bear a somewhat *equivocal* meaning, it is not easy to discover what *particular* theological or ecclesiastical maxims Montaigne could adduce.

‡ We use the term in deference to universal custom, although Francis Bacon protested 250 years ago that "Antiquity, as we call it, is the young state of the world ; for those times are ancient when the world is ancient, and not those we vulgarly account ancient by computing backwards—so that the present time is the real Antiquity."—*Advancement of Learning, I.* See also *Novum Organum.*

G

himself, exposed to injuries of every kind, should call itself master and lord of the universe, of which, so far from being lord of it, he knows but the smallest part ?   .
Who has given him this sealed charter ? Let him shew us the 'letters patent' of this grand commission. Have they been issued [*octroyées*] in favour of the wise only ? They affect but the few in that case. The fools and the wicked—are they worthy of so extraordinary a favour, and being the worst part of the world [*le pire pièce du monde*], do they deserve to be preferred to all the rest ? Shall we believe all this ?

"Presumption is our natural and original disease. The most calamitous and fragile of all creatures is man, and yet the most arrogant.* It is through the vanity of this same imagination that he equals himself to a god, that he attributes to himself divine conditions, that he picks himself out and separates himself from the crowd of other creatures, curtails the just shares of other animals his brethren [*confrères*] and companions, and assigns to them such portions of faculties and forces as seems to him good. How does he know, by the effort of his intelligence, the interior and secret movements and impulses of other animals ? By what comparison between them and us does he infer the stupidity [*la bêtise*] which he attributes to them ?

Montaigne quotes the example of his master, the just and benevolent Plutarch, who made it a matter of justice and conscience not to sell or send to the slaughter-house (according to the common selfish ingratitude) a Cow who had served him faithfully and profitably for so many years. With Plutarch and Porphyry he never wearies of denouncing the unreasoning opinions, or rather prejudices, prevalent amongst men as to the mental qualities of many of the non-human races, and, as we have already seen, insists that the difference between them and us is of *degree* and not of *kind*:—

"Plato, in his picture of the 'Golden Age,' reckons amongst the chief advantages of the men of that time the communication they had with other animals, by investigating and instructing themselves in whose nature they learned their true qualities and the differences between them, by which they acquired a ve·y perfect knowledge and intelligence, and thus made their lives more happy than we can make ours. Is a better test needed by which to judge of human folly in regard to other species ?

"I have said all this in order to bring us back and reunite ourselves to the crowd [*presse*]. We are [in the accidents of mortality] neither above nor below the rest. 'All who are under the sky,' says the Jewish sage, 'experience a like law and fate.' There is some difference, *there are orders and degrees*, but they are under the aspect of one and the same nature. Man must be constrained and ranged within the barriers of this police [*Il faut contraindre l'homme, et le ranger dans les barrières de cette police*]. The wretch has no right to encroach [*d'enjamber*] beyond these ; he is

---

* Compare Shakspere's eloquent indignation :—

> "Man, proud Man,
> Dressed in a little brief authority,
> Most ignorant of what he's most assured—
> His glassy essence—like an angry ape,
> Plays such fantastic tricks before high heaven," &c.

Measure for Measure.

fettered, entangled, he is subjected to like necessities with other creatures of his order, and in a very mean condition without any true and essential prerogative and pre-excellence. That which he confers upon himself by his own opinion and fancy has neither sense nor substance ; and if it be conceded to him that he alone of all animals has that freedom of imagination and that irregularity of thought representing to him what he is, what he is not, and what he wants, the false and the true, it is an advantage which has been very dearly sold to him, and of which he has very little to boast, for from that springs the principal source of the evils which oppress him—crime, disease, irresolution, trouble, despair."

Rejecting the still received prejudice which will not allow our humble fellow-beings the privilege of reason, but invents an imaginary faculty called "instinct," he repeats that—

"There is no ground for supposing that other beings do by *natural and necessary inclination* the same things that we do by choice, and while we are bound to infer from like effects like faculties—nay, from greater effects, greater faculties—we are forced to confess, consequently, that that same reason, that same method which we employ in action are also employed by the lower animals, or else that they have some still better reason or method. Why do we fancy in them that natural necessity or impulse [*contrainte*]—*we* who have no experience of that sort ourselves.*

"As for use in eating, it is with us as with them, natural and without instruction. Who doubts that a child, arrived at the necessary strength for feeding itself, could find its own nourishment ? The earth produces and offers to him enough for his needs without artificial labour, and if not for all seasons, neither does she for the other races—witness the provisions which we observe the ants and others collecting for the sterile seasons of the year. Those nations whom we have lately discovered [the peoples of Hindustan and of parts of America], so abundantly furnished with natural meat and drink without care and without labour, have just instructed us that bread is not our sole food, and that without toil our mother Nature has furnished us with every plant we need, to shew us, as it seems, how superior she is to all our *artificiality ;* while the extravagance of our appetite outruns all the inventions by which we seek to satisfy it."*

---

* With these just and common-sense arguments of Montaigne compare the very remarkable treatise (remarkable both by the profession and by the age of the author) of Hieronymus or Jerome Rorarius, published under the title—"That the [so-called] irrational animals often make use of reason better than men." (*Quod Animalia Bruta Sæpe Utantur Ratione Melius Homine.*) It was given to the world by the celebrated physician, Gabriel Naudé, in 1648, one hundred years after it was written, and, as pointed out by Lange, it is therefore earlier than the *Essais* of Montaigne. "It is distinguished," according to Lange, "by its severe and serious tone, and by the assiduous emphasising of just such traits of the lower animals as are most generally denied to them, as being products of the higher faculties of the soul. With their virtues the vices of men are set in sharp contrast. We can therefore understand that the MS., although written by a priest, who was a friend both of Pope and Emperor, had to wait so long for publication." (*Hist. of Materialism.* Vol. i., 225. Eng. Trans.) It is noteworthy that the title, as well as the arguments, of the book of Rorarius reveals its original inspiration—the Essay of Plutarch. Equally heterodox upon this subject is the *De La Sagesse* of Montaigne's friend, Pierre Charron.

* *Essais* de Michel de Montaigne, II., 12.

XIV.

GASSENDI.   1592—1655.

GASSENDI, one of the most eminent men, and, what is more to the purpose, the most meritorious philosophic writer, of France in the seventeenth century, claims the unique honour of being the first directly to revive in modern times the teaching of Plutarch and Porphyry. Other minds, indeed, of a high order, like More and Montaigne, had, as already shown, implicitly condemned the inveterate barbarism. But Gassendi is the writer who first, since the extinction of the Platonic philosophy, expressly and unequivocally attempted to enlighten the world upon this fundamental truth.

He was born of poor parents, near Digne, in Provence. In his earliest years he gave promise of his extraordinary genius. At nineteen he was professor of philosophy at Aix. His celebrated "Essays against the Aristotleians" (*Exercitationes Paradoxicœ Adversus Aristoteleos)* was his first appearance in the philosophic world. Written some years earlier, it was first published, in part, in the year 1624. It divides with the *Novum Organon* of Francis Bacon, with which it was almost contemporary, the honour of being the earliest effectual assault upon the old scholastic jargon which, abusing the name and authority of Aristotle, during some three or four centuries of mediæval darkness had kept possession of the schools and universities of Europe. It at once raised up for Gassendi a host of enemies, the supporters of the old orthodoxy, and, as has always been the case in the exposure of falsehood, he was assailed with a torrent of virulent invective. Five of the Books of the *Exercitationes,* by the advice of his friends, who dreaded the consequences of his courage, had been suppressed. In the Fourth Book, besides the heresy of Kopernik (which Bacon had not the courage or the penetration to adopt), the doctrine of the eternity of the Earth had been maintained, as already taught by Bruno ; while the Seventh, according to the table of contents, contained a formal recommendation of the Epicurean theory of morals, in which Pleasure and Virtue are synonymous terms.

In the midst of the obloquy thus aroused the philosopher devoted himself, by way of consolation, to the study of anatomy and astronomy, as well as to literary studies. "As the result of his anatomical researches he composed a treatise to prove that man was intended to live upon vegetables, and that animal food, as contrary to the human constitution, is baneful and unwholesome."* He was the first to observe the transit of the planet Mercury over the Sun's disc (1631), previously calculated by Kepler. He next appears publicly as the opponent of Descartes in his *Disquisitiones Anticartesianœ* (1643)—a work justly distinguished, according to the remark of an eminent German critic, as a model of controversial excellence. The philosophic world was soon divided between the two hostile camps. It is sufficient to observe here that Descartes, whatever merit may attach to him in other respects, by his equally absurd and mischievous paradox that the non-human species are possessed only of unconscious sensation and perception, had done as much as he well could to destroy his reputation for common sense and common reason with all the really thinking part of the world. Yet this "animated machine" theory, incredible as it may appear, has recently been revived by a well-known physiologist of the present day, in the very face of the most ordinary facts and experience—a theory about which it needs only to be said that it deserves to be classed with some of the most absurd and monstrous conceptions of mediævalism. As though, to quote Voltaire's admirable criticism, God had given to the lower animals reason and feeling to the end *that they might not feel and reason.* It was not thus, as the same writer reminds us, that Locke and Newton argued.†

In 1646 Gassendi became Regius Professor of Mathematics in the University of Paris, where his lecture-room was crowded with listeners of all classes. His *Life and Morals of Epikurus (De Vitâ et Moribus Epicuri),* his principal work, appeared in the year 1647. It is a triumphant refutation of the prejudices and false representations connected with the name of one of the very greatest and most virtuous of the Greek Masters, which had been prevalent during so many ages. Neither his European reputation, nor the universal respect extorted

---

* See Article in *English Cyclopœdia.*

† See *Elémens de la Philosophie de Newton.* The whole passage breathes the true spirit of humanity and philosophy, and deserves to be quoted in full in this place : "Il y a surtout dans l'homme une disposition à la compassion aussi généralement répandue que nos autres instincts. Newton avait cultivé ce sentiment d'humanité, et il l'etendait jusqu'aux animaux. Il était fortement convaincu avec Locke, que Dieu a donné aux animaux une mesure d'idées, et les mêmes sentiments qu'à nous. Il ne pouvait penser que Dieu, qui ne fait rien en vain, eût donné aux animaux des organes de sentiment, *afin qu'elles n'eussent point de sentiment.* Il trouvait une contradiction bien affreuse à croire que les animaux sentent, et à les faire souffrir. Sa morale s'accordait en ce point avec sa philosophie. *Il ne cédait qu'avec répugnance à l'usage barbare de nous nourrir du sang et de la chair des êtres semblables à nous,* que nous caressons tous les jours. Il ne permit jamais dans sa maison qu'on les fit mourir par des morts lentes et recherchées, pour en rendre la nourriture plus délicieuse. Cette compassion qu'il avait pour les animaux se tournait en vraie charité pour les hommes. En effet, *sans l'humanité—vertu qui comprend toutes les vertus — on ne mériterait guère le nom de philosophe."—Elémens* v. An expression of feeling in sufficiently striking contrast to the ordinary ideas. Compare *Essay on the Human Understanding,* ii., 2.

by his private as well as public merits, could corrupt the simplicity of Gassendi; and his sober tastes were little in sympathy with the luxurious or literary trifling of Paris :—

"He had only with difficulty resolved to quit his southern home, and being attacked by a lung complaint, he returned to Digne, where he remained till 1653. Within this period falls the greater part of his literary activity and zeal in behalf of the philosophy of Epikurus, and simultaneously the positive extension of his own doctrines. In the same period Gassendi produced, besides several astronomical works, a series of valuable biographies, of which those of Kopernik and Tycho Brahe are especially noteworthy. He is, of all the most prominent representatives of Materialism, the only one gifted with a historic sense, and that he has in an eminent degree. Even in his *Syntagma Philosophicum* he treats every subject, at first historically from all points of view. . . . Gassendi did not fall a victim to Theology, because he was destined to fall a victim to Medicine. Being treated for a fever in the fashion of the time, he had been reduced to extreme debility. He long, but vainly, sought restoration in his southern home. On returning to Paris he was again attacked by fever, and thirteen fresh blood-lettings ended his life. He died October 24th, 1655."

Lange, from whom we have quoted this brief notice, proceeds to vindicate his position as a physical philosopher :—

"The reformation of Physics and Natural Philosophy, usually ascribed to Descartes, was at least as much the work of Gassendi. Frequently, in consequence of the fame which Descartes owed to his Metaphysics, those very things have been credited to Descartes which ought properly to be assigned to Gassendi. It was also a result of the peculiar mixture of difference and agreement, of hostility and alliance, between the two systems that the influences resulting from them became completely interfused."[*]

Although of extraordinary erudition his learning did not, as too often happens, obscure the powers of original thought and reason. Bayle, writing at the end of the seventeenth century, has characterised him as "the greatest philosopher amongst scholars, and the greatest scholar amongst philosophers;" and Newton conceived the same high esteem for the great vindicator of Epikurus.[†]

It is in his celebrated letter to his friend Van Helmont, that Gassendi deals with the irrational assertions of certain physiologists, apparently more devoted to the defence of the orthodox diet than to the discovery of unwelcome truth, as to the character of the human teeth :—

---

[*] *History of Materialism.*—We may here observe that Descartes seems to have adopted his extraordinary theory as to the non-human races as a sort of *dernier resort*. In a letter to one of his friends (Louis Racine) he declares himself driven to his theory by the rigour of the dilemma, that (seeing the innocence of the victims of man's selfishness) it is necessary either that they should be insensible to suffering, or that God, who has made them, should be unjust. Upon which Gleïzès makes the following reflection : "This reasoning is conclusive. One must either be a Cartesian, or allow that man is very vile. Nothing is more rigorous than this consequence." —(*Thalysie Ou La Nouvelle Existence*). La Fontaine has well illustrated the absurdity of the animated machine theory in *Fables* x. 1.

[†] See "*Elémens de la Philosophie de Newton.*"

"I was contending," he writes to his medical friend, "that from the conformation of our teeth we do not appear to be adapted by Nature to the use of a flesh diet, since all animals (I spoke of terrestrials) which Nature has formed to feed on flesh have their teeth long, conical, sharp, uneven, and with intervals between them—of which kind are lions, tigers, wolves, dogs, cats, and others. But those who are made to subsist only on herbs and fruits have their teeth short, broad, blunt, close to one another, and distributed in even rows. Of this sort are horses, cows, deer, sheep, goats, and some others. And further—that men have received from Nature teeth which are unlike those of the first class, and resemble those of the second. It is therefore probable, since men are land animals, that Nature intended them to follow, in the selection of their food, not the carnivorous tribes, but those races of animals which are contented with the simple productions of the earth. . . . Wherefore, I here repeat that from the primæval institution of our nature, the teeth were destined to the mastication, not of flesh, but of fruits.

As for flesh, true, indeed, it is that man is sustained on flesh. But *how many things*, let me ask, *does man do every day which are contrary to, or beside, his nature?* So great, and so general, is the perversion of his mode of life, which has, as it were, eaten into his flesh by a sort of deadly contagion *(contagione veluti quâdam jam inusta est)*, that he appears to have put on another disposition. Hence, the whole care and concern of philosophy and moral instruction ought to consist in leading men back to the paths of Nature."

Helmont, it seems, had rested his principal argument for flesh-eating, not altogether in accordance with *Genesis,* and certainly not in accordance with Science, on the presumption that man was formed expressly for carnivorousness. To this Gassendi replied that, without ignoring theological argument, he still maintained comparative Anatomy to be a satisfactory and sufficient guide. He then applies himself to refute the physiological prejudice of Helmont about the teeth, &c. (as already quoted), and begins by warning his friend that he is not to wonder if the self-love of men is constantly viewed by him with suspicion.*

"For, in fact, we all, with tacit consent, conspire to extol our own nature, and we do this commonly with so much arrogance that, if people were to divest themselves of this traditional and inveterate prejudice, and seriously reflect upon it, their faces must be immediately suffused with burning shame."

He repeats Plutarch's unanswerable challenge :—

"Man lives very well upon flesh, you say, but, if he thinks this food to be natural to him, why does he not use it as it is, as furnished to him by Nature? But, in fact, he shrinks in horror from seizing and rending living or even raw flesh with his teeth, and lights a fire to change its natural and proper condition. Well, but if it were the intention of Nature that man should eat *cooked* flesh, she would surely have provided him with ready-made cooks ; or, rather, she would have herself cooked it as she is wont to do fruits, which are best and sweetest without the intervention of fire. Nature, surely, does not fail in providing necessary provision for her children, according to the common boast. But what is more necessary than to make food pleasurable? And, as she does in the case of sexual love by which she procures the preservation of the *species,* so would she procure the preservation of the *genus.*

---

* *Suspecta mihi semper fuerit* (he writes) *ipsa hominis* φιλαυτία.

"Nor let anyone say that Nature in this is corrected, since, to pass over other things, that is tantamount to convicting her of a blunder.   Consider how much more benevolent she would be proved to be, in that case, towards the savage beasts than towards us.   Again, since our teeth are not sufficient for eating flesh, even when prepared by fire, the invention of knives seems to me to be a strong proof Because, in fact, we have no teeth given us for rending flesh, and we are therefore forced to have recourse to those *non-natural* organs, in order to accomplish our purpose.   As if, forsooth, Nature would have left us destitute in so essential things ! I divine at once your ready reply : 'think that Nature has given man reason to supply defects of this kind.'   But this, I affirm, is always to accuse Nature, *in order to* defend our unnatural luxury.   So it is about dress—so it is about other things.

"What is clearer [he sums up] than that man is not furnished for hunting, much less for eating, other animals ?   In one word, we seem to be admirably admonished by Cicero that man was destined for other things than for seizing and cutting the throats of other animals.   If you answer that 'that may be said to be an industry ordered by Nature, by which such weapons are invented,' then, behold ! it is by the very same artificial instrument that men make weapons for mutual slaughter.   Do they this at the instigation of Nature ?   Can a use so noxious be called *natural ?*   Faculty is given by Nature, but it is our own fault that we make a perverse use of it."

He, finally, refutes the popular objection about the strength-giving properties of flesh-meat, and instances Horses, Bulls, and others.*

In his *Ethics* (affixed to his Books on *Physics*) he quotes and endorses the opinions of Epikurus on the slaughter of innocent life :—

"There is no pretence," he asserts, "for saying that any right has been granted us oy law to kill any of those animals which are not destructive or pernicious to the human race, for there is no reason why the innocent species should be allowed to increase to so great a number as to be inconvenient to us.   They may be restrained within that number' which would be harmless, and useful to ourselves."†

With   that   Great   Master   he   thus   rebukes   the   fashionable "hospitality" :—

"I, for my part, to speak modestly of myself, lived contented with the plants of my little garden, and have pleasure in that diet, and I wish inscribed on my doors : Guest, here you shall have good cheer ! here the *summum bonum* is Pleasure.   The guardian of this house, *humanely* hospitable, is ready to entertain you with pearl-barley *(polenta)*, and will furnish you abundantly with water.   These little gardens do not increase hunger, but extinguish it ; nor do they make thirst greater by the very potations themselves, but satisfy it by a natural and gratuitous remedy.' "‡

---

* See Gassendi's Letter, *Viro Clarissimo et Philosopho ac Medico Expertissimo Joanni Baptistœ Helmontio Amico Suo Singulari.*   Dated, Amsterdam, 1629.

† *Physics.*   Book II.   *De Virtutibus.*

‡ See *Philosophiæ Epicuri Syntagma.   De Sobrietate contra Gulam.*   ("View of the Philosophy of Epikurus : On Sobriety as opposed to Gluttony.")   Part III.   Florentiæ, 1727.   Folio. VoL III.

THERE is one name which, in reputation, occupies a pre-eminent position in philosophy, belonging to this period—Francis Bacon. But, for ourselves, for whom true ethical and humanitarian principles have a much deeper significance than mere mental force undirected to the highest aims of truth and of justice, the name of the modern assertor of the truths of Vegetarianism will challenge greater reverence than even that of the author of the *New Instrument*.

That Bacon should exhibit himself in the character of an advocate of the rights of the lower races is hardly to be expected from the selfish and unscrupulous promoter of his own private interests at the expense at once of common gratitude and common feeling. His remarks on Vivisection (where he questions whether experiments on human beings are defensible, and suggests the limitation of scientific torture to the non-human races)* are, in fact, sufficient evidence of his indifferentism to so unselfish an object as the advocacy of the claims of our defencelesss dependants. When we consider his unusual sagacity in exposing the absurd quasi-scientific methods of his predecessors, and of the prevailing (so-called) philosophical system and the many profound remarks to be found in his writings, it must be added that we are reluctantly compelled to believe that the opinions elsewhere which he publishes inconsistent with those principles were inspired by that notorious servility and courtiership by which he flattered the absurd and pedantic dogmatism of one of the most contemptible of kings.

One passage there is, however, in his writings which seems to give us hope that this eminent compromiser was not altogether insensible to higher and better feeling :—

"Nature has endowed man with a noble and excellent principle of compassion, which extends [? ought to extend] itself also to the dumb animals—whence this com-

---

* *Advancement of Learning*, iv., 2. Bacon's suggestion seems to imply that human beings were still vivisected, for the "good" of science, in his time. Celsus, the well-known Latin physician of the second century, had protested against this cold-blooded barbarity of deliberately cutting up a living human body. The wretched victims of the vivisecting knife were, it seems, slaves, criminals, and captives, who were handed over by the authorities to the physiological "laboratory." Harvey, Bacon's contemporary, is notorious (and, it ought to be added, infamous) for the number and the unrelenting severity of his experiments upon the [non-humam slaves, which, though constantly alleged by modern vivisectors to have been the means by which he discovered the "circulation of the blood," have been clearly proved to have served merely as demonstrations in physiology to his pupils. But we no longer wonder at Harvey's indifference to the horrible suffering of which he was the cause, when we read the similar atrocities of vivisection and "pathology" of our own time. From the cold-blooded cruelties of Harvey, who was accustomed to amuse Charles I. and his family with his demonstrations, it is a pleasant relief to turn to the better feeling of Shakspere on that subject. See his *Cymbeline* (i., 6), where the Queen, who is experimenting in poisons, tells her physician,

"I will try the force of these thy compounds on such creatures as
   We count not worth the hanging—but none human."

and is reminded that she would "from this practice but make hard her heart." Such a rebuke is in keeping with the true feeling which inspired the poet to picture the undeserved pangs of the hunted Deer in *As You Like It* ii., 1 passion has some resemblance to that of a prince towards his subjects. And it is certain that the noblest souls are the most extensively compassionate, for narrow and degenerate minds think that compassion belongs not to them ; but a great soul, the noblest part of creation, is ever compassionate. Thus, under the old laws, there were numerous precepts (not merely ceremonial) enjoining mercy—for example, the not eating of flesh with the blood, &c. So, also, the sects of the Essenes and Pythagoreans totally abstained from flesh, as they do also to this day, with an inviolate religion, in some parts of the empire of the Mogul [Hindustan]. Nay, the Turks, though a savage nation, both in their descent and discipline, give alms to the dumb animals, and suffer them not to be tortured."*

If Bacon had lived longer (he died in 1626) we may entertain the hope that the powerful arguments of his illustrious contemporary might have inspired him with more sound and satisfactory ideas on Dietetics than the somewhat crude ones which he published in his *De Augmentis* (iv., 2). As for Medicine, he had, reasonably enough, not conceived a high opinion of the methods of its ordinary professors. He says :—

"Medicine has been more professed than laboured, and more laboured than advanced ; rather circular than progressive ; for I find great repetition, and but little new matter in the writers of Physic."

XV.
RAY.   1627—1705.

JOHN RAY, the founder of Botanical and, only in little less degree, of Zoological Science, was an *alumnus* of the University of Cambridge. He was elected Fellow of Trinity College in 1649, and Lecturer in Greek in the following year. While at Cambridge he formed a collection of plants growing in the neighbourhood, a catalogue of which he published in 1660. Three years later, with his friend Francis Willoughby, he travelled over a large part of Europe, as during his academical life he had traversed the greater part of these islands, in pursuit of botanical and zoological science—an account of which tour he published in 1673.

He had been one of the first Fellows of the recently founded Royal Society. In 1682 appeared his *New Method of Plants*, which formed a new era in botany, or rather, which was the first attempt at making it a real science. It is the basis of the subsequent classification of Jussieu, which is still received ; and its author was the first to propose the division of plants into *monocotyledons* and *dicotyledons*.

His principal work is the *Historia Plantarum*, 1686—1704. "In it he collected and arranged all the species of plants which had been described by botanists. He enumerated 18,625 species. Haller, Sprengel, Adamson, and others speak of this work as being the produce of immense labour, and as containing much acute criticism."

* *Advancement of Learning.* viii., 2.

What, however, is more interesting to us is the fact that "in zoology Ray ranks almost as high as in botany, and his works on this subject are even more important, as they still, in great measure, preserve their utility. Cuvier says that 'they may be considered as the foundation of modern zoology, for naturalists are obliged to consult them every instant for the purpose of clearing up the difficulties which they meet with in the works of Linnæus and his copyists.'"

Between 1676—1686 appeared *Ornithologia* and *Historia Piscium*, the materials of which had been left him by his friend Willoughby. To his extraordinary erudition and industry the world was indebted for *A Methodical Synopsis of Quadrupeds* as well as a very valuable history of Insects. Conspicuous amongst his merits are his accuracy of observation and his philosophical method of classification. With others, Buffon is largely indebted to the most meritorious of the pioneers of zoological knowledge.

Ray has delivered his profession of faith in the superiority and excellence of the non-flesh diet in the following eloquent passage which has been quoted with approval by his friend John Evelyn :—

"The use of plants is all our life long of that universal importance and concern that we can neither live nor subsist with any decency and convenience, or be said, indeed, to live at all without them. Whatsoever food is necessary to sustain us, whatsoever contributes to delight and refresh us, is supplied and brought forth out of that plentiful and abundant store. And ah ! [he exclaims] how much more innocent, sweet, and healthful is a table covered with those than with all the reeking flesh of butchered and slaughtered animals. Certainly man by nature was never made to be a carnivorous animal, nor is he armed at all for prey and rapine, with jagged and pointed teeth and crooked claws sharpened to rend and tear, but with gentle hands to gather fruit and vegetables, and with teeth to chew and eat them.*

XVI.

EVELYN. 1620—1706.

JOHN EVELYN, the representative of the more estimable part of the higher middle life of his time, who has so eloquently set forth the praises of the vegetable diet, also claims with Ray the honour of having first excited, amongst the opulent classes of his countrymen, a rational taste for botanical knowledge. Especially meritorious and truly patriotic was his appeal to the owners of land, by growing trees to provide the country with useful as well as ornamental timber for the benefit of posterity. He was one of the first to treat gardening and planting in a scientific manner ; and his own cultivation of exotic and other valuable plants was a most useful example too tardily followed by ignorant or selfish land-

* See *Acetaria* (page 170). By John Evelyn.

lords of those and succceding times. It would have been well indeed for the mass of the peeple of these islands, had the owners of landed property cared to develope the teaching of Evelyn by stocking the country with various fruit trees, and so supplied at once an easy and wholesome food. *O fortunatos nimium, sua si bona nôrint, Agricolas!* . . *Fundit humo facilem victum justissima Tellus.**

The family of Evelyn was settled at Wooton, in Surrey. During the struggle between the Parliament and the Court he went abroad, and travelled for some years in France and in Italy, where he seems to have employed his leisure in ẚ more refined and useful way than is the wont of most of his travelling countrymen. He returned home in 1651. At the foundation of the Royal Society, some ten years later, Evelyn became one of its earliest Fellows. His first work was published in 1664, *Sylva; or, a Discourse of Forest Trees and the Propagation of Timber.* Its immediate cause was the application of the Naval Commissioners to the Royal Society for advice in view of the growing scarcity of timber, especially of oak, in England. A large quantity of the more valuable wood now existing is the practical outcome of his timely publication.

In 1675, appeared his *Terra: a Discourse of the Earth Relating to the Culture and the Improvement of it, to Vegetation and the Propagation of Plants.* The book by which he is most popularly known is his *Diary and Correspondence,* one of the most interesting productions of the kind. Besides its value as giving an insight into the manner of life in the fashionable society of the greater part of the seventeenth century, it is of importance as an independent chronicle of the public events of the day. The work which has the most interest and value for us is his *Acetaria* (Salads, or Herbs eaten with vinegar), in which the author professes his faith in the truth and excellence of the Vegetarian diet. Unfortunately, according to the usual perversity of literary enterprise, it is one of those few books which, representing some profounder truth, are nevertheless the most neglected by those who undertake to supply the mental and moral needs of the reading public.

Evelyn held many high posts under the varying Governments of the day; and being, by tradition and connexion, attached to the monarchical party, he attracted (contrary to the general experience) the grateful recognition of the restored dynasty.

---

* The tract of Samuel Hartlib, entitled, *A Design for Plenty, by a Universal Planting of Fruit Trees,* which appeared during the Commonwealth Government, no doubt suggested to Evelyn his kindred publication. Hartlib (of a distinguished German family) settled in this country somewhere about the year 1630. By his writings, in advocacy of better agriculture and horticulture, he has deserved a grateful commemoration from after-times. Cromwell gave him a pension of £300, which was taken away by Charles II., and he died in poverty and neglect. It was to him Milton dedicated his *Tractate on Education.*

Having adduced other arguments for abstinence from flesh, Evelyn continues :—

"And now, after all we have advanced in favour of the herbaceous diet, there still emerges another inquiry, viz., whether the use of crude herbs and plants is so wholesome as is alleged ? What opinion the prince of physicians had of them we shall see hereafter ; as also what the sacred records of olden times seem to infer, before there were any flesh-shambles in the world ; together with the reports of such as are often conversant among many nations and people, who, to this day, living on herbs and roots, arrive to an incredible age in constant health and vigour, which, whether attributable to the air and climate, custom, constitution, &c., should be inquired into."

Cardan—the pseudo-savant of the sixteenth century—had written, it seems, in favour of flesh-meat.   Evelyn informs us that :—

"This, [the alleged superiority of flesh] his learned antagonist, utterly denies. Whole nations—flesh devourers, such as the farthest northern—become heavy, dull, inactive, and much more stupid than the southern ; and such as feed more on plants are more acute, subtle, and of deeper penetration.   Witness the Chaldeans, Assyrians, Egyptians, &c. And he further argues from the short lives of most carnivorous animals, compared with grass feeders, and the ruminating kind, as the Hart, Camel, and the longævus Elephant, and other feeders on roots and vegetables.

"As soon as old Parr came to change his simple homely diet to that of the Court and Arundel House, he quickly sank and drooped away ; for, as we have shewn, the stomach easily concocts plain and familiar food, but finds it a hard and difficult task to vanquish and overcome meats of different substances.   Whence we so often see temperate and abstemious persons of a collegiate diet [of a distant age, we must suppose] very healthy ; husbandmen and laborious people more robust and longer-lived than others of an uncertain, extravagant habit."

He appeals to the biblical reverence of his readers, and tells them :—

"Certain it is, Almighty God ordaining herbs and fruit for the food of man, speaks not a word concerning flesh for two thousand years ; and when after, by the Mosaic constitution, there were distinctions and prohibitions about the legal uncleanness of animals, plants of what kind soever were left free and indifferent for everyone to choose what best he liked.   And what if it was held indecent and unbecoming the excellency of man's nature, before sin entered and grew enormously wicked, that any creature should be put to death and pain for him who had such infinite store of the most delicious and nourishing fruit to delight, and the tree of life to sustain him ? Doubtless there was no need of it.   Infants sought the mother's nipples as soon as born, and when grown and able to feed themselves, ran naturally to fruit, and still will choose to eat it rather than flesh, and certainly might so persist to do, did not Custom prevail even against the very dictates of Nature.*

"And now to recapitulate what other prerogatives the hortulan provision has been celebrated for besides its antiquity, and the health and longevity of the antediluvians— viz., that temperance, frugality, leisure, ease, and innumerable other virtues and advantages which accompany it, are no less attributable to it.   Let us hear our excellent botanist, Mr. Ray."

---

* Locke (one of the very highest names in Philosophy) had already exhorted English mothers to make their children abstain "wholly from flesh," at least until the completion of the fourth or fifth year.   He strongly recommends a very sparing amount of flesh for after years ; and thinks that many maladies may be traceable to the foolish indulgence of mothers in respect to diet.— See *Thoughts on Education*, 1690.

He then quotes the profession of faith of the father of English botany and zoology; and goes on eloquently to expatiate on the varied pleasures of a non-flesh and fruit diet :—

"To this might we add that transporting consideration, becoming both our veneration and admiration, of the infinitely wise and glorious Author of Nature, who has given to plants such astonishing properties ; such fiery heat in some to warm and cherish ; such coolness in others to temper and refresh ; such pinguid juice to nourish and feed the body ; such quickening acids to compel the appetite, and grateful vehicles to court the obedience of the palate ; such vigour to renew and support our natural strength ; such ravishing flavours and perfumes to recreate and delight us ; in short, such spirituous and active force to animate and revive every part and faculty to all kinds of human and, I had almost said, heavenly capacity.

" What shall we add more ?   Our gardens present us with them all : and, while the Shambles are covered with gore and stench, our Salads escape the insults of the summer-fly, purify and warm the blood against winter rage.   Nor wants there variety in more abundance than any of the former ages could show."

Evelyn produces an imposing array of the " Old Fathers " :—

" In short, so very many, especially of the Christian profession, advocate it [the bloodless food] that some even of the ancient fathers themselves have thought that the permission of eating flesh to Noah and his sons was granted them no other-wise than repudiation of wives was to the Jews—namely—for the hardness of their hearts and to satisfy a murmuring generation.*

He is " persuaded that more blood has been shed between Christians " through addiction to the sanguinary food than by any other cause :—

"Not that I impute it *only* to our eating blood ; but I sometimes wonder how it happened that so strict, so solemn, and famous a sanction—not upon a ceremonial account, but (as some affirm) a moral and perpetual one, for which also there seem to be fairer proofs than for most other controversies agitated amongst Christians—should be so generally forgotten, and give place to so many other impertinent disputes and cavils about superstitious fopperies which frequently end in blood and cutting of throats."†

I⟋ is opportune here to refer to the sentiments of Evelyn's contemporary and political and ecclesiastical opposite—the great Puritan poet and patriot—one of the very greatest names in all literature.   Milton's feeling, so far as he had occasion to express it, is quite in unison with the principles of dietetic reform, and in sympathy with aspirations after the more spiritual life.

In one of his earliest writings, on the eve of the production of one of the finest poems of its kind in the English language—the *Ode to Christ's*

---

* He quotes, amongst others, Tertullian *De Jejuniis* (On Fasting), cap, iv. ; Jerome *(Adv. Jovin) ;* Clemens of Alexandria (*Strom.* vii.) ; Eusebius, *Preparatio Evangelica* (Preparation for the Gospel), who cites several abstinents from amongst the philosophers of the old theologies.

† *Acetaria* (" A Discourse of Salads"). Dedicated to Lord Somers, of Evesham, Lord High Chancellor of England, and President of the Royal Society, London, 1699.

Nativity, composed at the age of twenty-one—he thus writes in Latin verse to his friend Charles Deodati, recommending the purer diet at all events to those who aspired to the nobler creations of poetry :—

> " Simply let those, like him of Samos, live :
> Let herbs to them a *bloodless* banquet give.
> In beechen goblets let their beverage shine,
> Cool from the crystal spring their sober wine !
> Their youth should pass in innocence secure
> From stain licentious, and in manners pure.
>
> .   .   .   .   .   .
>
> For these are sacred bards and, from above,
> Drink large infusions from the mind of Jove." *

To readers of his master-piece the *Paradise Lost,* it is perhaps a work of supererogation to point out the charming passages in which he sympathetically describes the food of the Age of Innocence :—

> " Savoury fruits, of taste to please
> True appetites."

In Raphael's discourse with his terrestrial entertainers, the ethereal messenger utters a prophecy (as we may take it) of the future general adoption by our race of "fruit, man's nourishment," and we may interpret his intimation :—

> " time may come when men
> With angels may participate, and find
> No inconvenient diet, nor too light fare.
> And from those corporal nutriments perhaps
> Your bodies may at last turn all to spirit,
> Improved by tract of time, and winged ascend
> Ethereal as we ; or may, at choice,
> *Here,* or in heavenly paradises, dwell,"

as a picture of the true earthly paradise to be—" the Paradise of Peace."

With these exquisite pictures of the life of bloodless feasts and ambrosial food we may compare the fearful picture of the Court of Death, displayed in prospective vision before the terror-stricken gaze of the traditional progenitor of our species, where, amongst the occupants, the largest number are the victims of "intemperance in meats and drinks, which on the earth shall bring diseases dire." In this universal lazar-house might be seen—

> " all maladies
> Of ghastly Spasm, or racking torture, Qualms
> Of heart-sick agony, all Feverous kinds,
> Convulsions, Epilepsies, fierce Catarrhs,
> Intestine Stone and Ulcer, Colic pangs,

---

* Translated by Cowper from the Latin poems of Milton. In a note to the original poem Thomas Warton justly remarks that "Milton's panegyrics on temperance both in eating and in drinking, resulting from his own practice, are frequent."

> Demoniac Phrensy, moping Melancholy,
> And moon-struck Madness, pining Atrophy,
> Marasmus, and wide-wasting Pestilence,
> Dropsies and Asthmas, and joint-racking Rheums." *

Very different, in other respects, from those of the author of the *History of the Reformation in England* the sentiments of his celebrated contemporary Bossuet, whose eloquence gained for him the distinguishing title of the "Eagle of Méaux," as to the degrading character of the prevalent human nourishment in the Western world, are sufficiently remarkable to deserve some notice. The *Oraisons Funèbres* and, particularly, his *Discours sur L'Histoire Universelle* have entitled him to a high rank in French literature. But a single passage in the last work, we shall readily admit, does more credit to his heart than his most eloquent efforts in oratory or literature do to his intellect. That, in common with other theologians, Catholic and Protestant, he has thought it necessary to assume the intervention of the Deity to sanction the sustenance of human life by the destruction of other innocent life, does not affect the weight of intrinsic evidence derivable from the natural feeling as to the debasing influence of the Slaughter-House. It is thus that he, impliedly at least, condemns the barbarous practice :—

"Before the time of the Deluge the nourishment which without violence men derived from the fruits which fell from the trees of themselves, and from the herbs which also ripened with equal ease, was, without doubt, some relic of the first innocence and of the gentleness (*douceur*) for which we were formed. Now to get food we have to shed blood in spite of the horror which it naturally inspires in us ; and all the refinements of which we avail ourselves, in covering our tables, hardly suffice to disguise for us the bloody corpses which we have to devour to support life. But this is but the least part of our misery. Life, already shortened, is still further abridged by the savage violences which are introduced into the life of the human species. Man, whom in the first ages we have seen spare the life of other animals, is accustomed henceforward to spare the life not even of his fellow-men. It is in vain that God forbade, immediately after the Deluge, the shedding of human blood ; in vain, in order to save some vestiges of the first mildness of our nature, while permitting the feeding on flesh did he prohibit consumption of the blood. Human murders multiplied beyond all calculation."

Bossuet, a few pages later, arrives at the necessary and natural consequence of the murder of other animals, when he records that "the brutalised human race could no longer rise to the true contemplation of intellectual things." †

---

* *Paradise Lost*, v. and xi. Cf. *Queen Mab*.

† *Le sang humain abruti ne pouvait plus s'élever aux choses intellectuelles.* See *Discours sur L'Histoire Universelle*, a historical sketch which, though necessarily infected by the theological prejudices of the bishop, is, for the rest, considering the period in which it was written, a meritorious production as one of the earliest attempts at a sort of "philosophy of history."

XVII.

BERNARD DE MANDEVILLE. 1670—1733.

THE most paradoxical of moralists, born at Dort, in Holland. He was brought up to the profession of medicine, and took the degree of M.D. He afterwards settled and practised in London.

It was in 1714 that he published his short poem called *The Grumbling Hive: or, Knaves Turned Honest,* to which he afterwards added long explanatory notes, and then republished the whole under the new and celebrated title of *The Fable of the Bees.* This work " which, however erroneous may be its views of morals and of society, is written in a proper style, and bears all the marks of an honest and sincere inquiry on an important subject, exposed its author to much obloquy, and met with answers and attacks. . . . It would appear that some of the hostility against this work, and against Mandeville generally, is to be traced to another publication, recommending the public licensing of " stews," the matter and manner of which are certainly exceptionable, though, at the same time, it must be stated that Mandeville earnestly and with seeming sincerity commends his plan as a means of diminishing immorality, and that he endeavoured, so far as lay in his power, by affixing a high price and in other ways, to prevent the work from having a general circulation." In fact, Mandeville is one of those injudicious but well-meaning reformers who, by their propensity to perverse paradox, have injured at once their reputation and their usefulness for after times.

A second part of *The Fable* appeared at a later period. Amongst other numerous writings were two entitled, *Free Thoughts on Religion, the Church, and National Happiness,* and *An Enquiry into the Origin of Honour,* and the *Usefulness of Christianity in War.* He appears to have been enabled to pursue his literary career in great measure by the liberality of his Dutch friends, and he was a constant guest of the first Earl of Macclesfield. " *The Fable of the Bees; or, Private Vices Public Benefits* may be received in two ways," says the writer in the *Penny Cyclopædia,* whom we have already quoted, " as a satire on men, and as a theory of society and national prosperity. So far as it is a satire, it is sufficiently just and pleasant, but received in its more ambitious character of a theory of society, it is altogether worthless. It is Mandeville's object to show that national greatness depends on the

H

prevalence of fraud and luxury; and for this purpose he supposes 'a vast hive of bees' possessing in all respects institutions similar to those of men; he details the various frauds, similar to those among men, practised by bees one upon another in various professions. . . . His hive of bees having thus become wealthy and great, he afterwards supposes a mutual jealousy of frauds to arise, and Fraud to be, by common consent, dismissed; and he again assumes that wealth and luxury immediately disappear, and that the greatness of the society is gone." For our part, in place of "greatness," we should have rather written *misery*, as far as concerns the mass of communities.

Strange, as it may appear, that views of this kind should be seriously put forth, "it is yet more so that they should come from one whose object always was, however strange the way in which he set about it, to promote good morals, for there is nothing in Mandeville's writings to warrant the belief that he sought to encourage vice."*

Mandeville, like Swift, in the piece entitled *An Argument against Abolishing Christianity;* or like De Foe, in his *Shortest Way with the Dissenters,* which were taken *au sérieux* almost universally at the time of their appearance, may have used the style of grave irony, so far as the larger portion of his Fable is concerned, for the purpose of making a stronger impression on the public conscience. If such were his purpose, the irony is so profound that it has missed its aim. Yet that his purpose was true and earnest is sufficiently evident in his opinion of the practice of slaughtering for food :—

" I have often thought [writes Mandeville] if it was not for the tyranny which Custom usurps over us, that men of any tolerable good nature could never be reconciled to the killing of so many animals for their daily food, so long as the bountiful Earth so plentifully provides them with varieties of vegetable dainties. I know that Reason excites our compassion but faintly, and therefore I do not wonder how men should so little commiserate such imperfect creatures as cray-fish, oysters, cockles, and, indeed, all fish in general, as they are mute, and their inward formation, as well as outward figure, vastly different from ours : they express themselves unintelligently to us, and therefore 'tis not strange that their grief should not affect our understanding which it cannot reach ; for nothing stirs us to pity so effectually as when the symptoms of misery strike immediately upon our senses, and I have seen people moved at the noise a live lobster makes upon the spit who could have killed half a dozen fowls with pleasure.

"'But in such perfect animals as Sheep and Oxen, in whom the heart, the brain, and the nerves differ so little from ours, and in whom the separation of the spirits from the blood, the organs of sense, and, consequently, feeling itself, are the same as they are in human creatures, I cannot imagine how a man not hardened in blood and massacre, is able to see a violent death, and the pangs of it, without concern.

---

* *Penny Cyclopædia*, Article Mandeville.

"In answer to this [he continues], most people will think it sufficient to say that things being allowed to be made for the service of man, there can be no cruelty in putting creatures to the use they were designed for,* but I have heard men make this reply, while the nature within them has reproached them with the falsehood of the assertion.

"There is of all the multitude not one man in ten but will own (if he has not been brought up in a slaughter-house) that of all trades he could never have been a *butcher;* and I question whether ever anybody so much as killed a chicken without reluctancy the first time. Some people are not to be persuaded to taste of any creatures they have daily seen and been acquainted with while they were alive; others extend their scruples no further than to their own poultry, and refuse to eat what they fed and took care of themselves; yet all of them feed heartily and without remorse on beef, mutton, and fowls when they are bought in the market. In this behaviour, methinks, there appears something like a *consciousness of guilt;* it looks as if they endeavoured to save themselves from the imputation of a crime (which they know sticks somewhere) by removing the cause of it as far as they can from themselves; and I discover in it some strong marks of primitive pity and innocence, which all the arbitrary power of Custom, and the violence of Luxury, have not yet been able to conquer."†

XVIII.

GAY. 1688—1732.

THE intimate friend of Pope and Swift is best known by his charming and instructive *Fables.* He was born at Barnstaple, in Devonshire, and belonged to the old family of the Le Gays of that county. His father, reduced in means, apprenticed him to a silk mercer in the Strand, London, in whose employment he did not long remain. The first of his poems, *Rural Sports,* appeared in 1711. In the following year he became secretary to the Duchess of Monmouth, and he served for a short time as secretary to the English embassy in Hanover. His next work was his *Shepherd's Week, in Six Pastorals,* in which he ridicules the sentimentality of the "pastorals" of his own and preceding age. It contains much naturalness as well as humour, and it was the precursor of Crabbe's rural sketches. In 1726 he published the most successful of his works, the *Beggars' Opera*—the idea of which had been suggested to him by

* Upon which Ritson aptly remarks: "The sheep is not so much 'designed' for the *man* as the *man* is for the *tiger*, this animal being naturally carnivorous, which man is not. But nature, and justice, and humanity are not always one and the same thing." To this remark we may add with equal force, that almost all the living beings upon whom our species preys have been so artificially changed from their natural condition for the gratification of its selfish appetite as to be with difficulty identified with the original stocks. So much for this theory of creative *design.*

† *Fable of the Bees,* i. 187, &c.

the Dean of St. Patrick's. It was received with unbounded applause, and it originated the (so-called) English opera, which for a time supplanted the Italian.

The *Fables* first appeared in 1726. They were supplemented afterwards by others, and the volume was dedicated to the young Duke of Cumberland, famous in after years by his suppression of the Highland rising of 1745. Gay's death, which happened suddenly, called forth the sincere laments of his devoted friends Swift and Pope. The former, in his letters, frequently refers to his loss with deep feeling; and Pope has characterised him as—

> "Of manners gentle, of affections mild—
> In wit a man, simplicity a child."

Of his *Fables*—the best in the language—one of the most interesting is the well-known *Hare and Many Friends*, in which he seems to record some of his own experiences. *The Court of Death*, suggested probably by Milton's fine passage in the *Paradise Lost*, is one of his most forcible. When the principal Diseases have severally advanced their claims to pre-eminence, Death calls upon *Intemperance :—*

> " All spoke their claim, and hoped the wand.
> Now expectation hushed the band,
> When thus the monarch from the throne :
> Merit was ever modest known—
> What ! no physician speak his right !
> None here ? But fees their toils requite.
> Let then Intemperance take the wand,
> Who fills with gold their jealous hand.
> You, Fever, Gout, and all the rest
> (Whom wary men as foes detest)
> Forego your claim. No more pretend—
> Intemperance is esteemed a friend.
> He shares their mirth, their social joys,
> And as a courted guest destroys.
> The charge on him must justly fall
> Who finds employment for you all."

It is in the following fable that Gay especially satirises the sanguinary diet :—

> " Pythagoras rose at early dawn,
> By soaring meditation drawn ;
> To breathe the fragrance of the day,
> Through flow'ry fields he took his way.
> In musing contemplation warm,
> His steps misled him to a farm :
> Where, on the ladder's topmost round,
> A peasant stood. The hammer's sound
> Shook the weak barn. 'Say, friend, what care
> Calls for thy honest labour there ? '

" The clown, with surly voice, replies :
' Vengeance aloud for justice cries.
This kite, by daily rapine fed,
My hens' annoy, my turkeys' dread,
At length his forfeit life hath paid.
See on the wall his wings displayed,
Here nailed, a terror to his kind.
My fowls shall future safety find,
My yard the thriving poultry feed,
And my barn's refuse fat the breed.'

" 'Friend,' says the Sage, 'the doom is wise—
For public good the murderer dies.
But if these tyrants of the air
Demand a sentence so severe,
Think how the glutton, man, devours;
What bloody feasts regale his hours !
O impudence of Power and Might !
Thus to condemn a hawk or kite,
When thou, perhaps, carnivorous sinner,
Had'st pullets yesterday for dinner.'

" 'Hold !' cried the clown, with passion heated,
' Shall kites and men alike be treated ?
When heaven the world with creatures stored,
Man was ordained their sovereign lord.'
' Thus tyrants boast,' the Sage replied,
' Whose murders spring from power and pride.
Own then this man-like kite is slain
Thy greater luxury to sustain—
For petty rogues submit to fate
That great ones may enjoy their state.' " *

---

* *Fable* xxxvi., *Pythagoras and the Countryman.* This fable of Gay may have been suggested
by that of Æsop—preserved by Plutarch—who represents a wolf watching a number of shepherds
eating a sheep, and saying to himself—" If *I* were doing what *you* are now about, what an uproar
you would make ! " See also the instructive fable of La Fontaine—*L'Homme et la Couleuvre,* one
of the finest in the whole twelve Books (*Livre* x., 2), in which the Cow and Ox accuse the base
ingratitude of Man for the cruel neglect, and, finally, for the barbarous slaughter of his fellow-
labourers. The Cow, appealed to by the Adder, replies :—

" Pourquoi dissimuler ?
Je nourris celui-ci depuis longues années :
Il n'a sans mes bienfaits passé nulles journées.
Tout n'est que pour lui seul : mon lait et mes enfants
Le font à la maison revenir les mains pleines.
Même j'ai rétabli sa santé, que les ans
    Avaient alterée ; et mes peines
Ont pour but son plaisir ainsi que son besoin.
Enfin me voilà vieille. *Il me laisse*
Sans herbe. S'il voulait encore me laisser paître !
Mais je suis attachée. . . . .
Force coups, peu de gré. Puis, quand il était vieux,
On croyait l'honorer chaque fois que les hommes
Achetaient de son sang l'indulgence des dieux."

This is not the only apologue in which the rhyming moralist exposes at once the inconsistency and the injustice of the human animal who, himself choosing to live by slaughter, yet hypocritically stigmatises with the epithets "cruel" and "bloodthirsty" those animals whom Nature has evidently *designed* to be predaceous.   In *The Shepherd's Dog and the Wolf* he represents the former upbraiding the ravisher of the sheepfolds for attacking " a weak, defenceless kind " :—

> " ' Friend,' says the Wolf, ' the matter weigh :
> Nature designed *us* beasts of prey.
> As such, when hunger finds a treat,
> 'Tis necessary wolves should eat.
> If, mindful of the bleating weal,
> Thy bosom burn with real zeal,
> Hence, and thy tyrant lord beseech—
> To *him* repeat thy moving speech.
> A wolf eats sheep but now and then—
> *Ten thousands are devoured by men !*
> An open foe may prove a curse,
> But a pretended friend is worse.' "

In *The Philosopher and the Pheasants* the same truth is conveyed with equal force :—

> " Drawn by the music of the groves,
> Along the winding gloom he roves.
> From tree to tree the warbling throats
> Prolong the sweet, alternate notes.
> But where he passed he terror threw ;
> The song broke short—the warblers flew :
> The thrushes chattered with affright,
> And nightingales abhorred his sight.
> All animals before him ran,
> To shun the hateful sight of man.
>   ' Whence is this dread of every creature ?
> Fly they our figure or our nature ? '
>   As thus he walked, in musing thought,
> His ear imperfect accents caught.
> With cautious step, he nearer drew,
> By the thick shade concealed from view.
> High on the branch a Pheasant stood,
> Around her all her listening brood :
> Proud of the blessings of her nest,
> She thus a mother's care expressed :—
>   ' No dangers here shall circumvent :
> Within the woods enjoy content.
> Sooner the hawk or vulture trust
> Than man, of animals the worst.
> In him ingratitude you find—
> A vice peculiar to the kind.

The Sheep, whose annual fleece is dyed
To guard his health and serve his pride,
Forćed from his fold and native plain,
Is in the cruel shambles slain.
The swarms who, with industrious skill,
His hives with wax and honey fill,
In vain whole summer days employed—
Their stores are sold, their race destroyed.
What tribute from the Goose is paid ?
Does not her wing all science aid ?
Does it not lovers' hearts explain,
And drudge to raise the merchant's gain ?
What now rewards this general use ?
He takes the quills and eats the Goose !' "

.   .   .   .

In another parable Gay, in some sort, gives the victims of the
Shambles their revenge :—

" Against an elm a Sheep was tied :
The butcher's knife in blood was dyed—
The patient flock, in silent fright,
From far beheld the horrid sight.
A savage Boar, who near them stood,
Thus mocked to scorn the fleecy brood :—
    ' All cowards should be served like you.
See, see, your murderer is in view :
With purple hands and reeking knife,
He strips the skin yet warm with life.
Your quartered sires, your bleeding dams,
The dying bleat of harmless lambs,
Call for revenge.   O stupid race !
The heart that wants revenge is base.'
    ' I grant,' an ancient Ram replies,
' We bear no terror in our eyes.
Yet think us not of soul so tame,
Which no repeated wrongs inflame—
Insensible of every ill,
Because we want thy tusks to kill—
Know, *those who violence pursue*
Give to themselves the vengeance due,
For in these massacres they find
The two chief plagues that waste mankind—
Our skin supplies the wrangling bar :
It wakes their slumbering sons to war.
And well Revenge may rest contented,
Since drums and parchment were invented.' " *

---

* *The Wild Boar and the Ram.* For admirable rebukes of human arrogance, see *The Elephant and the Bookseller* and *The Man and the Flea.*

XIX.

CHEYNE. 1671—1743.

ONE of the most esteemed of English physicians, and one of the first medical authorities in this country who expressly wrote in advocacy of the reformed diet, descended from an old Scottish family. He studied medicine at Edinburgh—then and still a principal school of medicine and surgery—where he was a pupil of Dr. Pitcairn. At about the age of thirty he removed to London, was elected a Fellow of the Royal Society, and took his M.D. degree, commencing practice in the metropolis.

The manner of life of a medical practitioner in the first half of the last century differed considerably from the present fashion. Not only personal inclination, but even professional interest, usually led him to frequent taverns and to indulge in all the excesses of "good living;" for in such boon companionship he most easily laid the foundation of his practice. Cheyne's early habits of temperance thus gave way to the double temptation, and soon by this indulgence he contracted painful disorders which threatened his life. An enormous weight of flesh, intermittent fevers, shortness of breath, and lethargy combined to enfeeble and depress him.

His first appearance in literature was the publication of his *New Theory of Fevers*, written in defence and at the suggestion of his old master Dr. Pitcairn, who was at war with his brethren on the nature of epidemics. The author, while in after life holding that it contained, though in a crude form, some valuable matter, wisely allowed it to fall into oblivion. The Mechanical or *Iatro-Mathematical* Theory, as it was called, of which Cheyne was one of the earliest and most distinguished expounders, by which it was attempted to apply the laws of Mechanics to vital phenomena, had succeeded to the principles of the old Chemical School. On the Continent the new theory had the support of the eminent authority of Boerhaave, Borelli, Sauvages, Hoffman, and others. The natural desire to discover some definite and simple *formulæ* of medical science lay at the root of this, as of many other hypotheses. Cheyne, himself, it is right to observe, ridiculed the notion that all vital processes can be explained on mechanical principles.

In 1705 he published his *Philosophical Principles of Natural Religion*, a book which had some repute in its day, apparently, since it was in use in the Universities. Between this and his next essay in literature a long interval elapsed, during which he had to pay the penalty of his old habits in apoplectic giddiness, violent headaches, and depression of spirits. Happily, it became for him the turning-point in his life, and eventually rendered him so useful an instructor of his kind. He had now arrived at a considerable amount of reputation in the profession. He seems to have been naturally of agreeable manners and of an amiable disposition, as well as of lively wit which, improved by study and reading, made him highly popular; and amongst his scientific and professional friends he was in great esteem. He had now, however—not too soon—determined to abandon his *bon-vivantism,* and speedily "even those who had shared the best part of my profusions," he tells us, "who, in their necessities had been relieved by my false generosity, and, in their disorders, been relieved by my care, did now entirely relinquish and abandon me." He retired into solitude in the country and, almost momentarily expecting the termination of his life, set himself to serious and earnest reflection on the follies and vices of ordinary living.

At this time it seems that, although he had reduced his food to the smallest possible amount, he had not altogether relinquished flesh-meat. He repaired to Bath for the waters and, by living in the most temperate way and by constant and regular exercise, he seemed to have regained his early health. At Bath he devoted himself to cases of nervous diseases which most nearly concerned his own state, and which were most abundant at that fashionable resort. About the year 1712, or in the forty-second year of his age, his health was fairly re-established, and he began to relax in the milk and vegetable regimen which he had previously adopted.

His next publication was *An Essay on the Gout and Bath Waters* (1720), which passed through seven editions in six years. In it he commends the vegetable diet, although not so radically as in his latest writings. His relaxation of dietetic reform quickly brought back his former maladies, and he again suffered severely. During the next ten or twelve years he continued to increase in corpulency, until he at last reached the enormous weight of thirty-two stones, and he describes his condition at this time as intolerable.* In 1725 he left Bath for London, to consult his friend Dr. Arbuthnot, whose advice probably renewed and confirmed his old inclination for the rational mode of living. At all events, within two years, by a strict adherence to the milk and vegetable regimen his maladies finally disappeared; nor did he afterwards suffer by any relapse into dietetic errors.

In the preceding year had appeared his first important and original

---

* He was at one time so corpulent that he could not get in and out of his carriage in visiting his patients at Bath.

work—his well-known *Essay of Health and a Long Life.* In the preface he declares that it is published for the benefit of those weakly persons who

"are able and willing to abstain from everything hurtful, and to deny themselves anything their appetites craved, to conform to any rules for a tolerable degree of health, ease, and freedom of spirits. It is for these, and these only," he proceeds, "the following treatise is designed. The robust, the luxurious, the pot-companions, &c., have here no business ; their time is not yet come."

It is generally acknowledged to be one of the best books on the subject. Haller pronounced it to be "the best of all the works bearing upon the health of sedentary persons and invalids." It went through several editions in the space of two years, and in 1726 was enlarged by the author and translated by his friend and pupil John Robertson M.A. into Latin, and three or four editions were quickly exhausted in France and Germany. In this book, while reducing flesh-meat to a *minimum,* and insisting upon the necessity of abstinence from grosser food and of the use of vegetables only, at the morning and evening meals, he had not advanced as yet so far as to preach the truth in its entirety. He arrived at it only by slow and gradual conviction. Expatiating on the follies and miseries of *bon-vivantism,* he proceeds to affirm that—

"All those who have lived long, and without much pain, have lived abstemiously, poor, and meagre. Cornaro prolonged his life and preserved his senses by almost starving in his latter days ; and some others have done the like. They have, indeed, thereby, in some measure, weakened their natural strength and qualified the fire and flux of their spirits, but they have preserved their senses, weakened their pains, prolonged their days, and procured themselves a gentle and quiet passage into another state. . . . All the rest will be insufficient without this [a frugal diet] ; and this alone, without these [medicines, &c.], will suffice to carry on life as long as by its natural flame it was made to last, and will make the passage easy and calm, as a taper goes out for want of fuel."

While the *Essay of Health* added greatly to his reputation with all thinking people, it also exposed him (as was to be expected) to a storm of small wit, ridicule, and misrepresentation :—

"Some good-natured and ingenious retainers to the Profession," he tells us, "on the publication of my book on *Long Life and Health,* proclaimed everywhere that I was turned mere enthusiast, advised people to turn monks, to run into deserts, and to live on roots, herbs, and wild fruits ! in fine, that I was, at bottom, a mere leveller, and for destroying order, ranks, and property, everyone's but my own. But that sneer had its day, and vanished into smoke. Others swore that I had eaten my book, recanted my *doctrine* and *system* (as they were pleased to term it), and was returned again to the devil, the world, and the flesh. This joke I have also stood. I have been slain again and again, both in prose and verse ; but, I thank God, I am still alive and well."

His next publication was his *English Malady : or, a Treatise of Nervous Diseases of all kinds,* which was also well received, going through four editions in two years. The incessant ridicule with which the *gourmands* had assailed his last work seems to have made him cautious in his next attempt to revolutionise dietetics; and he is careful to advertise the public that his milk and vegetable system was for those in weak health only. Denouncing the use of sauces and provocatives of unnatural appetite, " contrived not only to rouse a sickly stomach to receive the unnatural load, but to render a naturally good one incapable of knowing when it has enough," he asks, " Is it any wonder then that the diseases which proceed from idleness and fulness of meat should increase in proportion ?" He is bold enough by this time to affirm that, for the cure of many diseases an entire abstinence from flesh is indisputably necessary :—

"There are some cases wherein a vegetable and milk diet seems absolutely necessary, as in severe and habitual gouts, rheumatisms, cancerous, leprous, and scrofulous disorders ; extreme nervous colics, epilepsies, violent hysteric fits, melancholy, consumptions (and the like disorders, mentioned in the preface), and towards the last stages of all chronic distempers. In such distempers *I have seldom seen such a diet fail of a good effect at last.*"

Six years later, in 1740, appeared his *Essay on Regimen : together with Five Discourses Medical Moral and Philosophical, &c.* Since his last exhortation to the world Cheyne had evidently convinced himself, by long experience as well as reflection, of the great superiority of the vegetable diet for all—sound as well as sick ; and, accordingly, he speaks in strong and clear language of the importance of a general reform. As a consequence of this plain speaking, his new book met with a comparatively cold reception. Perhaps, too, its mathematical and somewhat abstruse tone may have affected its popularity. As regards its moral tone it was a new revelation, doubtless, for the vast majority of his readers. He boldly asserts :—

"The question I design to treat of here is, whether animal or vegetable food was, in the original design of the Creator, intended for the food of animals, and particularly of the human race. And I am almost convinced it *never was intended, but only permitted as a curse or punishment.* . . . At what time animal [flesh] food came first in use is not certainly known. He was a bold man who made the first experiment.

Illi robur et æs triplex
Circa pectus erat.

To see the convulsions, agonies, and tortures of a poor fellow-creature, whom they cannot restore nor recompense, dying to gratify luxury, and tickle callous and rank organs, must require a rocky heart, and a great degree of cruelty and ferocity. I cannot find any great difference, *on the foot of natural reason and equity only, between feeding on human flesh and feeding on brute animal flesh, except custom and example.*

I believe some [more] rational creatures would suffer less in being fairly butchered than a strong Ox or red Deer; and, in natural morality and justice, the *degrees of pain* here make the essential difference, for as to other differences, *they are relative only,* and can be of no influence with an infinitely perfect Being. Did not use and example weaken this lesson, and make the difference, reason alone could never do it."—*Essay on Regimen, &c.* 8vo. 1740. Pages 54 and 70.

Noble and courageous words ! Courageous as coming from an eminent member of a profession—which almost rivals the legal or even the clerical, in opposition to all change in the established order of things. In Dr. Cheyne's days such interested or bigoted opposition was even stronger than in the present time. From the period of the final establishment of his health, about 1728, little is known of his life excepting through his writings. Almost all we know is, that he continued some fifteen years to practise in London and in Bath with distinguished reputation and success. He had married a daughter of Dr. Middleton of Bristol by whom he had several children. His only son was born in 1712. Amongst his intimate friends was the celebrated Dr. Arbuthnot, a Scotchman like himself, and we find him meeting Sir Hans Sloane and Dr. Mead at the bedside of his friend and relative Bishop Burnet. Both Dr. Arbuthnot and Sir Hans Sloane, we may remark in passing, have given evidence in favour of the purer living. His own diet he thus describes in his *Author's Case*, written towards the end of his life :—

"My regimen, at present, is milk, with tea, coffee, bread and butter, mild cheese, salads, fruits and seeds of all kinds, with tender roots (as potatoes, turnips, carrots), and, in short, *everything that has not life*, dressed or not, as I like it, *in which there is as much or a greater variety than in animal foods*, so that the stomach need never be cloyed. I drink no wine nor any fermented liquors, and am rarely dry, most of my food being liquid, moist, or juicy.* Only after dinner I drink either coffee or green tea, but seldom both in the same day, and sometimes a glass of soft, small cider. The thinner my diet, the easier, more cheerful and lightsome I find myself ; my sleep is also the sounder, though perhaps somewhat shorter than formerly under my full animal diet ; but then I am more alive than ever I was. As soon as I wake I get up. I rise commonly at six, and go to bed at ten."

As for the effect of this regimen, he tells us that " since that time [his last lapse] I thank God I have gone on in one constant' tenor of diet, and enjoy as good health as, at my time of life (being now sixty), I or any man can reasonably expect." When we remember the complicity of maladies of which he had been the victim during his adhesion to the orthodox mode of living, such experience is sufficiently significant. Some ten years later he records his experiences as follow :—

"It is now about sixteen years since, for the last time, I entered upon a milk and vegetable diet. At the beginning of this period, this light food I took as my appetite directed, without any measures, and found myself easy under it. After some time, I

---

* One of the many excellences of the non-flesh dietary is this essential quality of fruits and vegetables, that they contain in themselves sufficient liquid to allow one to dispense with a large proportion of all extraneous drinks, and certainly with all alcoholic kinds. Hence it is at once the easiest and the surest preventive of all excessive drinking. Much convincing testimony has been collected to this effect by the English and German Vegetarian Societies.

found it became necessary to lessen this quantity, and I have latterly reduced it to one-half, at most, of what I at first seemed to bear ; and if it should please God to spare me a few years longer, in order to preserve, in that case, that freedom and clearness which by his presence I now enjoy, I shall probably find myself obliged to deny myself one-half of my present daily sustenance, which, precisely, is three Winchester pints of new milk, and six ounces of biscuit, made without salt or yeast, baked in a quick oven."*—[*Natural Method of Curing Diseases*, &c., page 298 ; see also Preface to *Essay on Regimen*].

The last production of Dr. Cheyne was his "*Natural Method of Curing the Diseases of the Body, and the Disorders of the Mind Depending on the Body.* In three parts. Part I.—General Reflections on the Economy of Nature in Animal Life.   Part II.—The Means and Methods for Preserving Life and Faculties ; and also Concerning the Nature and Cure of Acute, Contagious, and Cephalic Disorders.   Part III.—Reflections on the Nature and Cure of Particular Chronic Distempers.   8vo.   Strahan, London, 1742." It is dedicated to the celebrated Lord Chesterfield, who records his grateful recognition of the benefits he had experienced from his methods.   He writes : "I read with great pleasure your book, which your bookseller sent me according to your direction.   The physical part is extremely good, and the metaphysical part *may be* so too, for what I know, and I believe it is, for as I look upon all metaphysics to be guess work of imagination, I know no imagination likelier to hit upon the right than yours, and I will take your guess against any other metaphysician's whatsoever.   That part which is founded upon knowledge and experience I look upon as a work of public utility, and for which the present age and their posterity may be obliged to you, if they will be pleased to follow it." Lord Chesterfield, it will be seen below, was one of those more refined minds whose better conscience revoltd from, even if they had not the courage or self-control to renounce, the Slaughter House.

The *Natural Method* its author considers as a kind of supplement to his last book, containing " the practical inferences, and the conclusions drawn from [its principles], in particular cases and diseases, confirmed by forty years' experience and observation."   It is the most practical of all his works, and is full of valuable observations.   Very just and useful is his rebuke of that sort of John-Bullism which affects to hold "good living" not only as harmless but even as a sort of merit—

"How it may be in other countries and religions I will not say, but among us good Protestants, abstinence, temperance, and moderation (at least in eating), are so far from being thought a virtue, and their contrary a vice, that it would seem that not eating the fattest and most delicious, and *to the top*, were the only vice and disease

---

* It is neither necessary nor possible for everyone to practise so extreme abstemiousness ; but it is instructive to compare it for a moment with the ordinary and prevalent indulgence in eating.

known among us—against which our parents, relatives, friends, and physicians exclaim with great vehemence and zeal. And yet, if we consider the matter attentively we shall find there is no such danger in abstinence as we imagine, but, on the contrary, the greatest abstinence and moderation nature and its external laws will suffer us to go into and practise for any time, will neither endanger our health, nor weaken our just thinking, be it ever so unlimited or unrestrained. . . . And it is a wise providence that Lent time falls out at that season which, if kept according to its original intention, in seeds and vegetables well dressed and not in rich high-dressed fish, would go a great way to preserve the health of the people in general, as well as dispose them to seriousness and reflection—so true it is that 'godliness has the promise of this life, and of that which is to come,' and it is very observable that in all civil and established religious worships hitherto known among polished nations Lents, days of abstinence, seasons of fasting and bringing down the brutal part of the rational being, have had a large share, and been reckoned an indispensable part of their worship and duty, except among a wrong-headed part of our Reformation, where it has been despised and ridiculed into a total neglect. And yet it seems not only natural and convenient for health, but strongly commended both in the Old and New Testament, and might allow time and proper disposition for more serious and weighty purposes. And this 'Lent,' or times of abstinence, is one reason of the cheerfulness or serenity of some Roman Catholic or Southern countries, which would be still more healthy and long-lived were it not for their excessive use of aromatics and opiates, which are the worst kind of dry drams, and the cause of their unnatural and unbridled lechery and shortness of life."

Denouncing the general practice of the Profession of encouraging their patients in indulging vitiated habits and tastes, he reminds them:—

" That such physicians do not consider that they are accountable to the community, to their patients, to their conscience, and to their Maker, for every hour and moment they shorten and cut off their patients' lives *by their immoral and murderous indulgence:* and the patients do not duly ponder that suicide (which this is in effect) is the most mortal and irremissible of all sins, and neither have sufficiently weighed the possibility that the patient, if not quickly cut off by both these preposterous means, may linger out miserably, and be twenty or thirty years a-dying, under these heart and wheel-breaking miseries thus exasperated ; whereas, by the methods I propose, if they obtain not in time a perfect cure, yet they certainly lessen their pain, lengthen their days, and continue under the benign influence of 'the Sun of Righteousness, who has healing in His wings, and, at worst, soften and lighten the anguish of their dissolution, as far as the nature of things will admit."

Not the least useful and instructive portions of his treatise are his references to the proper regimen for mental diseases and disordered brains, which, he reasonably infers, are best treated by the adoption of a light and pure dietary. He despairs, however, of the general recognition, or at least adoption, of so rational a method by the " faculty " or the public at large,

"Who do not consider that *nine parts in ten* of the whole mass of mankind are necessarily confined to this diet (of farinacea, fruits, &c.), or pretty nearly to it, and yet live with the use of their senses, limbs, and faculties, without diseases or with but few, and those from accidents or epidemical causes ; and that there have been nations, and now are numbers of tribes, who voluntarily confine themselves to vegetables only, . . . and that there are whole villages in this kingdom whose inhabitants scarce eat animal food or drink fermented liquors a dozen times a year."

In regard to all nervous and brain diseases, he insists that the reformed diet would

"Greatly alleviate and render tolerable original distempers derived from diseased parents, and that it is absolutely necessary for the deep-thinking part of mankind, who would preserve their faculties ripe and pregnant to a green old age and to the last dregs of life ; and that it is the true and real antidote and preservative from wrong-headedness, irregular and disorderly intellect and functions, from loss of the rational faculties, memory, and senses, as far as the ends of Providence and the condition of mortality will allow."—(*Nat. Method*, page 90.)

This benevolent and beneficent dietetic reformer, according to the testimony of an eye-witness, exemplified by his death the value of his principles—relinquishing his last breath easily and tranquilly, while his senses remained entire to the end. During his last illness he was attended by the famous David Hartley, noticed below. He was buried at Weston, near Bath. His character is sufficiently seen in his writings which, if they contain some metaphysical or other ideas which our reason cannot always endorse, in their *practical* teaching prove him to have been actuated by a true and earnest desire for the best interests of his fellow-men. One of the merits of Cheyne's writings is his discarding the common orthodox *esoteric* style of his profession, who seem jealously to exclude all but the "initiated" from their sacred mysteries. One of his biographers has remarked upon this point that "there is another peculiarity about most of Dr. Cheyne's writings which is worthy of notice. Although there are many passages that are quite unintelligible to the reader unless he possesses a considerable knowledge, not only of medicine but also of mathematics, yet there is no doubt but that the greater part of his works were intended for popular perusal, and in this undertaking he is one of the few medical writers who have been completely successful. His productions, which were much read and had an extensive influence in their day, procured him a considerable degree of reputation, not only with the public, but also with the members of his own profession. If they present to the reader no great discoveries (?) they possess the merit of putting more prominently forward some useful but neglected truths ; and though now, probably, but little read, they contain much matter that is well worth studying, and have obtained for their author a respectable place in the history of medical literature."*

Our notice of the author of the *Essay on Regimen*, &c., would scarcely be

---

* *A Life of George Cheyne, M D.*, Parker and Churchill, 1843. See also *Biog. Britannica*

complete without some reference to his friendship with two distinguished characters—John Wesley and Samuel Richardson * the author of *Pamela*. It was to Dr. Cheyne that Wesley, as he tells us in his journals, was indebted for his conversion to those dietetic principles to which he attributes, in great measure, the invigoration of his naturally feeble constitution, and which enabled him to undergo an amount of fatigue and toil, both mentally and bodily, seldom or never surpassed. Of Cheyne's friendship for Richardson there are several memorials preserved in his familiar letters to that popular writer; and his free and naïve criticisms of his novels are not a little amusing. The novelist, it seems, was one of his patients, and that he was not always a satisfactory one, under the abstemious regimen, appears occasionally from the remonstrances of his adviser.

XX.

POPE.   1688—1744.

THE most epigrammatic, and one of the most elegant, of poets. He was also one of the most precocious. His first production of importance was his *Essay on Criticism*, written at the age of twenty-one, although not published until two years later. But he had composed, we are assured, several verses of an Epic at the age of twelve; and his *Pastorals* was given to the world by a youth of sixteen. Its division into the Four Seasons is said to have suggested to Thomson the title of his great poem. The MS. passed through the hands of some distinguished persons, who loudly proclaimed the merits of the boy-poet.

In the same year with his fine mock-heroic *Rape of the Lock* (1712) appeared *The Messiah*, in imitation of Isaiah and of Virgil (in his well-known *Eclogeu* IV.), both of whom celebrate, in similar strains, the advent of a "golden age" to be. The "Sybilline" prophecy, which Pope supposes the Latin poet to have read, existed, it need scarcely be added, only in the imagination of himself and of the authorities on whom he relied. *Windsor Forest* (1713) deserves special notice as one of the earliest of that class of poems which derive their inspiration directly from Nature. It was the precursor of *The Seasons*, although the anti-barbarous feeling is less pronounced in the former. We find, however, the germs of that higher feeling which appears more developed in

---

* Dr. Samuel Johnson gave up wine by the advice of Cheyne, and drank tea with Mrs. Thrale and Boswell till he died, æt. 75.

the *Essay on Man;* and the following verses, descriptive of the usual "sporting" scenes, are significant :—

> "See ! from the brake the whirring Pheasant springs,
> And mounts exulting on triumphant wings :
> Short is his joy ; he feels the fiery wound,
> Flutters in blood, and panting beats the ground.
> Ah, what avail his glossy, varying dyes,
> His purple crest and scarlet-circled eyes—
> The vivid green his shining plumes unfold,
> His painted wings, and breast that flames with gold ?
>
> .   .   .   .   .
>
> To plains with well-breathed beagles they repair,
> And trace the mazes of the circling Hare.
> Beasts, urged by us, their fellow-beasts pursue,
> And learn of man each other to undo.
> With slaughtering guns the unwearied fowler roves,
> When frosts have whitened all the naked groves,
> Where Doves, in flocks, the leafless trees o'ershade,
> And lonely Woodcocks haunt the watery glade—
> He lifts the tube, and level with his eye,
> Straight a short thunder breaks the frozen sky.
> Oft, as in airy rings they skim the heath,
> The clamorous Lapwings feel the leaden death :
> Oft, as the mounting Larks their notes prepare,
> They fall and leave their little lives in air."

His *Epistle of Eloisa to Abelard* (a romantic version of a very realistic story), *Temple of Fame, Imitations of Chaucer*, translation of the *Iliad* (1713-1720)—characterised by Gibbon as having "every merit but that of likeness to its original"—an edition of Shakspere, *The Dunciad* (1728), translation of the *Odyssey*, are some of the works which attest his genius and industry.   But it is with his *Moral Essays*—and in particular the *Essay on Man* (1732-1735), the most important of his productions— that we are especially concerned.

As is pretty well known, these *Essays* owe their conception, in great part, to his intimate friend St. John Bolingbroke.   Although the author by birth and, perhaps, still more from a feeling of pride which might make him reluctant to abandon an unfashionable sect (such it was at that time), belonged nominally to the Old Church, the theology and metaphysics of the work display little of ecclesiastical orthodoxy.   The pervading principles of the *Essay on Man* are natural theology or, as Warburton styles it, "Naturalism" (*i.e.*, the putting aside human assertion for the study of the attributes of Deity through its visible manifestations) and Optimism.*

---

* Bayle, the author of the great *Dictionnaire Historique et Critique* (1690), to whom belongs the lasting honour of having inaugurated the critical method in history and philosophy, which has

I

The merits of the *Essay*, it must be added, consist not so much in the philosophy of the poem as a whole as in the many fine and true thoughts scattered throughout it, which the author's epigrammatic terseness indelibly fixes in the mind. Of the whole poem the most valuable part, undoubtedly, is its ridicule of the common arrogant (pretended) belief that all other species on the earth have been brought into being for the benefit of the human race—an egregious fallacy, by the way, which, ably exposed as it has been over and over again, still frequently reappears in our popular theology and morals. To the writers and talkers of this too numerous class may be commended the rebukes of Pope :—

> " Nothing is foreign—parts relate to whole :
> One all-extending, all-preserving soul
> Connects each being, greatest with the least—
> Made beast in aid of man, and man of beast :
> All served, all serving—nothing stands alone.
>
> . . . . . .
>
> Has God, thou fool, worked solely for thy good,
> Thy joy, thy pastime, thy attire, thy food ?
>
> . . . . . .
>
> Is it for thee the Lark ascends and sings ?
> Joy tunes his voice, joy elevates his wings.
> Is it for thee the Linnet pours his throat ?
> Loves of his own and raptures swell the note.
> The bounding Steed you pompously bestride
> Shares with his lord the pleasure and the pride.
>
> . . . . . .
>
> Know Nature's children all divide her care,
> The fur that warms a monarch warmed a Bear.
> While Man exclaims, 'See all things for my use !'
> 'See Man for mine !' replies a pampered Goose.
> And just as short of reason he must fall,
> Who thinks *all made for one, not one for all.*"

He then paints the picture of the "Times of Innocence" of the Past, or rather (as we must take it) of the Future :—

> " *No murder clothed him, and no murder fed.*
> In the same temple—the resounding wood—
> All vocal beings hymned their equal God.

---

since led to such extensive and important results, seems also to have been the first explicitly to state the difficulties of that greatest *crux* of Theology—the problem of the existence, or rather dominance, of Evil. His rival Le Clerc, in his *Bibliothèque*, took up the orthodox cudgels. Lord Shaftesbury, the celebrated theologian and moralist, wrote his dialogue—*The Moralists* (1709)—in direct answer to Bayle, followed the next year by the *Theodike or Vindication of the Deity* of Leibnitz. Two of the most able and distinguished of the Anti-Optimists are Voltaire and Schópenhauer, the former of whom never wearies of using his unrivalled powers of irony and sarcasm on the *Tout est Bien* theory. As for the latter philosopher, he has carried his Anti-Optimism to the extremes of Pessimism

The shrine, with gore unstained, with gold undrest,
Unbribed, unbloody, stood the blameless priest.
Heaven's attribute was universal care,
And man's prerogative to rule but spare.
Ah, how unlike the man of times to come—
Of half that live the butcher and the tomb!
Who, foe to Nature, hears the general groan,
Murders their species, and betrays his own.
But just disease to luxury succeeds,
And every death its own avenger breeds:
The fury-passions from that blood began,
And turned on man a fiercer savage, man."

Again, depicting the growth of despotism and superstition, and speculating as to—

" Who first taught souls enslaved and realms undone
The enormous faith of Many made for One ?"

he traces the gradual horrors of sacrifice beginning with other, and culminating in that of the human, species :—

" She [Superstition] from the rending earth and bursting skies
Saw gods descend, and fiends infernal rise :
Here fixed the dreadful, there the blest, abodes—
Fear made her devils and weak Hope her gods—
Gods partial, changeful, passionate, unjust,
Whose attributes were rage, revenge, or lust—
Such as the souls of cowards might conceive,
And, formed like tyrants, tyrants would believe.

. . . . . .

Altars grew marble then, and reeked with gore ;
Then first the Flamen tasted living food,
Next his grim idol smeared with human blood.
With Heaven's own thunders shook the earth below,
And played the God an engine on his foe."

Whenever occasion arises, Pope fails not to stigmatise the barbarity of slaughtering for food ; and the *sæva indignatio* urges him to upbraid his fellows with the slaughter of—

" The lamb thy riot dooms to bleed,

. . . . . .

Who licks the hand just raised to shed his blood."

And, again, he expresses his detestation of the selfishness of our species who—

" Destroy all creatures for their sport or gust."

That all this was no mere affectation of feeling appears from his correspondence and contributions to the periodicals of the time :—

" I cannot think it extravagant," he writes, "to imagine that mankind are no less, in proportion, accountable for the ill use of their dominion over the lower ranks of beings, than for the exercise of tyranny over their own species. The more entirely the inferior creation is submitted to our power, the more answerable we must be for our mismanagement of them ; and the rather, as the very condition of Nature renders them incapable of receiving any recompense in another life for ill-treatment in this."*

Consistently with the expression of this true philosophy, he declares elsewhere that—

"Nothing can be more shocking and horrid than one of our kitchens sprinkled with blood, and abounding with the cries of expiring victims, or with the limbs of dead animals scattered or hung up here and there. It gives one the image of a giant's den in romance, bestrewed with scattered heads and mangled limbs."†

The personal character of Pope, we may add, has of late been subjected to minute and searching criticism. Some meannesses, springing from an extreme anxiety for fame with after ages, have undoubtedly tarnished his reputation for candour. His excessive animosity towards his public or private enemies may be palliated in part, if not excused, by his well-known feebleness of health and consequent mental irritability. For the rest, he was capable of the most sincere and disinterested attachments ; and not his least merit, in literature, is that in an age of servile authorship he cultivated literature not for place or pay, but for its own sake.

Amongst Pope's intimate friends were Dr. Arbuthnot, Dean Swift, and Gay. The first of these, best known as the joint author with Pope and Swift of *Martinus Scriblerus*, a satire on the useless pedantry prevalent in education and letters, and especially as the author of the *History of John Bull* (the original of that immortal personification of beef, beer, and prejudice), published his *Essay Concerning Aliments*, in which the vegetable diet is commended as a preventive or cure of certain diseases, about the year 1730. Not the least meritorious of his works was an epitaph on the notorious Colonel Chartres—one of

---

* Pope here is scarcely logical upon his own premiss. It seems impossible, upon any grounds of reason or analogy, to deny to the lower animals a posthumous existence while vindicating it for ourselves, inasmuch as the *essential* conditions of existence are identical for many other beings. To the serious thinker the question of a post-terrestrial state of existence must stand or fall for both upon the same grounds. Yet what can well be more weak, or more of a subterfuge, than the pretence of many well-meaning persons, who seek to excuse their indifferentism to the cruel sufferings of their humble fellow-beings by the expression of a belief or a hope that there is a future retributive state for them? It must be added that this idle speculation—whether the non-human races are capable of post-terrestrial life or no—might, to any serious apprehension, seem to be wholly beside the mark. But what can be more monstrously ridiculous (γέλοιον, in Lucian's language) than the inconsistency of those who would maintain the affirmative, and yet persist in *devouring* their clients? *Risum teneatis, amici!*

† *Spence's Anecdotes* and *The Guardian*, May 21, 1713. His indignation was equally aroused by the tortures of the vivisectors of the day. And he demands how do men know that they have "a right to kill beings whom they [at least, the vast majority] are so little above, for their own curiosity, or even for some use to them."

the few epitaphs which are attentive less to custom than to truth, and, we may add, in marked contrast with that typical one on his unhistorical contemporary Captain Blifil.

In the *Travels of Lemuel Gulliver* the reader will find the *sæva indignatio* of Swift—or, at all events, of the Houyhnhnms—amongst other things, launched against the indiscriminating diet of his countrymen :—

"I told him" [the Master-Horse], says Gulliver, "we fed on a thousand things which operated contrary to each other—that we eat when we are not hungry, and drink without the provocation of thirst . . . that it would be endless to give him a catalogue of all diseases incident to human bodies, for they could not be fewer than five or six hundred, spread over every limb and joint—in short, every part, external and intestine, having diseases appropriated to itself—to remedy which there was a sort of people bred up among us in the profession or pretence of curing the sick."

Among the infinite variety of remedies and prescriptions, in the human *Materia Medica,* the astounded Houyhnhnm learns, are reckoned "serpents, toads, frogs, spiders, dead men's flesh and bones, birds, beasts, fishes"—no mere travellers' tales (it is perhaps necessary to explain), but sober fact, as any one may discover for himself by an examination of some of the received and popular medical treatises of the seventeeth century, in which the most absurd "prescriptions," involving the most frightful cruelty, are recorded with all seriousness :—

"My master, continuing his discourse, said there was nothing that rendered the Yahoos more odious than their undistinguishing appetite to devour everything that came in their way, whether herbs, roots, berries, the *corrupted flesh of animals, or all mingled together;* and that it was peculiar in their temper that they were fonder of what they could get by rapine or stealth at a greater distance than much better food provided for them at home. If their prey held out, they would eat till they were ready to burst."

Although unaccustomed to the better living, and finding it "insipid at first," the human slave of the Houyhnhnm (a word which, by the way, in that language, means "the perfection of nature") records as the result of his experience, in the first place, how little will sustain human life ; and, in the second place, the fact of the superior healthfulness of the vegetable food. *

About this period or a little earlier, Philippe Hecquet, a French physician, published his *Traité des Dispenses du Carême* ("Treatise on Dispensations in Lent"), 1709, in which he gave in his adhesion to the principles of Vegetarianism—at all events, so far as health is concerned. He is mentioned by Voltaire, and is supposed to be the original of the doctor Sangrado of Le Sage. If this conjecture have

* See *Travels,* &c. Part IV.

† *Dict. Phil.,* in article *Viande,* where it is lamented that his book, as far as appeared, had made no more converts than had the Treatise of Porphyry fifteen centuries before.

any truth, the author of *Gil Blas* is open to the grave charge of misrepresentation, of sacrificing truth to effect, or (what is still worse and still more common) of pandering to popular prejudices·*

XXI.

THOMSON. 1700—1748.

In the long and terrible series of the Ages the distinguishing glory of the eighteenth century is its *Humanitarianism*—not visible, indeed, in legislation or in the teaching of the ordinarily-accredited guides of the public faith and morals, but proclaimed, nevertheless, by the great prophets of that era. As far as ordinary life was concerned, the last age is only too obnoxious to the charge of selfishness and heartlessness. Callousness to suffering, as regards the non-human species in particular, is sufficiently apparent in the common amusements and "pastimes" of the various grades of the community.

Yet, if we compare the tone of even the common-place class of writers with that of the authors of quasi-scientific treatises of the preceding century—in which the most cold-blooded atrocities on the helpless victims of human ignorance and barbarity are prescribed for the composition of their medical *nostrums*, &c., with the most unconscious audacity and ignoring of every sort of feeling—considerable advance is apparent in the slow onward march of the human race towards the goal of a true morality and religion.

To the author of *The Seasons* belongs the everlasting honour of being the first amongst modern poets earnestly to denounce the manifold wrongs inflicted upon the subject species, and, in particular, the savagery inseparable from the Slaughter-House—for Pope did not publish his *Essay on Man* until four years after the appearance of *Spring*.

---

* See the amusing scene of the gourmand Canon Sedillo and Dr. Sangrado, who had been called in to the gouty and fever-stricken patient : "'Pray, what is your ordinary diet?' [asks the physician.] 'My usual food,' replied the Canon, 'is broth and juicy meat.' 'Broth and juicy meat!' cried the doctor, alarmed. I do not wonder to find you sick ; such dainty dishes are poisoned pleasures and snares that luxury spreads for mankind, so as to ruin them the more effectually. . . . . What an irregularity is here! what a frightful regimen! You ought to have been dead long ago. How old are you, pray?' 'I am in my sixty-ninth year,' replied the Canon. 'Exactly,' said the physician ; 'an early old age is always the fruits of intemperance. If you had drunk nothing else than pure water all your life, and had been satisfied with simple nourishment—such as boiled apples, for example—you would not now be tormented with the gout, and all your limbs would perform their functions with ease. I do not despair, however, of setting you to rights, provided that you be wholly resigned to my directions.'" (*Adventures of Gil Blas*, ii., 2.) We may comment upon the satire of the novelist (for so it was intended), that irony or sarcasm is a legitimate and powerful weapon when directed against falsehood ; that there was, and is, only too much in the practice and principles of the profession open to ridicule ; but that the attempted ridicule of the better living does not redound to the penetration or good sense of the satirist.

James Thomson, of Scottish parentage, came to London to seek his fortune in literature, at the age of 25. For some time he experienced the poverty and troubles which so generally have been the lot of young aspirants to literary, especially poetic, fame. *Winter*—which inaugurated a new school of poetry—appeared in March, 1726. That the publisher considered himself liberal in offering three guineas for the poem speaks little for the taste of the time; but that a better taste was coming into existence is also plain from the fact of its favourable reception, notwithstanding the obscurity of the author. Three editions appeared in the same year. *Summer*, his next venture, was published in 1727, and the (Four) *Seasons* in 1730, by subscription—387 subscribers enrolling their names for copies at a guinea each.

Natural enthusiasm, sympathy, and love for all that is really beautiful on Earth (a sort of feeling not to be appreciated by vulgar minds) forms his chief characteristic. But, above all, his sympathy with suffering in all its forms (see, particularly, his reflections after the description of the snowstorm in *Winter*), not limited by the narrow bounds of nationality or of species but extended to all innocent life—his indignation against oppression and injustice, are what most honourably distinguish him from almost all of his predecessors and, indeed, from most of his successors. *The Seasons* is the forerunner of *The Task* and the humanitarian school of poetry. *The Castle of Indolence* in the stanza of Spenser, has claims of a kind different from those of *The Seasons*; and the admirers of *The Faerie Queen* cannot fail to appreciate the merits of the modern romance. Besides these *chefs-d'œuvre* Thomson wrote two tragedies, *Sophonisba* and *Liberty*, the former of which, at the time, had considerable success upon the stage. In the number of his friends he reckoned Pope and Samuel Johnson, both of whom are said to have had some share in the frequent revisions which he made of his principal production.

It is with his *Spring* that we are chiefly concerned, since it is in that division of his great poem that he eloquently contrasts the two very opposite diets. Singing the glories of the annual birth-time and general resurrection of Nature, he first celebrates

> " The living Herbs, profusely wild,
> O'er all the deep-green Earth, beyond the power
> Of botanist to number up their tribes,
> (Whether he steals along the lonely dale
> In silent search, or through the forest, rank
> With what the dull incurious weeds account,
> Bursts his blind way, or climbs the mountain-rock,
> Fired by the nodding verdure of its brow).
> With such a liberal hand has Nature flung
> Their seeds abroad, blown them about in winds,

Innumerous mixed them with the nursing mould,
The moistening current and prolific' rain.

But who their virtues can declare ?   Who pierce,
With vision pure, into those secret stores
Of health and life and joy—the food of man,
While yet he lived in innocence and told
A length of golden years, unfleshed in blood ?
A stranger to the savage arts of life—
Death, rapine, carnage, surfeit, and disease—
The Lord, and not the Tyrant, of the world."

And then goes on to picture the feast of blood :—

" And yet the wholesome herb neglected dies,
Though with the pure exhilarating soul
Of nutriment and health, and vital powers
Beyond the search of Art, 'tis copious blessed.
For, with hot ravin fired, ensanguined Man
Is now become the Lion of the plain
And worse.   The Wolf, who from the nightly fold
Fierce drags the bleating Prey, ne'er drank her milk,
Nor wore her warming fleece ; nor has the Steer,
At whose strong chest the deadly Tiger hangs,
E'er ploughed for him.   They, too, are tempered high,
With hunger stung and wild necessity,
Nor lodges pity in their shaggy breast.

But Man, whom Nature formed of milder clay,
With every kind emotion in his heart,
And taught alone to weep ; while from her lap
She pours ten thousand delicacies—herbs
And fruits, as numerous as the drops of rain
Or beams that gave them birth—shall he, fair form,
Who wears sweet smiles and looks erect on heaven,
E'er stoop to mingle with the prowling herd
And dip his tongue in gore ?   The beast of prey,
Blood-stained, deserves to bleed.   But you, ye Flocks,
What have you done ?   Ye peaceful people, what
To merit death ?   You who have given us milk
In luscious streams, and lent us your own coat
Against the winter's cold !   And the plain Ox,
That harmless, honest, guileless animal,
In what has he offended ?   He, whose toil,
Patient and ever ready, clothes the land
With all the pomp of harvest—shall he bleed,
And struggling groan beneath the cruel hands
E'en of the clowns he feeds, and that, perhaps,
To swell the riot of the autumnal feast
Won by his labour ?" *

---

* Compare the similar thoughts of the Latin poet, *Metam.* xv.

And again in denouncing the *amateur* slaughtering (euphemised by the mocking term of *Sport*) unblushingly perpetrated in the broad light of day :—

> " When beasts of prey retire, that all night long,
> Urged by necessity, had ranged the dark,
> As if their conscious ravage shunned the light,
> Ashamed. Not so [he reproaches] the steady tyrant Man,
> Who with the thoughtless insolence of Power,
> Inflamed beyond the most infuriate wrath
> Of the worst monster that e'er roamed the waste,
> For Sport alone pursues the cruel chase,
> Amid the beamings of the gentle days.
>
>   Upbraid, ye ravening tribes, our *wanton* rage,
> For hunger kindles *you*, and lawless want ;
> But lavish fed, in Nature's bounty rolled—
> To joy at anguish, and delight in blood—
> Is what your horrid bosoms never knew." *

We conclude these extracts from *The Seasons* with the poet's indignant reflection upon the selfish greed of Commerce, which barbarously sacrifices by thousands (as it does also the innocent mammalia of the seas) the noblest and most sagacious of the terrestrial races for the sake of a superfluous luxury :—

> " Peaceful, beneath primeval trees, that cast
>   Their ample shade o'er Niger's yellow stream,
>   And where the Ganges rolls his sacred waves ;
>   Or mid the central depth of blackening woods,
>   High raised in solemn theatre around,
>   Leans the huge Elephant, wisest of *brutes !*
>   O truly wise ! with gentle might endowed :
>   Though powerful, not destructive. Here he sees
>   Revolving ages sweep the changeful Earth,
>   And empires rise and fall : regardless he
>   Of what the never-resting race of men
>   Project. Thrice happy ! could he 'scape their guile
>   Who mine, from cruel avarice, his steps :
>   Or with his towering grandeur swell their state—
>   The pride of kings !—or else his strength pervert,
>   And bid him rage amid the mortal fray,
>   Astonished at the madness of mankind."†

---

* *Autumn.* Read the verses which immediately follow, describing, with profound pathos, the sufferings and anguish of the hunted Deer and Hare.

† *Summer.*

XXII.

HARTLEY.   1705—1757.

Celebrated as the earliest writer of the utilitarian school of morals. At the age of fifteen he entered Jesus College, Cambridge, of which he was afterwards elected a Fellow.   Scruples of conscience about the "Thirty-nine Articles" would not allow him to subscribe them and take orders, and he turned to the medical profession, in which he reached considerable eminence.

His *Observations on Man : his Frame, his Duties, and his Expectations,* appeared in 1748.   The principal interest in the book consists in the fact of its containing the germs of that school of moral philosophy of which Paley, Bentham, and Mill have been the most able expositors. He had imbibed the teaching of Locke upon the origin of ideas, which that first of English metaphysicians founded in Sensation and Reflection or Association, in contradiction to the old theory of *Innateness.* Although now universally received, it is hardly necessary to remark that at its first promulgation it met with as great opposition as all rational ideas experience long after their first introduction ; and Locke's controversy with the Bishop of Worcester is matter of history.

It has already been stated that David Hartley was the friend of Dr. Cheyne, whom he attended in his last illness, and he numbered amongst his acquaintances some of tke most eminent personages of the day.   His character appears to have been singularly amiable and disinterested.   His theology is, for the most part, of unsuspected orthodoxy.   The following sentences reveal the bias of his mind in the matter of *kreophagy* :—

"With respect to animal diet, let it be considered that taking away the lives of [other] animals in order to convert them into food, *does great violence to the principles of benevolence and compassion.*   This appears from the frequent hard-heartedness and cruelty found among those persons whose occupations engage them in destroying animal life, as well as from the uneasiness which others feel in beholding the butchery of [the lower] animals.   It is most evident, in respect to the larger animals and those with whom we have a familiar intercourse—such as Oxen, Sheep, and domestic Fowls, &c.—so as to distinguish, love, and compassionate individuals.   They resemble us greatly in the make of the body in general, and in that of the particular organs of circulation, respiration, digestion, &c. ; also in the formation of their intellects, memories, and passions, and in the signs of distress, fear, pain, and death. They often, likewise, win our affections by the marks of peculiar sagacity, by their instincts, helplessness, innocence, nascent benevolence, &c., &c. , and, if there be any glimmering of hope of an hereafter for them—if they should prove to be our *brethren and sisters* in this higher sense, in immortality as well as mortality—in the permanent principle of our minds as well as in the frail dust of our bodies—this ought to be still further reason for tenderness for them.

"This, therefore, seems to be nothing else," he concludes, " than an argument to stop us in our career, to make us sparing and tender in this article of diet, and put us upon consulting experience more faithfully and impartially in order to determine what is most suitable to the purposes of life and health, our compassion being made, by the foregoing considerations in some measure, a balance to our impetuous bodily appetites."*

Dr. Hartley is not the only theologian who has suggested the possibility or probability of a future life for all or some of the non-human races. This question we must leave to the theologians. All that we here remark is, that Hartley is one of the very few amongst his brethren who have had the consistency and the courage of their opinions to deduce the inevitable inference.

XXIII.

CHESTERFIELD.   1694—1773.

Notwithstanding his strange self-deception as to the "general order of nature," by which he attempted (sincerely we presume) to silence the better promptings of conscience, the remarkably strong feeling expressed by Lord Chesterfield gives him some right to notice here. His early *instinctive* aversion for the food which is the product of torture and murder is much better founded, we shall be apt to believe, than the fallacious sophism by which he seems eventually to have succeeded in stifling the voices of Nature and Reason in seeking refuge under the shelter of a superficial philosophy. At all events his example is a forcible illustration of Seneca's observation that the better feelings of the young need only to be evoked by a proper education to conduct them to a true morality and religion.†

As it is we have to lament that he had not the greater light (of

---

* *Observations on Man, II., 3.*

† Quam vehementes haberent tirunculi impetus primos ad optima quæque *si quis exhortaretur, si quis impelleret!* The general failure Seneca traces partly to the fault of the schoolmasters, who prefer to instil into the minds of their pupils a knowledge of *words* rather than of *things*—of *dialectics* rather than of *dietetics* (nos docent disputare non vivere), and partly to the fault of parents who expect a head in place of a heart training. (See *Letters to Lucilius*, cviii.) *Quis doctores docebit ?*

science) of the present time, if, indeed, the "deceitfulness of riches" would not have been for him, as for the mass of the rich or fashionable world, the shipwreck of just and rational feeling.

Philip Dormer, Earl of Chesterfield, succeeded to the family title in 1726. High in favour with the new king—George II.—he received the appointment of Ambassador-extraordinary to the Court of Holland in 1728, and amongst other honours that of the knighthood of the Garter. In 1745 he was appointed Lord Lieutenant of Ireland, in which post, during his brief rule, he seemed to have governed with more success than some of his predecessors or successors. He was soon afterwards a Secretary of State : ill-health obliged him to relinquish this office after a short tenure. He wrote papers for *The World*—the popular periodical of the time— besides some poetical pieces, but he is chiefly known as an author by his celebrated *Letters to his Son*, which long served as the text-book of polite society. It contains some remarks in regard to the relations of the sexes scarcely consonant with the custom, or at least with the outward code of sexual morals of the present day. His sentiments upon the subject in question are as follow :—

"I remember, when I was a young man at the University, being so much affected with that very pathetic speech which Ovid puts into the mouth of Pythagoras against the eating of the flesh of animals, that it was some time before I could bring myself to our college mutton again, with some inward doubt whether I was not making myself an accomplice to a murder. My scruples remained unreconciled to the committing of so horrid a meal, till upon serious reflection I became convinced of its legality* from the general order of Nature which has instituted the universal preying [of the stronger] upon the weaker as one of her first principles : though to me it has ever appeared an incomprehensible mystery that she, who could not be restrained by any want of materials from furnishing supplies for the support of her numerous offspring, should lay them under the necessity of devouring one another.†

"I know not whether it is from the clergy having looked upon this subject as too trivial for their notice, that we find them more silent upon it than could be wished ; for as slaughter is at present no branch of the priesthood, it is to be presumed that they have as much compassion as other men. The *Spectator* has exclaimed against the cruelty of roasting lobsters alive, and of whipping pigs to death, but the misfortune is the writings of an Addison are seldom read by cooks and butchers. As to the *thinking* part of mankind, it has always been convinced, I believe, that however conformable to the *general* rule of nature our devouring animals may be, we are nevertheless under indelible obligation to prevent their suffering any degree of pain more than is absolutely unavoidable.

---

* An instance of the common confusion af thought and logic. The too obvious fact that a large proportion of animals are carnivorous neither proves nor justifies the carnivorousness of the *human* species. The real question is, is the human race originally *frugivorous* or *carnivorous?* Is it allied to the Tiger or to the Ape?

† "Who is this female personification 'Nature'? What are 'her principles,' and where does she reside?" asks Ritson quoting this passage.

"But this conviction lies in such heads that I fear *not one poor creature in a million has ever fared the better for it,* and, I believe, never will : since people of condition, the only source from whence [effectual] pity is to flow, are so far from inculcating it to those beneath them, that a very few years ago they suffered themselves to be entertained at a public theatre by the performances of an unhappy company of animals who could only have been made actors by the utmost energy of whipcord and starving."[*]

The writer might have instanced still more frightful results of this insensibility on the part of the influential classes of the community : nor indeed, the better few always excepted, were he living now could he present a much more favourable picture of the morals (in this the most important department of them) of the ruling sections of society.

Ritson supplements the virtual adhesion of Lord Chesterfield to the principles of Humanity, with some remarks of Sir W. Jones, the eminent Orientalist, who (protesting against the selfish callousness of "Sportsmen" and even of "Naturalists" in the infliction of pain) writes : "I shall never forget the couplet of Ferdusi[†] for which Sadi,[‡] who cites it with applause, pours blessings on his departed spirit :—

"Ah ! spare yon emmet, rich in hoarded grain :
He lives with pleasure and he dies with pain."

To which creditable expression of feeling we would append a word of astonishment at that very common inconsistency, and failure in elementary logic, which permits men—while easily and hyperbolically commiserating the fate of an emmet, a beetle, or a worm—to ignore the necessarily infinitely greater sufferings of the highly-organised victims of the *Table*."

------ ◆◆ ------

XXIV.

VOLTAIRE. 1694—1778.

Of the life and literary productions of the most remarkable name in the whole history of literature—if at least we regard the extent and variety of his astonishing genius, as well as the immense influence, contemporary and future, of his writings—only a brief outline can be given here. Yet, as the most eminent humanitarian prophet of the eighteenth century, the principal facts of his life deserve somewhat larger notice than within the general scope of this work.

François Marie Arouet—commonly known by his assumed name of Voltaire—on his mother's side of a family of position recently ennobled, was born at Chatenay, near Paris. He was educated at the Jesuits' College of Louis XIV., where, it is said, the fathers already foretold his

------

[*] *The World* No. 190, as quoted by Ritson.
[†] Persian poets of the tenth and thirteenth centuries of our era.
[‡] *Asiatic Researches.* iv. 12

future eminence.   Like many other illustrious writers he was originally destined for the "Law," which was little adapted to his genius, and, like his great prototype, Lucian, and others, he soon abandoned all thought of that profession for letters and philosophy.   He had the good fortune, at an early age, to gain the favour of the celebrated Ninon de Lenclos, who left him a legacy of 2,000 livres for the purchase of a library—an important event which was doubtless the means of confirming his intellectual bias.

Voltaire's first literary conceptions were formed in the Bastile, that infamous representative of despotic caprice, to which some verses of which he was the reputed author, satirising the licentious extravagance of the Court of the late king, Louis XIV., had consigned him at the age of twenty.   Soon afterwards appeared the tragedy of *Œdipe* (founded upon the well-known dramas of Sophocles), the first modern drama in which the universal and traditional love scenes were discarded.   This contempt for the conventionalities, however, excited the indignation of the play-goers, and the *Œdipe* was, at its first representation, hissed off the stage.   The author found himself forced to sacrifice to the popular tastes, and his tragedy was received with applause.   Two memorable verses indicated the bias of the future antagonist of ecclesiastical ortho-doxy, and naturally provoked the hostility of the profession which he had dared so openly to assail :—

> " Nos prêtres ne sont pas ce qu'un vain peuple pense :
> Notre credulité fait toute leur science."

It was during this imprisonment, too, that he formed the first idea of the *Henriade* (or *The League*, as it was originally called), the only epic poem worthy of the name in the French language.   A chance quarrel with an insolent courtier was the cause of Voltaire's second incarceration in the Bastile with, at the end of six months, a peremptory order to absent himself from the capital.   These experiences of despotic caprice and of sophisticated society he long afterwards embodied in two of his best romances, *L'Ingénu* and *Micro-mégas* (the " Little-Big Man "), one of the most exquisite productions of Satire.

The youthful victim of these malicious persecutions determined upon seeking refuge in England, whose freer air had already inspired Newton, Locke, Shaftesbury, and other eminent leaders of Thought.   A flattering welcome awaited him—and subscriptions to the *Henriade*, better received here than in France, gratified his pride and filled his purse.   During his sojourn of three years in this country, he made the most of his time in studying its best literature, and cultivating the acquaintance of its most eminent living writers.   His tragedy of *Brutus* was followed by *La Mort de César* which, from its taint of liberalism, was not allowed to be printed in France. Upon his return to Paris he published his *Zaïre*—finished in eighteen days—the first tragedy in which, deserting the footsteps of Corneille and Racine, he ventured to follow the bent of his own genius. The plan of *Zaïre* has been pronounced to be one of the most perfect ever contrived for the stage.

More important, by its influence upon contemporary thought, was his famous *Letters on the English*—a work designed to inform his countrymen generally of the literature, thought, and political and theological parties of the rival nation, and, more especially, of the discoveries of Newton and Locke. Descartes, at this moment supreme in France, had succeeded to the vacant throne of the so-called Aristotelian Schoolmen. His system, a great advance upon the old, broached some errors in physics, amongst others the theory of "Vortices" to explain the planetary movements. A much more pernicious and reprehensible error was his absurd denial of conscious feeling and intelligence to the lower races, which was admirably exposed by Voltaire in his *Elémens de Newton* and elsewhere. In England, Newton's extraordinary discoveries had already made Descartes obsolete, as far as the *savans* were concerned at least, but the French scientific world still clung, for the most part, to the Cartesian principles. As for Locke, he had overturned the orthodox creed of 'innate ideas," supplying instead sensation and reflection. This advocacy of the new philosophy, added to the success of his tragedies for the theatre,

"Drew [says Voltaire in his *Mémoires*] a whole library of pamphlets down upon me, in which they proved I was a bad poet, an atheist, and the son of a peasant. A history of my life was printed in which this genealogy was inserted. An industrious German took care to collect all the tales of that kind which had been crammed into the libel, they had published against me. They imputed adventures to me with persons I never knew, and with others who never existed. I have found while writing this a letter from the Maréchal de Richelieu which informed me of an impudent lampoon where it was proved his wife had given me an elegant couch, with something else, at a time when he had no wife. At first I took some pleasure in making collections of these calumnies, but they multiplied to such a degree I was obliged to leave off. Such are the fruits I gathered from my labours. I, however, easily consoled myself, sometimes in my retreat at Cirey, and at other times in mixing with the best society."

Amongst other subjects the *Lettres* (a masterpiece of criticism and sort of essays, since often imitated but seldom or never, perhaps, equalled in their kind) contains an admirable essay upon the Quakers, to whom he did justice. He introduces one of them in conversation with him, thus apologising for his *eccentricities*:

"Confess that thou hast had some trouble to prevent thyself from laughing when I answered all thy civilities with my hat upon my head and with thouing and thee-ing thee (*en te tutoyant*). Yet thou seemest to me too well informed to be ignorant that, in the time of Christ, no nation fell into the ridiculousness of substituting the *plural* for the singular. They used to say to Cæsar-Augustus : ' I love thee,' 'I pray thee,' 'I thank thee.' He would not allow himself to be called 'Monsieur' (*dominus*). It was only a long time after him that men thought of causing themselves to be addressed as *you* in place of *thou*, as though they were double, and of usurping impertinent titles of grandeur, of eminence, of holiness, of divinity even, which earthworms give to other earthworms, while assuring them with a profound respect (and with an infamous falseness), they are their *very humble and very obedient servants*. It is in order to be upon our guard against this unworthy commerce of lies and of flatteries that we ' thee ' and ' thou ' equally kings and kitchen-maids : that we give the ordinary compliments to no one, having for men only charity, and reserving our respect for the laws. We wear a dress a little different from other men, in order that it may be for us a continual warning not to resemble them. Others wear marks of their dignities, we those of Christian humility. We never use *oaths*, not even in law courts : we think that the name of the *Most High* ought not to be pronounced in the miserable debates of men. When we are forced to appear before the magistrates on others' business (for we never have law suits ourselves), we affirm the truth by a ' yes ' or a ' no,' and the judges believe us upon our simple word, while so many other Christians perjure themselves upon the *Gospel*. We never go to war. It is not that we fear death, but it is because we are neither tigers, nor wolves, nor dogs, but men, but Christians. Our God, who has told us to love our enemies and to suffer without a murmur, doubtless would not have us cross the sea to go and cut the throats of our brothers, because assassins, clothed in red and in hats of two feet high, enrol citizens to the accompaniment of a noise produced by two little sticks upon the dried skin of an ass. And when, after battles won, all London is brilliant with illuminations, when the sky is in flames with musket shots, when the air re-echoes with sounds of thanksgiving, with bells, with organs, with cannons, we groan in silence over the murders which cause the public light-heartedness." (*Lettre II.*)

About this period, frequenting less the fashionable and trifling society of the capital, and contenting himself with the company of a few congenial minds, he formed amongst others a sympathetic friendship with the Marquise de Châtelet, a lady of extraordinary talents.

"I was tired [thus he begins his unfinished *Mémoires*], I was tired of the lazy and noisy life led at Paris, of the multitude of *petit-maîtres*, of bad books printed with the approbation of censors and the privilege of the king, of the cabals and parties among the learned, and of the mean arts of plagiarism and book-making which dishonour Literature."

The lady was the equal of Madame Dacier in knowledge of the Greek and Latin languages, and she was familiar with all the best modern writers. She wrote a commentary on Leibnitz. She also translated the *Principia*. Her favourite pursuits, however, were mathematics and metaphysics.

"She was none the less fond of the world and those amusements familiar to her age and sex. She determined to leave them all and bury herself in an old ruinous château on the borders of Champagne and Lorraine, situated in a barren and unhealthy soil. This old château she ornamented with sufficiently pretty gardens. I built a gallery, and formed a very good collection of natural history, added to which we had a library not badly furnished. We were visited by several of the *savans*, who came to philosophise in our retreat. . . I taught English to Madame de Châtelet, who, in about three months, understood it as well as I did, and read Newton, and Locke, and Pope, with equal ease. We read all the works of Tasso and Ariosto together, so that when Algerotti came to Cirey, where he finished his *Newtonianism for Women*, he found her sufficiently skilful in his own language to give him some very excellent information by which he profited."

Voltaire had already (1741) given to the world his *Elémens de Newton*— a work which, in conjunction with other parts of his writings, proves that had he chosen to apply himself wholly to natural philosophy or to mathematics he might have reached the highest fame in those departments of science. It is in the *Elémens* that Voltaire records his noble protest at the same time against the monstrous hypothesis of Descartes, to which we have already referred, and against the selfish cruelty of our species.

"There is in man a disposition to compassion as generally diffused as his other instincts. Newton had *cultivated* this sentiment of humanity, and he extended it to the lower animals. With Locke he was strongly convinced that God has given to them a proportion of ideas, and the same feelings which he has to us. He could not believe that God, who has made nothing in vain, would have given to them organs of feeling *in order that they might have no feeling.*

"He thought it a very frightful inconsistency to believe that animals feel and *at the same time to cause them to suffer.* On this point his morality was in accord with his philosophy. *He yielded but with repugnance to the barbarous custom of supporting ourselves upon the blood and flesh of beings like ourselves,* whom we caress, and he never permitted in his own house the putting them to death by slow and exquisite [*recherchées*] modes of killing for the sake of making the food more delicious. This compassion, which he felt for other animals, culminated in true charity for men. In truth, *without humanity, a virtue which comprehends all virtues,* the name of philosopher would be little deserved."*

At Cirey some of his best tragedies were composed—*Alzire, Mérope,* and *Mehemet* ; the *Discours sur l'Homme,* a moral poem in the style of Pope's Essays, pronounced to be one of the finest monuments of French poetry ; an *Essay on Universal History,* (for his friend's use, to correct as well as supplement Bossuet's splendid but little philosophic history), the foundation of perhaps his most admirable production the *Essai sur les Mœurs et l'Esprit des Nations,* and many lesser pieces, including a large correspondence. Besides these literary works, he engaged in mathematical and scientific studies, which resulted in some *brochures* of considerable value.

About this time (1740) news arrived of the death of Friedrich Wilhelm of Prussia. Most readers know the extraordinary character of this strange personage, who caned the women and his clergy in the streets of his capital, and who was with difficulty dissuaded from ordering his son's execution. Narrowly escaping with his life the prince had devoted

---

* *Elémens de la Philosophie de Newton,* v. Haller, the founder of modern physiology, assures us that "Newton, while he was engaged upon his *Optics,* lived almost entirely on bread, and wine, and water" (*Newtonus, dum* Optica *scribebat, solo pœnè vino pane et aquâ vixit).—Elements of Physiology,* vi. 198.

K

himself to literary pursuits, and had kept up a correspondence with the
leading men of letters of France, and above all with the author of *Zaïre*
whom he regarded as little less than divine. The new king set about
inspecting his territories, and proceeded *incognito* to Brussels, where the
first interview between the two future most eminent persons in Europe
took place. Repairing to his majesty's quarters—

"One soldier was the only guard I found. The Privy-Councillor and Minister
of State was walking in the court-yard blowing his fingers. He had on a large pair
of coarse ruffles, a hat all in holes, and a judge's old wig, one side of which
hung into his pocket and the other scarcely touched his shoulder. They informed me
that this man was charged with a state affair of great importance, and so indeed he
was. I was conducted into his majesty's apartments, in which I found nothing but
four bare walls. By the light of a taper I perceived a small truckle-bed two feet and
a half wide in a closet, upon which lay a little man wrapped in a morning dressing-
gown of blue cloth. It was his majesty who lay perspiring and shaking beneath a
miserable coverlet in a violent ague fit. I made my bow, and began my acquaintance
by feeling his pulse, as if I had been his first physician. The fit left him, and
he rose, dressed himself, and sat down to table with Algerotti, Maupertuis, the
ambassador of the States-general, and myself. At supper he treated most profoundly
of the soul, natural liberty, and the *Androgynes* of Plato. I soon found myself
attached to him, for he had wit, an agreeable manner, and moreover was a king,
which is a circumstance of seduction hardly to be vanquished by human weakness.
Generally speaking, it is the employment of men of letters to flatter kings, but in this
instance I was praised by a king from the crown of my head to the soles of my feet
at the same time that I was libelled at least once a week by the Abbé Desfontaines
and other Grub-street poets of Paris."

Voltaire received a pressing invitation to Berlin.

"But I had before given him to understand I could not come to stay with him ;
that I deemed it a duty to prefer friendship to ambition ; that I was attached to
Mdlle. de Châtelet, and that, between philosophers, I loved a lady better than a king.
He approved of the liberty I took, though, for his part, he did not love the ladies. I
went to pay him a visit in October, and the Cardinal de Fleury [the French premier]
wrote me a long letter, full of praises of the *Anti-Machiavel*, and of the author
[Friedrich], which I did not forget to let him see."

The French court wished to secure the alliance of Friedrich. No one
seemed a more fitting mediator than his early counsellor, who was
induced to accept the mission, and to set out for Berlin, where an
enthusiastic welcome awaited him, apartments in the palace being placed
at his disposal. Yet, in spite of the success of this and other public
services, his enemies in Paris remained in full possession of the field.
For the second time Voltaire sought admission into the *Académie*—an
empty honour, the granting or refusal of which could neither add to
nor detract from his fame. The prestige of that society, however, he
seemed to consider essential to his safety against the increasing violence
and formidable array of his enemies, who were bent on crushing him, by

whatever means. It was only by submitting to the mortification of qualifying some of his opinions that he at length succeeded in his object. Notwithstanding the address with which he manages his language, it were better, as his biographer—the Marquis de Condorcet—justly remarks, he had renounced the *Académie* than have had the weakness to submit to so evident a farce.

On succeeding to a vacant chair it was customary, besides a eulogy upon the deceased member, to speak in set terms of praise of Richelieu and Louis XIV. This traditional and servile practice the new Academician was the first to break through. Philosophy and literature were treated of in unaccustomed strains of freedom, and his good example has been influential on after generations.

"I was deemed worthy [writes Voltaire] to be one of the forty useless members of the *Académie*, was appointed historiographer of France, and created by the king one of the gentlemen in ordinary of his chamber. From this I concluded it was better, in order to make the most trifling fortune, to speak four words to a king's mistress, than to write a hundred volumes."

A sort of experience he has finely illustrated in his romance of *Zadig*.

Stanislaus, the ex-king of Poland, was keeping his Court at Luneville, not far from Cirey, where he divided his time between his mistress and his confessor. To this royal retreat the friends of Cirey were invited, and the whole of the year 1749 was passed there. Meanwhile Madame de Châtelet died, and Voltaire, much affected by his loss, returned to Paris. Friedrich redoubled his solicitation with new hope.

"I was destined to run from king to king, although I loved liberty to idolatry. . . He was well assured that in reality his verse and prose were superior to my verse and prose ; though as to the former, he thought there was a certain something that I, in quality of academician, might give to his writings, and there was no kind of flattery, no seduction, he did not employ to engage me to come."

The philosopher at length set out for Berlin, and his reception must have reached his highest expectations. We have no intention to repeat the account of this singular episode in his life, which has been so often narrated. Evenings of the most agreeable kind, abundance of wit, unrestrained conversation, the society of some of the most distinguished men of science of the time, the unbounded adoration of a royal host, eager, above all things, to retain so brilliant a guest—such were the pleasures of this palace of Alcina, as he calls it. But the imperious tempers of the two unequal friends soon proved the impossibility of a lasting *entente*, and rivalries amongst the literary courtiers hastened, if they did not effect, the final rupture.

After his escape from Berlin Voltaire passed a few weeks with the Duchess of Saxe-Gotha, "the best of princesses, full of gentleness, discretion, and equanimity, and who, God be thanked, did not make verses" (alluding to his late host's proclivities), and some days with the Landgrave of Hesse on his way to Frankfort. Literature had not suffered during the life at Berlin. Finishing touches were put to many of the tragedies—the *Âge de Louis XIV.* was completed, part of the *Essai sur les Mœurs et l'Esprit des Nations* written, *La Pucelle* (the least worthy of all his productions) corrected, and a poem, *Sur la Loi Naturelle,* composed (a work of a far better inspiration than the poem just mentioned, but which was publicly burned at Paris by the misdirected zeal of the bigots). In a later poem on the destruction of Lisbon, as well as in the romance of *Candide,* fired with indignation at the hypocrisies and mischiefs of the easy-going creed of Optimism (as generally understood), so welcome to self-complacent orthodoxy, he displayed all his vast powers of sarcasm in exposing its fatal absurdities. Leibnitz had been one of its most strenuous apologists. In the person of the wretched Pangloss the theory of "the best of all possible worlds," and of the "eternal fitness of things," is overwhelmed, indeed, with an excess of ridicule. It is to be lamented that the satirist allowed his *sœva indignatio* to overpower a proper sense of the proprieties of language and expression.

Voltaire was now become a potentate more dreaded than a sovereign-prince on his throne, an object of hatred and terror to political and other oppressors. After some hesitation he had chosen for his retreat the ever-memorable Ferney—a place within French territory, on the borders of Switzerland—and also a spot near Geneva, where he alternately resided, escaping at pleasure either from Catholic intolerance or from Puritanic rigour, with his niece—Madame Denis, who had anxiously attended him during a recent illness. From these retreats he made himself heard over all Europe in defence of reason and humanity. It was about this time (1756) that he employed his eloquence to save Admiral Byng, a victim to ministerial necessities, who was nevertheless condemned, as his advocate expresses it in *Candide,* "pour encourager les autres." A like philanthropic effort, equally vain, was made on behalf of the still more unfortunate Comte de Lally.

The year 1757 is memorable in literature as that in which he gave to the world an accurate edition of his already published works, enriched by one of his most meritorious productions, the *Essai sur les Mœurs et l'Esprit des Nations,* which now appeared in its complete form. History, the author justly complained, had hitherto been but a uniform chronicle of kings, courts, and court intrigues. The history of legislation, arts, sciences, commerce, morals, had been always, or almost always, neglected.

" We imagine [says Condorcet], while we read such histories, that the human race was created only to exhibit the political or military talents of a few individuals, and that the object of society is not the happiness of the Species but the pleasure of the Few."

If the best historical works of the present day are a considerable improvement upon those which were in fashion before Voltaire's *critiques*, the remarks of Condorcet are not altogether inapplicable to the popular and school manuals still in vogue. At all events this style of composing " history," ridiculed by the wit of Lucian sixteen centuries before, was the universal method down to the appearance of the celebrated *Essai.*

Beginning with Charlemagne, it presents, in a rapid, concise, and philosophic style, the most important and interesting features, not only of European but of the world's history, adorned with all the grace and ease of which he was always so consummate a master. Many there always are who conceive of philosophy and erudition only as enveloped in verbosity and obscurity. Dulness and learning in the common mind are convertible terms. The very transparency and clearness of his style were reproached to him as a sign of superficiality and want of exactness —the last faults which could be justly imputed to him. However, the influence of Voltaire became apparent in the productions of the English historical school, till then unknown, which soon afterwards arose. The Italian Vico, and Beaufort, in France, in the particular branch of Roman antiquity, and Bayle in general, had already contributed in some degree towards the founding of a critical school; but these attempts were partial only. To Voltaire belongs the honour of having applied the principles of criticism at once universally and popularly.

In reviewing the history and manners of the Hindus he repeatedly expresses his sympathy, more or less directly, with their aversion from the coarser living of the West :—

" The Hindus, in embracing the doctrine of the *Metémpsychosis*, had one restraint the more. The dread of killing a father or mother, in killing men and other animals, inspired in them a terror of murder and every other violence, which became with them a second nature. Thus all the peoples of India, whose families are not allied either to the Arabs or to the Tartars, are still at this day the mildest of all men. Their religion and the temperature of their climate made these peoples entirely resemble those peaceful animals whom we bring up in our sheep pens and our dove cotes for the purpose of cutting their throats at our good will and pleasure. . . .

" The Christian religion, which these *primitives* [the Quakers] alone follow out to the letter, is as great an enemy to bloodshed as the Pythagorean. But the Christian peoples have never practised *their* religion, and the ancient Hindu castes have always practised theirs. It is because Pythagoreanism is the only religion in the world which has been able to educe a religious feeling from the horror of murder and slaughter. . . .

" Some have supposed the cradle of our race to be Hindustan, alleging that the feeblest of all animals must have been born in the softest climate, and in a land which produces without culture the *most nourishing and most healthful fruits,* like dates and cocoa nuts. The latter especially easily affords men the means of existence, of clothing and of housing themselves—and of what besides has the inhabitant of that Peninsula need? . . . Our Houses of Carnage, which they call Butcher-Shops [*boucheries*], where they sell so many carcases to feed our own, would import the plague into the climate of India.

"These peoples need and desire **pure** and refreshing foods. Nature has lavished upon them forests of citron trees, orange trees, fig trees, palm trees, cocoa-nut trees, and plains covered with rice. The strongest man can need to spend but one or two sous a day for his subsistence.* Our workmen spend more in one day than a Malabar native in a month. . . .

"In general, the men of the South-East have received from Nature gentler manners than the people of our West. Their climate disposes them to abstain from strong liquors and from the flesh of animals—foods which excite the blood and often provoke ferocity—and, although superstition and foreign irruptions have corrupted the goodness of their disposition, nevertheless all travellers agree that the character of these peoples has nothing of that irritability, of that caprice, and of that harshness which it has cost much trouble to keep within bounds in the countries of the North."

In noticing the comparative progress of the various foreign religions in India, Voltaire observes that—

"The Mohammedan religion alone has made progress in India, especially amongst the richer classes, because it is the religion of the Prince, and because it teaches but the divine unity conformably to the ancient teaching of the first Brahmins. Christianity [he adds, only too truly] has not had the same success, notwithstanding the large establishments of the Portuguese, of the French, of the English, of the Dutch, of the Danes. It is, in fact, the conflict of these nations which has injured the progress of our Faith. As they all hate each other, and as several of them often make war one upon the other in their climates, what they teach is naturally hateful to the peaceful inhabitants. Their customs, besides, revolt the Hindus. Those people are scandalised at seeing us drinking wine and eating flesh, which they themselves abhor.†

This—one of the chief obstacles to the spread of Christian civilisation in the East, and especially in India, viz., the eating of flesh and the drinking of alcohol, its legitimate attendant—has been acknowledged by Christian missionaries themselves of late years.

Employed as he was in various literary undertakings he had been watching with great interest, not, perhaps, without a secret wish for vengeance, the important political and military complications of Europe. After some brilliant successes the Prussian king had been reduced to the last extremity. At this juncture the former friends agreed to forget, as far as possible, their old quarrel, and Voltaire enjoyed the satisfaction of having succeeded in dissuading Friedrich from suicide. The victories of Rosbach and Breslau not long afterwards changed the condition of things once again. From this time the prince and the philosopher resumed the

---

* A fact which brings out into strong relief the entirely superfluous luxuries of living of the English residents.

† *Essai sur les Mœurs et l'Esprit des Nations,* introduction section xvi., and chap. iii. and iv.

name, if not the cordiality, of friends. A curious accident put the arbitrament of peace and war for some weeks into the hands of Voltaire. The Prussian king, while inactive in his fortified camp, wrote, as his custom was, a quantity of verse and sent the packet to Ferney. Amongst the mass—good, bad, and indifferent—was a satire on Louis and his mistress. The packet had been opened before reaching its destination.

"Had I been inclined to amuse myself, it depended only on me to set the King of France and the King of Prussia to war in rhyme, which would have been a novel farce on earth. But I enjoyed another pleasure—that of being more prudent than Friedrich. I wrote him word that his Ode was beautiful, but that he ought not to publish it. . . . To make the pleasantry complete I thought it possible to lay the foundation of the peace of Europe on these poetical pieces. My correspondence with the Duc de Choiseul [the French Premier] gave birth to that idea, and it appeared so ridiculous, so worthy of the transactions of the times, that I indulged it, and had the satisfaction of proving on what weak and invisible pivots the destinies of nations turn."

Several letters passed between the three before the danger was averted.

The limited space at our disposal will allow us only rapidly to notice some of the remaining *chefs-d'œuvre* of Voltaire. The celebrated *Encyclopédie*, under the auspices of D'Alembert and Diderot, had been lately commenced. To this great work, to which he looked with some hope as promising a severe assault on ignorance and prejudice, Voltaire contributed a few articles. It is not the place here to narrate the history of the fierce war of words to which the *Encyclopédie* gave birth. It was completed in about fifteen years, in 1775—a memorable year in literature.

"Several men of letters [thus Voltaire briefly describes the project], most estimable by their learning and character, formed an association to compose an immense Dictionary of whatever could enlighten the human mind, and it became an object of commerce with the booksellers. The Chancellor, the Ministry, all encouraged so noble an enterprise. Seven volumes had already appeared, and were translated into English, German, Dutch, and Italian. This treasure, opened by the French to all nations, may be considered as what did us most honour at the time, so much were the excellent articles in the *Encyclopédie* superior to the bad, which also were tolerably numerous. One had little to complain of in the work, except too many puerile declamations unfortunately adopted by the editors, who seized whatever came to hand to swell the work. But all which those editors wrote themselves was good."

The article which was particularly selected by the prosecution was that on the Soul, "one of the worst in the work, written by a poor doctor of the Sorbonne, who killed himself with declaiming, rightly or wrongly, against materialism." The writers, as "encyclopédistes" and "philosophers" were long marked by those titles for the public opprobrium. This general persecution had the effect of uniting that party for common defence. For Voltaire himself an important advantage was secured. Most of the principal men of letters and science, up to this time either avowed enemies or coldly-distant friends, henceforward enrolled themselves under his undisputed leadership.

About the same period he published a number of pieces, prose and verse, directed against his enemies of various kinds, theatrical as well as theological. Amongst the latter, conspicuous by their attacks, but still more so by their punishment, were Fréron and Desfontaines, whose chastisement was such that, according to Macaulay's hyperbolic expression, " scourging, branding, pillorying would have been a trifle to it." It is more pleasing, however, to turn from this fierce war of retaliation, in which neither party was free from blame, to proofs of the real benevolence of his disposition. We can merely note the strenuous efforts he made, unsolicited, on behalf of Admiral Byng and the Comte de Lally, and the still more meritorious labours in the less well-known histories of Calas and Serven. Not by these public acts alone did the man, who has been accused of malignity, discover the humanity of his character : to whose ready assistance in money, as well as in counsel, the unfortunate of the literary tribe and others acknowledged their obligations.

His *Philosophie de l'Histoire*, the prototype of its successors in name at least, was designed to expose that long-established and prevailing idolatry of Antiquity, which received everything bequeathed by it with astounding credulity. The *Philosophie* called forth a numerous host of small critics, to which men who knew, or ought to have known better, allied themselves. Their curious way of maintaining the credit of Antiquity afforded, as may be imagined, the author of the *Defence of my Uncle*, under which title Voltaire chose to defend himself, full scope for the exercise of his unrivalled powers of irony. Warburton, the pedant Bishop of Gloucester, with his odd theories about the " Divine Legation," comes in for a share of this Dunciad sort of immortalisation.

A work of equal merit with the *Philosophie* are the *Questions*, addressed to the lovers of science, upon the *Encyclopœdia*, wherein, in the form of a dictionary, he treats, as the Marquis de Condorcet eloquently describes,

" Successively of theology, grammar, natural philosophy, and literature. At one time he discusses subjects of Antiquity ; at another questions of policy, legislation, and public economy. His style, always animated and seductive, clothed these various subjects with a charm hitherto known to himself alone, and which springs chiefly from the licence with which, yielding to his successive emotions, adapting his style less to his subject than to the momentary disposition of his mind, sometimes he spreads ridicule over objects which seem capable of inspiring only horror, and almost instantaneously hurried away by the energy and sensibility of his soul, he vehemently and eloquently exclaims against abuses which he had just before treated with mockery. His anger is excited by false taste; he quickly perceives that his indignation ought to be reserved for interests more important, and he finishes by laughing in his usual way. Sometimes he abruptly leaves a moral or political discussion for a literary criticism, and in the midst of a lesson on taste he pronounces abstract maxims of the profoundest philosophy, or makes a sudden and terrible attack on fanaticism and tyranny."

It is with his romances that we are here chiefly concerned, since it is in those lighter productions of his genius that he has most especially allowed us to see his opinions upon flesh-eating. In the charming tale of *The Princess of Babylon,* her attendant *Phœnix* thus accounts to his mistress for the silence of his brethren of the inferior races :—

"It is because men fell into the practice of eating us in place of holding converse with and being instructed by us. The barbarians ! Ought they not to have convinced themselves that, having the same organs as they, the same power of feeling, the same wants, the same desires, we have what they call *soul* as well as themselves, that we are their brethren, and that only the wicked and bad deserve to be cooked and eaten? We are to such a degree your brethren that the Great Being, the Eternal and Creative Being, having made a covenant with men*, expressly comprised us in the treaty. He forbad *you* to feed yourselves upon our blood, and *us* to suck yours. The fables of your Lokman, translated into so many languages, will be an everlasting witness of the happy commerce which you formerly had with us. It is true that there are many women among you who are always talking to their Dogs; but they have resolved never to make any answer, from the time that they were forced by blows of the whip to go hunting and to be the accomplices of the murder of our old common friends, the Deer and the Hares and the Partridges. You have still some old poems in which Horses talk and your coachmen address them every day, but with so much grossness and coarseness, and with such infamous words, that Horses who once loved you now detest you. . . . The shepherds of the Ganges, born all equal, are the owners of innumerable flocks who feed in meadows that are perpetually covered with flowers. They are never slaughtered there. It is a horrible crime in the country of the Ganges to kill and eat one's fellows [*semblables*]. Their wool, finer and more brilliant than the most beautiful silk, is the greatest object of commerce in the Orient."

A certain king had the temerity to attack this innocent people :—

"The king was taken prisoner with more than 600,000 men. They bathed him in the waters of the Ganges; they put him on the salutary *régime* of the country, which consists in vegetables, which are lavished by Nature for the support of all human beings. Men, fed upon carnage and drinking strong drinks, have all an empoisoned and acrid blood, which drives them mad in a hundred different ways. Their principal madness is that of shedding the blood of their brothers, and of devastating fertile plains to reign over cemeteries."

Her admirable instructor caused the princess to enter

"A dining-hall, whose walls were covered with orange-wood. The under-shepherds and shepherdesses, in long white dresses girded with golden bands, served

---

* See *Gen.* ix. and *Ecclesiastes* iii., 18, 19.—Note by Voltaire.

her in a hundred baskets of simple porcelain, with a hundred delicious meats, among which was seen no disguised corpse. The feast was of rice, of sago, of semolina, of vermicelli, of maccaroni, of omelets, of eggs in milk, of cream-cheeses, of pastries of every kind, of vegetables, of fruits of perfume and taste of which one has no idea in other climates, and a profusion of refreshing drinks superior to the best wines."

Having occasion to visit the land *par excellence* of flesh-eaters, and being entertained at the house of a certain English lord, the hero, the amiable lover of the princess, is questioned by his host

"Whether they ate 'good roast beef' in the country of the people of the Ganges. The Vegetarian traveller replied to him with his accustomed politeness that they did not eat their brethren in that part of the world. He explained to him the system and diet which was that of Pythagoras, of Porphyry, of Iamblichus ; whereupon *milord* went off into a sound slumber."[*]

Amabed, a young Hindu, writes from Europe to his affianced mistress his impressions of the Christian sacred books and, in particular, of Christian carnivorousness :—

"I pity those unfortunates of Europe who have, at the most, been created only 6,940 years ; while our era reckons 115,652 years [the Brahminical computation]. I pity them more for wanting pepper, the sugar-cane, and tea, coffee, silk, cotton, incense, aromatics, and everything that can render life pleasing. But I pity them still more for coming from so great a distance, among so many perils, to ravish from us, arms in hand, our provisions. It is said at Calicut they have committed frightful cruelties only to procure pepper. It makes the Hindu nature, which is in every way different from theirs, shudder ; their stomachs are carnivorous, they get drunk on the fermented juices of the vine, which was planted, they say, by their Noah. Father Fa-Tutto [one of the missionaries], polished as he is, has himself cut the throats of two little chickens ; he has caused them to be boiled in a cauldron, and has devoured them without pity. This barbarous action has drawn upon him the hatred of all the neighbourhood, whose anger we have appeased only with much difficulty. May God pardon me ! I believe that this stranger would have eaten our sacred Cows, who give us milk, if he had been allowed to do so. A promise has been extorted from him that he will commit no more murders of Hens, and that he will content himself with fresh eggs, milk, rice, and our excellent fruits and vegetables—pistachio nuts, dates, cocoa nuts, almond cakes, biscuits, ananas, oranges, and with everything which our climate produces, blessed be the Eternal ! "

In another letter to his old Hindu teacher from Rome, whither he had been induced to go by the missionaries, speaking of the feasts in that " citadel of the faith," he writes :—

"The dining-hall was grand, convenient, and richly ornamented. Gold and silver shone upon the sideboards. Gaiety and wit animated the guests. But, meantime, in the kitchens blood and fat were streaming in one horrible mass ; skins of quadrupeds, feathers of birds and their entrails, piled up pell-mell, oppressed the heart, and spread the infection of fevers."[†]

---

* See *Lettres d'Amabed à Shastasid.* See also article *Viande* in the *Dictionnaire Philosophique.*
† *La Princesse de Babylone.* Cf. *Dialogue du Chapon et de la Poularde.*

That one who hated and denounced injustice of all kinds, and who sympathised with the suffering of all innocent life, should thus characterise the cruelty of the Slaughter-House is what we might naturally look for; as also that he should denounce the kindred and even worse atrocity of the physiological Laboratory. And it is a strange and unaccountable fact that, amongst the humanitarians of his time, he stands apparently alone in condemnation of the secret tortures of the vivisectionists and pathologists—although, perhaps, the almost universal silence may be attributable, in part, to the very secrecy of the experiments which only recent vigilance has fully detected. Exposing the equally absurd and arrogant denial of reason and intelligence to other animals, and instancing the dog, he proceeds :—

" There are barbarians who seize this dog, who so prodigiously surpasses man in friendship, and nail him down to a table, and dissect him alive to shew you the mezaraic veins. You discover in him all the same organs of feeling as in yourself. Answer me, Machinist [*i.e.*, supporter of the theory of mere mechanical action], has Nature really arranged all the springs of feeling in this animal *to the end that he might not feel?* Has he nerves *that he may be incapable* of suffering? Do not suppose that mpertinent contradiction in Nature."*

To the final triumph which in Paris awaited this champion of the weak, at the advanced age of 84, and the unexampled enthusiam of the people, and the closing act of his eventful life, we can here merely refer. In Berlin, Friedrich ordered a solemn mass in the cathedral church in commemoration of his genius and virtues. A more enduring monument than any conventional mark of human vanity is the legacy which he left to posterity, which will last as long as the French language, and, still more, the humanity embodied in one of his later verses :—

" J'ai fait un peu de bien, c'est mon meilleur ouvrage."

The faults of his character and writings which, for the most part, lie on the surface (one of the most regretable of which was his sometimes servile flattery of men in power, and the only excuse for which was his eagerness to gain them over to moderation and justice) will be deemed by impartial criticism to have been more than counterbalanced by his real and substantial merits. That he allowed his ardent indignation to overmaster the sense of propriety in too many instances, in dealing with subjects which ought to be dealt with in a judicial and serious manner, is that fault in his writings which must always cause the greatest regret. In his discourse at his reception by the French Academy he remarks that " the art of instruction, when it is perfect, in the long run, succeeds better than the art of sarcasm, because Satire dies with those

---

* See article *Bêtes* in the *Dict. Phil.*

who are the victims of it; while Reason and Virtue are eternal." It would have been well, in many instances, had he practised this principle. But, however objectionably his convictions were sometimes expressed, his ardent love of truth and hatred of injustice have secured for him an imperishable fame; while Göthe's estimate of his intellectual pre-eminence—that he has the greatest name in all Literature—is not likely soon to be disputed by Posterity.

XXV.

HALLER. 1708—1777.

THE founder of Modern Physiology was born at Berne. In 1723 he went to Tübingen to study medicine, afterwards to Leyden, where the famous Boerhaave was at the height of his reputation. Twelve years later he received the appointment of physician to the hospital at Berne; but soon afterwards he was invited by George II., as Elector of Hanover, to accept the professorship of anatomy and surgery at the University of Göttingen.

His scientific writings are extraordinarily numerous. From 1727 to 1777 he published nearly 200 treatises. His great work is his *Elements of the Physiology of the Human Body* (in Latin), 1757—1766—the most important treatise on medical science—or at least on anatomy and surgery—up to that time produced. The *Icones Anatomicæ* ("Anatomical Figures") is "a marvellously accurate, well-engraved representation of the principal organs of the human . body." His writings are marked by unusual clearness of meaning, as well as by accurate and deep research.

We wish that we could here stop; but the force of truth compels us to affirm that, for us at least, his reputation, great as it is in science, has been for ever tarnished by his sacrifices—with frightful torture—of innocent victims on the altars of a selfish and sanguinary science.

One plea in extenuation of this callousness in regard to the suffering of other animals, and only one, can be offered in his defence. At this very moment, after all the humanitarian doctrine that has been preached during the century since the death of Haller, tortures of the most cold-blooded kind are being inflicted on tens of thousands of horses, deer, dogs, rabbits, and others, in all the "laboratories" of Europe; while he had neither the prolonged experience of the uselessness of all such unnatural experimentation, of which the vivisectors and pathologists of our day are in possession, nor the same indoctrination of a higher morality, which has been the heritage of these latter days. The scientific barbarity of Haller does not affect the nature of his physiological testimony, which, it might be presumed, ought to be of some weight with his disciples and representatives of the present day. He asserts :—

"This food, then, that I have hitherto described, in which flesh has no part, is salutary ; inasmuch as it fully nourishes a man, protracts life to an advanced period, and prevents or cures such disorders as are attributable to the acrimony or the grossness of the blood." *

XXVI.

COCCHI. 1695—1758.

It might justly provoke expression of feeling stronger than that of astonishment, when we have to record that in South Europe (where climate and soil unite to recommend and render a *humane* manner of living† still more easy than in our colder regions) the followers, or, at all events, the prophets of the Reformed Diet have been conspicuously few. Since, by the *à fortiori* argument, if abundant experience and teaching have proved it to be more conducive to health in higher latitudes, much more is it evident that it must be fitting for the people of those parts of the globe nearer to the Equator.

Italy, which has produced Seneca, Cornaro, and Cocchi, is less obnoxious to the reproach of indifferentism in this most vitally-important branch of ethics than the western peninsula. But the "paradise of Europe" has yet to deserve the more glorious title of "the paradise of Peace," and to atone (if, indeed, it be possible) for the cruel shedding of innocent, and in an especial degree superfluous, blood.

An eminent professor of medicine and of surgery, Antonio Cocchi distinguished himself also as a philologist. He was born at Benevento. Before giving himself up to the practice of medicine he devoted several years to the study of the old and the modern languages of Europe. His knowledge of English helped to bring him into contact with many

---

* *Elements of Physiology.*
† Cf. Virgil's "Magna parens frugum."

men of science in England, some of whom he met on his visit to London.
Returning to Italy he was named Professor of Medicine at Pisa. He
soon left that University for Florence, where he held the chair of
Anatomy as well as of Philosophy. To him Florence was indebted for
its Botanical Society, with which, in conjunction with Micheli, he
endowed it.

He was a voluminous writer.* His *Greek Surgical Books*† contain
valuable extracts from the Greek writers on medicine and surgery not
before published. Amongst other writings may be distinguished his
Treatise on the Use of Cold Baths by the Ancients.‡ The treatise which
gives him a place in this work was published at Florence under the title
of *The Pythagorean Diet: for the Use of the Medical Faculty.*§

Dr. Cocchi begins his little treatise with a eulogy and defence of the
great reformer of Samos, and of his radical revolution in food. He cites
the Greek and Latin writers, and especially the earlier Roman Laws,
the Fannian and the Licinian. He proceeds :—

"True and constant vigour of body is the effect of health, which is much better
preserved with watery, herbaceous, frugal, and tender food, than with *vinous*, abundant,
hard, and gross flesh (*che col carneo vinoso ed unto abundante e duro*). And in a sound
body, a clear intelligence, and desire to suppress the mischievous inclinations (*voglie
dannose*), and to conquer the irrational passions, produces true worth."

Cocchi cites the examples of the Greeks and of the Romans as proof
that the non-flesh diet does not diminish courage or strength :—

"The vulgar opinion, then, which, on health reasons, condemns vegetable food and
so much praises animal food, being so ill-founded, I have always thought it well to
oppose myself to it, moved both by experience and by that refined knowledge
of natural things which some study and conversation with great men have given me.
And perceiving now that such my constancy has been honoured by some learned and
wise physicians with their authoritative adhesion (*della autorevole sequela*), I have
thought it my duty publicly to diffuse the reasons of the Pythagorean diet, regarded
as useful in medicine, and, at the same time, as full of innocence, of temperance, and
of health. And it is none the less accompanied with a certain delicate pleasure,
and also with a refined and splendid luxury (*non è privo nemmeno d'una certa delicate
voluttà e d'un lusso gentile e splendido ancora*), if care and skill be applied in selection
and proper supply of the best vegetable food, to which the fertility and the natural
character of our beautiful country seem to invite us. For my part I have been so
much the more induced to take up this subject, because I have persuaded myself
that I might be of service to intending diet-reformers, there not being, to my
knowledge, any book of which this is the sole subject, and which undertakes exactly
to explain the origin and the reasons of it."

---

* See the *Nouvelle Biographie Universelle.* Didot, Paris.
† *Græcorum Chirurgici Libri.* Firenze, 1754.
‡ *Dissertazione sopra l'uso esterno appresso gli Antichi dell'acqua fredda sul corpo umano.*
Firenze, 1747.
§ *Del Vitto Pithagorico Per Uso Della Medicina: Discorso D'Antonio Cocchi.* Firenze, 1743. A
translation appeared in Paris in 1762 under the title of *Le Régime de Pythagore.*

His special motive to the publication of his treatise, however, was to vindicate the claims of the reformer of Samos upon the gratitude of men :—

" I wished to show that Pythagoras, the first founder of the vegetable regimen, was at once a very great physicist and a very great physician ; that there has been no one of a more cultured and discriminating humanity ; that he was a man of wisdom and of experience ; that his motive in commending and introducing the new mode of living was derived not from any extravagant superstition, but from the desire to improve the health and the manners of men." *

XXVII.

ROUSSEAU. 1712—1778.

FEW lives of writers of equal reputation have been exposed to our examination with the fulness and minuteness of the life of this the most eloquent name in French literature.   With the exception of the great Latin father, St. Augustine, no other leader of thought, in fact, has so entirely revealed to us his inner life, his faults and weaknesses (often sufficiently startling), no less than the estimable parts of his character, and we remain in doubt whether more to lament the infirmities or to admire the candour of the autobiographer.

Jean Jacques Rousseau, son of a Genevan tradesman, had the misfortune to lose his mother at a very early age.   It is to this want of maternal solicitude and fostering care that some of the errors in his after career may perhaps be traced.   After a short experience of school discipline he was apprenticed to an engraver, whose coarse violence must injuriously have affected the nervous temperament of the sensitive child.   Ill-treatment forced him to run away, and he found refuge with Mde. de Warens, a Swiss lady, a convert to Catholicism, who occupies a prominent place in the first period of his *Confessions*.   Influenced by her kindness, and by the skilful arguments of his preceptors at the college at Turin, where she had placed him, the young Rousseau (like Bayle and Gibbon, before and after him, though from a different motive) abjured Protestantism, and, for the moment, accepted, or at least professed, the tenets of the old Orthodoxy.   Dismissed from the college because he refused to take orders, he engaged himself as a domestic servant or valet.

---

* *Del Vitto Pithagorico.* Amongst the heralds and forerunners of Cocchi deserve to be mentioned with honour Ramazzini (1633-1714), who earned amongst his countrymen the title of Hippokrates the Third ; Lessio (in his *Hygiastricon*, or Treatise on Health), in the earlier part of the 17th century ; and Lemery, the French Physician and Member of the Académie, author of *A Treatise on all Sorts of Food*, which was translated into English by D. Hay, M.D., in 1745.

He did not long remain in this position, and he resought the protection of his friend Mde. de Warens at Chambéry. His connexion with his too indulgent patroness terminated in the year 1740. For some years after this his life was of a most erratic, and not always edifying, kind. We find him employed in teaching at Lyons, and at another time acting as secretary to the French Embassy at Venice. In 1745 he came to Paris. There he earned a living by copying music. About this time he met with Thérèse Levasseur, the daughter of his hostess, with whom he formed a lasting but unhappy connexion.

It was in 1748, at the age of ˙36, that he made the acquaintance, at the house of Mde. d'Epinay, of the editors of the *Encyclopédie*, D'Alembert and Diderot, who engaged him to write articles on music and upon other subjects in that first of comprehensive dictionaries. His first independent appearance in literature was in his essay on the question, " Whether the progress of science and of the arts has been favourable to the morals of mankind," in which paradoxically he maintains the negative. It was the eloquence, we must suppose, rather than the reasoning, which gained him the prize awarded by the Académie of Dijon. His next production— a more important one—was his *Discours sur l'Inegalité parmi les Hommes* (" Discourse upon Inequality amongst Men "). In this treatise—the prelude to his more developed *Contrat Social*—Rousseau affirms the paradox of the *natural* school, as it may be termed, which alleged the state of nature—the life of the uncivilised man—to be the ideal condition of the species. His thesis that all men are born with equal rights takes a much more defensible position. In this *Discours* diet is assigned its due importance in relation to the welfare of communities.

The romance of *Julie: ou la Nouvelle Héloise*, which excited an unusual amount of interest, appeared in 1759. *Emile: ou de l'Education* was given to the world three years later. It is the most important of his writings. In the education of Emile, or Emilius, he propounds his ideas upon one of the most interesting subjects which can engage attention—the right training of the young. The earlier part of the book is almost altogether admirable and useful. The later portion is more open to criticism, although not upon the grounds upon which was founded the hostility of the authorities of the day who unjustly condemned the book as irreligious and immoral. Rousseau begins with laying down the principles of a new and more rational method of rearing infants, agreeing, in many particulars, with the system of his predecessor, Locke. At least some of his protests against the unnatural treatment of children were not altogether in vain. Mothers in fashionable ranks of life began to recognise the mischief arising from the common practice of putting their infants out to nurse in place of suckling them themselves. They began also to abandon the absurd custom of confining their limbs in mummy-like bandages. Nor, though long in bearing adequate fruit, were his denunciations of the barbarous severity of parents and schoolmasters without some result. He insists upon the incalculable evils of inoculating the young, according to the almost uuiversal custom, with superstitious beliefs and fancies which grow with the growth of the recipient until they become radically fixed in the mind as by a natural development. Most important of all his innovations in education, and certainly the most heretical, is his recommendation of a pure dietary.

The publication of his treatise on education brought down a storm of persecution and opprobrium upon the author. The *Contrat Social* (in which he seemed to aim at subverting the political and social traditions, as he had in *Emile* the educational prejudices of the venerated Past) appearing soon afterwards added fuel to the flames. Rousseau found himself forced to flee from Paris, and he sought shelter in the territory of Geneva. But the authorities, unmindful of the old reputation of the land of freedom, refusing him an asylum, he proceeded to Neuchâtel, then under Prussian rule, where he was well received. From this retreat he replied to the attacks of the Archbishop of Paris, and addressed a letter to the magistrates of Geneva renouncing his citizenship. He also published *Letters Written from the Mountain*, severely criticising the civil and church government of his native canton. These acts did not tend to conciliate the goodwill of the rulers of the people with whom he had taken refuge. At this moment an object of dislike to all the Continental sovereign powers, he gladly embraced the offer of David Hume to find him an asylum in England. The social and political revolutionist arrived in London in 1766, and took up his residence in a village in Derbyshire. He did not remain long in this country, his irritable temperament inducing him too hastily to suspect the sincerity of the friendship of his host.

The next eight years of his life were passed in comparative obscurity, and in migrating from one place to another in the neighbourhood of Paris. In his solitude gardening and botanising occupied a large part of his leisure hours. It was at this period he made the acquaintance of Bernardin St. Pierre, his enthusiastic disciple, and immortalised as the author of *Paul et Virginie*. His end came suddenly. He had been settled only a few months in a cottage given him by one of his numerous aristocratic friends and admirers, when one morning, feeling unwell, he requested his wife to open the window that he " might once more look on

L

the lovely verdure of the fields," and as he was expressing his delight at the exquisite beauty of the scene and of the skies he fell forward and instantly breathed his last. At his special request his place of burial was chosen on an island in a lake in the Park of Ermondville, a fitting resting-place for one of the most eloquent of the high priests of Nature.

His character (as we have already remarked) is revealed in his *Confessions*—which was written, in part, during his brief exile in England. It, as well as his other productions, shews him to us as a man of extraordinary sensibility, which, in regard to himself, occasionally degenerated into a sort of disease or, in popular language, *morbidness* (a word, by the way, constantly abused by the many who seem to excuse their own insensibility to surrounding evils by stigmatising with that vague expression the acuter feeling of the few), which sometimes assumed the appearance of partial unsoundness of mind. This it was that caused him to suspect and quarrel with his best friends, and which, we may suppose, led him, in his minute dissection of himself, to exaggerate his real moral infirmities.

In summing up his personal character we shall perhaps impartially judge him to have been, on the whole, amiable rather than admirable, of good impulses, and of a naturally humane disposition, cultivated by reading and reflection, but to have been wanting in firmness of mind and in that virtue so much esteemed in the school of Pythagoras—self control. His philosophy is distinguished rather by refinement than by vigour or depth of thought.

It is in the education of the young that Rousseau exerts his eloquence to enforce the importance of a non-flesh diet :—

" One of the proofs that the taste of flesh is not natural to man is the indifference which children exhibit for that sort of meat, and the preference they all give to vegetable foods, such as milk-porridge, pastry, fruits, &c. It is of the last importance not to *denaturalise* them of this primitive taste (*de ne pas dénaturer ce goût primitif*), and not to render them carnivorous, if not for health reasons, at least *for the sake of their character*. For, however the experience may be explained, it is certain that great eaters of flesh are, in general, more cruel and ferocious than other men. This observation is true of all places and of all times. English coarseness is well known.* The Gaures, on the contrary, are the gentlest of men. All savages are cruel, and it is not their morals that urge them to be so ; this cruelty proceeds *from their food* They go to war as to the chase, and treat men as they do bears. Even in England the butchers

---

* Rousseau adds in a note: "I know that the English boast loudly of their humanity and of the good disposition of their nation, which they term ' good nature,' but it is in vain for them to proclaim this far and wide. Nobody repeats it after them. ' Gibbon, in the well-known passage in his xxvith chapter, in which he speculates upon the influence of flesh-eating in regard to the savage habits of the Tartar tribes, quoting this remark of Rousseau, in his ironical way, says : " Whatever we may think of the general observation *we* shall not easily allow the truth of his example."— *Decline and Fall of the Roman Empire*, **xxvi.**

are not received as legal witnesses any more than surgeons.* Great criminals harden themselves to murder by drinking blood.† Homer represents the *Cyclopes*, who were flesh-eaters, as frightful men, and the Lotophagi [Lotus-eaters] as a people so amiable that as soon as one had any dealing with them one straightway forgot everything, even one's country, to live with them."

Rousseau, in a free translation, here quotes a considerable part of Plutarch's *Essay*. He insists, especially, that children should be early accustomed to the pure diet :—

"The further we remove from a natural mode of living the more do we lose our natural tastes ; or rather habit makes a *second* nature, which we substitute to such a degree for the first that none among us any longer knows what the latter is. It follows from this that the most simple tastes must also be the most natural, for they are those which are most easily changed, while by being sharpened and by being irritated by our whims they assume a form which never changes. The man who is yet of no country will conform himself without trouble to the customs of any country whatever, but the man of one country never becomes that of another. This appears to me true in every sense, and still more so applied to taste properly so-called. Our first food was milk. We accustom ourselves only by degrees to strong flavours. At first they are repugnant to us. Fruits, vegetables, kitchen herbs, and, in fine, often broiled dishes, without seasoning and without salt, composed the feasts of the first men. The first time a savage drinks wine he makes a grimace and rejects it ; and even amongst ourselves, whoever has lived to his twentieth year without tasting fermented drinks, cannot afterwards accustom himself to them. We should all be abstinents from alcohol if we had not been given wines in our early years. In fine, the more simple our tastes are the more universal are they, and the most common repugnance is for made-up dishes. Does one ever see a person have a disgust for water or bread ? Behold here the impress of nature ! Behold here, then, our rule of life. Let us preserve to the child as long as possible his primitive taste ; let its nourishment be common and simple ; let not its palate be familiarised to any but natural flavours, and let no exclusive taste be formed. . . . I have sometimes examined those people who attached importance to *good living*, who thought, upon their first awaking, of what they should eat during the day, and described a dinner with more exactitude than Polybius would use in describing a battle. I have thought that all these so-called men were but children of forty years without vigour and without consistence—*fruges consumere nati*.‡ Gluttony is the vice of souls that have no solidity (*qui n'ont point d'étoffe*). The soul of a gourmand is in his palate. He is brought into the world but to devour. In his stupid incapacity he is at home only at his table. His powers of judgment are limited to his dishes. Let us leave him in his employment without regret. Better that for him than any other, as much for our own sakes as for his."§

---

* He corrects this mistake in a note : "One of my English translators has pointed out this error, and both [of my translators] have rectified it. Butchers and surgeons are received as witnesses, but the former are not admitted as jurymen or peers in *criminal* trials, while surgeons are so." Even this amended statement needs revision.

† How the French apostle of humanitarianism and refinement of manners, if he were living, would regard the recently reported practice of French and other physicians of sending their patients to the slaughter-houses to drink the blood of the newly-slaughtered oxen may be more easily imagined than expressed.

‡ Rather *carnes consumere nati*—"born simply to devour."—See *Hor.*, Ep. I., 2.

§ *Emile : ou de l'Education*, II.

In the *Julie: ou la Nouvelle Heloise* he describes his heroine as preferring the innocent feast :—

"Although luxurious in her repasts she likes neither flesh-meat nor ragoûts. Excellent vegetable dishes, eggs, cream, fruits—these constitute her ordinary food ; and, excepting fish, which she likes as much, she would be a true Pythagorean."*

Although he was not a thorough or consistent abstainer, Rousseau speaks with enthusiasm of the pleasures of his frugal repasts, in which, it seems, when he was not seduced by the sumptuous dinners of his fashionable admirers, flesh, as a rule, had no part :—

"Who shall describe, who shall understand, the charm of these repasts, composed of a quartern loaf, of cherries, of a little cheese, and of a half-pint of wine, which we drank together. Friendship, confidence, intimacy, sweetness of soul, how delicious are your seasonings ! "†

XXVIII.

LINNÉ. 1707—1778.

KARL VON LINNÉ, or (according to the antiquated fashion of *Latinising* eminent names still retained) Linnæus, the distinguished Swedish naturalist, and the most eminent name in botanical literature, in a notable manner arrived at his destined immortality in spite of friends and fortune.   Prophecies do not always fulfil themselves, and the estimate of his teachers that he was a hopeless " blockhead," and the prediction that he would be of no intellectual worth in the world (they had advised his parents to apprentice him to a handicraft trade), are a conspicuous instance of the falsification of prophecy.   After one year's course of study at the University of Lund—where he had access to a good library and collections of natural history—he proceeded to the University of Upsala.   There, upon an allowance by his father of £8 a year to meet all his expenses of living, he struggled desperately against the almost insuperable obstacles of extreme poverty, which forced him often to reduce his diet to one meal during the day.   He was then at the age of 20.   At length, by the hospitable friendship of the professor of botany, and a small income derived from a few pupils, Linné found himself free to devote himself to the great labour of his life.   It was in the house of his host (Rudbeck) that he sketched the subject-matter of the important works he afterwards published.   In 1731 he was

---

* *Julie* IV., *Lettre* 10. See also her protests against shooting and fishing.

† *Confessions.* One of his friends, Dussault, surprised him, it seems, on one occasion eating a "cutlet." Rousseau, conscious of the betrayal of his principles, "blushed up to the whites of his eyes." (See Gleïzè's *Thalysic.*) In truth, as we have already observed, his principles on the subject of *dietetics*, as on some other matters, were better than his practice. His sensibility was always greater than his strength of mind

commissioned by his university to explore the vegetable life of Lapland. Within the space of five months he traversed alone, and with slender provision, some 4,000 miles. The result of this laborious expedition was his *Flora Laponica.*

Three years later, with the sum of fifteen pounds, which he had with great difficulty gathered together, he set out in search of some university where he might obtain the necessary degree of doctor in medicine at the least outlay, in order to gain a living by the practice of physic. He found the object of his search in Holland. In that country he met with a hospitable reception. During his residence in Holland he came over to England, and visited the botanical collections at Oxford and Eltham, with which the Swedish *savant,* it seems, had not much reason to be satisfied. Returning to Sweden, he began practice as a physician at the age of 31, and he lectured, by Government appointment, upon botany and mineralogy at Stockholm. His fame had now become European. He was in correspondence with some of the most eminent scientific men throughout the world. Books and collections were sent to him from every quarter, and his pupils supplied him with the results of their explorations in the three continents. He was elected to the Professorship of Medicine at Upsala, and (a vain addition to his real titles) he was soon afterwards " ennobled."

The productions of his genius and industry during the twenty years from 1740 were astonishingly numerous. Besides his *Systema Naturæ* and *Species Plantarum,* his two most considerable works, he wrote a large number of dissertations, afterwards collected under the title of *Amœnitates Academicæ*—" Academic Delights." Everything he wrote was received with the greatest respect by the scientific world. Upon his death the whole University of Upsala united in showing respect to his memory ; sixteen doctors of medicine, old pupils, bearing the " pall," and a general mourning was ordered throughout the land of his birth.

The scientific merits of Linné are his exactness and conciseness in classification. He reduced to something like order the chaotic and pedantic systems of his predecessors, which were prolix and overladen with names and classes. If the science still labours under the stigma of needless pedantry, the fault lies not with himself, but with his successors. Linné's evidence to the scientific truth of Vegetarianism is brief but *pregnant :*—

" This species of food [fruits and farinacea] is that which *is most suited* to man, as is proved by the series of quadrupeds, analogy, wild men, apes, the structure of the mouth, of the stomach, a·d of the hands."*

———

* *Amœnitates Academicæ,* x., 8.

XXIX.

BUFFON. 1707—1788.

An eminent instance of perversity of logic—of which, by the way, the history of human thought supplies too many examples—is that of the well-known author of the *Histoire Naturelle*, a work which (highly interesting as it is, and always will be, by reason of the detailed and generally accurate delineation of the characters and habits of the various forms of animated nature, and by reason of the graces of style of that French classic) is, from a strictly scientific point of view, of not always the most reliable authority. Although Buffon has depicted as forcibly as well can be conceived the low position in Nature of the carnivorous tribes, and not a few of the evils arising from human addiction to carnivorousness, yet, by a strange perversion of the facts of comparative physiology, he has chosen to enlist himself amongst the apologists of that degenerate mode of living. But facts are stronger than prejudices, and his very candid *admissions*, which we shall here quote, speak sufficiently for themselves :—

"Man [says he] knows how to use, as a master, his power over [other] animals. He has selected those whose flesh *flatters his taste.* He has made domestic slaves of them. He has multiplied them more than Nature could have done. He has formed innumerable flocks, and by the cares which he takes in propagating them he *seems** to have acquired the right of sacrificing them for himself. But he extends that right *much beyond* his needs. For, independently of those species which he has subjected, and of which he disposes at his will, he makes war also upon wild animals, upon birds upon fishes. He does not even limit himself to those of the climate he inhabits. He seeks at a distance, even in the remotest seas, new meats, and entire Nature seems scarcely to suffice for his intemperance and the inconsistent variety of his appetites.

"*Man alone consumes and engulfs more flesh than all other animals put together. He is, then, the greatest destroyer, and he is so more by abuse than by necessity.* Instead of enjoying with moderation the resources offered him, in place of dispensing them with equity, in place of repairing in proportion as he destroys, of renewing in proportion as he annihilates, the rich man makes all his boast and glory in *consuming,* all his splendour in destroying, in one day, at his table, more material (*plus de biens*) than would be necessary for the support of several families. He abuses equally other animals and his own species, the rest of whom live in famine, languish in misery, and work only to satisfy the immoderate appetite and the still more insatiable vanity of this human being who, *destroying others by want, destroys himself by excess.*

"And yet Man might, like other animals, live upon vegetables. *Flesh is not a better nourishment than grains or bread.* What constitutes true nourishment, what contributes

---

* This little word " seems " here, as in very many other controversies, has a vast importance and needs a double emphasis.

to the nutrition, to the development, to the growth, and to the support of the body, is not that brute matter which, to our eyes, composes the texture of flesh or of vegetables, but it is those organic molecules which both contain ; since the ox, in feeding on grass, acquires as much flesh as man or as animals who live upon flesh and blood. . . . The essential source is the same ; it is the same matter, it is the same organic molecules which nourish the Ox, Man, and all animals. . . . It results from what we have just said that Man, whose stomach and intestines are not of a very great capacity relatively to the volume of his body, could not live simply upon grass. Nevertheless *it is proved by facts that he could well live upon bread. vegetables, and the grains of plants*, since we know entire nations and classes of men to whom religion forbids to feed upon anything that has life."

To the ordinary apprehension all this might seem *primâ facie* conclusive evidence of the non-necessariness of the food of the richer classes of the community. But, unhappily, Buffon seems to have considered himself as holding a brief to defend his clients, the flesh-eaters, in the last resort, and, accordingly, in spite of these admissions, which to an unbiassed mind might appear conclusive argument for the relinquishment of flesh as food, he proceeds to contradict himself by adding :—

"But these examples, supported even by the authority of Pythagoras [and he might have added many later names of equal authority], and recommended by some physicians too friendly to a reformed diet (*trop amis de diète !)*, appear to me not sufficient to convince us that it would be for the advantage of human health (*qu'il y eût à gagner pour la santè des hommes*) and for the multiplication of the human species to live upon vegetables and bread only, for so much the stronger reason, that the poor country people, whom the luxury of the cities and towns and the extravagant waste of tables reduce to this mode of living, languish and die off sooner than persons of the middle class, to whom inanition and excess are equally unknown !" *

In stigmatising, in the following sentence, the cruel rapacity of the lower carnivorous tribes, Buffon consciously or unconsciously stamps the same stigma upon the carnivorous human animal :—

"*After Man*, the animals who live only upon flesh are the greatest destroyers. They are at once the enemies of Nature and the rivals of Man."†

---

* Buffon here entirely ignores the true cause of the "inanition" of the poor classes of the community. It is not the want of *flesh*-meats, but the want of all solid and nutritious *meat* of any kind, which is to be found amply in the abundant stores supplied by Nature at first hand in the various parts of the vegetable world. Were the poor able to procure, and were they instructed how best to use, the most nourishing of the various *farinacea*, fruits, and kitchen herbs, supplied by the home and foreign markets, we should hear nothing or little of the scandalous scenes of starvation which are at present of daily occurrence in our midst. The example of the Irish living upon a few potatoes and buttermilk, or of the Scotch peasantry, instanced by Adam Smith, proves how all-sufficient would be a diet judiciously selected from the riches of the vegetable world. For, *à fortiori*, if the Irish, living thus meagrely, not only support life, but exhibit a *physique* which, in the last century, called forth the admiration of the author of *The Wealth of Nations*, might not our English poor thrive upon a richer and more substantial vegetable diet which could easily be supplied but for the astounding indifference of the ruling classes?

† *Hist. Naturelle, Le Bœuf.*

XXX.

HAWKESWORTH. 1715—1773.

BEST known as the editor of *The Adventurer*—a periodical in imitation of the *Spectator, Rambler,* &c.—which appeared twice a week during the years 1752-54. Johnson, Warton, and others assisted him in this undertaking, which has the honour of being one of the first periodicals which have ventured to denounce the cruel barbarism of "Sport," and the papers by Hawkesworth upon that subject are in striking contrast with the usual tone and practice of his contemporaries and, indeed, of our own times.

In 1761 he published an edition of Swift's writings, with a life which received the praise of Samuel Johnson (in his *Lives of the Poets*), and it is a passage in that book which entitles him to a place here. In 1773 he was entrusted by the Government of the day with the task of compiling a history of the recent voyages of Captain Cook. He also translated the *Aventures de Télémaque* of Fénélon. The coarseness and repulsiveness of the dishes of the common diet seldom have been stigmatised with greater force than by Dr. Hawkesworth. His expressions of abhorrence are conceived quite in the spirit of Plutarch :—

" Among other dreadful and disgusting images which Custom has rendered familiar, are those which arise from eating animal food. He who has ever turned with abhorrence from the skeleton of a beast which has been picked whole by birds or vermin, must confess that *habit* alone could have enabled him to endure the sight of the mangled bones and flesh of a dead carcase which every day cover his table. And he who reflects on the *number* of lives that have been sacrificed to sustain his own, should enquire by *what* the account has been balanced, and whether his life is become proportionately of more value by the exercise of virtue and by the superior happiness which he has communicated to [more] reasonable beings." *

---

* Edition of Swift's Works. Canon Sydney Smith, equally celebrated as a *bon-vivant* and as a wit, at the termination of his life writes thus to his friend Lord Murray: " You are, I hear, attending more to diet than heretofore. If you wish for anything like happiness in the *fifth* act of life *eat and drink about one-half what you could eat and drink.* Did I ever tell you my calculation about eating and drinking ? Having ascertained the weight of what I could live upon, so as to preserve health and strength, and what I did live upon, I found that, between ten and seventy years of age, I had eaten and drunk *forty-four horse wagon-loads of meat and drink more than would have preserved me in life and health !* The value of this mass of nourishment I considered to be worth seven thousand pounds sterling. It occurred to me *that I must, by my voracity, have starved to death fully a hundred persons.* This is a frightful calculation, but irresistibly true." Commentary upon this candid statement is superfluous. *Ab uno disce omnes.* If amongst the richer classes the ordinary liver may consume a somewhat smaller quantity of life during his longer or shorter existence, at all events the *sum total* must be a sufficiently startling one for all who may have the courage and candour to reflect upon this truly appalling subject. Another thought irresistibly suggests itself. What *proportion* of human lives thus supported is of any real value in the world ?

XXXI.

PALEY.   1743—1805.

WITH the exception of Joseph Butler, perhaps the ablest and most interesting of English orthodox theologians. As one of the very few of this numerous class of writers who seem seriously to be impressed with the difficulty of reconciling orthodox *dietetics* with the higher moral and religious instincts, Paley has for social reformers a title to remembrance, and it is as a moral philosopher that he has a claim upon our attention.

The son of a country curate, Paley began his career as tutor in an academy in Greenwich. He had entered Christ's College, Cambridge, as "sizar." Being senior wrangler of his year, he was afterwards elected a Fellow of his college. His lectures on moral philosophy at the University contained the germs of his most useful writing. After the usual previous stages, finally he received the preferment of the Archdeaconry of Carlisle. The failure of the most eminent of the modern apologists of dogmatic Christianity to attain the highest rewards of ecclesiastical ambition, and the refusal of George III. to promote "pigeon" Paley when it was proposed to that reactionary prince to make so skilful a controversialist a bishop—a refusal founded on the famous apology for monarchy in the *Moral and Political Philosophy*—is well known.

The most important, by far, of his writings, is the *Elements of Moral and Political Philosophy* (1785). He founds moral obligation upon principles of utility. In politics he asserts the grounds of the duties of rulers and ruled to be based upon the same far-reaching consideration, and upon this principle he maintains that as soon as any Government has proved itself corrupt or negligent of the public good, whatever may have been the alleged legitimacy of its original authority, the right of the governed to put an end to it is established. "The final view of all national politics," he affirms, "is [ought to be] to produce the greatest quantity of happiness." The comparative boldness, indeed, of certain of his disquisitions on Government alarmed not a little the political and ecclesiastical dignitaries of the time. His adhesion to the programme of Clarkson and the anti-slavery "fanatics" (as that numerically insignificant band of reformers was styled) did not tend, it may be presumed, to counteract the damaging effects of his political philosophy.

In his *Natural Theology* (1802), his best theological production, he labours to establish the fact of benevolent design from observation of the various phenomena of nature and life. Whatever estimate may be formed of the success of this undertaking, there can be no question of the ability and eloquence of the accomplished pleader; and the book proves him, at least, to have acquired a surprising amount of physiological and anatomical knowledge.   It is justly described by Sir J. Mackintosh as " the wonderful work of a man who, after sixty, had studied anatomy in order to write it."   Of the *Evidences* (1790-94)—the most popularly known of his writings—the considerable literary merit is in somewhat striking contrast, in regard to clearness and simplicity of style, with the ordinary productions of the evidential school.

We are concerned now with the *Moral and Political Philosophy*. It has been already stated that it is based upon the principles of utilitarianism.   As for personal moral conduct, he justly considered it to be vastly influenced by early custom ; or, as he expresses it, the art of life consists in the right "setting of our habits."

On the subjoined examination of the question of the lawfulness or otherwise of flesh-eating, his ultimate refuge in an alleged biblical authority (forced upon him, apparently, by the necessity of his position rather than by personal inclination) confirms rather than weakens his preceding candid *admissions*, which sufficiently establish our position :—

"A right to the flesh of animals. This is a *very different claim* from the former ['a right to the fruits or vegetable produce of the earth'].   *Some* excuse seems necessary for the pain and loss which we occasion to [other] animals by restraining them of their liberty, mutilating their bodies, and, at last, putting an end to their lives for our pleasure or convenience.

" The reasons alleged in vindication of this practice are the following—that the several species of animals being created to prey upon one another* affords a kind of analogy to prove that the human species were intended to feed upon them ; that, if let alone, they would overrun the earth, and exclude mankind from the occupation of it ;† that they are requited for what they suffer at our hands by our care and protection.

" Upon which reasons I would observe that the analogy contended for *is extremely lame*, since [the carnivorous] animals have no power to support life by any other means, and *since we have, for the whole human species might subsist entirely upon fruit, pulse, herbs, and roots, as many tribes of Hindus‡ actually do*. The two other reasons may be valid reasons, as far as they go, for, no doubt, if men had been supported entirely by vegetable food a great part of those animals who die to furnish our tables would never have lived § but they by no means justify our right over the lives of other animals to the extent to which we exercise it. What danger is there, *e.g.*, of fish

---

* In reply to this sort of apology it is obvious to ask—" Have the *frugivorous* races, who form no inconsiderable proportion of the *mammals*, no claim to be considered ?"

† To this very popular fallacy it is necessary only to object that Nature may very well be supposed able to maintain the proper balance for the most part. For the rest, man's proper duty is to harmonise and regulate the various conditions of life, as far as in him lies, not indeed by satisfying his selfish propensities, but by assuming the part of a benevolent and beneficent superior. To this we may add with some force, that man appeared on the scene within a comparatively very recent geological period, so that the Earth fared, it seems, very well without him for countless ages.

‡ And, in point of fact, two-thirds at least of the whole human population of our globe.

interfering with us in the occupation of their element, or what do we contribute to their support or preservation ?

" *It seems to me that it would be difficult to defend this right by any arguments which the light and order of Nature afford,* and that we are beholden for it to the permission recorded in Scripture (*Gen.* ix., 1, 2, 3). To Adam and his posterity had been granted, at the creation, ' every green herb for meat,' and nothing more. In the last clause of the passage now produced the old grant is recited and extended to the flesh of animals—' even as the green herb, have I given you all things.' But this was not until after the Flood. The inhabitants of the antediluvian world had therefore no such permission that we know of. Whether they actually refrained from the flesh of animals is another question. Abel, we read, was a keeper of sheep, and for what purpose he kept them, except for food, is difficult to say (unless it were sacrifice). Might not, however, some of the stricter sects among the antediluvians be scrupulous as to this point ? And might not Noah and his family be of this description ? For, it is not probable that God should publish a permission to authorise a practice which had never been disputed."‖

Thus far as regards the *moral* aspect of the subject. Dealing with the social and economical view, Paley, untrammelled by professional views, is more decided. In his chapter, *Of Population and Provision, &c.,* he writes :—

" The natives of Hindustan being confined, by the laws of their religion, to the use of vegetable food, and requiring little except rice, which the country produces in plentiful crops ; and food, in warm climates, composing the only want of life, these countries are populous under all the injuries of a despotic, and the agitations of an unsettled, Government. If any revolution, or what would be called perhaps *refinement of manners (!),* should generate in these people a taste for the flesh of animals, similar to what prevails amongst the Arabian hordes—should introduce flocks and herds into grounds which are now covered with corn—should teach them to account a certain portion of this species of food amongst the necessaries of life—the population from this single change would suffer in a few years a great diminution, and this diminution would follow in spite of every effort of the laws, or even of any improvement that might take place in their civil condition. In Ireland the simplicity of

---

§ This popular excuse is perhaps the feeblest and most disingenuous of all the defences usually made for flesh-eating. Can the mere gift of life compensate for all the horrible and frightful sufferings inflicted, in various ways, upon their victims by the multiform selfishness and barbarity of man? To what unknown, as well as known, tortures are not every day the victims of the slaughter-house subjected? From their birth to their death, the vast majority—it is too patent a fact—pass an existence in which freedom from suffering of one kind or other—whether from insufficient food or confined dwellings on the one hand, or from the positive sufferings endured *in transitu* to the slaughter-house by ship or rail, or by the brutal savagery of cattle-drivers, &c.— is the exception rather than the rule.

‖ *Moral and Political Philosophy,* i., 2. It is deeply to be deplored that Dr. Paley is in a very small minority amongst christian theologians, of candour, honesty, and feeling sufficient to induce them to dispute at all so orthodox a thesis as the right to slaughter for food. That he is compelled, by the force of truth and honesty, to abandon the popular pretexts and subterfuges, and to seek refuge in the *supposed* authority of the book of *Genesis,* is significant enough. Of course, to all reasonable minds, such a course is tantamount to giving up the defence of kreophagy altogether ; and, if it were not for theological necessity, it would be sufficiently surprising that Paley's intelligence or candour did not discover that if flesh-eating is to be defended on biblical grounds, so, by parity of reasoning, are also to be defended—slavery, polygamy, wars of the most cruel kind, &c.

living alone maintains a considerable degree of population under great defects of police, industry, and commerce. . . . Next to the mode of living, we are to consider ' the quantity of provision suited to that mode, which is either raised in the country or imported into it,' for this is the order in which we assigned the causes of population and undertook to treat of them. Now, if we measure the quantity of provision by the number of human bodies it will support in due health and vigour, this quantity, the extent and quality of the soil from which it is raised being given, will depend greatly upon the *kind.* For instance, a piece of ground capable of supplying animal food sufficient for the subsistence of ten persons *would sustain, at least, the double of that number with grain, roots, and milk.*

" The first resource of savage life is in the flesh of wild animals. Hence the numbers amongst savage nations, compared with the tract of country which they occupy, are universally small, because this species of provision is, of all others, supplied in the slenderest proportion. The next step was the invention of pasturage, or the rearing of flocks and herds of tame animals. This alteration added to the stock of provision much. But the last and *principal improvement was to follow, viz., tillage, or the artificial production of corn, esculent plants, and roots.* This discovery, whilst it changed the quality of human food, augmented the quantity in a vast proportion.

" So far as the state of population is governed and limited by the quantity of provision, perhaps there is no single cause that affects it so powerfully as the kind and quality of food which chance or usage hath introduced into a country. In England, notwithstanding the produce of the soil has been of late considerably increased by the enclosure of wastes and the adoption, in many places, of a more successful husbandry, yet we do not observe a corresponding addition to the number of inhabitants, the reason of which appears to me to be the more general consumption of animal food amongst us. Many ranks of people whose ordinary diet was, in the last century, prepared almost entirely from milk, roots, and vegetables, now require every day a considerable portion of the flesh of animals. *Hence a great part of the richest lands of the country are converted to pasturage.* Much also of the bread-corn, which went directly to the nourishment of human bodies, now only contributes to it by fattening the flesh of sheep and oxen. *The mass and volume of provisions are hereby diminished,* and what is gained in the amelioration of the soil is lost in the quality of the produce

" This consideration teaches us that tillage, as an object of national care and encouragement, is universally preferable to pasturage, because the kind of provision which it yields goes much farther in the sustentation of human life. Tillage is also recommended by this additional advantage—that it *affords employment to a much more numerous peasantry.* Indeed pasturage seems to be the art of a nation, either imperfectly civilised, as are many of the tribes which cultivate it in the internal parts of Asia, or of a nation, like Spain, declining from its summit by luxury and inactivity."*

Elsewhere Paley asserts that " luxury in dress or furniture is universally preferable to luxury *in eating,* because the articles which constitute the one are more the production of human art and industry than those which supply the other."

---

* *The Principles of Moral and Political Philosophy,* xii., 11. See, amongst others, the philosophical reflections of Mr. Greg in his *Enigmas of Life,* Appendix. But the subject has been most fully and satisfactorily dealt with by Professor Newman in his various Addresses.

XXXII.

ST. PIERRE. 1737—1814.

PRINCIPALLY known as the author of the most charming of all idyllic romances—*Paul et Virginie*. Beginning his career as civil engineer he afterwards entered the French army. A quarrel with his official superiors forced him to seek employment elsewhere, and he found it in the Russian service, where his scientific ability received due recognition.

Encouraged by the esteem in which he was held, he formed the project of establishing a colony on the Caspian shores, which should be under just and equal laws. St. Pierre submitted the scheme to the Russian Minister, who, as we should be apt to presume, did not receive it too favourably. He then went to Poland in the vain expectation of aiding the people of that hopelessly distracted country in throwing off the foreigners' yoke. Failing in this undertaking, and despairing, for the time, of the cause of freedom, we next find him in Berlin and in Vienna. He had also previously visited Holland, in which great refuge of freedom he had been received with hospitality. In Paris, upon his return to France, his project of a free colony found better reception than in St. Petersburg—owing, perhaps, to the not altogether disinterested sympathy of the Government with the recently revolted American colonies. To further his plans he accepted an official post in the Ile de France, intending eventually to proceed to Madagascar, where was to be realised his long-cherished idea. On the voyage he discovered that his associates had formed a very different design from his own—to engage in the slave traffic. Separating from these nefarious speculators, he landed in the Ile de France, where he remained two years. It is to the experiences of this part of his life that we owe his *Paul et Virginie*, the scenes of which are laid in that tropical island.

Returning home once again, he made the acquaintance of D'Alembert and of other leading men of letters in Paris, and, particularly, of Rousseau, his philosophical master. At the period of the Great Revolution of 1789, St. Pierre lost his post as superintendent of the Royal Botanical Gardens under the old Bourbon Government, and he found himself reduced to poverty; and although his sympathies were with the party of constitutional, though not of radical, reform, the supremacy of the extreme revolutionists (1792-1794) exposed him to some hazard by reason of his known deistic convictions. Upon the establishment of the reactionary revolution of the Empire, St. Pierre recovered his former post, and, with the empty honour of the Imperial Cross, he received the more solid benefit of a pension and other emoluments.

His writings have been collected and published in two quarto volumes (Paris, 1836). Of these, after his celebrated romance, perhaps the most popular is *La Chaumière Indienne* ("The Indian or Hindu Cottage"). His principal productions are *Etudes de la Nature* ("Studies of Nature"), *Vœux d'un Solitaire* ("Aspirations of a Recluse"), *Voyage à L'Ile de France* ("Voyage to Mauritius"), and *L'Arcadie* ("Arcadia"). His merits consist in a certain refinement of feeling, in charming eloquence in description of natural beauty, and in the humane spirit which breathes in his writings. Of the *Paul et Virginie* he tells us—

"I have proposed to myself great designs in that little work. . . . I have desired to reunite to the beauty of Nature, as seen in the tropics, the moral beauty of a small society of human beings. I proposed to myself thereby to demonstrate several great truths ; amongst others this—that our happiness consists in living according to Nature and Virtue."

He assures us that the principal characters and events he describes are by no means only the imaginings of romance. In truth, it seems difficult to believe that the genius of the author alone could have impressed so wonderful an air of reality upon merely fictitious scenes. The popularity of the story was secured at once in the author's own country, and it rapidly spread throughout Europe. *Paul et Virginie* was successively translated into English, Italian, German, Dutch, Polish, Russian, and Spanish. It became the fashion for mothers to give to their children the names of its hero and heroine, and well would it have been had they also adopted for them that method of innocent living which is the real, if too generally unrecognised, secret of the fascinating power of the book.

It is thus that he eloquently calls to remembrance the *natural* feasts of his young heroine and hero :—

"Amiable children ! thus in innocence did you pass your first days. How often in this spot have your mothers, pressing you in their arms, thanked Heaven for the consolation you were preparing for them in their old age, and for the happiness of seeing you enter upon life under so happy auguries ! How often, under the shadow of these rocks, have I shared, with them, your out-door repasts *which had cost no animals their lives.* Gourds full of milk, of newly-laid eggs, of rice cakes upon banana leaves, baskets laden with potatoes, with mangoes, with oranges, with pomegranates, with bananas, with dates, with ananas, offered at once the most wholesome meats, the most beautiful colours, and the most agreeable juices. The conversation was as refined and gentle as their food."

The humaneness of their manners had attracted to the charming arbour, which they had formed for themselves, all kinds of beautiful birds, who sought there their daily meals and the caresses of their human protectors. Our readers will not be displeased to be reminded of this charming scene :—

"Virginie loved to repose upon the slope of this fountain, which was decorated with a pomp at once magnificent and wild. Often would she come there to wash the household linen beneath the shade of two cocoa-nut trees. Sometimes she led her goats to feed in this place ; and, while she was preparing cheese from their milk, she pleased herself in watching them as they browsed the herbage upon the precipitous sides of the rocks, and supported themselves in mid-air upon one of the jutting points as upon a pedestal. Paul, seeing that this spot was loved by Virginie, brought from the neighbouring forest the nests of all sorts of birds. The fathers and mothers of these birds followed their little ones, and came and established themselves in this new colony. Virginie would distribute to them from time to time grains of rice, maize, and millet. As soon as she appeared, the blackbirds, the *bengalis*, whose flight is so gentle, the cardinals, whose plumage is of the colour of fire, quitted their bushes ; parroquets, green as emerald, descended from the neighbouring lianas, partridges ran along under the grass—all advanced pell-mell up to her feet like domestic hens. Paul and she delighted themselves with their transports of joy, with their eager appetites, and with their loves."

In his views upon national education, St. Pierre invites the serious attention of legislators and educators to the importance of accustoming the young to the nourishment prescribed by Nature :—

"They [the true instructors of the people] will accustom children to the vegetable *régime.* The peoples living upon vegetable foods, are, of all men, the handsomest, the most vigorous, the least exposed to diseases and to passions, and they whose lives last longest. Such, in Europe, are a large proportion of the Swiss. The greater part of the peasantry who, in every country, form the most vigorous portion of the people, eat very little flesh-meat. The Russians have multiplied periods of fasting and days of abstinence, from which even the soldiers are not exempt ; and yet they resist all kinds of fatigues. The negroes, who undergo so many hard blows in our colonies, live upon manioc, potatoes, and maize alone. The Brahmins of India, who frequently reach the age of one hundred years, eat only vegetable foods. It was from the Pythagorean sect that issued Epaminondas, so celebrated by his virtues ; Archytas, by his genius for mathematics and mechanics ; Milo of Crotona, by his strength of body. Pythagoras himself was the finest man of his time, and, without dispute, the most enlightened, since he was the father of philosophy amongst the Greeks. Inasmuch as the non-flesh diet introduces many virtues and excludes none, it will be well to bring up the young upon it, since it has so happy an influence upon the beauty of the body and upon the tranquility of the mind. This regimen prolongs childhood, and, by consequence, human life.*

"I have seen an instance of it in a young Englishman aged fifteen, and who did not appear to be twelve years of age. He was of a most interesting figure, of the most robust health, and of the most sweet disposition. He was accustomed to take very long walks. He was never put out of temper by any annoyance that might happen. His father, Mr. Pigott, told me that he had brought him up entirely upon the Pythagorean regimen, the good effects of which he had known by his own experience. He had formed the project of employing a part of his fortune, which was considerable, in establishing in English America a society of dietary reformers who should be engaged in educating, under the same regimen, the children of the colonists in all the arts which bear upon agriculture. Would that this educational scheme, worthy of

---

* Compare the similar observation of Flourens, Secretary of the French Academy of Sciences, in his *Treatise on the Longevity of Man* (Paris, 1812). He quotes Cornaro, Lessio, Haller, and other authorities on the reformed regimen.

the best and happiest times of Antiquity, might succeed ! Physically, it suits a warlike people no less than an agricultural one. The Persian children, of the time of Cyrus, and by his orders, were nourished upon bread, water, and vegetables. . . . It was with these children, become men, that Cyrus made the conquest of Asia. I observe that Lycurgus introduced a great part of the physical and moral regimen of the Persian children into the education of those of the Lacedemonians."(*Etudes.*)*

Of the many practical witnesses of this period, more or less interesting, for the sufficiency, or rather superiority, of the reformed regimen, four names stand out in prominent relief—Franklin, Howard, Swedenborg, Wesley—prominent either for scientific ability or for philanthropic zeal. To his early resolution to betake himself to frugal living, Benjamin Franklin, then in a printer's office in Boston, attributes mainly his future success in life.†

It was to his pure dietary that the great Prison Reformer assigns his immunity, during so many years, from the deadly jail-fever, to the infection of which he fearlessly exposed himself in visiting those hotbeds of *malaria*—the filthy prisons of this country and of continental Europe. (See the correspondence of John Howard—*passim.*) Equally significant is the testimony of the eminent founder of Methodism whose almost unexampled energy and endurance, both of mind and body, during some fifty years of continuous persecution, both legal and popular, were supported (as he informs us in his *Journals*) mainly by abstinence from gross foods ; while, in regard to Emanuel Swedenborg, if abstinence does not assume so prominent a place in his theological or other various writings as might have been expected from his special opinions, the cause of such silence must be referred not to personal addiction to an *anti-spiritualistic* nourishment (for he himself was notably frugal) but to preoccupation of mental faculties which seem to have been absorbed in the elaboration of his well-known spiritualistic system.

---

* He well exposes the fatal mischief of *emulation* (in place of love of truth and of love of know ledge, for its own sake) in schools which tends to intensify, if not produce, the *selfism* dominant in all ranks of the community. Not the least meritorious of his exhortations to Governments is his desire that they would employ themselves in such useful works as the general planting of trees, producing nourishing foods, in place of devastating the earth by wars, &c.

† The reason, as given by himself, for his abandonment in after years of his self-imposed reform, is worthy neither of his philosophic acumen nor of his ordinary judgment. It seems that on one occasion, while his companions were engaged in sea-fishing, he observed that the captured fish, when opened, revealed in its interior the remains of another fish recently devoured. The young printer seemed to see in this fact the ordinance of Nature, by which living beings live by slaughter, and the justification of human carnivorousness. (See *Autobiography.*) This was, how- ever, to use the famous Sirian's phrase, "to reason badly ; " for the sufficient answer to this alleged justification of man's flesh-eating propensity is simply that the fish in question was, by natural organisation, *formed* to prey upon its fellows of the sea, whereas man is *not formed* by Nature for feeding upon his fellows of the land ; and, further, that the larger proportion of *terrestrials* do not live by slaughter.

The limits of this work do not permit us to quote all the many writers of the eighteenth century whom philosophy, science, or profounder feeling urged *incidentally* to question the necessity or to suspect the barbarism of the Slaughter-House. But there are two names, amongst the highest in the whole range of English philosophic literature, whose expression of opinion may seem to be peculiarly noteworthy—the author of the *Wealth of Nations* and the historian of the *Decline and Fall of the Roman Empire.*

" It may, indeed, be doubted [writes the founder of the science of Political Economy] whether butchers' meat is *anywhere* a necessary of life. Grain and other vegetables, with the help of milk, cheese, and butter, or oil (where butter is not to be had), it is known from experience, can, *without any butchers' meat, afford the most plentiful, the most wholesome, the most nourishing, and the most invigorating diet.*" †

As for the reflections of the first of historians, who seems always carefully to guard himself from the expression of any sort of emotion not in keeping with the character of an impartial judge and unprejudiced spectator, but who, on the subject in question, cannot wholly repress the *natural* feeling of disgust, they are sufficiently significant. Gibbon is describing the manners of the Tartar tribes :—

" The thrones of Asia have been repeatedly overturned by the shepherds of the North, and their arms have spread terror and devastation over the most fertile and warlike countries of Europe. On this occasion, as well as on many others, the sober historian is forcibly awakened from a pleasing vision, and is compelled, with some reluctance, to confess that the pastoral manners, which have been adorned with the fairest attributes of peace and innocence, are much better adapted to the fierce and cruel habits of a military life.

" To illustrate this observation, I shall now proceed to consider a nation of shepherds and of warriors in the three important articles of (1) their diet, (2) their habitations, and (3) their exercises. 1. The corn, or even the rice, which constitutes the ordinary and wholesome food of a civilised people, can be obtained only by the patient toil of the husbandman. Some of the happy savages who dwell between the tropics are plentifully nourished by the liberality of Nature ; but in the climates of the North a nation of shepherds is reduced to their flocks and herds. The skilful practitioners of the medical art will determine (if they are able to determine) how far the temper of the human mind may be affected by the use of animal or of vegetable food ; and

---

† *Wealth of Nations* iii., 341. See, too, Sir Hans Sloane (*Natural History of Jamaica*, i., 21, 22), who enumerates almost every species of vegetable food that has been, or may be, used for food, in various parts of the globe ; the philosophic French traveller, Volney (*Voyages*), who, in comparing flesh with non-flesh feeders, is irresistibly forced to admit that the "habit of shedding blood, or even of seeing it shed, corrupts all sentiment of humanity ; " the Swedish traveller Sparrman, the disciple of Linné, who corrects the astonishing physiological errors of Buffon as to the human digestive apparatus ; Anquetil (*Récherches sur les Indes*), the French translator of the *Zend-Avesta* who, from his sojourn with the vegetarian Hindus and Persians, derived those more refined ideas which caused him to discard the coarser Western living ; and Sir F. M. Eden (*State of the Poor*).

M

whether the common association of carnivorous and cruel deserves to be considered in any other light than that of an innocent, perhaps a salutary, prejudice of humanity Yet if it be true that the sentiment of compassion is imperceptibly weakened by the sight and practice of domestic cruelty, we may observe that *the horrid objects which are disguised by the arts of European refinement* are exhibited in their naked and most disgusting simplicity in the tent of a Tartar shepherd. The Oxen or the Sheep are slaughtered by the same hand from which they were accustomed to receive their daily food, and the bleeding limbs are served, with very little preparation, on the table of their unfeeling murderers."*

To the poets, who claim to be the interpreters and priests of Nature, we might, with justness, look for celebration of the anti-materialist living. Unhappily we too generally look in vain. The prophet-poets—Hesiod, Kalidâsa, Milton, Thomson, Shelley, Lamartine—form a band more noble than numerous. Of those who, not having entered the very sanctuary of the temple of humanitarianism, have been content to officiate in its outer courts, Burns and Cowper occupy a prominent place. That the latter, who felt so keenly

> " The persecution and the pain
> That man inflicts on all inferior kinds
> Regardless of their plaints,"

and who has denounced with so eloquent indignation the pitiless wars " waged with defenceless innocence," and the protean shapes of human selfishness, should yet have stopped short of the *final* cause of them all, would be inexplicable but for the blinding influence of habit and authority. Nevertheless, his picture of the savagery of the Slaughter-House, and of some of its associated cruelties, is too forcible to be omitted :

> "To make him sport,
> To justify the phrensy of his wrath,
> *Or his base gluttony*, are causes good
> And just, in his account, why bird and beast
> Should suffer torture, and the stream be dyed
> With blood of their inhabitants impaled.
> Earth groans beneath the burden of a war
> Waged with defenceless Innocence : while he,
> *Not satisfied to prey on all around,*
> *Adds tenfold bitterness to death by pangs*
> *Needless, and first torments ere he devours.*
> Now happiest they who occupy the scenes
> The most remote from his abhorred resort.

.   .   .   .   .   .

---

* *History of the Decline and Fall of the Roman Empire*, xxvi. Notwithstanding Gibbon's expression of horror, we shall venture to remark that the "unfeeling murderers" of the Tartar steppes, in slaughtering each for himself, are more just than the *civilised* peoples of Europe, with whom a pariah-class is set apart to do the cruel and degrading work of the community.

Witness at his feet
The Spaniel dying for some venial fault,
Under dissection of the knotted scourge :
Witness the patient Ox, with stripes and yells
Driven to the slaughter, goaded as he runs
To madness, while the savage at his heels
Laughs at the frantic sufferer's fury spent
Upon the heedless passenger o'erthrown.
He, too, is witness—noblest of the train
Who waits on Man—the flight-performing Horse :
With unsuspecting readiness he takes
His murderer on his back, and, pushed all day,
With bleeding sides, and flanks that heave for life,
To the far-distant goal arrives, and dies !
So little mercy shows, who needs so much !
Does Law—so jealous in the cause of Man [?]—
Denounce no doom on the delinquent ? None." *

XXXIII.

OSWALD. 1730—1793.

AMONGST the less known prophets of the new Reformation the author of the *Cry of Nature*—one of the most eloquent appeals to justice and right feeling ever addressed to the conscience of men—deserves an honourable place. Of the facts of his life we have scanty record. He was a native of Edinburgh. At an early age he entered the English army as a private soldier, but his friends soon obtained for him an officer's commission. He went to the East Indies, where he distinguished himself by his remarkable courage and ability. He did not long remain in the military life; and, having sold out, he travelled through Hindustan to inform himself of the principles of the Brahmin and Buddhist religions of the peninsula, whose dress as well as milder manners he assumed upon his return to England.

During his stay in this country he uniformly abstained from all flesh meats, and so great, we are told, was his abhorrence of the Slaughter-House, that, to avoid it or the butcher's shop, he was accustomed to make a long *détour*. His children were brought up in the same way.

---

* *The Task.* When Cowper wrote this (in 1782) the Law was entirely silent upon the rights of the lower animals to protection, It was not until nearly half a century later that the British Legislature passed the first Act (and it was a very partial one) which at all considered the rights of any non-human race. Yet Hogarth's *Four Stages of Cruelty*—to say nothing of literature—had been several years before the world. It was passed by the persistent energy and courage of one man—an Irish member—who braved the greatest amount of scorn and ridicule, both within and without the Legislature, before he succeeded in one of the most meritorious enterprises ever undertaken. Martin's Act has been often amended or supplemented, and always with no little opposition and difficulty.

In 1790, like some others of the more enthusiastic class of his country-men, he espoused the cause of the Revolution, and went to Paris. By introducing some useful military reforms he gained distinction amongst the Republicans, and he received an important post. He seems to have fallen, with his sons, fighting in La Vendée for the National Cause.

The author, in his preface, tells us that—

"Fatigued with answering the inquiries and replying to the objections of his friends with respect to the singularity of his mode of life, he conceived that he might consult his ease by making, once for all, a public apology for his opinions. . . . The author is very far from entertaining a presumption that his slender labours (crude and imperfect as they are now hurried to the press) will ever operate an effect on the public mind ; and yet, when he considers the natural bias of the human heart to the side of mercy,* and observes, on all hands, the barbarous governments of Europe giving way to a better system of things, he is inclined to hope that the day is beginning to approach when the growing sentiment of peace and goodwill towards men will also embrace, in a wide circle of benevolence, the lower orders of life.

"At all events, the pleasing persuasion that his work may have contributed to *mitigate* the ferocities of prejudice, and to *diminish*, in some degree, the great mass of misery which oppresses the lower animal world, will, in the hour of distress, convey to the author's soul a consolation which the tooth of calumny will not be able to empoison."

A noble and true inspiration nobly and eloquently used! The arguments, by which he attempts to reach the better feeling of his readers, are drawn from the deepest source of morality. Having given a beautiful picture of the tempting and alluring character of Fruits, he exclaims in his poetic-prose :—

"But far other is the fate of animals. For, alas ! when they are plucked from the tree of Life, suddenly the withered blossoms of their beauty shrink to the chilly hand of Death. Quenched in his cold grasp expires the lamp of their loveliness, and struck by the livid blast of loathed putrefaction, their comely limbs are involved in ghastly horror. Shall we leave the living herbs to seek, in the den of death, an obscene aliment ? Insensible to the blooming beauties of Pomona—unallured by the fragrant odours that exhale from her groves of golden fruits—unmoved by the nectar of Nature, by the ambrosia of innocence—shall the voracious vultures of our impure appetites speed along those lovely scenes and alight in the loathsome sink of putre-faction to devour the remains of other creatures, to load with cadaverous rottenness a wretched stomach ?"

He repeats Porphyry's appeal to the consideration of human interests themselves—

---

* The term "Mercy," it is important to observe, is one of those words of ambiguous meaning, which are liable, in popular parlance, to be misused. It seems to have a double origin—from *misericordia,* "Pity" (its better parentage), and *merces,* "Gain," and, by deduction, "Pardon" granted for some consideration. It is in this latter sense that the term seems generally to be used in respect of the non-human races. But it is obvious to object that "pardon," applicable to *criminals,* can have no meaning as applied to the innocent. *Pity* or *Compassion,* still more *Justice*—these are the terms properly employed.

"And is not the human race itself highly interested to prevent the habit of spilling blood? For, will the man, habituated to violence, be nice to distinguish the vital tide of a quadruped from that which flows from a creature with two legs? Are the dying struggles of a Lamb less affecting than the agonies of any animal whatever? Or, will the ruffian who beholds unmoved the supplicatory looks of innocence itself, and, reckless of the Calf's infantine cries, pitilessly plunges in her quivering side the murdering knife, will he turn, I say, with horror from human assassination?

> ' What more advance can mortals make in sin,
> So near perfection, who with blood begin?
> Deaf to the calf who lies beneath the knife,
> Looks up, and from the butcher begs her life.
> Deaf to the harmless kid who, ere he dies,
> All efforts to procure thy pity tries,
> And imitates, in vain, thy children's cries.
> Where will he stop?'

"From the practice of slaughtering an innocent animal of another species to the murder of man himself the steps are neither many nor remote. This our forefathers perfectly understood, who ordained that, in a cause of blood, no butcher should be permitted to sit in jury. . . . .

"But from the nature of the very human heart arises the strongest argument in behalf of the persecuted beings. Within us there exists a rooted repugnance to the shedding of blood, a repugnance which yields only to Custom, and which even the most inveterate custom can seldom entirely overcome. Hence the ungracious task of shedding the tide of life (for the gluttony of the table) has, in every country, been committed to the lowest class of men, and their profession is, in every country, an object of abhorrence.

"They feed on the carcass without remorse, because the dying struggles of the butchered victim are secluded from their sight—because his cries pierce not their ears —because his agonising shrieks sink not into their souls. But were they forced, with their own hands, to assassinate the beings whom they devour, who is there among us who would not throw down the knife with detestation, and, rather than embrue his hands in the murder of the lamb, consent for ever to forego the accustomed repast? What then shall we say? Vainly planted in our breast is this abhorrence of cruelty— this sympathetic affection for innocence? Or do the feelings of the heart point to the command of Nature more unerringly than all the elaborate subtlety of a set of men who, at the shrine of science, have sacrificed the dearest sentiments of humanity?"

This eloquent vindicator of the rights of the oppressed of the non-human races here addresses a scathing rebuke to the torturers of the vivisection-halls, as well as to those who abuse Science by attempting to enlist it in the defence of slaughter.

"You, the sons of modern science, who court not Wisdom in her walks of silent meditation in the grove—who behold her not in the living loveliness of her works, but expect to meet her in the midst of obscenity and corruption—you, who dig for knowledge in the depths of the dunghill, and who expect to discover Wisdom enthroned amid the fragments of mortality and the abhorrence of the senses—you, that with cruel violence interrogate trembling Nature, who plunge into her maternal bosom the butcher-knife, and, in quest of your nefarious science, delight to scrutinise the fibres of agonising beings, you dare also to violate the human form, and holding up the entrails of men, you exclaim, 'Behold the bowels of a carnivorous animal!' Barbarians! to these very bowels I appeal against your cruel dogmas—to these bowels which Nature hath sanctified to the sentiments of pity and of gratitude, to the yearnings of kindred, to the melting tenderness of love.

'Mollissima corda
Humano generi dare se Natura fatetur,
Quæ *lachrymas* dedit: hæc nostri pars optima sensus.'*

" Had Nature intended man to be an animal of prey, would she have implanted in his breast an instinct so adverse to her purpose? . . . Would she not rather, in order to enable him to brave the piercing cries of anguish, have wrapped his ruthless heart in ribs of brass, and with iron entrails have armed him to grind, without shadow of remorse, the palpitating limbs of agonising life? But has Nature winged the feet of men with fleetness to overtake the flying prey? And where are his fangs to tear asunder the beings destined for his food? Does the lust of carnage glare in his eye-balls? Does he scent from afar the footsteps of his victim? Does his soul pant for the feast of blood? Is the bosom of men the rugged abode of bloody thoughts, and from the den of Death rush forth, at sight of other animals, his rapacious desires to slay, to mangle, and to devour?

"But come, men of scientific subtlety, approach and examine with attention this dead body. It was late a playful Fawn, who skipping and bounding on the bosom of parent Earth, awoke in the soul of the feeling observer a thousand tender emotions. But the butcher's knife has laid low the delight of a fond mother, and the darling of Nature is now stretched in gore upon the ground. Approach, I say, men of scientific subtlety, and tell me, does this ghastly spectacle whet your appetite? But why turn you with abhorrence? Do you then yield to the combined evidence of your senses, to the testimony of conscience and common sense; or with a show of rhetoric, pitiful as it is perverse, will you still persist in your endeavour to persuade us that to murder an innocent being is not cruel nor unjust, and that to feed upon a corpse is neither filthy nor unfitting?"

Amid the dark scenes of barbarism and cold-blooded indifferentism to suffering innocence, there are yet the glimmers of a better nature, which need but the life-giving impulse of a true religion and philosophy :—

"And yet those channels of sympathy for inferior animals, long—a very long—custom has not been able altogether to stifle. Even now, notwithstanding the narrow, joyless, and hard-hearted tendency of the prevailing superstitions ; even now we discover, in every corner of the globe, some good-natured *prejudice* in behalf of [certain of] the persecuted animals ; we perceive, in every country, certain privileged animals, whom even the ruthless jaws of gluttony dare not to invade. For, to pass over unnoticed the vast empires of India and of China, where the lower orders of life are considered as relative parts of society, and are protected by the laws and religion of the natives,† the Tartars abstain from several kinds of animals ; the Turks are

---

* The observation of a *non-Christian* moralist *(Juvenal*, xv.) It is the motto chosen by Oswald for his title page.

‡ In the Hindu sacred scriptures, and especially in the teaching of the great founder of the most extensive religion on the globe, this regard for non-human life, however originating, is more obvious than in any other sacred books. But it is most charmingly displayed in that most interesting of all Eastern poetry and drama—*Sakuntala; or The Fatal Ring*, of the Hindu Kalidâsa, the most frequently translated of all the productions of Hindu literature. We may refer our readers also to *The Light of Asia*, an interesting versification of the principal teaching of Sakya-Muni or Gautama.

charitable to the very dog, whom they abominate; and even the English peasant pays towards the *red-breast* an inviolable respect to the rights of hospitality.

"Long after the perverse practice of devouring the flesh of animals had grown into inveterate habit among peoples, there existed still in almost every country, and of every religion, and of every sect of philosophy, a wiser, a purer, and more holy class of men who preserved by their institutions, by their precepts, and by their example, the memory of primitive innocence [?] and simplicity. The Pythagoreans abhorred the slaughter of any animal life; Epicurus and the worthiest part of his disciples bounded their delights with the produce of their garden; and of the first Christians several sects abominated the feast of blood, and were satisfied with the food which Nature, unviolated, brings forth for our support. . . . . .

"Man, in a state of nature, is not, apparently, much superior to other animals. His organisation is, without doubt, extremely happy; but then the dexterity of his figure is counterpoised by great advantages in other beings. Inferior to the Bull in force, and in fleetness to the Dog, the *os sublime*, or erect front, a feature he bears in common with the Monkey, could scarcely have inspired him with those haughty and magnificent ideas which the pride of human refinement thence endeavours to deduce. Exposed, like his fellow-creatures, to the injuries of the air, urged to action by the same physical necessities, susceptible of the same impressions, actuated by the same passions, and equally subject to the pains of disease and to the pangs of dissolution, the simple savage never dreams that his nature was so much more noble, or that he drew his origin from a purer source or more remote than the other animals in whom he saw a resemblance so complete.

"Nor were the simple sounds by which he expressed the singleness of his heart at all fitted to flatter him into that fond sense of superiority over the beings whom the unreasoning insolence of cultivated ages absurdly styles *mute*. I say absurdly styles *mute;* for with what propriety can that name be applied, for example, to the little sirens of the groves, to whom Nature has granted the strains of ravishment—the soul of song? Those charming warblers who pour forth, with a moving melody which human ingenuity vies with in vain, their loves, their anxiety, their woes. In the ardour and delicacy of his amorous expressions, can the most impassioned, the most respectful, human lover surpass the 'glossy kind,' as described by the most beautiful of all our poets?

"And, indeed, has not Nature given to almost every being the same spontaneous signs of the various affections? Admire we not in other animals whatever is most eloquent in man—the tremor of desire, the tear of distress, the piercing cry of anguish, the pity-pleading look—expressions which speak to the soul with a feeling which words are feeble to convey?"

The whole of the little book of which the above extracts are properly representative, breathes the spirit of a true religion. We shall only add that it exhibits almost as much learning and valuable research as it exhibits justness of thought and sensibility—enriched, as it is, by copious illustrative notes.*

---

* *The Cry of Nature: an Appeal to Mercy and to Justice on behalf of the Persecuted Animals.* By John Oswald. London, 1791.

XXXIV.

HUFELAND. 1762—1836.

NOT entitled to rank among the greater prophets who have had the penetration to recognise the *essential* barbarism, no less than the unnaturalness, of Kreophagy (disguised, as it is, by the arts of civilisation), this most popular of all German physicians, with the Cornaros and Abernethys, may yet claim considerable merit as having, in some degree, sought to stem the tide of unnatural living, which, under less gross forms indeed than those of the darker ages of dietetics, and partially concealed in the refinements of Art, is more difficult to be resisted by reason of its very disguise. If the renaissance of Pythagorean dietetics had already dawned for the deeper thinkers, the age of science and of reason, as regards the mass of accredited teachers, was yet a long way off; and to all pioneers, even though they failed to clear the way entirely, some measure of our gratitude is due.

Christian Wilhelm Hufeland is one of the most prolific of medical writers. Having studied medicine at Jena and at Gottingen he took the degree of doctor in 1783. At Jena he occupied a professorial chair (1793), and came to Berlin five years later, where he was entrusted with the superintendence of the Medical College. Both as practical physician and as professor, Hufeland attained a European reputation. The French Academy of Sciences elected him one of its members. His numerous writings have been often reprinted in Germany. Among the most useful are : (1) *Popular Dissertations upon Health* (Leipsig, 1794) ; (2) *Makrobiotik : oder die Kunst das Menschliche Leben zu Verlängern* (Jena, 1796), a celebrated work which has been translated into all the languages of Europe* ; (3) *Good Advice to Mothers upon the most Important Points of the Physical Education of Children in the First Years* (Berlin, 1799); (4) *History of Health, and Physical Characteristics of our Epoch* (Berlin, 1812)†. Of Hufeland's witness to the general superiority of the *Naturgemässe Lebensweise* the following sentences are sufficiently representative :

" The more man follows Nature and obeys her laws the longer will he live. The further he removes from them *(je weiter er von ihnen abweicht)* the shorter will be his duration of existence. . . . Only inartificial, simple nourishment promotes health and long life, while mixed and rich foods but shorten our existence. . . . We frequently find a very advanced old age amongst men who from youth upwards have lived, for the most part, upon the vegetable diet, and, perhaps, have never tasted flesh."‡

---

* *Long Life, or the Art of Prolonging Human Existence.*
† See the *Nouvelle Biographie Universelle* for complete enumeration of his writings.
‡ *Makrobiotik.*

XXXV.

RITSON. 1761—1830.

KNOWN to the world generally as an eminent antiquarian and, in particular, as one of the earliest and most acute investigators of the sources of English romantic poetry, for future times his best and enduring fame will rest upon his at present almost forgotten Moral Essay upon Abstinence—one of the most able and philosophical of the ethical expositions of anti-kreophagy ever published.

His birthplace was Stockton in the county of Durham. By profession a conveyancer, he enjoyed leisure for literary pursuits by his income from an official appointment. During the twenty years from 1782 to 1802 his time and talents were incessantly employed in the publication of his various works, antiquarian and critical. His first notable critique was his *Observations* on Warton's *History of English Poetry*, in the shape of a letter to the author (1782), in which his critical zeal seems to have been in excess of his literary amenity. Of other literary productions may be enumerated his *Remarks on the Commentators of Shakspere ; A Select Collection of English Songs, with a Historical Essay on the Origin and Progress of National Songs* (1783) *; Ancient Songs from the Time of King Henry III. to the Revolution* (1790), reprinted in 1829—perhaps the most valuable of his archæological labours ; *The English Anthology* (1793) *; Ancient English Metrical Romances*, and *Bibliographia Poetica*, a catalogue of English poets from the 12th to the 16th century, inclusive, with short notices of their works. These are only some of the productions of his industry and genius.

We give the origin of his adhesion to the Humanitarian Creed as recorded by himself in one of the chapters of his Essay, in which, also, he introduces the name of an ardent and well-known humanitarian reformer :—

" Mr. Richard Phillips,* the publisher of this compilation, a vigorous, healthy, and well-looking man, has desisted from animal food for upwards of twenty years ; and the compiler himself, induced to serious reflection by the perusal of Mandeville's *Fable of the Bees*, in the year 1772, being the 19th year of his age, has ever since, to the revisal of these sheets [1802], firmly adhered to a milk and vegetable diet; having, at least, never tasted, during the whole course of those thirty years, any flesh, fowl, or fish, or anything, to his knowledge, prepared in or with those substances or any extract from them, unless, on one occasion, when tempted by wet, cold, and hunger

---

* Afterwards Sir Richard Phillips, whose admirable exposition of his reasons for abandoning flesh-eating, published in the *Medical Journal*, July 1811, is quoted in its due place.

in the south of Scotland, he ventured to eat a few potatoes dressed under roasted flesh, nothing less repugnant to his feelings being obtainable ; or, except by ignorance or imposition, unless, it may be, in-eating eggs, which, however, deprives no animal of life, although it may prevent some from coming into the world to be murdered and devoured by others."*

Ritson begins his Essay with a brief review of the opinions of some of the old Greek and Italian philosophers upon the origin and constitution of the world, and with a sketch of the position of man in Nature relatively to other animals.   Amongst others he cites Rousseau's Essay *Upon Inequality Amongst Men*. He then demonstrates the unnaturalness of flesh-eating by considerations derived from Physiology and Anatomy, and from the writings of various authorities ; the fallacy of the prejudice that flesh-meats are necessary or conducive to strength of body, a fallacy manifest as well from the examples of whole nations living entirely, or almost entirely, upon non-flesh food, as from those of numerous individuals whose cases are detailed at length. He quotes Arbuthnot, Sir Hans Sloane, Cheyne, Adam Smith, Volney, Paley, and others. Next he insists upon the ferocity or coarseness of mind directly or indirectly engendered by the diet of blood :—

"That the use of animal food disposes man to cruel and ferocious actions is a fact to which the experience of ages gives ample testimony, The Scythians, from drinking the blood of their cattle, proceeded to drink that of their enemies. The fierce and cruel disposition of the wild Arabs is supposed chiefly, if not solely, to arise from their feeding upon the flesh of camels : and as the gentle disposition of the natives of Hindustan is probably owing, in great degree, to temperance and abstinence from animal food, so the common use of this diet, with other nations, has, in the opinion of M. Pagès, intensified the natural tone of their passions ; and he can account, he says, upon no other principle, for the strong, harsh features of the Mussulmen and the Christians compared with the mild traits and placid aspect of the Gentoos. 'Vulgar and uninformed men,' it is observed by Smellie, 'when pampered with a variety of animal food, are much more choleric, fierce, and cruel in their tempers, than those who live chiefly upon vegetables.' This affection is equally perceptible in other animals—'An officer, in the Russian service, had a bear whom he fed with bread and oats, but never gave him flesh. A young hog, however, happening to stroll near his cell, the bear got hold of him and pulled him in ; and, after he had once drawn blood and tasted flesh, he became unmanageable, attacking every person who came near him, so that the owner was obliged to kill him.'—[*Memoirs of P. H. Bruce.*] It was not, says Porphyry, from those who lived on vegetables that robbers, or murderers, or tyrants have proceeded, but from flesh-eaters.† Prey being almost the sole object of quarrel

---

* *Abstinence from Animal Food a Moral Duty*, IX. Ritson, in a note, quotes the expression of surprise of a French writer, that whereas abstinence "from blood and from things strangled" is especially and solemnly enjoined by the immediate successors of Christ, in a well-known pro-hibition, yet this sacred obligation is daily "made of none effect" by those calling themselves *Christians.*

† "I have known," says Dr. Arbuthnot, "more than one instance of irascible passions having been much subdued by a vegetable diet."—Note by Ritson.

amongst carnivorous animals, while the frugivorous live together in constant peace and harmony, it is evident that if men were of this latter kind, they would find it much more easy to subsist happily."

"The barbarous and unfeeling sports (as they are called) of the English—their horse-racing, hunting, shooting, bull and bear baiting, cock-fighting, * prize-fighting, and the like, all proceed from their immoderate addiction to animal food. Their natural temper is thereby corrupted, and they are in the habitual and hourly commission of crimes against nature, justice, and humanity, from which a feeling and reflective mind, unaccustomed to such a diet, would revolt, but in which they profess to take delight. The kings of England have from a remote period, been devoted to hunting; in which pursuit one of them, and the son of another lost his life. James I., according to Scaliger, was merciful, except at the chase, where he was cruel, and was very much enraged when he could not catch the Stag. 'God,' he used to say, 'is enraged against me, so that I shall not have him.' Whenever he had caught his victim, he would put his arm all entire into his belly and entrails. This anecdote may be paralleled with the following of one of his successors : 'The hunt on Tuesday last, (March 1st, 1784), commenced near Salthill, and afforded a chase of upwards of fifty miles. His Majesty was present at the death of the stag near Tring, in Herts. It is the first deer that has been ran to death for many months ; and when opened, the heart strings were found to be quite rent, as is supposed, with the force of running.'†

Siste, vero, tandem carnifex ! The slave trade, that abominable violation of the rights of Nature, is most probably owing to the same cause, as well as a variety of violent acts, both national and personal, which usually are attributed to other motives. In the sessions of Parliament, 1802, a majority of the members voted for the continuance of bull-baiting, and some of them had the confidence to plead in favour of it."‡

---

* Written in 1802. Since that time the "pastime" of worrying bulls and bears, has in this country become illegal and extinct. Cock-fighting, though illegal, seems to be still popular with the "sporting" classes of the community.

† *General Advertiser*, March 4th, 1784. Since Ritson quoted this from the newspaper of his day, 80 years ago, the same scenes of equal and possibly of still greater barbarity have been recorded in our newspapers, season after season, of the royal and other hunts, with disgusting monotony of detail. Voltaire's remarks upon this head are worthy of quotation : "It has been asserted that Charles IX. was the author of a book upon hunting. It is very likely that if this prince had cultivated less the art of torturing and killing other animals, and had not acquired in the forests the habit of seeing blood run, there would have been more difficulty in getting from him the order of St. Bartholomew. The chase is one of the most sure means for blunting in men the sentiment of pity for their own species ; an effect so much the more fatal, as those who are addicted to it, placed in a more elevated rank, have more need of this bridle."—*Œuvres* LXXII., 213. In Flaubert's remarkable story of *La Légende de St. Julien* the hero "developes by degrees a propensity to bloodshed. He kills the mice in the chapel, the pigeons in the garden, and soon his advancing years gave him opportunity of indulging this taste in hunting. He spends whole days in the chase, caring less for the 'sport' than for the slaughter." One day he shoots a Fawn, and while the despairing mother, "looking up to heaven, cried with a loud voice, agonising and human," St. Julien remorselessly kills her also. Then the male parent, a noble-looking Stag, is shot last of all ; but, advancing, nevertheless, he comes up to the terrified murderer, and "stopped suddenly, and with flaming eyes and solemn tone, as of a just judge, he spoke three times, while a bell tolled in the distance, 'Accursed one ! ruthless of heart ! thou shalt slay thy father and mother also,' and tottering and closing his eyes he expired." The blood-stained man on one occasion is followed closely by all the victims of his wanton cruelty, who press around him with avenging looks and cries. He fulfils the prophecy of the Stag, and murders his parents.—See *Fortnightly Review*, April, 1878.

‡ It is scarcely necessary to remind our readers that a quarter of a century later (1827), when Martin had the courage to introduce the first bill for the prevention of cruelty to certain of the domesticated animals (a very partial measure after all), the humane attempt was greeted by an almost universal shout of ridicule and derision, both in and out of the Legislature.

Ritson enforces his observations upon this head by citing Plutarch, Cowper, and Pope (in the *Guardian, No. 61*—a most forcible and eloquent protest against the cruelties of "sport" and of gluttony).* In his fifth chapter he traces the origin of human sacrifices to the practice of flesh eating :—

"Superstition is the mother of Ignorance and Barbarity. Priests began by persuading people of the existence of certain invisible beings, whom they pretended to be the creators of the world and the dispensers of good and evil ; and of whose wills, in fine, they were the sole interpreters. Hence arose the necessity of sacrifices [ostensibly] to appease the wrath or to procure the favour of imaginary gods, but in reality to gratify the gluttonous and unnatural appetites of *real* demons. Domestic animals were the first victims. These were immediately under the eye of the priest, and he was pleased with their taste. This satisfied for a time ; but he had eaten of the same things so repeatedly, that his luxurious appetite called for variety. He had devoured the sheep, and he was now desirous of devouring the shepherd. The anger of the gods—testified by an opportune thunderstorm, was not to be assuaged but by a sacrifice of uncommon magnitude. The people tremble, and offer him their enemies, their slaves, their parents, their children, to obtain a clear sky on a summer's day, or a bright moon by night. When, or upon what particular occasion, the first human being was made a sacrifice is unknown, nor is it of any consequence to enquire. Goats and bullocks had been offered up already, and the transition was easy from the 'brute' to the man. The practice, however, is of remote antiquity and universal extent, there being scarcely a country in the world in which it has not, at some time or other, prevailed."

He supports this probable thesis by reference to Porphyry, the most erudite of the later Greeks, who repeats the accounts of earlier writers upon this matter, and by a comparison of the religious rites of various nations, past and present. Equally natural and easy was the step from the use of non-human to that of human bodies :—

"As human sacrifices were a natural effect of that superstitious cruelty which first produced the slaughter of other animals, so is it equally natural that those accustomed to eat the 'brute' should not long abstain from the man. More especially as, when roasted or broiled upon the altar, the appearance, savour, and taste of both, would be nearly, if not entirely the same. But, from whatever cause it may be deduced, nothing can be more certain than that the eating of human flesh has been a practice in many parts of the world from a very remote period, and is so, in some countries, at this day. That it is a consequence of the use of other animal food there can be no doubt, as it would be impossible to find an instance of it among people who were accustomed solely to a vegetable diet. The progress of cruelty is rapid. Habit renders it familiar, and hence it is deemed *natural*.

"The man who, accustomed to live on roots and vegetables, first devoured the flesh of the smallest mammal, committed a greater violence to his own nature than the most beautiful and delicate woman, accustomed to other animal flesh, would feel in shedding the blood of her own species for sustenance ; possessed as they are of exquisite feel-

---

* See Appendix.

ings, a considerable degree of intelligence, and even, according to her own religious system, of a *living soul.* That this is a principle in the social disposition of mankind, is evident from the deliberate coolness with which seamen, when their ordinary provisions are exhausted, sit down to devour such of their comrades as chance or contrivance renders the victim of the moment; a fact of which there are but too many, and those too well-authenticated instances. Such a crime, which no necessity can justify, would never enter the mind of a starving Gentoo, nor, indeed, of anyone who had not been previously accustomed to other animal flesh. Even among the Bedouins, or wandering Arabs of the desert—according to the observation of the enlightened Volney—though they so often experience the extremity of hunger, the practice of devouring human flesh was never heard of."

In the two following chapters Ritson traces a large proportion of human diseases and suffering, physical and mental, to indulgence in unnatural living. He cites Drs. Buchan, Goldsmith, Cheyne, Stubbes *(Anatomy of Abuses,* 1583*)*, and Sparrman the well-known pupil of Linné *(Voyages).*

In his ninth chapter, he gives a copious catalogue of "nations and of individuals, past and contemporary, subsisting entirely upon vegetable foods"—not the least interesting part of his work. Some of the most eminent of the old Greek and Latin philosophers and historians are quoted, as well as various modern travellers, such as Volney and Sparrman. Especially valuable are the enquiries of Sir F. M. Eden *(State of the Poor)*, who, in a comparison of the dietary of the poor, in different parts of these islands, proves that flesh has, or at all events *had,* scarcely any share in it—a fact which is still true of the agricultural districts, manifest not only by the commonest observation, but also by scientific and official enquiries of late years.

Of individual cases, two of the most interesting are those of John Williamson of Moffat, the discoverer of the famous chalybeate spring, who lived almost to the age of one hundred years, having abstained from all flesh-food during the last fifty years of his life,* and of John Oswald,

---

* Quoted from an article in the *Gentleman's Magazine,* (August, 1787), signed *Etonensis,* who, amongst other particulars, states of the hero of his sketch that he was "one of the most original geniuses who have ever existed. . . . He was well skilled in natural philosophy, and might be said to have been a moral philosopher, not in *theory* only, but in strict and uniform *practice.* He was remarkably humane and charitable; and, though poor, was a bold and avowed enemy to every species of oppression. . . . Certain it is, that he accounted the murder (as he called it) of the meanest animal, except in self defence, a very criminal breach of the laws of nature; insisting that the creator of all things had constituted man not the *tyrant,* but the lawful and limited *sovereign,* of the inferior animals, who, he contended, answered the ends of their being better than their little despotic lord. . . . He did not think it

'Enough
In this late age, advent'rous to have touched
Light on the precepts of the Samian Sage,'

for he acted in strict conformity with them. . . . His vegetable and milk diet afforded him, in particular, very sufficient nourishment; for when I last saw him, he was still a tall, robust, and rather corpulent man, though upwards of fourscore." He was reported it seems, to be a the author of *The Cry of Nature.* It is in this part of his work that Ritson narrates the history of his own conversion and dietetic experiences, and of his well-known publisher, Mr. R. Phillips.

—————◆•◆—————

XXXVI.
NICHOLSON. 1760—1825.

AMONG the least known, but none the less among the most estimable, of the advocates of the rights of the oppressed species and the heralds of the dawn of a better day, the humble Yorkshire printer, who undertook the unpopular and unremunerative work of publishing to the world the sorrows and sufferings of the non-human races, claims our high respect and admiration. He has also another title (second only to his humanitarian merit) to the gratitude of posterity as having been the originator of cheap literature of the best class, and of the most instructive sort, which, alike by the price and form, was adapted for wide circulation.

George Nicholson was born at Bradford. He early set up a printing press, and began the publication of his *Literary Miscellany,* "which is not, as the name might lead one to suppose, a magazine, but a series of choice anthologies, varied by some of the gems of English literature. The size is a small 18mo., scarcely too large for the waistcoat pocket. The printing was a beautiful specimen of the typographic art, and for the illustrations he sought the aid of the best artists. He was one of the patrons of Thomas Bewick, some of whose choicest work is to be found in the pamphlets issued by Nicholson. He also issued 125 cards, on which were printed favourite pieces, afterwards included in the *Literary Miscellany.* This 'assemblage of classical beauties for the parlour, the closet, the carriage, or the shade,' became very popular, and extended to twenty volumes. The plan of issuing them in separate numbers enabled individuals to make their own selection, and they are found bound up in every possible variety. Complete sets are now rare, and highly prized by collectors."

Of his many useful publications may be enumerated—*Stenography: The Mental Friend and Rational Companion, consisting of Maxims and Reflections relating to the Conduct of Life.* 12mo. *The Advocate and Friend of Woman.* 12mo. *Directions for the Improvement of the Mind.* 12mo. *Juvenile Preceptor.* Three vols., 12mo. The books which

---

believer in the *Metempsychosis.* "It was probably so said," remarks Ritson, "by ignorant people who cannot distinguish justice or humanity from an absurd and impossible system. The compiler of the present book, like Pythagoras and John Williamson, abstains from flesh-food, but he does not believe in the *Metempsychosis,* and much doubts whether it was the *real* belief of either of those philosophers."—*Abstinence from Animal Food a Moral Duty,* by Joseph Ritson. R. Phillips, London, 1802.

concern us now are—*On the Conduct of Man to Inferior Animals* (Manchester, 1797 : this was adorned by a woodcut from the hand of Bewick). And his *magnum opus*, which appeared in the year 1801, under the title of *The Primeval Diet of Man : Arguments in Favour of Vegetable Food; with Remarks on Man's Conduct to [other] Animals* (Poughnill, near Ludlow).

The value of *The Primeval Diet* was enhanced by the addition, in a later issue, of a tract *On Food* (1803), in which are given recipes for the preparation of "one hundred perfectly palatable and nutritious substances, which may easily be procured at an expense much below the price of the limbs of our fellow animals. . . . Some of the recipes, on account of their simple form, will not be adopted even by those in the middle rank of life. Yet they may be valuable to many of scanty incomes, who desire to avoid the evils of want, or to make a reserve for the purchasing of books and other mental pleasures." He also published a tract *On Clothing*, which contains much sensible and practical advice on an important subject.

Nicholson resided successively in Manchester, Poughnill, and Stourport, and died at the last-named place in the year 1825. " He possessed," says a writer in *The Gentleman's Magazine* (xcv.), " in an eminent degree, strength of intellect, with universal benevolence and undeviating upright-ness of conduct." The learned bibliographer, to whom we are indebted for this brief notice, thus sums up the character of his labours : " In all his writings the purity and benevolence of his intentions are strikingly manifest. Each subject he took in hand was thought out in an indepen-dent manner, and without reference to current views or prejudices."*

In his brief preface the author thus expresses his sad conviction of the probable futility of his protests :—

"The difficulties of removing deep-rooted prejudices, and the inefficiency of reason and argument, when opposed to habitual opinions established on general approbation, are fully apprehended. Hence the cause of humanity, however zealously pleaded, will not be materially promoted. Unflattered by the hope of exciting an impression on the public mind, the following compilation is dedicated to the sympathising and generous Few, whose opinions have not been founded on implicit belief and common acceptation : whose habits are not fixed by the influence of false and pernicious maxims or corrupt examples : who are neither deaf to the cries of misery, pitiless to suffering innocence, nor unmoved at recitals of violence, tyranny, and murder."

In the whole literature of humanitarianism, nothing can be more impressive for the sympathising reader than this putting on record by these nobler spirits their profound consciousness of the moral torpor of the world around them, and their sad conviction of the prematureness

---

*In a sketch of the life of George Nicholson, contributed to a Manchester journal, by Mr. W E. A. Axon.

of their attempt to regenerate it. In both his principal works, he judiciously chooses, for the most part, the method of compilation, and of presenting in a concise and comprehensive form the opinions of his humane predecessors, of various minds and times, rather than the presentation of his own individual sentiments. He justly believed that the large majority of men are influenced more by the authority of great names than by arguments addressed simply to their conscience and reason. He intersperses, however, philosophic reflections of his own, whenever the occasion for them arises. Thus, under the head of "Remarks on Defences of Flesh-eating," he well disposes of the common excuses :—

"The reflecting reader will not expect a formal refutation of common-place objections, which *mean nothing*, as, 'There would be more unhappiness and slaughter among animals did we not keep them under proper regulations and government. Where would they find pasture did we not manure and enclose the land for them? &c. The following objection, however, may deserve notice :—'Animals must die, and is it not better for them to live a short time in plenty and ease, than be exposed to their enemies, and suffered in old age to drag on a miserable life?' The lives of animals in *a state of nature* are very rarely miserable, and it argues a barbarous and savage disposition to cut them *prematurely* off in the midst of an agreeable and happy existence ; especially when we reflect on the *motives* which induce it. Instead of a friendly concern for promoting their happiness, your aim is the gratification of your own sensual appetites. How inconsistent is your conduct with the fundamental principle of pure morality and true goodness (which some of you ridiculously profess)— *whatsoever you would that others should do to you, do you even so to them.* No man would willingly become the food of other animals ; he ought not therefore to prey on *them.* Men who consider themselves members of universal nature, and links in the great chain of Being, ought not to usurp power and tyranny over others, beings naturally free and independent, however such beings may be inferior in intellect or strength. . . . . It is argued that 'man has a permission, proved by the practice of mankind, to eat the flesh of other animals, and consequently to kill them ; and as there are many animals which subsist wholly on the bodies of other animals, the practice is sanctioned among mankind.' By reason of the at present very low state of morality of the human race, there are many evils which it is the duty and business of enlightened ages to eradicate. The various refinements of civil society, the numerous improvements in the arts and sciences, and the different reformations in the laws, policy, and government of nations, are proofs of this assertion. That mankind, in the present stage of *polished* life, act in direct violation of the principles of justice, mercy, tenderness, sympathy, and humanity, in the practice of eating flesh, is obvious. To take away the life of any happy being, to commit acts of depredation and outrage, and to abandon every refined feeling and sensibility, is to degrade the human kind beneath its professed dignity of character ; but to *devour* or eat any animal is an additional violation of those principles, because it is the *extreme* of brutal ferocity. Such is the conduct of the most savage of wild beasts, and of the most uncultivated and barbarous of our own species. Where is the person who, with calmness, can hear himself compared in disposition to a lion, a hyæna, a tiger or a wolf? And yet, how exactly similar is his disposition.

"Mankind affect to revolt at murders, at the shedding of blood, and yet eagerly, and without remorse, feed on the corpse after it has undergone the culinary process. What mental blindness pervades the human race, when they do not perceive that every feast of blood is a *tacit encouragement* and licence to the very crime their pretended delicacy abhors ! I say *pretended* delicacy, for that it is pretended is most evident. The profession of sensibility, humanity, &c., in such persons, therefore, is egregious folly. And yet there are respectable persons among everyone's acquaintance, amiable in other dispositions, and advocates of what is commonly termed the cause of humanity, who are weak or prejudiced enough to be satisfied with such arguments, on which they ground apologies for their practice ! Education, habit, prejudice, fashion, and interest, have blinded the eyes of men, and seared their hearts.

"Opposers of compassion urge : 'If we should live on vegetable food, what shall we do with our *cattle ?* What would become of them ? They would grow so numerous they would be prejudicial to us—they would eat us up if we did not kill and eat them.' But there is abundance of animals in the world whom men do not kill and eat ; and yet we hear not of their injuring mankind, and sufficient room is found for their abode. Horses are not usually killed to be eaten, and yet we have not heard of any country overstocked with them. The raven and redbreast are seldom killed, and yet they do not become too numerous, If a decrease of cows, sheep, and others were required, mankind would readily find means of reducing them. Cattle are at present an article of trade, and their numbers are *industriously* promoted. If cows are kept solely for the sake of milk, and if their young should become too numerous, let the evil be nipped in the bud. Scarcely suffer the innocent young to feel the pleasure of breathing. Let the least pain possible be inflicted ; let its body be deposited entire in the ground, and let a sigh have vent for the calamitous necessity that induced the painful act. . . . Self-preservation justifies a man in putting noxious animals to death, yet cannot warrant the least act of cruelty to any being. By suddenly despatching one when in extreme misery, we do a kind office, an office which reason approves, and which accords with our best and kindest feelings, but which (such is the force of custom) we are denied to show, though solicited, to our own species. When they can no longer enjoy happiness, they may perhaps be deprived of life. Do not suppose that in this reasoning an intention is included of *perverting* nature. No ! some animals are savage and unfeeling ; but let not *their* ferocity and brutality be the standard and pattern of the conduct of *man.* Because *some* of them have no compassion, feeling, or reason, are *we* to possess no compassion, feeling, or reason ? "

In another section of his book Nicholson undertakes to expose the inconsistencies of flesh-eaters, and the strange illogicalness of the position of many protestors against various forms of cruelty, who condone the greatest cruelty of all—the (necessary) savagery of the butchers :—

"The inconsistencies of the conduct and opinions of mankind in general are evident and notorious ; but when ingenious writers fall into the same glaring errors, our regret and surprise are justly and strongly excited. Annexed to the impressive remarks by Soame Jenyns, to be inserted hereafter, in examining the conduct of man to [other] animals, we meet with the following passage :—

"'God has been pleased to create numberless animals intended for our sustenance, and that they are so intended, the agreeable flavour of their flesh to our palates, and the wholesome nutriment which it administers to our stomachs, are suffici-nt proofs ; these, as they are formed for our nse, propagated by our culture, and fed by our care,

N

we have certainly a right to deprive of life, because it is given and preserved to them on that *condition*.'

"Now, it has already been argued that the bodies of animals are *not* intended for the sustenance of man; and the decided opinions of several eminent medical writers and others sufficiently disprove assertions in favour of the wholesomeness of the flesh of animals. The *agreeable taste* of food is not always a proof of its *nourishing* or *wholesome* properties. This truth is too frequently experienced in mistakes, ignorantly or accidentally made, particularly by children, in eating the fruit of the deadly nightshade, the taste of which resembles black currants, and is extremely inviting by the beauty of its colour and shape.*

"That we have a right to make attacks on the existence of any being *because* we have assisted and fed such being, is an assertion opposed to every established principle of justice and morality. A 'condition' cannot be made without the mutual consent of parties, and, therefore, what this writer terms 'a condition,' is nothing less than an unjust, arbitrary, and deceitful imposition. 'Such is the deadly and stupifying influence of habit or custom,' says Mr. Lawrence, 'of so poisonous and brutalising a quality is prejudice, that men, perhaps no way inclined by nature to acts of barbarity, may yet live insensible of the constant commission of the most flagrant deeds.' . . . A cook-maid will weep at a tale of woe, while she is skinning a living eel, and the devotee will mock the Deity by asking a blessing on food supplied by murderous outrages against nature and religion! Even women of education, who readily weep while reading an affecting moral tale, will clear away clotted blood, still warm with departed life, cut the flesh, disjoint the bones, and tear out the intestines of an animal, without sensibility, without sympathy, without fear, without remorse. What is more common than to hear this *softer* sex talk of, and assist in, the cookery of a deer, a hare, a lamb or a calf (those acknowledged emblems of innocence) with perfect composure? Thus the female character, by nature soft, delicate, and susceptible of tender impressions, is debased and sunk. It will be maintained that in other respects they still possess the characteristics of their sex, and are humane and sympathising. The inconsistency then is the more glaring. To be virtuous in some instances does not constitute the moral character, but to be uniformly so."

We can allow ourselves space only for one or two further quotations from this excellent writer. The remarks upon the common usage of language, by which it is vainly thought to conceal the true nature of the dishes served up upon the tables of the rich, are particularly noteworthy, because the inaccurate expression condemned is almost universal, and that even, from force of habit, amongst reformed dietists themselves:—

"There is a natural horror at the shedding of blood, and some have an aversion to the practice of devouring the carcase of an innocent sufferer, which bad habits

---

*Perhaps the fallacy of this line of apology, on the part of the ordinary dietists, cannot be better illustrated than by the example of the man-eating tribes of New Zealand, Central Africa, and other parts of the world, who confessedly are (or were) *hominivorous*, and who have been by travellers quoted as some of the finest races of men on the globe. The "wholesome nutriment" of their human food was as forcible an argument for their stomach as the "agreeable flavour" was attractive for their palates. Such glaring fallacy might be illustrated further by the example of the man-eating tiger who, we may justly imagine, would use similar apologies for his practice.

improper education, and silly prejudices have not overcome. This is proved by their affected and absurd refinement of calling the dead bodies of animals *meat*. If the meaning of words is to be regarded, this is a gross mistake ; for the word *meat* is a universal term, applying equally to all nutritive and palatable substances. If it be intended to express that all other kinds of food are comparatively not meat, the intention is ridiculous. The truth is that the proper expression, *flesh*, conveys ideas of murder and death. Neither can it easily be forgotten that, in grinding the body of a fellow animal, substances which constitute *human* bodies are masticated. This reflection comes somewhat home, and is recurred to by eaters of flesh in spite of themselves, but recurred to *unwillingly*. They attempt, therefore, to pervert language in order to render it agreeable to the ear, as they disguise animal flesh by cookery in order to render it pleasing to the taste."

His reflections upon the essential injustice (to use no stronger term) of delegating the work of butchering to a particular class of men (to which frequent reference has already been made in these pages) are equally admirable :—

"Among butchers, and those who qualify the different parts of an animal into food, it would be easy to select persons much further removed from those virtues which should result from reason, consciousness, sympathy, and animal sensations, than any savages on the face of the earth ! In order to avoid all the generous and spontaneous sympathies of compassion, the office of shedding blood is committed to the hands of a set of men who have been educated in inhumanity, and whose sensibility has been blunted and destroyed by early habits of barbarity. Thus men *increase* misery in order to avoid the sight of it, and because they cannot endure being obviously cruel themselves, or commit actions which strike painfully on their senses, they commission those to commit them who are formed to delight in cruelty, and to whom misery, torture, and shedding of blood is an amusement ! They appear not once to reflect that *whatever we do by another we do ourselves*."

" When a large and gentle Ox, after having resisted a ten times greater force of blows than would have killed his murderers, falls stunned at last, and his armed head is fastened to the ground with cords ; as soon as the wide wound is made, and the jugular veins are cut asunder, what mortal can, without horror and compassion, hear the painful bellowings, intercepted by his flow of blood, the bitter sighs that speak the sharpness of his anguish, and the deep-sounding groans with loud anxiety, fetched from the bottom of his strong and palpitating heart. Look on the trembling and violent convulsions of his limbs ; see, whilst his reeking gore streams from him, his eyes become dim and languid, and behold his strugglings, gasps, and last efforts for life.

"When a being has given such convincing and undeniable proofs of terror and of pain and agony, is there a disciple of Descartes so inured to blood, as not to refute, by his commiseration, the philosophy of that vain reasoner ?" *

In his previous essay, *On the Conduct of Men to Inferior Animals*, Nicholson has collected from various writers, both humane and inhumane,

---

*On the Conduct, &c., and *The Primeval Diet of Man*, &c., by George Nicholson, Manchester and London, 1797, 1801. The author assumes as his motto for the title-page the words of Rousseau— *Hommes, soyez humains ! C'est votre premier devoir. Quelle sagesse y a-t-il pour vous hors de l'humanité?* "Humans, be *humane* ! It is your first duty. What wisdom is there for you without humanity ?"

a fearful catalogue of atrocities of different kinds perpetrated upon his helpless dependants by the being who delights to boast himself (at least in civilised countries) to be made "in the image and likeness of God." Among these the hellish tortures of the vivisectionists and "pathologists" hold, perhaps, the bad pre-eminence, but the cruel tortures of the Slaughter-House come very near to them in wanton atrocity.

XXXVII.
ABERNETHY. 1763—1831.

DISTINGUISHED as a practical surgeon and as a physiologist, Abernethy has earned his lasting reputation as having been one of the first to attack the old prejudice of the profession as to the origin of diseases, and as having sought for such origin, not in mere local and accidental but, in general causes—in the constitution and habits of the body.

A pupil of John Hunter, in 1786 he became assistant surgeon at St. Bartholomew's Hospital, and shortly afterwards he lectured on anatomy and surgery at that institution, which to his ability and genius owes the fame which it acquired as a school of surgery. As a lecturer he had a reputation and popularity seldom or perhaps never before so well earned in the medical schools—founded, as they were, upon a rare penetration and logical method, united with clearness and perspicuity in communicating his convictions. In honesty, integrity, and in the domestic virtues his character was unimpeachable, but the gentleness of deportment for which he was noted in his home he was far from exhibiting in public and towards his patients. His roughness and even coarseness of manner in dealing with capricious valetudinarians, indeed, became notorious.

The Constitutional Origin and Treatment of Local Diseases—his principal work—in comparison with the vast mass of medical literature up to that time put forth, stands out in favourable relief. In it two great principles are laid down—that "local diseases are symptoms of a disordered constitution, not primary and independent maladies, and that they are to be cured by remedies calculated to make a salutary impression on the *general frame*, not by local treatment, nor by any mere manipulations of surgery." This single principle changed the aspect of the entire field of surgery, and elevated it from a manual art into the rank of a science. And to this first principle he added a second, the range of which is, perhaps, less extensive, but the practical importance of which is scarcely inferior to that of the first—namely, that "this disordered state of the constitution either originates from, or is rigorously allied with, derangement of the stomach and bowels, and that it can only be reached by remedies which first exercise a curative influence upon these organs." It will not detract from the merit of Abernethy to add to this account that his predecessor, Dr. Cheyne, and his contemporary, Dr. Lambe, have most satisfactorily and radically carried out into practice these just principles; or to remark that great public reputations ought not to be allowed, as too often is the fact, to overwhelm less known but not therefore less meritorious labours.

As to *dietetics*, the theory of Abernethy seems to have been better than his practice. When reproached with the inconsistency that the reformed diet which he so forcibly commended to others he himself failed to follow, he is related to have used the well-known simile of the sign-post with his usual readiness of repartee.

It was while Dr. Lambe was at the Aldersgate Street Dispensary that Abernethy formed the acquaintance of that unostentatious but true reformer—an acquaintance which was destined to have no unimportant influence upon the medical theories of the great surgeon. Abernethy was at that time writing his *Observations on Tumours*, and he had intrusted to his friend one of his cancer patients to be treated by the non-flesh and distilled water regimen. He carefully watched the effects, and he has thus given us the results of his observations :—

"There can be no subject which I think more likely to interest the mind of a surgeon than that of an endeavour to amend and alter the state of a cancerous constitution. The best timed and best conducted operation brings with it nothing but disgrace if the diseased propensities of the constitution are active and powerful. It is after an operation that, in my opinion, we are most particularly concerned to regulate the constitution, lest the disease should be revived or renewed by its disturbance. In addition to that attention, to tranquillise and invigorate the nervous system, and keep the digestive organs in as healthy a state as possible (which I have recommended in my first volume), I believe general experience sanctions the recommendation of a more vegetable because less stimulating diet, with the addition of so much milk, broth, and eggs, as seems necessary to prevent any declension of the patient's strength.

"Very recently Dr. Lambe has proposed a method of treating cancerous diseases. which is *wholly* dietetic. He recommends the adoption of a strict vegetable regimen, to avoid the use of fermented liquors, and to substitute water purified by distillation in the place of common water as a beverage, and in all parts of diet in which common water is used, as tea, soups, &c. The grounds upon which he founds his opinion of the propriety of this advice, and the prospects of benefit which it holds out, may be seen in his *Reports on Cancer*, to which I refer my readers.

"My own experience on the effects of this regimen is of course very limited. Nor does it authorise me to speak decidedly on the subject. But I think it right to observe that, in one case of cancerous ulceration in which it was used, the symptoms of the disease were, in my opinion, rendered more mild, the erysipelatous inflammation surrounding the ulcer was removed, and the life of the patient was, in my judgment, considerably prolonged. The more minute details of the facts constitute the sixth case of Dr. Lambe's *Reports*. It seems to me very proper and desirable that the powers of the regimen recommended by Dr. Lambe should be fairly tried, for the following reasons :—

"Because I know some persons who, whilst confined to such diet, have enjoyed very good health ; and further, I have known several persons, who did try the effects of such a regimen, declare that it was productive of considerable benefit. They were not, indeed, afflicted with cancer, but they were induced to adopt a change of diet to allay a state of nervous irritation and correct disorder of the digestive organs, upon which medicine had but little influence.

"Because *it appears certain, in general, that the body can be perfectly nourished by vegetables.*

"Because all great changes of the constitution are more likely to be effected *by alterations of diet and modes of life than by medicine.*

"Because it holds out a source of hope and consolation to the patient in a disease in which medicine is known, to be unavailing, and in which surgery affords no more than a temporary relief." *

"The above opinion of Mr. Abernethy," remarks an experienced authority upon the subject, "is most valuable, for he watched the case for three and a half years under Dr. Lambe's regimen, which is directly opposed to the system of diet which he had advocated, before he met Dr. Lambe, in the first volume of his work on *Constitutional Diseases,* and from his rough honesty there is no doubt that had Dr. Abernethy lived to publish a second edition he would have corrected his mistake." As it is, the candour by which so distinguished an authority was impelled to alter or modify opinions already put f rth to the world, claims our respect as much as the too general want of it deserves censure.

XXXVIII.

LAMBE. 1765—1847.

ONE of the most distinguished of the hygeistic and scientific promoters of the reformed regimen, Dr. Lambe, occupies an eminent position in the medical literature of vegetarianism, and he divides with his predecessor, Dr. Cheyne, the honour of being the founder of scientific *dietetics* in this country.

His family had been settled some two hundred years in the county of Hereford, in which they possessed an estate that descended to Dr. William Lambe; and is now held by his grandson. He early gave promise of his future mental eminence. Head boy of the Hereford Grammar School, he proceeded, in due course, to St. John's College,

---

* *Surgical Observations on Tumours.* John Abernethy, M.D., F.R.C.S.

Cambridge. In 1786, being then in the twenty-first year of his age, he graduated as fourth wrangler of his year. As a matter of course, he soon was elected a Fellow of his college, where he continued to reside until his marriage in 1794. During this period of learned leisure he devoted his time to the study of medicine, and the MS. notes in the possession of his biographer, Mr. Hare, " prove the diligence with which he studied his profession, and there we see the origin of his enlarged views of the causes of disease, so much insisted on by these fathers of medicine, and so much neglected by modern physicians in their search for chemical remedies." After his marriage he went to reside and practise in Warwick, where he was the intimate friend of Parr, the well-known Greek critic, and of Walter Savage Landor, who writes of him as " very communicative and good humoured. I had enough talk with Lambe to assure myself that he is no ordinary man." It was to the discoveries of Dr. Lambe, and to his publications reporting the curative value of its mineral waters, that Leamington owed its fame and popularity ; and Dr. Jefferson, in his address to the British Medical Association a few years ago, thus eulogises him :—

" It was not until the end of the last century that any really scientific research ever was recorded on this subject [impure water]. About this period Dr. Lambe was engaged in practice in Warwick. Somewhat eccentric in some of his practical views, Dr. Lambe was not the less a scientific man, an intelligent observer of nature, and an accomplished physician, and was, moreover, one of the most elegant medical writers of his day. The springs of the neighbouring village of Leamington did not escape his observation, and, having carefully studied and analysed the waters, he published an account of them, in 1797, in the fifth volume of the *Transactions of the Philosophical Society of Manchester*, a society embracing the respected names of Priestley, Dalton, Watt, and others, and not inferior, perhaps, to any contemporary association in Europe."

Like many other seceders from orthodox dietetics both before and after him, Dr. Lambe found himself impelled to experiment in the non-flesh diet by ill-health. His bodily disorders, indeed, were so complicated and of such a nature, as to excite astonishment that not only he greatly mitigated their violence, but that also he survived to an advanced age. In an exceedingly minute and conscientious narrative of his own case in his *Additional Reports* (writing in the third person), he informs us, that having during several years—from his eighteenth year—suffered greatly and with constantly aggravated symptoms :—

" He resolved, therefore, finally to execute what he had been contemplating for some time—to abandon animal food altogether, and everything analogous to it, and to confine himself wholly to vegetable food. This determination he put in execution the second week of February, 1806, and he has adhered to it with perfect regularity to the present time. His only subject of repentance with regard to it has been that it had not been adopted much earlier in life. He never found the smallest real ill-consequence from this change. He sank neither in strength, flesh, nor in spirits. He was at all times of a very thin and slender habit, and so he has continued to be, but upon the whole he has rather gained than lost flesh. He has experienced neither indigestion nor flatulence even from the sort of vegetables which are commonly thought to produce flatulence, nor has the stomach suffered from any vegetable matter, though unchanged by culinary art or uncorrected by condiments. The only unpleasant consequence of the change was a sense of emptiness of stomach, which continued many months. In about a year, however, he became fully reconciled to the new habit, and felt as well satisfied with his vegetable meal as he had been formerly with his dinner of flesh. He can truly say that since he has acted upon this resolution no year has passed in which he has not enjoyed better health than in that which preceded it. But he has found that the changes introduced into the body by a vegetable regimen take place with extreme slowness ; that it is in vain to expect any *considerable* amendment in successive weeks or in successive months. We are to look rather to the intervals of *half-years or years.*"

With extreme candour as well as carefulness, this patient and philosophic experimentalist details every particular circumstance of his own *diagnosis.* After a minute report of the various symptoms of his maladies and his gradual subjugation of them, he deduces the only just inference :—

"Granting this representation of facts to be correct, and the nature of this case to, be truly determined, I must be permitted to ask, What other method than that which has been adopted would have produced the same benefit? If such methods exist, I confess my ignorance of them. . . . But though these pains [in the head] still recur in a trifling degree, the relief given to the brain in general has been decided and most essential. It has appeared in an increased sensibility of all the organs, particularly of the senses—the touch, the taste, and the sight, in greater muscular activity, in greater freedom and strength of respiration, greater freedom of all the secretions, and in increased intellectual power. It has been extended to the night as much as to the day. The sleep is more tranquil, less disturbed by dreams, and more refreshing. Less sleep, upon the whole, appears to be required ; but the loss of quantity is more than compensated by its being sound and uninterrupted. . .

"The hypochondriacal symptoms continued to be occasionally very oppressive during the second year, particularly during the earlier part of it, but they afterwards very sensibly declined, and at present he enjoys more uniform and regular spirits than he had done for many years upon the mixed diet. From the whole of these facts it follows that all the organs, and indeed every fibre of the body, are simultaneously affected by the matters habitually conveyed into the stomach, and that it is the incongruity of these matters to the system, which gradually forms that morbid diathesis which exists alike both in apparent health and in disease. I might illustrate this fact still more minutely by observations on the teeth, on the hair, and on the skin. I might show that by a steady attention to regimen, the skin of the palm of the hand becomes of a firmer and stronger texture, that even an excrescence which had for twenty years and upwards been growing more fixed, firm, and deep, had, first, its habitudes altered, and, finally, was softened and disappeared. But, perhaps, enough has been said already to give a pretty clear idea both of the kind of change introduced into the habit by diet, and of the extent to which it may be carried. I proceed, therefore, to relate some new phenomena which took place during the course of this regimen, which are both curious in themselves and lead to important conclusions."

The author then goes on to record further gradual diminution of painful symptoms. From long and careful observation of himself, amongst other important deductions, Dr. Lambe infers that :—

"We may conclude that it is the property of this regimen, and, in particular, of the vegetable diet, to transfer diseased action from the *viscera* to the exterior parts of the body—from the central parts of the system to the periphery. Vegetable diet has often been charged with causing cutaneous diseases ; in common language, they are, in these cases, said to proceed from poorness of blood.* In some degree the charge is probably just, and the observation I have already made may give us some insight into the causes of it. But this charge, instead of being a just cause of reproach, is *a proof of the superior salubrity of vegetable diet.* Cutaneous eruptions appear, because disease is translated from the internal organs to the skin."

For all brain disease abandonment of the gross and stimulating flesh-meats is shown to be of the first importance. At the same time, that it involves any loss of actual bodily strength is a fallacy :—

"We see, then, how ill-founded is the notion that inaction and loss of power are induced by a vegetable diet. In fact, all the observations that have been made have shewn the very reverse to be the truth. Symptoms of plenitude and oppression have continued in considerable force for at least five years ; and the consequence of this peculiar regimen has been an increase of strength and power, and not a diminution. In the subject of this case the pulse, which may be deemed, perhaps, the best idea of the condition of all the other functions, is at present much more strong and full than under the use of animal food. It is also perfectly calm and regular."

His personal experience of satisfaction derivable from vegetables and fruits as affording, for the most part, sufficient liquids in themselves, without use of extraneous drinks, is of importance :—

"He had, when living on the common diet, been habitually thirsty, and, like most persons inclined to studious and sedentary habits, was much attached to tea-drinking. But for the last two or three years he has almost wholly relinquished the use of liquids, and by the substitution of fruit and recent vegetables he has found that th sensation of thirst has been in a manner abolished. Even tea has lost its charms, and he very rarely uses it. He is therefore certain, from his own experience, that the habit of employing liquids is an artificial habit, and not necessary to any of the functions of the animal economy."

Whatever may be thought of the theory of the possibility of entire abstinence from all *extraneous* liquids, there is not the least doubt that a judicious use of vegetable foods reduces to a *minimum* the feeling of thirst and craving for artificial drinks, an experience, we imagine, almost universal with abstinents from flesh-dishes.

Dr. Lambe concludes the first part of his valuable *diagnosis* with the assurance, " that if those for whose service these labours are principally designed, I mean persons suffering under habitual and chronic illness, are able to go along with me in my argument to form a general correct

* Excessive poverty of blood, it is obvious to remark, is caused, not by abstaining from flesh but by abstaining from a *sufficient* amount of *nutritious* non-flesh foods.

notion of what they are to expect from [a reformed] regimen, and, above all, to arm their minds with firmness, patience, and perseverance, I shall not readily be induced to think that I have written one superfluous line."*

In 1805, at the age of forty, we find him established in practice in London. Five years later he was physician to the General Dispensary, Aldersgate Street. He was also elected Fellow and Censor of the College of Physicians, whose meetings he regularly attended. His peculiar opinions did not tend to secure popularity for him, and the adhesion of such men as Dr. Abernethy, Dr. Pitcairn, Lord Erskine, and of Mr. Brotherton, M.P. (one of the earliest members of the Vegetarian Society), served only to make the indifference of the mass of the community more conspicuous.

Not the least interesting fact in his life is his share in the conversion of Shelley, and his friendship with J. F. Newton and his interesting family, at whose house these earlier pioneers of the New Reformation were accustomed to meet, and celebrate their charming *réunions* with vegetarian feasts. A cardinal part of the dietetic system of Dr. Lambe was his insistance upon the use of *distilled* water. In his *Reports on Regimen* he writes of the Newton family : "I am well acquainted with a family of young children who have scarcely ever touched animal food, and who now for three years have drunk only distilled water. For clearness and beauty of complexion, muscular strength, fulness of habit free from grossness, hardiness, healthiness, and ripeness of intellect these children are unparalleled."†

We have already mentioned Lord Erskine as one of the many eminent friends of Dr. Lambe. That more humane and distinguished lawyer, in a letter to his friend acknowledging the receipt of the *Reports*, writes as follows : "I am of opinion that both this work and the other referred to in it are deserving of the highest consideration. I read them both with more interest and attention from the abuse of the *British Critic* [one of the periodicals of the day] mentioned in the preface, as no periodical criticism ever published in this country is so uniformly unjust, ignorant, and impudent." Dr. Abernethy's testimony to the

* *Additional Reports*, 1814. Amongst valuable diagnoses of this kind the reader may be referred in particular to the highly interesting one of the Rev. C. H. Collyns, M.A., Oxon, which originally appeared in the *Times* newspaper, and which twice has been republished by the Vegetarian Society. The success of the pure regimen in first mitigating and, finally, in altogether subduing long-inherited gouty affections, was complete and certain. The recently published evidence of the President of the newly-formed French Society, Dr. A. H. de Villeneuve, is equally satisfactory. (See *Bulletin de la Société Végétarienne* of Paris, as quoted in *Nature*, Jan., 1881.)

† See, too, the testimony of Newton, *Return to Nature*, and of Shelley in his *Essay on the Vegetable Diet*, in which he describes these children as "the most beautiful and healthy beings it is possible to conceive. The girls are the most perfect models for a sculptor. Their dispositions, also, are the most gentle and conciliating."

efficacy of abstinence in cases of cancer will be found in the notice of that eminent practitioner. Amongst the most interesting correspondence of his later years is his interchange of ideas with Sylvester Graham— the first of the American prophets of the reformed regimen. The letter to the celebrated American vegetarian is, as Dr. Lambe's latest biographer justly observes, "a most valuable relic, because it continues the result of Dr. Lambe's diet up to September, 1837—twenty-three years after the last notice of his health in the account of his own case, which he published in November, 1814. It is, besides, an admirable proof of his truthful and philosophic mind, which was slow to arrive at conclusions, and willing rather to exaggerate than otherwise the traces of disease which he still felt." He proves, also, in this letter, how slow and yet sure are the effects of diet, and it supplies an answer to those objectors who complain that they have tried the diet (perhaps for a few weeks only) without any good result. After complimenting his transatlantic fellow-worker in the cause of truth upon his zeal and industry, Dr. Lambe proceeds :—

"My book, entitled *Additional Reports on Regimen*, has now been before the world three and twenty years. That it has attracted little notice, and still less popular favour—though it may have excited in the writer some mortification—has not occasioned much surprise. The doctrine it seeks to establish is in direct opposition to popular and deep-rooted prejudice. It is thought (most erroneously) to attack the best enjoyments and most solid comforts of life ; and, moreover, it has excited the bitter hostility of a numerous and influential body in society—I mean that body of medical practitioners who exercise their profession for the sake of its profits merely, and who appear to think that disease was made for the profession and not the profession for disease.

"To drop, however, all idle complaints of public neglect, let us go to the more useful inquiry whether or not the principles propounded in these *Reports* have been confirmed by subsequent and more extensive experience. To this inquiry I answer directly and fearlessly, that in the interval between the present time and the year 1815 (the date of that publication) the practice recommended has succeeded in cases very numerous and of extreme variety, and I can promise the practitioner who will try it fairly and judge with candour that he will experience no disappointment. I say, *let him try it fairly.* I do not assert that it will succeed in cases where the powers of life are sunk, in confirmed hectic fever, in ulcerated cancer, in established chronic disease, or in the decrepitude of old age. I may have attempted the relief of such cases in an early stage of my experiments, but experience speedily demonstrated the hopelessness of such attempts. But let subjects be taken not far advanced in life, let them be *tabid* children (for example) with tumid abdomen, swelled joints, and depraved appetites, or with obstinate cutaneous diseases, erythema, *scabus*, rickets, epileptic convulsions (not grown habitual by long continuance). But a practitioner in moderate practice will find no difficulty in selecting proper subjects, if he is himself actuated by a regard to humanity united to principles of honour.

"Moreover, let not the patient, particularly if arrived at mature age, expect to receive a perfect cure. In many cases the consequences are rather preventive than curative. This I hold to be no objection. It is enough, surely, if a disease which, from its nature, might be expected to be continually on the increase, is obviously checked in its progress, if the symptoms become more and more mild, and if a human being is preserved in comfortable existence who would otherwise have been consigned to the grave."

He devoted his great medical knowledge and experience particularly to the cure or mitigation of cancer. In the letter, from which we have already quoted, he informs his correspondent of this interesting fact :—

"My most ardent wish was to attempt the relief of cases of cancer. This object 1 have steadily pursued (from the year 1803) to the present day. The case—the particulars of which 1 briefly mentioned to you in my former communication—has hitherto succeeded so perfectly that I should myself suspect an error in the *diagnosis,* if it were not for the strongly-marked constitutional symptoms, which are such as, in my mind, put it out of doubt. There does not now remain what I expected, and what I have called a *nucleus,* for the resolution is *complete.* Now, this is contrary to most of my former observations, and would furnish, as I have said, some ground of suspicion. But still it is not wholly unsupported by corroborative facts. I have observed, particularly in one case, that the whole extreme edge of a schirrous tumour has been restored, whilst the portion has remained unchanged ; not, indeed, speedily, as in the former case, but after having used the diet for a very considerable time, Now, if a portion of a true schirrous tumour can be resolved, there can be no reason why a resolution of the whole—taken very early and under favourable circumstances—shall be deemed impossible. The truth is, that at present we are not advanced enough to form general conclusions, but ought to content ourselves with *accumulating* facts for the use of our successors."

If the experience of the benefits of a reasonable living in the cases of his patients was thus satisfactory, he himself afforded, in his own person, perhaps the best testimony to its revivifying and invigorating qualities. One of his visitors gives his impressions of the now famous *doctor* (a title, in the present instance, of real meaning) as follow : "Agreeably to your request, I submit to your perusal a short account of the friendly interview I had with Dr. Lambe in London. I first called on him in February. I found him to be very gentlemanly in manners and venerable in appearance. He is rather taller than the middle height His hair is perfectly white, for he is now seventy-two years of age. He told me he had been on the vegetable diet thirty-one years, and that his health was better now than at forty, when he commenced his present system of living. He considers himself as likely to live thirty years longer as to have lived to his present age. . . . . Although he is seventy-two years of age he walks into town, a distance of three miles from his residence, every morning, and back at night. Dr. Lambe, I am told, has spent large sums of money in making experiments and publishing their results to the world." In his earlier life he had been conspicuously thin and attenuated. In later years he seems to have acquired even a certain amount of robustness, and he is described as being active and strong at an advanced age. Some instances of extraordinary energy and endurance have been put on record by his family; and his feats of pedestrianism, when he was verging on his eightieth year, are, we imagine, rarely to be paralleled.

His hope of attaining the age of one hundred years, unhappily, was not to be fulfilled. " Our bodies," his biographer justly remarks, " are but machines adapted to perform a definite amount of work, and Dr. Lambe's originally weak constitution had been severely tried by sickness and wrong diet during the first forty years of his life. At the age of eighty his strength began to fail, but his grandson writes, ' up to a very short time before his death there were no outward signs of ill-health, only the marks of old age.' " Existence had its enjoyment for him up to almost the last days, and his intellectual powers remained to the end. He calmly expired in his eighty-third year.

Of contemporary and posthumous eulogies of his personal, as well as scientific, worth, the following may suffice : "A man of learning, a man of science, a man of genius, a man of distinguished integrity and honour." Such is the testimony of his friend Dr. Parr, as quoted by Samuel Johnson. In the Anniversary Harveian Oration before the College of Physicians, by Dr. Francis Hawkins, in the year 1848, the representative of the Faculty thus recalls his memory : " Nor can I pass over in silence the loss we have sustained in Dr. William Lambe—an excellent chemist, a learned man, and a skilful physician. His manners were simple, unreserved, and most modest. His life was pure. Farewell, therefore, gentle spirit, than whom no one more pure and innocent has passed away ! "

XXXIX.

NEWTON. 1770—1825.

JOHN FRANK NEWTON, the friend and associate of Dr. Lambe, Shelley, and the little band who met at the house of the former to share his vegetarian repasts, appears to have been one of the earliest converts of Dr. Lambe, to whom he dedicated his *Return to Nature*, in gratitude for the recovery of his health through the adoption of the reformed regimen.

He published his little work, as he informs us in his preface, to impart to others the benefits which he himself had experienced; and

* The Life of William Lambe, M.D., Fellow of the Royal College of Physicians. By E. Hare, C.S.I., Inspector-General of Hospitals, to which valuable biography we are indebted for the present sketch. In Mr. Hare's memoir will be found, among other testimonies to the truths of Vegetarianism, a highly-interesting letter, written to him by his friend Dr. H. G. Lyford, an eminent physician of Winchester.

especially to make known to the heads of households the fact that his whole family of himself, wife, and four children under nine years of age, with their nurse, had been living, at the date of his publication, for two years upon a non-flesh diet, during which time the apothecary's bill, he tells us, had amounted to the sum of sixpence; and that charge had been incurred by himself.

The ever-memorable meetings of the reformers at the house of Newton, where Shelley was a constant guest, have been thus recorded by one of the biographers of the great poet :—" Shelley was intimate with the Newton family, and was converted by them in 1813, and he began then a strict vegetable diet.   His intimate association with the amiable and accomplished votaries of a *Return to Nature* was perhaps the most pleasing portion of his poetical, philosophical, and lovely life. . . . For some years I was in the thick of it ; for I lived much with a select and most estimable society of persons (the Newtons), who had ' returned to Nature,' and I heard much discussion on the topic of vegetable diet. Certainly their vegetable dinners were delightful, elegant, and excellent repasts ; flesh, fowl, fish, and 'game' never appeared—nor eggs nor butter *bodily*, but the two latter were admitted into cookery, but as sparingly as possible, and under protest, as not approved of and soon to be dispensed with.   We had soups in great variety, that seemed the more delicate from the absence of flesh-meat.

" There were vegetables of every kind, plainly stewed or scientifically disguised. Puddings, tarts, confections and sweets abounded.   Cheese was excluded.   Milk and cream might not be taken unreservedly, but they were allowed in puddings, and sparingly in tea.   Fruits of every kind were welcomed.   We luxuriated in tea and coffee, and sought variety occasionally in cocoa and chocolate.   Bread and butter, and buttered toast were eschewed ; but bread, cakes, and plain seed-cakes were liberally divided among the faithful."*

The cause of the publication of his book Newton thus states :—

" Having for many years been an habitual invalid, and having at length found that relief from regimen which I had long and vainly hoped for from drugs, I am anxious, from sympathy with those afflicted, to impart to others the knowledge of the benefit I have experienced, and to dispel, as far as in me lies, the prejudices under which I conceive mankind to labour on points so nearly connected with their health and happiness.

" The particulars of my case I have already related at the concluding pages of Dr. Lambe's *Reports on Cancer*. To the account there given I have little to add, but that, by continuing to confine myself to the regimen advised in that work, I continue to experience the same benefit ; that the winter which has just elapsed has been passed

* *Life of Shelley*, by Jefferson Hogg, quoted by Mr. Hare in Life of Dr. Lambe. Hogg adds that he conformed for good fellowship, and found the purer food an agreeable change.

much more comfortably than that which preceded it, and that, if my habitual disorder is not completely eradicated, it is so much subdued as to give but little inconvenience ; that I have suffered but a single day's confinement for several months ; and, upon the whole, that I enjoy an existence which many might envy who consider themselves to be in full possession of the blessings of health.

" All that I have to regret in my present undertaking is the imperfect way in which it is executed. The adepts in medicine have gained their knowledge originally from the experience of the sick. I have taken my own sensations for my guide, and am myself alone responsible for the conclusions which I have drawn from them, the manuscript of this volume having been neither corrected nor looked over by any individual. While I make no pretensions to medical science, I cannot consent to be reasoned or ridiculed out of my feelings ; nor to believe that to be an illusion, the truth of which has been confirmed to me by long-continued and repeated observation."

The use of distilled water was a cardinal article in the dietary creed of his friend Dr. Lambe, and upon this point Newton particularly insists. He appeals with much fervour, as we have just stated, to parents to have recourse to the natural means of prevention and cure, in place of vainly trying every available *artificial* method by medicine and drugs. He instances, with minute particularity, the regimen of his children, whom he asserts to have been, up to the moment of his writing, perfectly free from any sort of malady or disorder, and to be—

" So remarkably healthy that several medical men who have seen and examined them with a scrutinizing eye, all agreed in the observation that they knew nowhere a whole family which equals them in robustness. Should the success of this experiment, now of three years' standing, proceed as it has begun, there is little doubt, [he ventures to flatter himself] that it must at length have some influence with the public, and that every parent who finds the illness of his family both afflicting and expensive, will say to himself ' Why should I any longer be imprudent or foolish enough to have my children sick ? ' All hail to the resolution which that sentiment implies ! But until it becomes general, I feel it necessary to exhort, in the warmest language I can think of, those who have the young in their charge to institute an experiment which I have made before them with the completest success. To those parents especially do I address myself who, aware that temperance in enjoyment is the best warrant of its duration, feel how dangerous and how empty are all the feverous amusements of our assemblies, our dinners, and our theatres, compared with the genuine and tranquil pleasures of a happy circle at home."

He presents an alluring picture of the health-producing results for the young of the natural regimen. He promises that

" They will become not only more robust but more beautiful ; that their carriage will be erect, their step firm ; that their development at a critical period of youth, the prematurity of which has been considered an evil, will be retarded ; that, above all, the danger of being deprived of them will in every way diminish ; while by these light repasts their hilarity will be augmented, and their intellects cleared in a degree which shall astonishingly illustrate the delightful effects of this regimen. . . . . . I will beg here to attempt an answer in this place to that trite and specious objection to Dr. Lambe's opinions that ' what is suitable to one constitution may be not so to another.' If there be a single person existing, whose health would not be improved by the vegetable diet and distilled water, then the whole system falls at once to the ground. The question is simply, whether fruits and other vegetables be not the natural sustenance of man, who would have occasion for no other drink than these afford, and whose thirst is at present excited by an unnatural flesh diet, which causes his disorders bodily and mentally. . . . . Another objection sometimes urged is this : ' If children, brought up on a vegetable regimen, should at a future period of their lives adopt a flesh diet, they will certainly suffer more from the change than they otherwise would have done.' The very contrary of this, I conceive, would happen. The stomach is so fortified by the general increase of health, that a person thus nourished is enabled to bear what one whose humours are less impaired would sink under. The children of our family can each of them eat a dozen or eighteen walnuts for supper without the most trifling indigestion, an experiment which those who feed their children in the usual manner would consider it adventurous to attempt. So also the Irish porters in London bear these alterations of diet success-fully, and owe much of their actual vigour to the vegetable food of their forefathers, and to their own, before they emigrated from Ireland, where, in all probability, they did not taste flesh half-a dozen times in the year."

As to another well-known pretext, that the propensity to flesh-eating, and the relish with which it is evidently enjoyed by the majority of flesh eaters, is proof of its fitness, Newton justly objects the various unnatural and disgusting foods of many savage peoples which are eaten with equal relish, so that " the argument of the agreeable flavour proves nothing, I apprehend, by proving too much." He exhorts the medical faculty generally, and those members of it who are in charge of hospitals, in-firmaries, or workhouses, to try the effect of the pure regimen on the sufferers and patients—in particular, in the cases of the victims of cancer. Amongst others of his personal acquaintance who had derived the greatest benefit from the regimen, he instances Dr. Adam Ferguson, the historian of the Roman Republic, who lived strictly on a vegetable diet. He was in the habit of accompanying Mr. Newton, in the year 1794, in rides through the environs of Rome. He was still living in 1811, and he died, in fact, at the age of ninety, holding a professorship in the University of Edinburgh.

XL.

GLEÏZÈS.   1773—1843.

OF all the enlightened and humane spirits to which the philosophic eighteenth century gave birth, and who were quickened into activity by the great movement which originated in France in its last quarter, not one, assuredly, was actuated by a purer and more exalted feeling than Jean Antoine Gleïzès—the most *enthusiastic*, perhaps, of all the apostles of humanity and of refinement. He was born at Dourgne, in the (present) department of the Tarn. His father was advocate to the old provincial parliament. His mother's name was Anna Francos. After attending preliminary schools, he applied himself to the study of medicine—urged, says his biographer, more by love of his species than by predilection for the profession. His intense horror of the vivisectional experiments in the physiological torture-dens soon compelled him to abandon his intended career : the experience, however, gained during his brief medical course he was able to utilize more than once in his after life for the benefit of his neighbours.

The earlier period of the Revolution had been hailed by him, still very young as he then was, as the hopeful beginning of a new era ; when its direction, unhappily, fell into the hands of fanatical leaders, who, following too much the examples of the old *régimes,* thought, by wholesale executions, to clear the way for the establishment of a universal republic and of lasting peace. The youthful enthusiast, whose whole soul revolted from the very idea of bloodshed and of suffering, withdrew despairing into solitude, and devoted himself to scientific and literary studies, and to calm contemplation of Nature.

In 1794, at the age of 21, Gleïzès married Aglae de Baumelle, daughter of a writer of some repute. At this time he seems to have entertained the hope of instructing his countrymen, by engaging in public teaching; but, disappointed in a scheme for the inauguration of a course of historical lectures in the central school of his department, he retired altogether from the active business of the world, and settled down in a happy and peaceful home, in a small château belonging to his wife, at the foot of the Pyrenees near Mezières. It was here, amidst the magnificent solitudes of Nature, that in 1798, in his twenty fifth year, he determined upon abandoning for ever the diet of blood and slaughter. Until the moment of his death, forty-five years later, his diet consisted solely of milk, fruits, and vegetables.

So great was his scrupulousness, that there might be no possibility or mistake Gleïzès prepared his own food ; and he always ate alone (his wife being unable or unwilling to follow his loftier aims), since he could not endure either the smell or the sight of the ordinary dishes. And this intense aversion it was, indeed, that compelled him to forego in great measure his intercourse with the world, or, at all events, to shun the ordinary celebrations of social " festivity."

Full of enthusiastic belief that the transparent truth and sublimity of his creed could not fail to commend themselves to the better spirits of the age amongst his countrymen, Gleïzès addressed himself to some of the more thoughtful of his contemporaries ; amongst others to Lamartine,

o

Lamennais, and Chateâubriand.   Lamartine—the author of the *Fall of an Angel,* in which he gives expression to his akreophagistic sympathies—respended, if not with the enthusiasm that might justly have been expected from the author of that poem, at least in a friendly spirit. The others kept silence.   This indifferentism of those who should have been the first to lend the support of their names naturally affected him ; and made much more sensible the intellectual and moral isolation of his existence.   He was not left quite alone, however.   There were found three or four minds of a loftier reach who had the courage of their convictions, and followed them out to their logical conclusion.   These were Anquetil (the author of *Recherches sur les Indes)*; Charles Nodier, Girod de Chantrans, and Cabantous, dean of the Faculty of Letters at Toulouse.   His brother, Colonel Gleïzès, a member of the Academy of Sciences of the same university, also declared for the reformation.   It is superflous to say that these converts were all men of superior moral calibre to their contemporaries, however high they might be exalted by popular estimates of worth.

Deeply sensible as he was of the profound selfishness and indifferentism of the world surrounding him upon the subject which to him had all the interest and importance of a new religion, he yet constantly displayed the benevolence of his disposition, and the beneficence of his morality, in his efforts for the good of all with whom he came in contact, and particularly in respect to his domestics and his tenants, amongst whom his memory was long held in reverence.   " His exalted nature," states his brother, " glowed with enthusiasm for everything true and good."   His " life-sorrow " seems to have been the want of sympathy on the part of his wife, to whom, nevertheless, he proved an indulgent husband.

His first book, *Les Mélancolies d'un Solitaire,* appeared in the year 1794, in 1800 his *Nuits Elysiennes,* and four years later his *Agrestes;* all more or less advocating the truth.   A long interval elapsed before he again essayed an appeal to the world.   His *Christianisme Expliqué : ou l'Unité de Croyance pour tous les Chrétiens* (Christianity Explained : or, Unity of Belief for all Christians) was published in 1830.   Seven years later it appeared under the title of " Christianity Explained : or, the True Spirit of that Religion Misinterpreted up to the Present Day." In this work, says his estimable editor and translator Herr Springer, "he sought to prove, from the standing-point of a protestant christian, that Christ's mission had for its end the abolition of the murder of animals *(Thiermord),* and that the whole significance of his teaching lay in the words spoken at the institution of the ' Supper,' that is to say, the substitution of bread instead of flesh, and wine instead of blood." This undertaking, it is needless to remark, admirable as was its motive, could hardly, from the nature of the case, be successful.

His last work was his *Thalysie: ou La Nouvelle Existence*, the first part of which was published at Paris in 1840, the second in 1842. He survived this his final appeal to the world on behalf of the new reformation but a few months. He had reached the proverbial limit of human existence; but that his life was shortened by disappointment and the bitter weariness of hope deferred, "by that sorrow which perpetually gnaws at the heart of the unrecognised reformer" (as his biographer well expresses it), we have too much reason to believe. The *Thalysie*—his *magnum opus*—excited, it appears, little interest, or even notice, upon its first appearance. It found one sympathising critic in M. Cabantous, to whom reference has been already made, who delivered a course of lectures upon it from his professorial chair. A few years later a Parisian advocate, M. Blot-Lequène, wrote a treatise in terms of strong recommendation of its principles; and Eugène Stourm, editor of *The Phalanx*, also eloquently advocated its claims upon the public notice. At length it was criticised in the *Révue des Deux Mondes* by Alphonse Esquiros, known to English readers by his contributions to that Review on English life and manners. We are hardly surprised that the criticism was conceived in the usual supercilious and prejudiced spirit.

No attempt appears to have been made to re-publish the *New Existence* until Herr Springer undertook the task for his countrymen. His German version, with an interesting notice of the life and labours of Gleïzès, was published at Berlin in 1872. Criticising a flippant article in *The Food Journal* in the same year, Herr Springer eloquently rebukes the easy and arrogant tone—so successful in appealing to popular prejudices—and observes : " Gleïzès at last published his eminent work, which, as Weilhaüser says, he has written with the blood of his own heart. If it be eccentric, as Mr. Jerrold asserts, it has only *the eccentricity of a gospel of humanity.* Gleïzès was so eccentric as to write the following lines, which were found amongst his posthumous papers : 'God, pure Source of Light, in order to obey thy commands I wrote this book. Be gracious to protect and to support my efforts ; for the humble creature which raises its voice from its grain of sand may, perhaps, be speechless to-morrow, and deep silence reign in the desert.' Yes ; Mr. Jerrold is right : that theory was to its author a religion. In the *Thalysie* we are instructed in the highest questions concerning the health and happiness of mankind. Surpassing all naturalists and philosophers, he explained to us the great mystery of Nature—that robbery and murder [in its full meaning] arose only by corruption, and by alienation from the original laws of creation, and that man, instead of favouring the corruption, as he has done till now, would be able to abolish it. In this way, and in contradiction to the hollow phrases of optimism and the depressing contemplation of pessimism, Gleïzès restores the peace of our mind, and bestows upon us the hope for a future reign of Wisdom and Love."*

In the preface to the *Thalysie* Gleïzès thus expresses his convictions, his hopes, and the general purpose of his labours :—

"The system which I now publish to the world is not, as the usual acceptation of that word might seem to indicate, a collection of principles more or less probable, and of which it depends upon each one to admit or reject the consequences. It is a chain of principles, rigorously true and just, from which man cannot depart without incurring penalties proportionate to his deviation. But, in spite of these penalties which he has suffered, and which he still suffers, he is not aware of his lost condition [*égarement*]. His fate is that of the slave, born in servitude, who plays with his chains, sometimes insults the freemen, and carries his madness to the point of refusing freedom when it is offered to him, and of choosing slavery.

"It is not that *all* men have allowed themselves to be carried willingly down the fatal descent : a large number have struggled against the press, but their diverse and scattered efforts have resembled the eddies of the flood, which ends with forcing together all the diverging waters and hurrying away with them into the gulf of the ocean. Or, if some few have raised and kept themselves above the rapid current, no permanent advantage has resulted from it to the human race, which has been none the less abandoned to itself."

We know that the greatest intellects amongst the Greeks† had taught the better way; but they failed, says Gleïzès, inasmuch as their doctrine was too exclusive and esoteric.

"The condition of the human race is a plain witness of its error. This condition, in fact, is so alarming that it might seem desperate, if it were certain that men had acquired *all* their knowledge. But, happily, there is one branch of it—the most essential of all, and without which the rest is scarcely of any account—which is yet entirely ignored. This knowledge is precisely that of which these great men had glimpses, and of which they reserved to themselves the sole enjoyment ;‡ and it is this knowledge, or, rather, this wisdom (and we know that with the Greeks these two things were comprised under the same denomination) which I publish. I shall give

---

* See the *Dietetic Reformer and Vegetarian Messenger*, August, 1873.

† *Pythagoran, Anytique reum, doctumque Platona* : "Pythagoras and the Man accused by Anytus [Socrates] and the learned Plato."—*Satires* of Horace.

‡ This is, perhaps, scarcely just to Pythagoras and his school. It is, without doubt, deeply to be lamented that they did not more widely promulgate a doctrine of such vital importance to the world ; but the reasons of their reserve and partial reticence have been indicated already in our notice of the founder of *Akreophagy*. In a word—like the Founder of Christianity in a later age— they had many things to say which the world could not then learn. Moreover, as Gleïzès remarks, the teachers themselves could not have from the nature of the case, the full knowledge of late: times.

it an extension which it was not possible for *them* to perceive or to give ; because Nature refuses its life-giving spirit [*esprit de vie*] to solitary and isolated seeds, and makes those only to fructify which enter into the common heritage of mankind.

" With such support, the most feeble must have an advantage over the strongest without it. I have, besides, another advantage. Men feeling to-day, more than ever, the privation of what is wanting to them, invoke on all sides new principles, and demand a higher civilisation. It is not the first time, doubtless, that such a state of things has been manifested. It has been seen to supervene after all the moral revolutions that have left man greater than they have found him. But that of which we have been the witnesses [the revolution in France of 1789—the reforms of 1830] seems to have something more remarkable, more complete—one would almost be tempted to believe that it must be the last, and terminate that long sequence of vain disputes across which the human kind has painfully advanced, seeing it rise in the midst of the *débris* of all the old-world ideas which have expired or are expiring at one's feet. What a moment for rebuilding ! No more favourable one could exist ; and it is urged on, so to speak, by the breeze of these happy circumstances that I offer to the meditation of men the following propositions. . . .

" I shall add but a few words. The principles which I have laid down are absolute —they cannot bend [*fléchir*]. But there are *steps* on the route which conduct to the heights which they occupy ; and were there but a single step made in that direction, that single step could not be regarded as indifferent and unimportant. Thus this work—guide of those whom it shall convince—will be useful also to the rest of the world as, at least, a moderator and a check ; and, I shall avow it, my hopes do not extend beyond this latter object. I should feel myself even perfectly satisfied, if this book should inspire in my contemporaries enough of esteem and favour to prevent them from arresting and impeding it at its start, and to allow it to follow its course towards a generation, I will not say more worthy, but better prepared than the present to receive it."

Gleïzès divides his great work into twelve Discourses, in two volumes, supplemented by a third volume which he entitles *Moral Proofs*. It is an almost exhaustive, as well as eloquent, *résumé* of the history and ethics of the subject. The only fault of this, perhaps, most heartfelt appeal to the reason and conscience of mankind ever published is its too great discursiveness. The manifest anxiety of the author to meet, or to anticipate, every possible objection or subterfuge on the part of the hostile or the indifferent, may well excuse this apparent blemish ; and the slightest acquaintance with his *New Existence* can hardly fail to extort, even from the most prejudiced reader, a tribute of admiration to a spirit so noble and so pure, devoting all its energies to the furtherance of an exalted and refined morality.

In the earlier portion of his book he reviews the dietetic habits and practices of the various peoples of the younger world, and notices the various philosophic and other writers who have left any record of their opinions upon flesh-eating. He next treats of modern authorities, and, after quoting a large number of anti-kreophagistic testimonies, in his fifth Discourse he applies himself to answer the sophisms of the chief opponents, and particularly of its arch-enemy—his countryman, Buffon, in his well-known *Histoire Naturelle* — and he may be said effectually to have disposed of his astonishing fallacies.*

"What most strikes the observer when he throws an attentive glance over the earth, is the *relative* inferiority of man, considered as what he is, in regard to what he ought to be : it is the feebleness of the work compared with the aptitude of the workman. All his inspirations are good, and all his actions bad ; and it is to this singular fact that must be attributed, without doubt, the universal contempt that man exhibits towards his fellows. . . . We must remount to the source, and see if there is not in man's existence some essential act which, reflecting itself on all the rest, would communicate to them its fatal influence. Let us consider, above everything, the *distinctive* quality of man—that which raises him above all other beings. It is clear that it is Pity,† source of that intelligence which has placed him at the head of that fine moral order, invincible in the midst of the catastrophes of Nature. His utter failure to exhibit this feeling of pity towards his humble fellow-beings, as well as to his own kind, engages us to inquire what is the *permanent* cause of such failure ; and we find it, at first, in that unhappy facility with which man receives his *impressions* of the beings by whom he is surrounded. These impressions, transmitted with life and cemented by habit, have formed a creation apart and separate from himself, which is consequently beyond the domain of his conscience, or, if you prefer it, of the ordinary jurisprudence of men. Thus men continue to accuse themselves of being unjust, violent, cruel, and treacherous to one another, but they do not accuse themselves of cutting the throats of other animals and of feeding upon their mangled limbs, which, nevertheless, is the single cause of that injustice, of that violence, of that cruelty, and of that treachery.

"Although all have not these vices to the same degree, and it is exactly this fact which aids the self-deception, I shall clearly prove that all have the *germs* of them ; and that, if they are not equally developed, we must thank the circumstances only which have failed them.

"It is thus that many Europeans, whom their destiny conducts to the cannibal countries, after some months of sojourn with the natives, make no difficulty of seating themselves at their banquet, and of sharing their horrible repast, which at first had excited their horror and disgust. They begin with devouring a dog : from the dog to the man the space is soon cleared.

"Men believe themselves to be just, provided that they fulfil, in regard to their fellows, the duties which have been prescribed to them. But it is goodness which is the justice of man ; and it is impossible, I repeat it, to be good towards one's fellow without being so towards other existences. Let us not be the dupes of *appearances.* Seneca, who lived only on the herbs of his garden, to which he owed those last gleams of philosophy which enlightened, so to speak, the fall of the Roman Empire, also

---

* The eloquence and style of Buffon, it need scarcely be remarked, are more indisputable than his scientific accuracy. Amongst his many errors, none, however, is more surprising than his assertion of the carnivorous anatomical organisation of man, which has been corrected over and over again by physiologists and *savants* more profound than Buffon.

† " *Lachrymas—nostri pars optima sensûs.*"

hinks that crime cannot be circumscribed : *Nullum intrà se ma et vitium.* And if, as Ovid affirms, the sword struck men only after having been first dyed in the blood of the lower animals, what interest have we not in respecting si ch a barrier ? Like Æolus, who held in his hands the bag in which the winds were confined, we may at our will, according as we live upon plants or upon animals, tran uillize the earth or excite terrible tempests upon it.

" I am too well aware that a subterfuge will be found in excusing the crime by necessity, and calumniating Providence. According to the pretended belief of the greatest number of people, if other animals were not put to death, they would deprive men of the empire of the earth. But it is easy to reply to this objection by the examples of people who, holding in horror the effusion of blood, and robbing no being of life—even the vilest or most hateful—are by no means disturbed in the exercise of their sovereignty.* And it would result from the examples of these people, if one had not other proofs besides, that man is absolutely master of the means of increasing or limiting the multiplication of the species which are more or less in dependence upon him. And it is not less evident that the earth, in this latter hypothesis, would support an infinitely greater number of the human species. Thus will the vegetable regimen be *necessarily* adopted one day over the whole earth, when the multiplication of our species shall have reached a certain number fixed and pre-established by that imperious and irrevocable law which is intimately connected, for the most part, with humanity, justice, and virtue—the number at which it is slowly arriving, arrested by the very causes which I am striving to destroy, and which, for that single reason, ought to arm against them all generous beings who appreciate the benefit of existence.†

Amongst other pretexts by which men seek to excuse selfishness, is the assertion that its victims have little or no consciousness of suffering, and that their death is so unexpected that it cannot excite their terror. This monstrous fiction is eloquently exposed by Gleïzès, as it is, indeed, by the commonest everyday experience :—

" The instinct of life among animals generally gives them a presentiment and fear of death—that is to say *violent* death ; for as for natural death it inspires in them no alarm, for the simple reason that it is in the course of nature. And it is the same with man. He is not afflicted with the thought of dying when he knows his hour is come ; he resigns himself to that fate as to any other imposed upon him by necessity. The sensations of other beings differ in no respect from those of men ; and when the horse, for example, is condemned to death by the lion, that is to say, when he hears the confused roar of that terrible beast which fills space, while the precise spot from which it emanates cannot be determined, which takes from the victim all hope of escape by flight, the perspiration rolls down all his limbs, he falls to the earth

---

* In newly-discovered countries, no decided predominance of one species over another has been found ; and the reason is, that qualities are pretty nearly equally divided, and that the strongest animal is not at the same time the most agile or the most intelligent.—*Note* by Gleïzès.

† Upon this, not the least interesting and important of the side-views of Vegetarianism, we refer our readers, amongst numerous authorities, to the opinions of Paley, Adam Smith, Prof. Newman, Liebig, and W. R. Greg (in *Social Problems*).

as if he had just been struck by a thunderbolt, and would die of terror alone if the lion did not run up to terminate the tragedy."[*]

"There exists so great an analogy, so strong a resemblance, between the life of man and that of other animals who surround him, that a simple return to himself—simple reflection—ought to suffice to make him respect the latter ; and if he were condemned by Nature to rend it from them, he might justly curse the order of things which, on the one hand, should have implanted in his heart the source of feeling so gentle, and, on the other, should have imposed on him a necessity so cruel. . . . . And if this man have children, if he bear in his heart objects which are so dear to him, how can he unceasingly surround himself with images of death—of that death which must deprive him one day of those whom he loves, or snatch himself away from their love ?   And if he be just, if he be good, how will he not have repugnance for acts which will continually recall to him ideas of ingratitude, of cruelty, and of violence ? There exists in the East a tree which, by a mechanical movement, inclines its branches towards the traveller, whom it seems to invite to repose under its shade.   This simple image of hospitality, which is revered in that part of the world, makes them regard it as sacred, and they would punish with death him who should dare to apply a hatchet to its trunk.   Our humble fellow-beings, should *they* be less sacred because they represent, not by mechanical movements, but by actions resembling our own, feelings the dearest to our hearts ?   Ah ! let us respect them, not alone because they aid us to bear the burdens of the world, which would overwhelm us without them but *because they have the same right with ourselves to life.* . . . A reason which is without reply, at least for generous souls, is the trust and confidence reposed in man by other animals.   Nature has not taught them to distrust him.   He is the only enemy whom she has not pointed out to them.   Is it not evident proof that he was not intended to be so ?   For can one believe that Nature, who holds so just a balance, could have been willing to deceive all other beings in favour of man alone ?   It has been observed that birds of the gentle species express certain cries when they perceive the fox, the weasel, &c., although they have nothing to fear from them, without doubt, by reason of the analogy which they offer.   They are the cries of hatred rather than of fear, whilst they utter these latter at sight of the eagle, of the hawk, &c. Now, it is certain that in all the islands on which man has landed, the native animals have not fled before them.   They have been able to take even birds with the hand."

Gleïzès rejects the common fallacy that, because men have *acquired* a lust for flesh, *therefore* it is natural or proper for them.

"It is a specious but very false reason to allege that, since man has acquired this taste, he ought to be permitted to indulge it—in the first place because Nature has not given him *cooked* flesh, and because several ages must have rolled away before fire was used. It is very well known that there are many countries in which it was not known at the

---

[*] That the victims of the Slaughter-House have, in fact, a full presentiment of the fate in store for them, must be sufficiently evident to every one who has witnessed a number of oxen or sheep driven towards the scene of slaughter—the frantic struggles to escape and rush past the horrible locality, the exertions necessary on the part of the drovers or slaughtermen to force them to enter as well as the frequent breaking away of the maddened victim—maddened alike by the blows and clamours of its executioners and the presentiment of its destiny—who frantically rushes through the public streets and scatters the terrified human passengers—all this abundantly proves the transparent falsity of the assertion of the unconsciousness or indifference of the victims of the shambles. See a terribly graphic description of a scene of this kind in *Household Words*, No. 14, quoted in *Dietetic Reformer* (1852), in *Thalysie,* and in the *Dietetic Reformer, passim.*   Also in *Animal World,* &c., &c.

period of their discovery. Nature, then, could have given man only *raw* or *living* flesh, and we know that it is repugnant to him over the whole extent of the earth. Now it is exactly this character which essentially distinguishes animals of prey from others. The former, those at least of the larger species, have generally an extreme repugnance, not only for cooked flesh, but even for that which has lost its freshness. Man, then, is not carnivorous but under certain abnormal conditions ; and his senses, to which he appeals in support of his carnivorousness, are perverted to such a degree, that he would devour his fellow-man without perceiving it, if they served him up in place of veal, the flesh of which is said to have the same taste. Thus Harpagus ate, without knowing it, the corpse of his son."

Gleïzès instances the case of Cows and of Reindeer who, in Norway, have been denaturalised so far as to feed on fish, and readily to take to that unnatural food.

" It would be too long to enumerate here all the causes which may have produced so great an aberration. This will be the matter of another Discourse. I shall content myself for the moment with saying some words upon that which perpetuates it. It is essentially that lightness of mind, or, rather, that sort of stupidity, which makes all reflection upon anything which is opposed to their habits painful to the generality of mankind. They would turn their head aside with horrror if they saw what a single one of their repasts costs Nature. They eat animals as some amongst them launch a bomb into the midst of a besieged town, without thinking of the evils which it must bring to a crowd of individuals, strangers to war— women, children, and old men—evils the near spectacle of which they could not support, in spite of the hardness of their hearts. . . . . To-day, when every- thing is calculated with so much precision [he remarks with bitterness], there will not be wanting persons with sufficient assurance to attempt to prove that there is more of advantage for the domesticated animals to be born and live on condition of having their throats cut, than if they had remained in ' nothingness,' or in the natural state. As for the word ' nothingness,' I confess that I do not understand it, but I understand the other very well ; and I have never conceived how man could have had the barbarity to accumulate all the calamities of the earth upon a single in- dividual ; that is to say, to slaughter it in return for having caused its degeneracy. But if he thinks himself to escape from the influence of an action so dastardly and so infamous, he would be in a very great error. . . . .

" I shall finish these prolegomena with an important remark. I have known a large number of good souls who offered up the most sincere wishes for the establish- ment of this doctrine of humaneness, who thought it just and true in all its aspects, who believed in all that it announces ; but who, in spite of so praiseworthy a disposition, dared not be the first to give the example. They awaited this movement from minds stronger than their own. Doubtless they are the minds which give the impulse to the world ; but is it necessary to await this movement when one is convinced of one's self ? Is it permissible to temporise in a question of life or death for innocent beings whose sole crime is *to have been born*, and is it in a case like this that strength of mind should fail justice ? No ! Well-doing is, happily, not so difficult. Ah ! what is your excuse, besides, pusillanimous souls ? I blush for you at the miserable pretexts which keep you back. It would be necessary, say you, to separate one's self from the world ; to renounce one's friends and neighbours. I see no such necessity, and I think, on the contrary, that if you truly loved the world and your neighbours, you would hasten to give them an example which must have so powerful an influence upon their present happiness and upon their future destiny."*

We have reason once again to lament the perversity of literary or publishing enterprise which will produce and reproduce, *ad infinitum*, books of no real and permanent value to the world, and altogether neglect its true luminaries. This is, in an especial manner, the case with Gleïzès. The *Nouvelle Existence* has never been republished, we believe, in the author's own country; while it has never found a translator, perhaps scarcely a reader, in this country outside the Vegetarian ranks. Germany, as we have already noticed, alone has the honour of attempting to preserve from oblivion one of the few who have deserved immortality.

<hr />

XLI.

SHELLEY.   1792—1822.

THAT a principle of profound significance for the welfare of our own species in particular, and for the peaceful harmony of the world in general—that a true spiritualism, of which some of the most admirable of the poets of the pre-Christian ages proved themselves not unconscious, has been, for the most part, altogether overlooked or ignored by modern aspirants to poetic fame is matter for our gravest lament. Thomson, Pope, Shelley, Lamartine—to whom Milton, perhaps, may be added—these form the small band who almost alone represent, and have developed the earlier inspiration of a Hesiod, Ovid, or Virgil, the prophet-poets who, faithful to their proper calling,† have sought to *unbarbarise* and elevate human life by arousing, in various degree, feelings of horror and aversion from the prevailing materialism of living.

Of this illustrious band, and, indeed, of all the great intellectual and moral luminaries who have shed a humanising influence upon our planet—who have left behind them "thoughts that breathe and words that burn"—none can claim more reverence from humanitarians than the poet of poets—the influence of whose life and writings, considerable even now, and gradually increasing, doubtless in a not remote future is destined to be equal to that of the very foremost of the world's teachers, and of whom our sketch, necessarily limited though it is, will be extended beyond the usual allotted space.

<hr />

* *Thalysie: ou La Nouvelle Existence*: Par J. A. Gleïzès. Paris, 1840. in 3 vols., 8vo. See also preface to the German version of R. Springer, Berlin, 1872. Our English readers will be glad to learn that a translation by the English Vegetarian Society is now being contemplated.

† *Poeta*, in its original Greek meaning, marks out a *creator* of new, and, therefore, (it is presumable) true ideas.

Percy Bysshe Shelley descended from an old and wealthy family long settled in Sussex. At the age of 13 he was sent to Eton, where (such was the spirit of the public and other schools at that time, and, indeed, of long afterwards) he was subjected to severe trials of endurance by the rough and rude manners of the ordinary schoolboy, and the harsh and unequal violence of the schoolmaster. Of an exceptionally refined and sensitive temperament, he was none the less determined in resistance to injustice and oppression, and his refusal to submit tamely to their petty tyrannies seems to have brought upon him more than the common amount of harsh treatment. It penetrated into his inmost soul, and inspired the opening stanzas of "The Revolt of Islam," in intensity of feeling seldom equalled. Some alleviation of these sufferings of childhood he found in his own mental resources. For his amusement he translated, we are assured, several books of the *Natural History* of Pliny. Of Greek writers he even then (in an English version) read Plato, who afterwards, in his own language, always remained one of his chief literary companions, and he applied himself also to the study of French and of German. In natural science, Chemistry seems to have been his especial pursuit.

In 1810, at the age of seventeen, he entered University College, Oxford. There he studied and wrote unceasingly. With a strong predilection for metaphysics, he devoted himself in particular to the great masters of dialectics, Locke and Hume, and to their chief representatives in French philosophy. Ardent and enthusiastic in the pursuit of truth, he sought to enlarge his knowledge and ideas from every possible quarter, and he engaged in correspondence with distinguished persons, suggested to him by choice or chance, with whom he discussed the most interesting philosophical questions. Like all truly fruitful minds, the youthful inquirer was not satisfied with the *dicta* of mere authority, or with the *consensus*, however general, of past ages, and he hesitated not, in matters of opinion in which every well-instructed intelligence is capable of judging for itself, to bring to the test of right reason the most widely-received dogmas of Antiquity. Actuated by this spirit, rather than by any matured convictions, and wishing to elicit sincere as well as exhaustive argument on the deepest of all metaphysical inquiries, in an unfortunate moment for himself, he caused to be printed an abstract of anti-theistic speculations, drawn from David Hume and other authorities, presented in a series of mathematically-expressed propositions. Copies of this modest thesis of two pages were sent either by the author, or by some other hand, to the heads of his College. The clerical dignitaries, listening to the dictates of outraged authority, rather than influenced by calm reflection, which would have, perhaps, shewn them the useless injustice of so extreme a measure, proceeded at once to expel him from the University.*

That in spite of this impetuous attack upon the stereotyped presentations of Theism, Shelley had an eminently religious temperament has been well insisted upon by a recent biographer :—

"Brimming over with love for men, he was deficient in sympathy with the conditions under which they actually think and feel. Could he but dethrone the anarch, Custom, the 'Millennium,' he argued, would immediately arrive ; nor did he stop to think how different was the fibre of his own soul from that of the unnumbered multitudes around him. In his adoration of what he recognised as *living*, he retained no reverence for the ossified experience of past ages. . . . For he had a vital faith, and this faith made the ideals he conceived seem possible—faith in the duty and desirability of overthrowing idols ; faith in the gospel of liberty, fraternity, equality ; faith in the divine beauty of Nature ; faith in the perfectibility of man ; faith in the omnipresent soul, whereof our souls are atoms ; faith in love, as the ruling and co-ordinating substance of morality. The man who lived by this faith was in no vulgar sense of the word 'atheist.' When he proclaimed himself to be one he pronounced his hatred of a gloomy religion which had been the instrument of kings and priests for the enslavement of their fellowbeings. As he told his friend Trelawney, he used the word *Atheism* ' to express his abhorrence of superstition : he took it up, as a knight took up a gauntlet, in defiance of injustice.' '†

So thorough was his contempt for mere received and routine thought, that even Aristotle, the great idol of the mediæval schoolmen, and still an object of extraordinary veneration in the elder University, became for him a kind of synonym for despotic authority—

"Tomes
Of reasoned Wrong glozed on by Ignorance "—

and was, accordingly, treated with undue neglect. As for politics, as represented in the parliament and public Press of his day, he was indignantly impatient of the too usual trifling and unreality of public life. He seldom read the newspapers ; nor could he ever bring himself to mix with the "rabble of the House."

Thus, forced into antipathy to the ordinary and orthodox business of life around him, the poet withdrew himself more and more from it into his own thoughts, and hopes, and aspirations, which he communicated to his familiar friends. Some of those, however, into whose, society he chanced to be thrown, were not of a sort of mind most congenial to his own. Yet they all bear witness to his surpassing moral

---

* Compare the fate of Gibbon, who, at the same age, found himself an outcast from the University for a very opposite offence—for having embraced the dogmas of Catholicism. (See *Memoirs of my Life and Writings*, by Edw. Gibbon.) The future historian of *The Decline and Fall*, it may be added, speedily returned to Protestantism, though not to that of his preceptors.

† *Shelley*. By J. A. Symonds. Macmillan, 1887.

no less than mental, constitution. " In no individual, perhaps, was the moral sense ever more completely developed than in Shelley," says one of his most intimate acquaintances ; " in no being was the perception of right and wrong more acute."

"As his love of intellectual pursuits was vehement, and the vigour of his genius almost celestial, so were the purity and sanctity of his life most conspicuous. . . . I have had the happiness to associate with some of the best specimens of gentleness ; but (may my candour and preference be pardoned), I can affirm that Shelley was almost the only example I have yet found that was never wanting, even in the most minute particular, of the infinite and various observances of pure, entire, and perfect gentility." This is the voluntary testimony of a friend who was not inclined to excess of praise.*

The sudden end of his career at Oxford had estranged him from his father, who was of a temperament the very opposite to that of the enthusiastic reformer—harsh, intolerant, and bigoted in his prejudices ; and the young Shelley's marriage, shortly afterwards, to Harriet Westbrook, a young girl of much beauty, but of little cultivation of mind, and in a position of life different from his own, incensed him still further. The marriage, happy enough in the beginning, proved to be an ill-assorted one, and various causes contributed to the inevitable *dénouement*. After a union of some three years, the marriage, by mutual consent, was dissolved. Two years later—not, it seems, in consequence of the divorce, as sometimes has been suggested—the young wife put an end to her existence—a terrible and tragic termination of an ill-considered attachment, which must have caused him the deepest pangs of grief, and which seems always, and justly, to have cast a gloomy shadow upon his future life.

Brief as his career was, we can refer only to the most interesting events in it. Of these, his enthusiastic effort to arouse a bloodless revolution in Ireland, such as, if effected, might have prevented the continued miseries of that especially neglected portion of the three kingdoms, is not the least noteworthy. With his lately-married wife and her sister he was living at Keswick, when, by a sudden inspiration, he resolved to 'cross the Channel, and engage in the work of propagating his principles of political and social reform. This was in the early part of 1812. In Dublin, where they established their head-quarters, he printed an *Address to the Irish People*, which, by his own hands, as well as by other agency, was distributed far and wide. In this wonderfully well-considered and reasonable manifesto, the principles laid down as necessary

---

* Hogg's *Life of Shelley*. Moxon (1858).

to success in attempting deliverance from ages of bad laws and mis-government, are as sound as the ardour and sincerity of his hopeless undertaking are unmistakeable. The cosmopolitan scope of the *Address* appears in such passages as these:—

"Do not inquire if a man be a heretic, if he be a Quaker, a Jew, or a Heathen, but if he be a virtuous man, if he love liberty and truth, if he wish the happiness and peace of human kind. If a man be ever so much 'a believer,' and love not these things, he is a heartless hypocrite and a knave. . . . It is not a merit to tolerate, but it is a crime to be intolerant. . . . Be calm, mild, deliberate, patient. . . . Think, and talk, and discuss. . . . Be free and be happy, but *first be wise and good.* . . . Habits of sobriety, regularity, and thought must be entered into and firmly resolved upon."

Truer in his perception of the radical causes and cure of national evils than most party politicians, he urged the essential need of ethical and social change, without which mere political change of parties, or increase in material wealth of some sections in the community, must be valueless in any true estimate of a nation's prosperity. Shelley also issued, in pamphlet form, *Proposals for an Association*—a plan for the formation of a vast society of Irish Catholics, to enforce their "emanci-pation"—a measure which was not brought about until twenty years later after long and vehement opposition.

Two months were devoted to this generous but futile work; the people of Ireland did not move, and the young reformer returned to England, but without abandoning his *propaganda* of the principles of liberty and justice. While residing in Somersetshire he published a paper entitled a *Declaration of Rights*, to circulate which recourse was had to ingenious methods. Four years later, in 1817, he published *A Proposal for putting Reform to the Vote throughout the Kingdom.* "He saw that the House of Commons did not represent the country; and acting upon his principle that Government is the servant of the Governed, he sought means for ascertaining the real will of the nation with regard to its Parliament, and for bringing the collective opinions of the population to bear upon its rulers. The plan proposed was that a large network of committees should be formed, and that by their means every individual man should be canvassed. We find here the same method of advancing reform by peaceable associations as in Ireland." At the same time, in presence of the incalculable amount of ignorance, destitution, and consequent venality of the great mass of the community—the necessary outcome of long ages of bad and selfish legislation—Universal Suffrage for the present appeared to him to be not a safe experiment. Evidence of controversial power, is his "grave and lofty" Letter to Lord Ellenborough, who had recently sentenced to imprisonment the printers of the *Age of Reason*, "an eloquent argument in favour of toleration and the freedom of the intellect, carrying the matter beyond the instance of legal tyranny, which occasioned its composition, and treating it with philosophical if impassioned, seriousness."* Before his visit to Ireland, he had been engaged (as he tells his corrrespondent, William Godwin) in writing *An Inquiry into the Causes of the Failure of the French Revolution to Benefit Mankind.* We have to lament that this Essay seems never to have been completed, since it is hardly doubtful that it would have been of unusual interest. Such was the force and activity of Shelley's intellect, as displayed in the regions of practical philosophy, at the age of twenty, and before he had given to the world his first productions in poetry.

Queen Mab, written in part two years before, was finished and printed in 1813. Although it may have some of the defects of immaturity of genius, it has the charm of a genuine poetic inspiration. Intense hatred of selfish injustice and untruth in all their shapes, equally intense sympathy with all suffering, sublime faith in the ultimate triumph of Good, clothed in the language of entrancing eloquence and sublimity, are the characteristics of this unique poem. The author's depreciation of his earliest poetic attempt in after years, in a letter addressed to the *Examiner,* only a month before his death, strikes us as scarcely sincere, and as having been a sort of necessary sacrifice on the altar of Expediency.

In this exquisitely beautiful prophecy of a "Golden Age" to be, the fairy Queen Mab, the unembodied being who acts as his instructress and guide through the Universe, displays to his affrighted vision, in one vast panorama, the horrors of the Past and the Present. She afterwards, in a glorious apocalypse, relieves his despair by revealing to him the "new heavens and the new earth," which eventually will displace the present evil constitution of things on our planet. On the redeemed and regenerated Globe :—

> "Ambiguous Man ! he that can know
> More misery, and can dream more joy than all :
> Whose keen sensations thrill within his heart,
> To mingle with a loftier instinct there,
> Lending their power to pleasure and to pain,
> Yet raising, sharpening, and refining each :
> Who stands amid the ever-varying world
> The burden of the glory of the Earth—
> He chief perceives the change : his being notes
> The gradual renovation, and defines
> Each movement of its progress on his mind.
> *   *   *   *   *

* *Shelley.* By J. A. Symonds.

Here now the human being stands, adorning
This loveliest Earth with taintless body and mind.
Blest from his birth with all bland impulses,
Which gently in his truthful bosom wake
All kindly passions and all pure desires.
Him (still from hope to hope the bliss pursuing,
Which from the exhaustless store of human weal
Draws on the virtuous mind), the thoughts that rise
In time-destroying infiniteness, gift
With self-enshrined eternity, that mocks
The unprevailing hoariness of age :
And Man, once fleeting o'er the transient scene,
Swift as an unremembered vision, stands
Immortal upon Earth. *No longer now*
He slays the Lamb who looks him in the face,
And horribly devours his mangled flesh,
Which, still avenging Nature's broken law,
Kindled all putrid humours in his frame—
All evil passions and all vain belief—
Hatred, despair, and loathing in his mind,
The germs of misery, death, disease, and crime.
    No longer now the wingèd habitants,
That in the woods their sweet lives sing away,
Flee from the form of Man.

    *   *   *   *   **

All things are void of terror.   Man has lost
His terrible prerogative, and stands
An equal amidst equals.   Happiness
And Science dawn, though late, upon the Earth.
Peace cheers the mind, Health renovates the frame.
Disease and pleasure cease to mingle here,
Reason and passion cease to combat there ;
Whilst each, unfettered, o'er the Earth extends
Its all-subduing energies, and wields
The sceptre of a vast dominion there ;
Whilst every shape and mode of matter lends
Its force to the omnipotence of Mind,
Which from its dark mine drags the gem of Truth
To decorate its paradise of Peace."

In rapt vision the prophet-poet apostrophises the " New Earth ":

" O happy Earth ! reality of Heaven,
To which those restless souls, that ceaselessly
Throng through the human universe, aspire.

    *   *   *   *   **

Of purest spirits, thou pure dwelling-place,
Where care and sorrow, impotence and crime,
Languor, disease, and ignorance dare not come.

O happy Earth ! reality of Heaven.
  Genius has seen thee in her passionate dreams ;
And dim forebodings of thy loveliness,
Haunting the human heart, have there entwined
Those rooted hopes of some sweet place of bliss.

    ✸   ✻   ✻   ✻   ✻

                 and the souls
That, by the paths of an aspiring change,
Have reached thy haven of perpetual Peace,
There rest from the eternity of toil,
That framed the fabric of thy perfectness.''

From the Essay, in the form of a note, which he subjoined to the passage we have quoted, we extract the principal arguments :——

"Man, and the other animals whom he has afflicted with his malady or depraved by his dominion, are *alone diseased.* The Bison, the wild Hog, the Wolf, are perfectly exempt from malady, and invariably die either from external violence or from mature old age. But the domestic Hog, the Sheep, the Cow, the Dog, are subject to an incredible variety of distempers, and, like the corruptors of their nature, have physicians who thrive upon their miseries. The super-eminence of man is, like Satan's, the super-eminence of pain ; and the majority of his species, doomed to penury, disease, and crime, have reason to curse the untoward event that, by enabling him to communicate his sensations, raised him above the level of his fellow-animals. But the steps that have been taken are irrevocable. The whole of human science is comprised in one question : How can the advantages of intellect and civilisation be reconciled with the liberty and pure pleasures of natural life ? How can we take the benefits and reject the evils of the system which is now interwoven with the fibre of our being ? I believe that abstinence from animal food and spirituous liquors would, in a great measure, capacitate us for the solution of this important question.

"It is true that mental and bodily derangements are attributable, in part, to other deviations from rectitude and nature than those which concern diet. The mistakes cherished by society respecting the connexion of the sexes, whence the misery and diseases of unsatisfied celibacy, unenjoyed prostitution, and the premature arrival of puberty, necessarily spring. The putrid atmosphere of crowded cities, the exhalations of chemical processes, the muffling of our bodies in superfluous apparel, the absurd treatment of infants—all these, and innumerable other causes, contribute their mite to the mass of human evil.

"Comparative Anatomy teaches us that man resembles the frugivorous animals in everything, the carnivorous in nothing. He has neither claws wherewith to seize his prey, nor distinct and pointed teeth to tear the living fibre. A mandarin of the first class, with nails two inches long, would probably find them alone inefficient to hold even a hare. After every subterfuge of gluttony, the bull must be degraded into the "ox," and the ram into the "wether," by an unnatural and inhuman operation, that the flaccid fibre may offer a fainter resistance to rebellious nature. It is only by softening and disguising dead flesh by culinary preparation that it is rendered susceptible of mastication or digestion, and that the sight of its bloody juice and raw horror does not excite loathing and disgust.

"Let the advocate of animal food force himself to a decisive experiment on its fitness, and, as Plutarch recommends, tear a living lamb with his teeth and, plunging

P

his head into its vitals, slake his thirst with the streaming blood. When fresh from this deed of horror, let him revert to the irresistible instinct of nature that would rise in judgment against it and say, 'Nature formed me for such work as this '. Then, and then only would he be consistent.

"Man resembles no carnivorous animal. There is no exception, unless man be one, to the rule of herbivorous animals having cellulated colons.

"The orang-outang perfectly resembles man both in the order and in the number of his teeth. The orang-outang is the most anthropomorphous of the ape tribe, all of whom are strictly frugivorous. There is no other species of animals, which live on different food, in which this analogy exists.* In many frugivorous animals the canine teeth are more pointed and distinct than those of man. The resemblance also of the human stomach to that of the orang-outang is greater than to that of any other animal.

"The structure of the human frame, then, is that of one fitted to a pure vegetable diet in every essential particular. It is true that the reluctance to abstain from animal food, in those who have been long accustomed to its stimulus, is so great in some persons of weak minds as to be scarcely overcome. But this is far from bringing any argument in its favour. A Lamb, who was fed for some time on flesh by a ship's crew, refused her natural diet at the end of the voyage. There are numerous instances of Horses, Sheep, Oxen, and even Wood-Pigeons having been taught to live upon flesh until they have loathed their natural aliment. Young children evidently prefer pastry, oranges, apples, and other fruit, to the flesh of animals, until, by the gradual depravation of the digestive organs, the free use of vegetables has, for a time, produced serious inconveniences—*for a time*, I say, since there never was an instance wherein a change from spirituous liquors and animal food to vegetables and pure water has failed ultimately to invigorate the body by rendering its juices bland and consentaneous, and to restore to the mind that cheerfulness and elasticity which not one in fifty possesses on the present system. A love of strong liquors also is with difficulty taught infants. Almost every one remembers the wry faces which the first glass of port produced. Unsophisticated instinct is invariably unerring, but to decide on the fitness of animal food from the *perverted* appetites which its continued adoption produces, is to make the criminal a judge of his own cause. It is even worse, for it is appealing to the infatuated drunkard in a question of the salubrity of brandy.

"Except in children, there remain no traces of that instinct which determines, in all other animals, what aliment is *natural* or otherwise ; and so perfectly obliterated are they in the reasoning adults of our species, that it has become necessary to urge considerations drawn from comparative anatomy to prove that we are *naturally* frugivorous.

"Crime is madness. Madness is disease. Whenever the cause of disease shall be discovered, the root from which all vice and misery have so long overshadowed the Globe will be bare to the axe. All the exertions of man, from that moment, may be considered as tending to the clear profit of his species. No sane mind in a sane body resolves upon real crime. . . . The system of a simple diet promises no Utopian advantages. It is no mere reform of legislation, whilst the furious passions and evil propensities of the human heart, in which it had its origin, are still unassuaged. It *strikes at the root of all evil*, and is an experiment which may be tried with success, not

---

* Cuvier's *Leçons d'Anatomie Comp.*, Tom. III., pages 169, 373, 443, 465, 480   Rees' *Cyclop.*, Art Man.

alone by nations, but by small societies, families, and even individuals. In no cases has a return to vegetable diet produced the slightest injury ; in most it has been attended with changes undeniably beneficial. Should ever a physician be born with the genius of Locke, I am persuaded that he might trace all bodily and mental derangements to our unnatural habits as clearly as that philosopher has traced all knowledge to sensation. . . .

"By all that is sacred in our hopes for the human race, I conjure those who love happiness and truth to give a fair trial to the vegetable system. Reasoning is surely superfluous on a subject whose merits an experience of six months would set for ever at rest. But it is only among the enlightened and benevolent that so great a sacrifice of appetite and prejudice can be expected, even though its ultimate excellence should not admit of dispute. It is found easier by the short-sighted victims of disease to *palliate* their torments by medicine than to *prevent* them by regimen. The vulgar of all ranks are invariably sensual and indocile, yet I cannot but feel myself persuaded that when the benefits of vegetable diet are mathematically proved ; when it is as clear that those who live naturally are exempt from premature death as that one is not nine, the most sottish of mankind will feel a preference towards. a long and tranquil, contrasted with a short and painful, life. On the average, out of sixty persons four die in three years. Hopes are entertained that, in April, 1814, a statement will be given that sixty persons, all having lived more than three years on vegetables and pure water, are then in *perfect health.* More than two years have now elapsed—*not one of them has died.* No such example will be found in any sixty persons taken at random.

"Seventeen persons of all ages (the families of Dr. Lambe and Mr. Newton) have lived for seven years on this diet without a death, *and almost without the slightest illness.* . . . In proportion to the number of proselytes, so will be the weight of evidence, and when a thousand persons can be produced living on vegetables and distilled water,* who have to dread no disease but old age, the world will be compelled to regard flesh and fermented liquors as slow but certain poisons."

Shelley next insists on the incalculable benefits of a reformed diet economically, socially, and politically :—

"The monopolising eater of flesh would no longer destroy his constitution by devouring an acre at a meal ; and many loaves of bread would cease to contribute to gout, madness, and apoplexy, in the shape of a pint of porter or a dram of gin, when appeasing the long-protracted famine of the hard-working peasant's hungry babes. The quantity of nutritious vegetable matter consumed in fattening the carcase of an ox would afford ten times the sustenance, undepraved, indeed, and incapable of generating disease, if gathered immediately from the bosom of the earth. The most fertile districts of the habitable globe are now actually cultivated by men for [other] animals, at a delay and waste of aliment absolutely incapable of calculation. It is only the wealthy that can, to any great degree, even now, indulge the unnatural craving for dead flesh, and they pay for the greater licence of the privilege by

* Inasmuch as at this moment there are in this country more than two thousand persons of all classes, very many for thirty or forty years strict abstinents from flesh-meat, enrolled members of the Vegetarian Society (not to speak of a probably large number of isolated individual abstinents scattered throughout these islands, who, for whatever reason, have not attached themselves to the Society), and that there have long been Anti-flesh eating Societies in America and in Germany, the *à fortiori* argument in the present instance will be allowed to be of *double* weight.

subjection to supernumerary diseases. Again, the spirit of the nation, that should take the lead in this great reform, would insensibly become *agricultural.*

" The advantage of a reform in diet is obviously greater than that of any other. It strikes at the *root* of the evil. To remedy the abuses of legislation, before we annihilate the propensities by which they are produced, is to suppose that by taking away the *effect* the *cause* will cease to operate. . . .

" Let not too much, however, be expected from this system. The healthiest among us is not exempt from hereditary disease. The most symmetrical, athletic, and long-lived is a being inexpressibly inferior to what he would have been, had not the unnatural habits of his ancestors accumulated for him a certain portion of malady and deformity. In the most perfect specimen of civilised man, something is still found wanting by the physiological critic. Can a return to Nature, then, instantaneously eradicate predispositions that have been slowly taking root in the silence of innumerable Ages? Undoubtedly not. All that I contend for is, that from the moment of relinquishing all *unnatural* habits no new disease is generated ; and that the predisposition to hereditary maladies gradually perishes for want of its accustomed supply. In cases of consumption, cancer, gout, asthma, and scrofula, such is the invariable tendency of a diet of vegetables and pure water. . . .

He concludes this philosophic discourse with an earnest appeal to the various classes of society :—

" I address myself not to the young enthusiast only, to the ardent devotee of truth and virtue—the pure and passionate moralist, yet unvitiated by the contagion of the world. He will embrace a pure system from its abstract truth, its beauty, its simplicity, and its promise of wide-extended benefit. Unless custom has turned poison into food, he will hate the brutal pleasures of the chase by instinct. It will be a contemplation full of horror and disappointment to his mind that beings, capable of the gentlest and most admirable sympathies, should take delight in the deathpangs and last convulsions of dying animals.

" The elderly man, whose youth has been poisoned by intemperance, or who has lived with apparent moderation, and is afflicted with a variety of painful maladies, would find his account in a beneficial change, produced without the risk of poisonous medicines. The mother, to whom the perpetual restlessness of disease, and unaccountable deaths incident to her children, are the causes of incurable unhappiness, would, on this diet, experience the satisfaction of beholding their perpetual health and natural playfulness.* The most valuable lives are daily destroyed by diseases that it is dangerous to palliate, and impossible to cure, by medicine. How much longer will man continue to pimp for the gluttony of Death—his most insidious, implacable, and eternal foe ? "

---

* " See Mr. Newton's Book [*Return to Nature.* Cadell, 1811.] His children are the most beautiful and healthy creatures it is possible to conceive. The girls are perfect models for a sculptor ; their dispositions also are the most gentle and conciliating. The judicious treatment they receive may be a correlative cause of this. In the first five years of their life, of 18,000 children that are born, 7,500 die of various diseases—and how many more that survive are rendered miserable by maladies not immediately mortal ! The quality and quantity of a mother's milk are materially injured by the use of dead flesh. On an island, near Iceland, where no vegetables are to be got, the children invariably die of *tetanus* before they are three weeks old, and the population is supplied from the mainland.—Sir G. Mackenzie's *History of Iceland*—note by Shelley."

Some time after the melancholy death of his first wife, Shelley married Mary Wolstoncroft, the daughter of William Godwin, author of *Political Justice*—perhaps the most revolutionary of all pleas for a change in the constitution of society that has ever proceeded from a prosaic tradesman, such as, in the ordinary intercourse of life and interchange of ideas, his biography and correspondence (lately published) prove him to have been. Her mother was the celebrated and earliest advocate of the rights of women. Previously, the lovers had travelled through France and part of Germany, and an account of their six weeks' tour was afterwards printed by Mrs. Shelley.

In 1815 appeared his *Alastor; or the Spirit of Solitude.* In 1817 he again left England for Geneva. While in Switzerland he made the acquaintance of Byron, which was renewed during his stay in Italy. In the same year he returned to this country and, after a short sojourn with Leigh Hunt, he settled at Great Marlow, one of the most picturesque parts of the Thames. There, in spite of his own ill-health, he showed the active benevolence of his character, not only in the easier form of alms-giving but also in frequent visits to the sick and destitute, at the risk of aggravating symptoms of consumption now alarmingly apparent. There, too, he composed the *Revolt of Islam*, or, as it was originally more fitly entitled, *Laon and Cythna*. In this poem, by the mouth of Laone, he again expresses his humanitarian convictions and sympathies. She calls upon the enfranchised nations :—

> " ' My brethren, we are free ! The fruits are glowing
> Beneath the stars, and the night-winds are flowing
> O'er the ripe corn ; the Birds and Beasts are dreaming—
> Never again may blood of bird or beast
> Stain with his venomous stream a human feast,
> To the pure skies in accusation steaming.
> Avenging poisons shall have ceased
>   To feed disease, and fear, and madness.
>   The dwellers of the earth and air
>   Shall throng around our steps in gladness,
>   Seeking their food or refuge there.
> Our toil from Thought all glorious forms shall cul.
> To make this earth, our home, more beautiful,
> And Science, and her sister Poesy,
> Shall clothe in light the fields and cities of the Free .

>   *   *   *   *   *   *   *   *   *

> " Their feast was such as Earth, the general Mother,
> Pours from her fairest bosom, when she smiles
> In the embrace of Autumn—to each other
> As when some parent fondly reconciles

> Her warring children, *she* their wrath beguiles
> With her own sustenance ; *they*, relenting, weep—
> Such was this Festival, which, from their isles,
> And continents, and winds, and oceans deep,
> All shapes might throng to share, that fly, or walk, or creep :

> " Might share in peace and innocence, for *gore*,
> *Or poison none this festal did pollute.*
> But, piled on high, an overflowing store
> Of pomegranates, and citrons—fairest fruit,
> Melons, and dates, and figs, and many a root
> Sweet and sustaining, and bright grapes, ere yet
> Accursed fire their mild juice could transmute
> Into a mortal bane ; and brown corn set
> In baskets : with pure streams their thirsting lips they wet."*

While he was yet residing in Marlow, the Princess Charlotte, daughter of the Prince of Wales (afterwards George IV.,) died; and, since her character had been in strong contrast with her father's and with royal persons' in general, her early death seems to have caused, not only ceremonial mourning, but also genuine regret amongst all in the community having any knowledge of her exceptional amiability. The poet seized the opportunity of so public an event, and published *An Address to the People on the Death of the Princess Charlotte. By the Hermit of Marlow*, in which he inscribed the motto—" We pity the plumage, but forget the dying bird." In this pamphlet, while paying due tribute of regret for the death of an amiable girl, and fully appreciating the sorrow caused by death as well among the destitute and obscure (with whom, indeed, the too usual absence of the care and sympathy of friends intensifies the sorrow) as among the rich and powerful, he invited, in studiously moderate language, attention to the many just reasons for national mourning in the interests of the poor no less than of princes ; and, in particular, invited the nation to express its indignant grief for the fate of the Lancashire mechanics who, missing the happier fate of their brethren slaughtered at Peterloo, were subjected to an ignominious death by a government which had, by its neglect, encouraged the growth of a just discontent.

In 1818 Shelley left England never to return. At this time was composed the principal part of his masterpiece—*Prometheus Unbound*, the most finished and carefully executed of all his poems. While in Rome (1819) he published *The Cenci*, which had been suggested to him by the famous picture of Guido, until lately supposed to be that of Beatrice Cenci, and by the traditions, current even in the poet's time,

---

* *Revolt of Islam*, v. 51, 55, 56.

of the cruel fate of his heroine. Shakspere's four great dramas excepted, *The Cenci* must take rank as the finest tragic drama since the days of the Greek masters. It is worked up to a degree of pathos unsurpassed by anything of the kind in. literature. "The Fifth Act," remarks Mrs. Shelley, his editor and commentator, "is a masterpiece. Every character has a voice that echoes truth in its tones." *The Cenci* was followed in quick succession by the *Witch of Atlas*, *Adonais* (an elegy on the death of Keats), the most exquisite "In Memoriam"—not excepting Milton's or Tennyson's—ever written; and *Hellas*, which was inspired by his strong sympathy with the Greeks, who were then engaged in the war of independence.

Of his lesser productions, the *Ode to the Skylark* is of an inspiration seldom equalled in its kind. With the "blythe spirit," whom he apostrophises, the poet rises in rapt ecstasy "higher still and higher." For the rest of his productions (the *Letters from Italy* and criticisms or rather eulogies on Greek art have an especial interest) and for the other events in his brief remaining existence we must refer our readers to the complete edition of his works.* The last work upon which he was engaged was his *Triumph of Life*, a poem in the *terza rima* of the *Divine Comedy*. It breaks off abruptly—it is peculiarly interesting to note—with the significant words, "Then what is Life, I cried?"

The manner of his death is well known. While engaged in his usual recreation of boating he was drowned in the bay of Spezia. His body was washed on to the shore and, according to regulations then in force by the Italian governments of the day, in guarding against possible infection from the plague, it was burned where it lay, in presence of his friends Byron and Trelawney, and the ashes were entombed in the Protestant cemetery in Rome—a not unfitting disposal of the remains of one the most spiritualised of human beings.

The following just estimate of the character of his genius and writings, by a thoughtful critic, is worth reproduction here :—"No man was more essentially a poet—'glancing from earth to heaven.' He was, indeed, ' of imagination all compact.' . . . In all his poems he uniformly denounces vice and immorality in every form; and his descriptions of love, which are numerous, are always refined and delicate, with even less of sensuousness than in many of our most admired writers. It is true that he decried marriage, but not in favour of libertinism; and the evils he depicts, or laments, are those arising from the indissolubility of the bond,

* Lately given to the world by Mr. Forman who has carefully collated and printed from Shelley's MSS.

or from the opinions of society as to its necessity—opinions to which he himself submitted by marrying the woman to whom he was attached. . . . His reputation as a poet has gradually widened since his death, and has not yet reached its culminating point. He was the poet of the future—of an ideal futurity—and hence it was that his own age could not entirely sympathise with him. He has been called the 'poet of poets,' a proud title, and, in some respects, deserved."*

Of his creed, the article which he most firmly held, and which, perhaps, most distinguishes him from ordinary thinkers, was the *Perfectibility* of his species, and his firm faith in the ultimate triumph of Good. " He believed," says the one authority who had the best means of knowing his thought and feeling, "that mankind had only to *will* that there should be no evil, and there would be none. It is not my part in these notes to criticise the arguments that have been urged against this opinion, but to mention the fact that he entertained it, and was, indeed, attached to it with fervent enthusiasm. That man could be so perfectionised as to be able to expel Evil from his own nature, and from the greater part of the world, was the cardinal point of his system. And the subject he liked best to dwell upon was the image of One warring with an evil principle, oppressed not only by it but by all, even the good, who were deluded into considering evil a *necessary* portion of humanity—a victim full of gratitude and of hope and of the spirit of triumph emanating from a reliance in the ultimate omnipotence of Good." Such was the conviction which inspired his greatest poem *The Prometheus Unbound.*

A principal charm of his poetry is that which repels the common class of readers : " He loved to *idealise* reality, and this is a task shared by few. We are willing to have our passing whims exalted into passions, for this gratifies our vanity. But few of us understand or sympathise with the endeavour to ally the love of abstract beauty and adoration of abstract Good with sympathies with our own kind."† Of so rare a spirit it is peculiarly interesting to know something of the outward form :—

" His features [describes one of his biographers] were not symmetrical—the mouth, perhaps, excepted. Yet the effect of the whole was extremely powerful. They breathed an animation, a fire, an enthusiasm, a vivid and preternatural intelligence, that I never met with in any other countenance. Nor was the moral expression less beautiful than the intellectual : for there was a softness, a delicacy, a gentleness, and especially (though this will surprise many) that air of profound religious veneration that characterises the best works, and chiefly the frescoes, of the great Masters of Florence and of Rome.

---

* *English Cyclopædia.*
† *Works of Percy Bysshe Shelley.* Edited by Mrs. Shelley. Moxon.

" His eyes were blue, unfathomably dark and lustrous. His hair was brown : but very early in life it became grey, while his unwrinkled face retained to the last a look of wonderful youth. It is admitted on all sides that no adequate picture was ever painted of him. Mulready is reported to have said that he was too beautiful to paint. And yet, although so singularly lovely, he owed less of his charm to regularity of feature, or to grace of movement, than to an indescribable personal fascination.''

As to his voice, impressions varied :—

" Like all finely-tempered natures, he vibrated in harmony with the subjects of his thought. Excitement made his utterance shrill and sharp. Deep feeling, or the sense of beauty, lowered its tone to richness ; but the *timbre* was always acute, in sympathy with his intense temperament. All was of one piece in Shelley's nature. This peculiar voice, varying from moment to moment, and affecting different sensibilities in diverse ways, corresponds to the high-strung passion of his life, his finedrawn and ethereal fancies, and the clear vibrations of his palpitating verse. Such a voice, far-reaching, penetrating, and unearthly, befitted one who lived in rarest ether on the topmost heights of human thought."*

If the physical characteristics of a great Teacher or of a sublime Genius excite a natural curiosity, it is the principal *moral* characteristics which most reasonably and profoundly interest us. To the supremely amiable disposition of the creator of *The Cenci* and *Prometheus Unbound* brief reference has been made ; and we shall fitly supplement this imperfect sketch of his humanitarian career with the vivid impressions left on the mind of the friend who best knew him. Love of truth and hatred of falsehood and injustice were not, in his case, limited to the pages of a book, and forgotten in the too often deadening influence of intercourse with the world—they permeated his whole life and conversation.

" The qualities that struck any one newly introduced to Shelley were, first, a gentle and cordial goodness that animated his discourse with warm affection and helpful sympathy ; the other, the eagerness and ardour with which he was attached to the cause of human happiness and improvement, and the fervent eloquence with which he discussed such subjects. His conversation was marked by its happy abundance, and the beautiful language in which he clothed his poetic ideas and philosophical notions. To defecate life of its misery and its evil was the ruling passion of his soul ; he dedicated to it every power of his mind, every pulsation of his heart. He looked on political freedom as the direct agent to effect the happiness of mankind ; and thus any new-sprung hope of liberty inspired a joy and even exultation more intense and wild than he could have felt for any personal advantage. Those who have never experienced the workings of passion on general and unselfish subjects cannot understand this ; and it must be difficult of comprehension to the younger generation rising around, since they cannot remember the scorn and hatred with which the partisans of reform were regarded some few years ago, nor the persecution to which they were exposed.

" Many advantages attended his birth ; he spurned them all when balanced with what he considered his duties. He was generous to imprudence—devoted to heroism. These characteristics breathe throughout his poetry. The struggle for human weal ; the resolution firm to martyrdom ; the impetuous pursuit ; the glad triumph in good ;

---

* *Shelley.* By J. A. Symonds.

the determination not to despair . . . Perfectly gentle and forbearing in manner, he suffered a great deal of internal irritability, or rather excitement, and his fortitude to bear was almost always on the stretch ; and thus, during a short life, he had gone through more experience of sensation than many whose existence is protracted. ' If I die to-morrow,' he said, on the eve of unanticipated death, ' I have lived to be older than my father.' The weight of thought and feeling burdened him heavily. You read his sufferings in his attenuated frame, while you perceived the mastery he held over them in his animated countenance and brilliant eyes.

"He died, and the world showed no outward sigh ; but his influence over mankind, though slow in growth, is fast augmenting ; and in the ameliorations that have taken place in the political state of his country we may trace, in part, the operation of his arduous struggles. . . . He died, and his place among those who knew him intimately has never been filled up. He walked beside them like a spirit of good to comfort and benefit—to enlighten the darkness of life with irradiations of genius, to cheer with his sympathy and love."*

WITH the name of Shelley is usually connected that of his more popular contemporary, Byron (1788-1824). The brother poets, it already has been noted, met in Switzerland ; and, afterwards, they had some intercourse in Italy during Shelley's last years. Excepting surpassing genius, and equal impatience of conventional laws and usages they had little in common. The one was first and above all a reformer, the other a satirist. To assert, however, the author of *Childe Harold* to have been inspired solely by cynical contempt for his species is unjust. A large part of his poems is pervaded apparently with an intense conviction of the evils of life as produced by human selfishness and folly. But what distinguishes the author of *Prometheus Unbound* from his great rival (if he may be so called) is the sure and certain hope of a future of happiness for the world. Thus, that belief in the all-importance of humane dietetics, as a principal factor in the production of weal or woe on earth, is far less apparent in Byron is matter of course.

Yet, that in moments of better feeling, Byron revolted from the gross materialism of the banquets, of which, as he expresses it, England

> " Was wont to boast—as if a Glutton's tray
> Were something very glorious to behold."†

---

* See preface to *The Poetical Works of Percy Bysshe Shelley.* Edited by Mrs. Shelley. New edition. London, 1869. The increasing reputation of Shelley is proved, at the present time, by the increasing number of editions of his writings, and by the increasing number of thoughtful criticisms and biographies of the poet, by some of the most cultured minds of the day. Since the time, indeed, when a popular writer but sometimes rash critic, with condemnable want of discernment and still more condemnable prejudice, so egregiously misrepresented to his readers the character as well of the poet as of his poems—which latter, nevertheless, he was constrained to admit to be the most "melodious" of all English poetry excepting Shakespere, and (their " utopian " inspiration apart) the most "perfect "—(*Thoughts on Shelley and Byron,* by Rev. C. Kingsley, " Fraser," 1853,) the pre-eminence of the poet, both morally and æsthetically, has been sufficiently established.

and that, had he not been seduced by the dinner-giving propensity of English society, he would have retained his early preference for the refined diet, we are glad to believe. In a letter to his mother, written in his early youth, he announces that he had determined upon relinquishment of flesh-eating, and his clearer mental perceptions in consequence of his reformed living; ‡ and he seems even to have advanced to the extreme frugality of living, at times, upon biscuits and water only.

It would have been well for him had he, like Shelley, abstained from gross eating and drinking upon *principle*; and had he uniformly adhered to the resolution formed in his earlier years, we should, in that case, not have to lament his too notorious sexual intemperance.

XLII.

PHILLIPS. 1767—1840.

It is an obvious truth—in vain demonstrated seventeen centuries since by the best moral teachers of non-Christian antiquity—that abolition of the slaughter-house, with all the cruel barbarism directly or indirectly associated with it, by a necessary and logical corollary, involves abolition of every form of injustice and cruelty. Of this truth the subject of the present article is a conspicuous witness. During his long and active career, in social and political as well as in literary life, Sir Richard Phillips was a consistent *philanthropist*; and few, in his position of influence, have surpassed him in real beneficence. In the face of rancorous obloquy and opposition from that too numerous proportion of communities which systematically resist all "innovation" and deviation from the "ancient paths," he fearlessly maintained the cause of the oppressed; and, as a prison reformer, he claims a place second only to that of Howard.

Of his life we have fuller record than we have of some others of the prophets of dietetic reformation. Yet there is uncertainty as to his birthplace. One account represents him to have been born in London, and to have been the son of a brewer. Another statement, which appears to

---

† In another place he indulges his ironical wit at the expense of the beef-eaters, in representing a certain Cretan personage in Greek story to have

        "Promoted breeding cattle,
      To make the Cretans bloodier in battle;
      For we all know that English people are
      Fed upon beef   .   .   .   .   .
      We know, too, *they are very fond of war*—
      A pleasure—like all pleasures—rather dear.

‡ See *Life and Letters*. Murray.

be more authentic, reports his place of birth to have been in the neighbourhood of Leicester, and his father to have been a farmer. What is of more permanent interest is the account preserved of the reason of his first revolt from the practice of kreophagy. Disliking the business of farming, it seems, while yet quite young, not without the acquiescence of his parents, he had adventurously sought his living, on his own account, in the metropolis. What, if any, plans had been formed by him is not known; but it is certain that he soon found himself in imminent danger of starvation, and, after brief trial, he gladly re-sought his home. Upon his return to the farm, he found awaiting him the welcome of the " Prodigal Son"—although, happily, he had no just claim to the title of that well-known character. A "fatted calf" was killed, and the boy shared in the dish with the rest of the family. It was not until after the feast that he learned that the slaughtered calf had been his especial favourite and playmate. So revolting to his keener sensibility was the consciousness of this fact, that he registered a vow never again to live upon the products of slaughter. To this determination he adhered during the remainder of his long life.*

His next venture, and first choice of a profession, while he was still quite young, led him to engage in teaching. As an advertisement he placed a flag at the door of a house in which he rented a room, where he gave elementary instruction to such children as were entrusted to his tuition by the townspeople of Leicester. The experiment proved not very successful, and at the end of a twelvemonth he tried his fortune elsewhere. He next turned to commerce—at first in a humble fashion. His business prospered, and his next important undertaking was the establishment of a newspaper—the *Leicester Herald*. This journal was what is now called a " Liberal " paper. Yet by those who affected to identify the welfare of England with the continued existence of rotten boroughs and other corruptions, it was held up to opprobrium as revolutionary and "incendiary." Phillips himself had the reputation of an able political writer; but the chief support of the journal was the celebrated Dr. Priestley, whose name and contributions gave it a reputation it otherwise might not have gained. The responsible editor did not escape the perils that then environed the denouncers of legal or social iniquity, and Phillips, convicted of a "misdemeanour," was sentenced to three years' imprisonment in the Leicester jail. During his imprisonment he displayed the beneficence of his disposition in relieving the miseries of some of his more wretched companions. Upon his release, he sold his interest in the *Leicester Herald*, and for some time confined himself altogether to his business.

* *Memoirs of the Public and Private Life of Sir R. Phillips.* London, 1808.

Leaving Leicester he migrated to London and set up a hosiery establishment, which, however, he soon converted into the more congenial bookshop. It was the success of the *Leicester Herald* that, probably, led him to think of starting a new periodical. Upon consultation with Priestley and other friends he was encouraged to proceed, and the *Monthly Magazine* was the result. It commenced in July 1795 and proved to be a most decided success. At first conducted by Priestley, it was afterwards partly under the editorship of Dr. Aikin, author of the *Country Around Manchester*. The proprietors shared in the management of the magazine, but to what extent it is difficult to ascertain. Amongst the contributors was "Peter Pindar," so well known as the author, amongst other satirical rhymes, of the verses upon George III., perplexed by the celebrated "apple dumpling." The monthly receipts from the sale amounted to £1,500. A quarrel with Aikin was followed by the resignation of the editor. Increase of business soon led to a removal of the publishing-house from St. Paul's Churchyard to a much larger establishment in Blackfriars. His home was at Hampstead where, in a beautiful neighbourhood and in an elegant villa, the opulent publisher enjoyed the refined pleasures which his humaneness of living, as well as beneficent industry, had justly deserved. At this time he began a correspondence with C. J. Fox, on the subject of the History of James II., upon which the famous Whig statesman was then engaged. Four letters addressed to him by Fox have been printed, but they have no special importance. He was already married, and the story of his courtship has more than the mere gossiping interest of ordinary biography. Upon his first arrival in London, he had taken lodgings in the house of a milliner. One of her assistants was a Miss Griffiths, a beautiful young Welsh girl, who, learning the unconquerable aversion of their guest from the common culinary barbarism, had amiably volunteered to prepare his dishes on strictly anti-kreophagist principles. This incident induced a sympathy and friendship which speedily resulted in a proposal of marriage. They were a handsome pair; and a somewhat precipitate matrimonial alliance was followed by many years of unmixed happiness for both.

In 1807 the "Livery" of London elected him to the office of High Sheriff of the City and County of Middlesex for the ensuing year. This responsible post put to the proof the sincerity of his professions as a reformer. Nor did he fail in the trial. During his term of power he effected many improvements in the treatment of the real or pretended criminals who, as occupants of the jails, came under his jurisdiction. No one who has read Howard's *State of the Prisons*, published thirty years before Phillips' entrance upon his office, or even general accounts of them, needs to be told that they were the very nurseries of disease, vice, misery, and crime of all kinds—one of the many everlasting disgraces of the governments and civilisation of the day. Nor had they been appreciably improved during the interval of thirty years.

The new Sheriff daily visited Newgate and the Fleet prisons and, by personal inquiry, made himself acquainted with the actual state of the occupants, and in many ways was able to ameliorate their condition. By his direction several collecting boxes were conspicuously displayed, and the alms collected were applied to the relief of the families of destitute debtors. He further insisted that persons, whose indictments had been ignored by the grand jury, should not be detained in the foul and ,pestilential atmosphere, as was then the case, but should be immediately released.

In his admirable *Letter to the Livery of London*, he begins with an appeal to the common sentiments of humanity which ought to have some influence with those in authority. He reminds his readers that:—

"It is too much the fashion to exclude *feeling* from the business of public life, and a total absence of it is considered as a necessary qualification in a public man. Among statesmen and politicians he is considered as weak and incompetent who suffers natural affection to have any influence on his political calculations."

In a note to this passage he adds :—

"It appears to me that political errors of all kinds arise, in a great degree, from the studied banishment of feeling from the consideration of statesmen. Reasoning frequently fails us from a false estimate of the premises on which our deductions are founded. But *feeling*, which, in most respects, is synonymous with conscience, is almost always right. Statesmen are apt to view society as a machine, the several parts of which must be made by them to perform their respective functions for the success of the whole. The comparison is often made, but the analogy is not perfect. The parts of the social machine are made up of sensitive beings, each of whom (though in the obscurest situation) is equal, in all the affections of our nature, to those in the most conspicuous places. The harmony and happiness of the whole will depend on the *degree* of feeling exercised by the directors and prime movers.

After this preliminary exhortation, he presents to their contemplation an appalling revelation of the stupid cruelties of the criminal law and its administration. He gives a graphic account of the jail of Newgate—both of the felons' and the debtors' division. The dimensions of the entire building were 105 yards by 40 yards, of which only one-fourth part was used by the prisoners. Into this space were crowded sometimes seven or eight hundred, never less than four or five hundred, human beings of both sexes and of all ages. "Felons" and debtors seem to have fared pretty much the same, and filth, fever, and starvation prevailed in all parts of the jail alike. The women prisoners he describes as pressed together so closely as, upon lying down, to leave no atom of space between their bodies. As for the results of this neglect on the part of the State, he finds it impossible to draw an adequa.e picture of them, and is at a loss to imagine how the whole city is not carried off by a plague. By persevering energy he obtained some reformation, although he failed in his proposal for a new building.

As to the individual occupants of these pest-houses, he found a large number whose offences were comparatively of an innocent kind, but who were herded with the most savage criminals. He espoused the cause of several of these prisoners—especially of the women—who, after some years of incarceration, were frequently drifted off to Botany Bay, which, besides its other terrors, was for almost all of them a perpetual separ ation from their homes, their husbands, and families. Twice he vainly addressed a memorial to the Secretary of State (Lord Hawkesbury) on their behalf. The traditions and routine of office were too powerful even for his persistent energy.

Romilly had lately introduced his measure for amendment of the barbarous and bloody penal code of this country. Sir Richard Phillips addressed to him also a thoughtful letter, in which were pointed out some of the more glaring abuses in the administration of the laws, with which his official experience as High Sheriff had made him familiar. When Mansfield was Lord Chief Justice, and Thurlow Lord Chancellor, the hangings were so numerous that, as he informs us, on one "hanging holiday" he saw nineteen persons on the gallows, the eldest of whom was not twenty-two years of age. The larger number, probably, had been sentenced to this barbarous death for theft of various kinds. Three hundred years had passed away since the animadversions of More (*before* his accession to office) in the *Utopia,* and some half-century since Beccaria and Voltaire had protested against this monstrous iniquity of criminal legislation, without effect, in England, at least. As far as their contemporaries and their successors for long afterwards were concerned these philanthropists had written wholly in vain.

In the letter to Romilly Phillips insists particularly upon the following reforms: (1) No prisoner to be placed in irons before trial. (2) None to be denied free access of friends or legal advisers. (3) None to be deprived of adequate means of subsistence—14 ounces of bread then being the *maximum* of allowance of food. (4) Every prisoner to be discharged as soon as the grand jury shall have thrown out the bill of indictment. (5) Abolition of payment to jailors by exactions forced from the most destitute prisoners, and of various other exorbitant or illegal fines and extortions. (6) Separation of lunatic from other occupants of the jails. (7) That counsel be provided for those too poor to pay for themselves.

In 1811 Phillips published his *Treatise on the Powers and Duties of Juries, and on the Criminal Laws of England*. Three years later *Golden Rules for Jurymen*, which he afterwards expanded into a book entitled *Golden Rules of Social Philosophy* (1826), in which he lays down rules of conduct for the ordinary business of life—lawyers, clergymen, schoolmasters, and others being the objects of his admonitions. It is in this work that the civic dignitary—so "splendidly false" to the habits of his class—sets forth at length the principles upon which his unalterable faith in the truth of humanitarian dietetics was founded. The reasons of this "true confession" are fully and perspicuously specified, and the first forms the key-note of the rest :—*

"1. *Because*, being mortal himself, and holding his life on the same uncertain and precarious tenure as all other sensitive beings, he does not find himself justified by any supposed superiority or inequality of condition in destroying the enjoyment of existence of any other mortal, except in the necessary defence of his own life.

"2. *Because* the desire of life is so paramount, and so affectingly cherished in all sensitive beings, that he cannot reconcile it to his feelings to destroy or become a voluntary party in the destruction of any innocent living being, however much in his power, or apparently insignificant.

"3. *Because* he feels the same abhorrence from devouring flesh in general that he hears carnivorous men express against eating human flesh, or the flesh of Horses, Dogs, Cats, or other animals which, in some countries, it is not customary for carnivorous men to devour.

"4. *Because* Nature seems to have made a superabundant provision for the nourishment of [frugivorous] animals in the saccharine matter of Roots and Fruits, in the farinaceous matter of Grain, Seed, and Pulse, and in the oleaginous matter of the Stalks, Leaves, and Pericarps of numerous vegetables.

"5. *Because* he feels an utter and unconquerable repugnance against receiving into his stomach the flesh or juices of deceased animal organisation.

"6. *Because* the destruction of the mechanical organisation of vegetables inflicts no sensible suffering, nor violates any moral feeling, while vegetables serve to sustain his health, strength, and spirits above those of most carnivorous men.

"7. *Because* during thirty years of rigid abstinence from the flesh and juices of deceased sensitive beings, he finds that he has not suffered a day's serious illness, that his animal strength and vigour have been equal or superior to that of other men, and that his mind has been fully equal to numerous shocks which he has had to encounter from malice, envy, and various acts of turpitude in his fellow-men.

"8. *Because* observing that carnivorous propensities among animals are accompanied by a total want of sympathetic feelings and gentle sentiments—as in the Hyæna, the Tiger, the Vulture, the Eagle, the Crocodile, and the Shark—he conceives that the practice of these carnivorous tyrants affords no worthy example for the imitation or justification of rational, reflecting, and *conscientious* beings.

"9. *Because* he observes that carnivorous men, unrestrained by reflection or sentiment, even refine on the most cruel practices of the most savage animals [of other species], and apply their resources of mind and art to prolong the miseries of the

---

* They had been published by him several years earlier in the *Medical Journal* for July 27 1811.

victims of their appetites—bleeding, skinning, roasting, and boiling animals alive, and torturing them without reservation or remorse, if they thereby add to the variety or the delicacy of their carnivorous gluttony.

"10. *Because* the natural sentiments and sympathies of human beings, in regard to the killing of other animals, are generally so averse from the practice that few men or women could devour the animals whom *they might be obliged themselves to kill;* and yet they forget, or affect to forget, the living endearments or dying sufferings of the being, while they are wantoning over his remains.

"11. *Because* the human stomach appears to be naturally so averse from receiving the remains of animals, that few could partake of them if they were not disguised and flavoured by culinary preparation; yet rational beings ought to feel that the prepared substances are not the less what they truly are, and *that no disguise of food, in itself loathsome,* ought to delude the unsophisticated perceptions of a considerate mind.

"12. *Because* the forty-seven millions of acres in England and Wales *would maintain in abundance as many human inhabitants,* if they lived wholly on grain, fruits, and vegetables; but they sustain only twelve millions [in 1811] *scantily,* while animal food is made the basis of human subsistence.

"13. *Because* animals do not present or contain the substance of food in mass, like vegetables; every part of their economy being subservient to their mere existence, and their entire frames being solely composed of blood necessary for life, of bones for strength, of muscles for motion, and of nerves for sensation.

"14. *Because* the practice of killing and devouring animals can be justified by no moral plea, by no physical benefit, nor *by any just allegation of necessity in countries where there is abundance of vegetable food,* and where the arts of gardening and husbandry are favoured by social protection, and by the genial character of the soil and climate.

"15. *Because* wherever the number and hostility of predatory land animals might so tend to prevent the cultivation of vegetable food as to render it necessary to destroy and, perhaps, to eat them, there could in that case exist no necessity for destroying the animated existences of the distinct elements of air and water; and, as in most civilised countries, there exist no land animals besides those which are properly bred for slaughter or luxury, of course the destruction of mammals and birds in such countries must be ascribed either to unthinking wantonness or to carnivorous gluttony.

"16. *Because* the stomachs of locomotive beings appear to have been provided for the purpose of conveying about with the moving animal nutritive substances, analogous in effect to the soil in which are fixed the roots of plants and, therefore, nothing ought to be introduced into the stomach for digestion and for absorption by the *lacteals,* or roots of the animal system, but the natural bases of simple nutrition—as the saccharine, the oleaginous, and the farinaceous matter of the vegetable kingdom.*

Perhaps his most entertaining book is his *Morning Walk from London to Kew* (1817). In it he avails himself of the various objects on his road for instructive moralising—as, for example, when he meets with a mutilated soldier, on the frightful waste and cruelty of war; or with a horse struggling up a precipitous hill in agony of suffering from the torture of the bearing-rein, on the common forms of selfish cruelty; or again, when he deplores the incalculable waste of food resources, by the

* *Golden Rules of Social Philosophy : being a System of Ethics.* 1826.

Q

careless indifferentism of owners of land and of the State in allowing the country to remain encumbered with useless, or comparatively useless, timber, in place of planting it with valuable fruit trees of various sorts according to the nature of the soil.

His next publication of importance was his *Million of Facts and Correct Data and Elementary Constants in the entire Circle of the Sciences, and on all Subjects of Speculation and Practice* (1832) 8vo. It is this work by which, perhaps, Phillips is now most known—an immense collection and, although many of the "Constants" may be open to criticism or have already become obsolete, it may still be examined with interest. The plan of the work is that of a classified collection of scraps of information on all the arts and sciences. It was so popular that five large editions were published in seven years. His preface to the stereotyped edition is dated 1839. He remarks that "his pretensions for such a task are a prolonged and uninterrupted intercourse with books and men of letters. He has, for forty-nine years, been occupied as the literary conductor of various public journals of reputation; he has superintended the press in the printing of many hundred books in every branch of human pursuit, and he has been intimately associated with men celebrated for their attainments in each of them." In the facts concerning anatomy and physiology will be found references to scientific and other authorities upon the subject of flesh-eating.

Occasionally we meet with biographical facts of special interest. Thus, he says that, early in 1825, he suggested the first idea of the Society for the Diffusion of Useful Knowledge to Dr. Birkbeck and then, by his advice, to Lord Brougham. His idea was the establishment of a fund for selling or giving away books and tracts, after the manner of the Religious Tract Society. As regards his astronomic paradoxes, his theory, in opposition to the Newtônian, that the phenomena attributed to gravitation are, in reality, the "proximate effects of the orbicular and rotatory motions of the earth" (for which he was severely criticised by Professor De Morgan), exhibits at least the various activity, if not the invariable infallibility, of his mental powers.

A work of equal interest with a *Million of Facts* is his next compilation—*A Dictionary of the Arts of Life and Civilisation* (1833). Under the article *Diet* he well remarks :—

"Some regard it as a purely *egotistical* question whether men live on flesh or on vegetables. But others mix with it moral feelings towards animals. If theory prescribed *human* flesh, the former party would lie in wait to devour their brethren; but the latter, regarding the value of life to all that breathe, consider that, even in a balance of argument, feelings of sympathy ought to turn the scale. . . . . We see all the best animal and social qualities in mere vegetable-feeders. . . . Beasts of prey are necessarily solitary and fearful, even of one another. Physiologists, themselves carnivorous, differ on the subject, but they never take into account *moral* considerations.

"Though it is known that the Hindus and other Eastern peoples live wholly on rice—that the Irish and Scotch peasantry subsist on potatoes and oatmeal—and that the labouring poor of all countries live on the food, of which an acre yields one hundred times more than of flesh, while they enjoy unabated health and long life—yet an endless play of sophistry is maintained about the alleged necessity of killing and devouring animals.

"At twelve years of age the author of this volume was struck with such horror in accidentally seeing the barbarities of a London slaughter-house, that since that hour he has never eaten anything but vegetables. He persevered, in spite of vulgar forebodings, with unabated vigorous health ; and at sixty-six finds himself more able to undergo any fatigue of mind and body than any other person of his age. He quotes himself because the case, in so carnivorous a country, is uncommon—especially in the grades of society in which he has been accustomed to live. . . . On principle he does not abstain from any *vegetable* luxuries or from fermented liquors ; but any indulgence in the latter requires (he hastens to add) the correction of carbonate of soda. He is always in better health when water is his sole beverage ; and such is the case with all who have imitated his practice." *

Under the article " Farming," he observes that " a man who eats 1lb. of flesh eats the exact equivalent of 6lbs. of wheat, and 128lbs. of potatoes." That is, that he, in such proportion, wastes the national resources of a country.

The High Sheriff, on the occasion of some petition to the King, had been knighted (to the affected scandal of his political enemies, who, apparently, wished to reserve all titular or other recognition for their own party), and the conspicuous beneficence of his career, while in office, had gained for him an honourable popularity. But fortune, so long favourable, now for a time showed itself adverse. In 1809 his affairs became embarrassed, and recourse to the bankruptcy court inevitable. Happily his friends aided him in saving from the general wreck the copyright of the *Monthly Magazine*. Its management was a chief occupation of his remaining years ; and his own contributions, under the signature of " Common Sense," attracted marked attention. In his publishing career, the most curious incident was the refusal of the MSS. of *Waverley*. The author's demands seem to have been in excess of the value placed upon the novel by the publisher. It had been advertised in the first instance (he tells us) as the production of Mr. W. Scott. The name was then withdrawn, and the famous novel came before the world anonymously.

A Dictionary of the Arts of Life and Civilisation. 1833. London : Sherwood & Co. It will be seen that the origin of his revolt from orthodox dietetics, given by himself, differs from that narrated in the Life from which we have quoted above. It is possible that both incidents may have equally affected him at the moment, but that the spectacle of the London slaughter-house remained most vividly impressed upon his mind.

Besides the writings already noticed, Phillips compiled or edited a large number of school books. He tells us that all the elementary books, published under the names of Goldsmith, Blair and others, were his own productions—between the years 1798 and 1815. Nor was his mental activity confined to literary work; mechanical and scientific inventions largely occupied his attention. To prevent the enormous expenses of railway viaducts, embankments, and removals of streets, he proposed suspension roads, ten feet above the housetops, with inclined planes of 20° or 30°, and stationary engines to assist the rise and fall at each end. Cities, he maintained, might be traversed in this way on right lines, with intermediate points for ascent and descent. This bold and ingenious idea seems to be very like an anticipation of the elevated railways of New York, although even these have not yet reached the height Phillips thought to be desirable.

He interested himself, also, in steam navigation. When Fulton was in England he was in frequent communication with his English friend, to whom he despatched a triumphant letter on the evening of his first voyage on the Hudson. This letter, having been shown to Earl Stanhope and some eminent engineers, was treated by them with derision as describing an impossibility. Sir R. Phillips then advertised for a company, to repeat on the Thames what had become an accomplished fact on the American rivers. After expenditure of a large sum of money in advertising he obtained only two ten-pound conditional subscribers. He then printed, with commendation, Fulton's letters in the *Monthly Magazine*, and his credulity was almost universally reprobated. It is worth recording that, in the first steam voyage from the Clyde to the Thames, Phillips, three of his family, and five or six others, were the only passengers who had the courage to test the experiment. To allay the public alarms he published a letter in the newspapers, and before the end of that summer he saw the same packet set out on its voyage with 350 passengers. *

In 1840, the year following the final edition of his most popular book, he died at Brighton in the seventy-third year of his age. During his busy life if, by his reforming energy, he had raised up some bitter enemies and detractors, he had made, on the other hand, some valuable friendships. Amongst these—not the least noteworthy—is his intimate friendship with that most humane-minded lawyer, Lord Erskine, one of those who have best adorned the legal profession in this country.

---

* *Million of Facts*, p. 176. For the substance of the greater part of this biography, our acknowledgments are due to the researches of Mr. W. E. A. Axon, F.R.S.L., F.S.S.

XLIII.

LAMARTINE. 1790—1869.

OF aristocratic descent, and educated at the college of the "Fathers of the Faith" (Pères de la Foi), Du Prat—such was the name of his family—imbibed in his youth principles very different from those of his great literary contemporary Michelet. Happily, Nature seems to have endowed his mother with a rare refinement and humaneness of feeling; and from her example and instruction he derived, apparently, the germs of those loftier ideas which, in maturer age, characterise a great part of his writings. While the first Napoléon was still emperor, he entered the army, from which he soon retired to employ his leisure in the more congenial amusement of travel.

In 1820 he first came before the world as the author of *Meditations Poétiques*, of which, within four years, 45,000 copies were sold, and the new poet was eagerly welcomed by the party of Reaction, who thought to find in him a future successor to the brilliant author of the *Génie du Christianisme*, the literary hope of their party, and the champion of the Church and royalty—the political counterbalance to Béranger, the poet of the Revolution—for Hugo had not yet raised the standard of revolt. Yet this remarkable volume with the greatest difficulty found its way into print. "A young man, [writes one of his biographers] his health scarcely re-established from a cruel malady, his face pale with suffering and covered with a veil of sadness, through which could be read the recent loss of an adored being, went about from publisher to publisher, carrying a small packet of verses dyed with tears. Everywhere the poetry and the poet were politely bowed out. At length, a bookseller, better advised, or seduced by the infinite grace of the young poet, decided to accept the manuscript so often rejected." It was published without a name and without recommendation. The melancholy beauty of the style, and the melody of the rhythm, could not fail to attract sympathy from readers of taste and feeling, even from those opposed to his political prejudices—"A rhythm of a celestial melody, verse supple, cadenced, and sonorous, which softly vibrates as an Æolian harp sighing in the evening breeze."

Its political, rather than its poetical, recommendations, we may presume, gained for the writer from the Government of Louis XVIII. a diplomatic post at Florence, which he held until the dynastic revolution of 1830. For some short time he acted as secretary to the French

Embassy in London, and during his stay in England he made the acquaintance of a rich Englishwoman, whom he afterwards married at Florence. A legacy of valuable property from an uncle, upon the condition of his assuming the name of Lamartine, still further enriched him.

In 1829 appeared the collection of *Harmonies Poétiques et Réligieuses*, in which, as in all his poetry up to this time, one of the most character-istic features is his devotion to Legitimacy and the Church. The *renversement* of 1830 considerably modified his political and ecclesiastical ideas. "I wish," he declared at this turning-point in his career, "to enter the ranks of the people ; to think, speak, act, and struggle with them." One of the first proofs of his advanced opinions was his pam-phlet advocating abolition of "capital" punishment. He failed to to obtain a seat in the Chambre des Députés of Louis Philippe, whether in consequence of this advocacy or by reason of his antecedent politics. His enforced leisure he employed in travelling, and in 1832, with his English wife and their young daughter Juliette (whose death at Beyrout caused him inconsolable grief), he set sail for the East in a vessel equipped and armed at his own expense. A narrative of these travels he published in his *Voyage en Orient* (1835). In the following year appeared his *Jocelyn*, a poem of charming tenderness and eloquence, and, in 1838, *La Chute d'un Ange* ("The Fall of an Angel"), in which he, for the first time, gives expression to his feeling of revolt from the bar-barisms of the Slaughter-House. In this strikingly original poem, one of the most remarkable of its kind in any language, Lamartine discovers to us that he no longer views human institutions, the customs of society, and the consecrated usages of nations through the rose-coloured medium of traditional prejudice. It is penetrated with a deep consciousness of the injustice and falseness of a large proportion of those things which are tolerated, and even approved, under the sanction of religious or social law, and with ardent indignation against cruelty and selfishness. In the frightful representation of the practices of the early tyrants of the world saved from the "universal deluge," he allows us to see his own feeling. One of more humane race thus addresses his charming heroine Daïdha :—

> " Ces hommes, pour apaiser leur faim,
> N'ont pas assez des fruits que Dieu mit sous leur main.
> Par un crime envers Dieu dont frémit la Nature,
> Ils demandent au sang une autre nourriture.
> Dans leur cité fangeuse il coule par ruisseaux !
> Les cadavres y sont étalés en monceaux.
> *Ils traînent par les pieds des fleurs de la prairie,*
> *L'innocente brebis que leur main a nourrie,*

Et sous l'œil de l'agneau l'égorgeant sans remords,
Ils savourent ses chairs et vivent de la mort!

*   *   *   *   *   *

De cruels aliments incessamment repus,
Toute pitié s'efface en leurs cœurs corrompus.
Et leur œil, qu'au forfait le forfait habitue,
Aime le sang qui coule et l'innocent qu'on tue.
Ils aiguisent le fer en flèches, en poignard;
Du métier de tuer ils ont fait le grand art:
Le meurtre par milliers s'appelle une victoire,
C'est en lettres de sang que l'on écrit la Gloire."

From the pages of the "Primitive Book," which he imagines to have been originally delivered to men, their hermit-host reads to Daïdha and her celestial, but incarnate, lover the true divine revelation, which is thus sublimely prefaced :—

"Hommes! ne dites pas, en adorant ces pages,
Un Dieu les écrivit par la main de ses sages.

*   *   *   *   *   *

La langue qu'il écrit chante éternellement—
Ses lettres sont ces feux, mondes du firmament
Et, par delà ces cieux, des lettres plus profondes—
Mondes étincelants voilés par d'autres mondes.
Le seul livre divin dans lequel il écrit
Son nom toujours croissant, homme, c'est Ton Esprit!
C'est ta Raison, miroir de la Raison suprême,
Où se peint dans ta nuit quelque ombre de lui-même.
Il vous parle, ô Mortel, mais c'est par ce seul sens.
Toute bouche de chair altère ses accents."

In pronouncing the following code of morality, the voice of conscience and of reason coincides with the divine voice in our hearts :—

"Tu ne leveras point la main contre ton frère :
Et tu ne verseras aucun sang sur la terre,
Ni celui des humains, ni celui des troupeaux
Ni celui des animaux, ni celui des oiseaux :
Un cri sourd dans ton cœur défend de le répandre,
Car le sang est la vie, et tu ne peux la rendre.
Tu ne te nourriras qu'avec les épis blonds
Ondoyant comme l'onde aux flancs de tes vallons,
Avec le riz croissant en roseaux sur tes rives—
Table que chaque été renouvelle aux convives,
Les racines, les fruits sur la branche mûris,
L'excédant des rayons par l'abeille pétris,
Et tous ces dons du sol où la séve de vie
Vient s'offrir de soi-même à ta faim assouvie.
La chair des Animaux crierait comme un remord,
Et la Mort dans ton sein engendrerait la Mort!"

Not only is the human animal sternly forbidden to imbrue his hands in the blood of his innocent earth-mates : it is also enjoined upon him to respect and cultivate their undeveloped intelligence and reason :—

> " Vous ferez alliance avec les 'brutes' même :
> Car Dieu, qui les créa, veut que l'homme les aime.
> D'intelligence et d'âme, à différents degrés,
> Elles ont eu leur part, vous la reconnaîtrez :
> Vous livez dans leurs yeux, douteuse comme un rêve,
> L'aube de la raison qui commence et se lève.
> Vous n'étoufferez pas cette vague clarté,
> Présage de lumière et d'immortalité :
> Vous la respecterez.
> La chaîne à mille anneaux va de l'homme à l'insecte :
> Que ce soit le premier, le dernier, le milieu,
> N'en insultez aucun, car tous tiennent à Dieu ! "

From such more rational estimate should follow, necessarily, just treatment :—

> " Ne les outragez pas par des noms de colère :
> Que la verge et le fouet ne soient pas leur salaire.
> Pour assouvir par eux vos brutaux appétits,
> Ne leur dérobez pas le lait de leurs petits :
> Ne les enchaînez pas serviles et farouches :
> Avec des mors de fer ne brisez pas leurs bouches ·
> Ne les écrasez pas sous de trop lourds fardeaux :
> Comprenez leur nature, adoucissez leur sort :
> *Le pacte entre eux et vous, hommes, n'est pas la Mort.*
> À sa meilleure fin façonnez chaque engeance,
> Prêtez-leur un rayon de votre intelligence :
> Adoucissez leurs mœurs en leur étant plus doux,
> Soyez médiateurs et juges entre eux tous.
>
> *   *   *   *   *   *
>
> *Le plus beau don de l'homme, c'est la Miséricorde.*"

Consistently with, and consequently from, such just human relations with the lower species are the admonitions to break down the walls of partition between the various human races, and to the proper cultivation of the Earth, the common mother of all :—

> " Vous n'établirez pas ces séparations
> En races, en tribus, peuples ou nations.
>
> *   *   *   *   *   *
>
> Vous n'arracherez pas la branche avec le fruit :
> *Gloire à la main qui sème, honte à la main qui nuit !*
> Vous ne laisserez pas le terre aride et nue,
> Car vos pères par Dieu la trouvèrent vêtue.

Que ceux qui passeront sur votre trace un jour
Passent en bénissant leurs pères à leur tour.
Vous l'aimerez d'amour comme on aime sa mère,
Vous y posséderez votre place éphémère,
Comme au soleil assis les hommes, tour à tour,
Possedènt le rayon tant que dure le jour.

*   *   *   *   *   *

Par un inconcevable et maternel mystère,
L'homme en la fatiguant fertilise la Terre.
Nulle bouche ne sent sa tendresse tarir :
Tout ce qu'elle a porté, son flanc peut le nourrir.

*   *   *   *   *   *

Vous vous assisterez dans toutes vos misères,
Vous serez l'un à l'autre enfants, pères, et mères :
Le fardeau de chacun sera celui de tous,
La Charité sera la justice entre vous.
Votre ombre ombragera le passant, votre pain
Restera sur le seuil pour quiconque aura faim :
Vous laisserez toujours quelques fruits sur la branche
Pour que le voyageur vers ses lèvres la penche.
Et vous n'amasserez jamais que pour un temps,
Car la Terre pour vous germe chaque printemps,
Et Dieu, qui verse l'onde et fait fleurir ses rives,
Sait au festin des champs le nombre des convives.*

It is hardly necessary to record that *The Fall of an Angel* was far from receiving, from the world of fashion, the applause of his earlier and more conventional productions.

Lamartine was still in the East (we refer to an earlier period), when news of his election to the Chambre des Deputés by a Legitimist constituency brought him back to Paris. Among the prominent political leaders of the day he figured "as a progressive Conservative, strongly blending reverence for the antique with a kind of philosophical democracy. He spoke frequently on social and philanthropic questions." In 1838 he became deputy for Macon, his native town. During the Orleanist régime he refused to hold office, professing aversion for the "vulgar utility" of the government of Guizot and the Bourgeois King, and in 1845 he openly joined the Liberal opposition. His *Histoire des Girondins* (1847) probably contributed to the expulsion of the Orleanist dynasty in the next year.

In the scenes of the Revolution of February, 1848, he occupied a prominent position as mediator between the two opposite parties; and

---

* *La Chute d'un Ange. Huitième Vision.*

the retention of the tricolour, in place of the Red flag, is attributed to his intervention. Elected a member of the Provisional Government, Lamartine served as Foreign Minister of the Republic. In this capacity he published his well-known *Manifesto à l'Europe*. But, in spite of the fact that ten departments had elected him as representative in the Assemblée Constituante, and that he was also made one of the five members of the Executive Commission, his popularity was short-lived. With all his, apparently, sincere sympathy with the cause of the Oppressed, traditionary associations and strong family attachments (sufficiently manifest in his *Mémoirs*) impeded him in his political course ; and his compromising attitude provoked the distrust of more advanced political reformers. In competition with Louis Napoléon and Cavaignac, he was nominated for the presidency ; but he received the support of few votes. From this period he withdrew into private life and devoted himself entirely to literature. His *Histoire de la Révolution* (1849), *Histoire de la Restauration*, *Histoire de la Russie*, *Histoire de la Turquie*, *Raphael* (a narrative of his childhood and youth) *Confidences* (1849-1851), a further autobiography—one of the most interesting of all his prose productions— and various other writings, most of them appearing, in the first instance, in the periodicals of the day, attested the activity and versatility of his genius. He also for some time conducted a journal—*Conseiller du Peuple*. In 1860 he collected his entire writings into forty-one volumes. Of them his *Histoire des Girondins* is, probably, the most widely known. But, next to *The Fall of an Angel*, it is his own Memoirs which will always have most interest and instruction for those who know how to appreciate true refinement of soul, and, making due deductions from political or traditionary prejudice, can discern essential worth of mind. In *Les Confidences* he allows us to see the natural sensibility and superiority of his disposition in his deep repugnance to the orthodox table—none the less real because he seems, unhappily, to have deemed himself forced to comply with the universal or, rather, fashionable barbarism. Writing of his early education, he tells us :—

"Physically it was derived (*découlait*) in a large measure from Pythagoras and from the *Emile*. Thus it was based upon the greatest simplicity of dress and the most rigorous frugality with regard to food. My mother was convinced, as I myself am, that killing animals for the sake of nourishment from their flesh and blood, is one of the infirmities of our human condition ; that it is one of those curses imposed upon man either by his fall, or by the obduracy of his own perversity. She believed, as I do still, that the habit of hardening the heart towards the most gentle animals, our companions, our helpmates, our brothers in toil, and even in affection, on this earth ; that the slaughtering, the appetite for blood, the sight of quivering flesh are the very things to *have the effect (sont faits pour)* to brutalise and harden the instincts of the heart. She believed, as I do still, that such nourishment, although, apparently, much more succulent and active *(énergique)* contains within itself irritating and putrid principles which embitter the food and shorten the days of man.

"To support these ideas she would instance the numberless refined and pious people of India who abstain from everything that has had life, and the hardy, robust pastoral race, and even the labouring population of our fields, who work the hardest, live the longest and most simply, and who do not eat meat ten times in their lives. She never allowed me to eat it until I was thrown into the rough-and-tumble *(pêle-mêle)* life of the public schools. To wean me from the liking for it she used no arguments, but availed herself of that instinct in us which reasons better than logic. I had a lamb, which a peasant of Milly had given me, and which I had trained to follow me everywhere, like the most attached and faithful dog. We loved each other with that first love *(première passion)* which children and young animals naturally have for each other. One day the cook said to my mother in my presence "Madame, the lamb is fat, and the butcher has come for it ; must I give it him ?" I screamed and threw myself on the lamb, asking what the butcher would do with it, and what was a 'butcher.' The cook replied that he was a man who gained his living by killing lambs, sheep, calves and cows. I could not believe it. I besought my mother and readily obtained mercy for my favourite. A few days afterwards my mother took me with her to the town and led me, as by chance, through the shambles. There I saw men with bared and blood-stained arms felling a bullock. Others were killing calves and sheep, and cutting off their still palpitating limbs. Streams of blood smoked here and there upon the pavement. I was seized with a profound pity, mingled with horror, and asked to be taken away. The idea of these horrible and repulsive scenes, the necessary preliminaries of the dishes I saw served at table, made me hold animal food in disgust, and butchers in horror.

"Although the necessity of conforming to the customs of society has since made me eat what others eat, I shall preserve a rational *(raisonnée)* dislike to flesh dishes, and I have always found it difficult not to consider the trade of a butcher almost on a par with that of the executioner. I lived, then, till I was twelve on bread, milk-products, vegetables and fruit. My health was not the less robust, nor my growth the less rapid ; and perhaps it is to that *regimen* that I owed the beauty of feature, the exquisite sensibility, the serene sweetness of character and temper that I preserved till that date."*

Some years before the publication of his *Fall of an Angel*, Lamartine, from the height of the National Tribune, had given significant expression to the feeling of all the more thoughtful minds, vague though it was, of the urgent need of some new and better principle to inspire and govern human actions than any hitherto tried :—

"I see [he exclaimed] men who, alarmed by the repeated shocks of our political commotions, await from providence a social revolution, and look around them for some man, a philosopher, to arise—*a doctrine* which shall come to take violent possession of the government of minds *(une doctrine qui vienne s'emparer violemment du gouvernement des esprits)*, and reinvigorate the staggered *(ébranlé)* world. They hope, they invoke, they look for this power, which shall impose itself by inherent right *(de son plein droit)* as the Arbitrator and Supreme Ruler of the Future."

---

* *Les Confidences*, par Alphonse de Lamartine, Paris, 1849-51, quoted in *Dietetic Reformer*, August, 1881. It is in this book, too, that he commemorates some of the many atrocities perpetrated by schoolboys with impunity, or even with the connivance of their masters, for their amusement, upon the helpless victims of their unchecked cruelty of disposition.

But a few years earlier, in the same place, a still more positive protest —not the less noteworthy because futile—was heard upon the occasion of a discussion as to the introduction into France of foreign "Cattle," when one of the Deputies, Alexandre de Laborde, maintained that flesh-meat is but an *object of luxury ;* and was supported, at least, by one or two other thoughtful deputies who had the courage of their better convictions. It deserves to be noted that while the Left seemed not unfavourable to the humaner feeling, the Centre apathetic, and the Right derisively antagonistic, the minister of the King (Charles X.) threw all the weight of his position into the materialistic side of the scales. Thus this feeble and last public attempt in France to stop the torrent of Materialism proved abortive.*

XLIV.

MICHELET.   1797—1874.

The early life of this most original and eloquent of French historians passed amidst much hardship and difficulty. His father, who was a printer, had been employed by the government of the Revolution period (1790-1794), and at the political reaction, a few years later, he found himself reduced to poverty. From the experiences of his earlier life Jules Michelet doubtless derived his contempt for the common rich and luxuriant manner of living. Until his sixteenth year, flesh-meat formed no part of his food ; and his diet was of the scantiest as well as simplest kind.

Naturally sensitive and contemplative, and averse from the rough manners and petty tyranny of his schoolfellows, the young student found companionship in a few choice books, of which A'Kempis' *Imitation of Christ* seems to have been at that time one of the most read. At the Sorbonne Michelet carried away some of the most valued prizes, which were conferred with all the *éclat* of the public awards of the *Académie.* At the age of 24, having graduated as doctor in philosophy, he obtained

---

† The question of kreophagy and anti-kreophagy had already been mooted, it appears, in the *Institut,* at the period of the great Revolution of 1789, as a legitimate consequence of the apparent general awakening of the human conscience, when slavery also was first publicly denounced. What was the result of the first raising of this question in the French Chamber of Savans does not appear, but, as Gleïzès remarks, we may easily divine it. One interesting fact was published by the discussion in the Deputies' Chamber—viz., that in the year 1817, in Paris, the consumption of flesh was less than that of the year 1780 by 40,000,000lb., in proportion to the population (see Gleïzès, *Thalysie, Quatrième Discours*), a fact which can only mean that the rich, who support the butchers, had been *forced* by reduced means to live less *carnivorously.*

the chair of History in the Rollin College.   His manner, original and full of enthusiasm, though wanting often in method and accuracy, possessed an irresistible fascination for his readers ; and all, who had the privilege of listening to him, were charmed by his earnest eloquence.

His first principal work was his *Synopsis of Modern History* (1827). His version of the celebrated *Scienza Nuova* of Vico, of whom he regarded himself as the especial disciple, appeared soon after.   Upon the revolution of July, Michelet received the important post of Keeper of the Archives, by which appointment he was enabled to prosecute his researches in preparation for his *magnum opus* in history, *L'Histoire de la France*, the successive volumes of which appeared at long intervals. It contains some of the finest passages in French prose, the episode of *La Pucelle d'Orleans* being, perhaps, the finest of all.   Having previously held a professorship in the Sorbonne (of which he was deprived by Guizot, then minister), he was afterwards invited to fill the chair of History in the Collège de France.

In 1847 his advanced political views deprived him once more of his professorial post and income, in which the Revolution of the next year, however, reinstated him.   The *coup d'état* of 1851 finally banished him from public life—at least as far as teaching was concerned—for being too conscientious to subscribe the oath of allegiance to the new Empire. Michelet, like an eminent writer of the present day, upon principle, elected to be his own publisher ; a fact which, in conjunction with the unpopularity of his opinions, considerably lessened the sale and circulation of his books ; and, by this independency of action, the historian was a pecuniary loser to a great extent.

Deprived of the means of subsistence by his conscientiousness, he left Paris almost penniless, and sought an asylum successively in the Pyrenees and on the Normandy coast.   In 1856 appeared the book with which the name of Michelet will hereafter be most worthily associated— the one which may be said to have been written with his heart's blood. That the taste of the reading world was not entirely corrupt, was proved by the rapid sale of this the most popular of all his productions.   A new edition of *L'Oiseau* came from the press each year for a long period of time, and it has been translated into various European languages.   How far the attractiveness of the book, through the illustrative genius of Giacomelli, influenced the buying public ; how far the surpassing merits of the style and matter of the work—we will not stay to determine ; but it is certain that *The Bird* at once established his popularity as a writer, and relieved his pecuniary needs.   *L'Oiseau* was followed by several

other eloquent interpretations of Nature.   But the first—there can be no
question with persons of taste—remains the masterpiece.   It is, indeed,
unique in its kind in literature—by the intense sympathy and love for
the subject which inspired the writer.   It is the only book which treats
the Bird as something more than an object of interest to the mere classi-
fier, to the natural-history collector, or to the "sportsman."   It considers
the winged tribes—those of the non-raptorial kinds—as possessed of a
high intelligence, of a certain moral faculty, of devoted maternal affection—of a soul, in fine.

Of his remaining writings, *La Bible de l'Humanité* (1863) is one of the
most notable, characteristic as it is of the author's method of treatment
of historical and ethnographical subjects.

The calamities of his native land he so greatly loved, through
the corrupt government which had brought upon it the devastations of a
terrible war, ending, by a natural sequence, in the fearful struggle
of the suffering proletariat, deeply affected the aged champion of the
rights of humanity.   Almost broken-hearted, he withdrew from his accustomed haunts and went to Switzerland, and afterwards to Italy.   He
died at Hyères, in 1874, in the 77th year of his age.   A public funeral,
attended by great numbers of the working classes, awaited him in
the capital.

In the following passage Michelet *virtually* subscribes to the creed of
Vegetarianism.   The saving clause, in which he seems to suppose the
diet of blood to be imposed upon our species by the "cruel fatalities" of
life, it is pretty certain he would have been the first to wish to cancel,
had he enjoyed the opportunity of investigating the scientific basis of
dietetic reform :—

"There is no selfish and exclusive salvation. Man merits his salvation only *through
the salvation of all*. The animals below us have also their rights before God. 'Animal
life, sombre mystery ! Immense world of thoughts and of dumb sufferings ! But
signs too visible, in default of language, express those sufferings. All Nature protests
against the barbarity of man, who misapprehends, who humiliates, who tortures his
inferior brethren.' This sentence, which I wrote in 1846, has recurred to me very
often. This year (1863), in October, near a solitary sea, in the last hours of the
night, when the wind, the wave were hushed in silence, I heard the voices of our
humble domestics. From the basement of the house, and from the obscure depths,
these voices of captivity, feeble and plaintive, reached me and penetrated me with
melancholy—an impression of no vague sensibility, but a serious and positive one.
    "The further we advance in knowledge, the more we apprehend the true meaning
of realities, the more do we understand simple but very serious matters which the
hurry *(entrainement)* of life makes us neglect. Life ! Death ! The daily murder, which
feeding upon other animals implies—those hard and bitter problems sternly placed
themselves before my mind. Miserable contradiction ! Let us hope that there

may be another globe in which the base, the cruel fatalities of this may be spared to us."*

Extolling the greater respect of the Hindus for other life, as exhibited in their sacred scriptures, Michelet vindicates the pre-eminently beneficent character of the Cow, in Europe so ungratefully treated by the recipients of her bounty :—

"Let us name first, with honour, his beneficent nurse—so honoured and beloved by him—the sacred Cow, who furnished the happy nourishment—favourable intermediate between insufficient herbs and flesh, which excites horror. The Cow, whose milk and butter has been so long the sacred offering. She alone supported the primitive people in the long journey from Bactria to India. By her, in face of so many ruins and desolations—by this fruitful nurse, who unceasingly renovates the earth for him, he has lived and always lives."†

In his *Bird* he constantly preaches the faith that can remove mountains—the faith that regards the regeneration and pacification of earth as the proper destiny of our species :—

"The devout faith which we cherish at heart, and which we teach in these pages, is that man will peaceably subdue the whole earth, when he shall gradually perceive that every adopted being, accustomed to a domesticated life, or at least to that degree of friendship and companionship of which his nature is susceptible, will be a hundred times more useful to him than he can be with his throat cut (*qu'il ne pourrait l'être egorgé*). Man will not be truly man until he shall labour seriously for that which the Earth expects from him—the pacification and harmonious union (*ralliement*) of all living Nature. Hunt and make war upon the lion and the eagle if you will, but not upon the Weak and Innocent."

This Michelet never wearies of repeating, and he returns again and again to a truth which is scorned by the modern self-seeking and money-getting, as it was by the fighting, wholly barbarous, world :—

"Conquerors have never failed to turn into derision this gentleness, this tenderness for animated Nature. The Persians, the Romans in Egypt, our Europeans in India, the French in Algeria, have often outraged and stricken these innocent brothers of man—the objects of his ancient reverence. Cambyses slew the sacred Cow ; a Roman the Ibis who destroyed unclean reptiles. But what means the Cow ? The fecundity of the country. And the Ibis ? Its salubrity. Destroy these animals, and

---

* In the same strain an eminent *savan*, Sir D. Brewster, has given expression to his feeling of aversion from the slaughter-house—a righteous feeling which (strange perversion of judgment) is so constantly repressed in spite of all the most forcible promptings of conscience and reason ! These are his words : "But whatever races there be in other spheres, we feel sure that there must be one amongst whom there are no man-eaters—no heroes with red hands—no sovereigns with bloody hearts—and no statesmen who, leaving the people untaught, educate them for the scaffold. In the Decalogue of that community will stand pre-eminent, in letters of burnished gold, the highest of all social obligations—'*Thou shalt not kill*, neither for territory for fame, for lucre, *nor for food, nor for raiment, nor for pleasure.*' The lovely forms of life, and sensation, and instinct, so delicately fashioned by the Master-hand, shall no longer be destroyed and trodden under foot, but shall be the objects of increasing love and admiration, the study of the philosopher, the theme of the poet, and the companions and auxiliaries of Man."—*More Worlds than One.*

† *Bible de l' Humanité—Redemption de la Nature, VI.*

the country is no longer habitable. That which has saved India and Egypt through so many misfortunes and preserved their fertility, is neither the Nile nor the Ganges. It is respect for other life, the mildness and the [comparatively] gentle heart of man.

"Profound in meaning was the speech of the Priest of Saïs to the Greek Herodotus—'You shall be children always.'

"We shall always be so—we men of the West—subtle and graceful reasoners, so long as we shall not have comprehended, with a simple and more exhaustive view, the *motive* of things. To be a child, is to seize life only by partial glimpses. To be a man is to be fully conscious of *all its harmonious unity.* The child disports himself, shatters and destroys; he finds his happiness in *undoing.* And science, in its childhood, does the same. It cannot study unless it kills. The sole use which it makes of a living mind, is, in the first place, to dissect it. None carry into scientific pursuits that tender reverence for life which Nature rewards by unveiling to us her mysteries."*

Like Shelley, he firmly believed in the indefinite amelioration of our world by the ultimate triumph of principles of *humaneness,* so that the " sting of death " and of pain might almost, if not entirely, be removed :—

To prevent death is, undoubtedly, impossible; but we may *prolong* life. We may eventually render pain rarer, less cruel, and *almost suppress* it. That the hardened old world laughs at our expression is so much the better. We saw quite such a spectacle in the days when our Europe, barbarised by war, centered all medical art in surgery, and made the knife its only means of cure, while young America discovered the miracle of that profound dream in which all pain is annihilated.

He upbraids the sportsman no less than he does the scientist, and finds sufficient cause for the too general sterility of the intellect in the habituation to slaughter, and in disregard for the subject species :—

"Woe to the ungrateful! By this phrase I mean the sporting crowd, who, unmindful of the numerous benefits we owe to other animals, exterminate innocent life. A terrible sentence weighs upon the tribes of 'sportsmen'—*they can create nothing.* They originate no art, no industry. They have added nothing to the hereditary patrimony of the human species. . . . . .

"Do not believe the axiom, that huntsmen gradually develope into agriculturalists. It is not so—they kill or die. Such is their whole destiny. We see it clearly through experience. He who has killed will kill—he who has created will create.

"In the want of emotion, which every man suffers from his birth, the child who satisfies it habitually by murder, by a miniature ferocious drama of surprise and treason, of the torture of the weak, will find no great enjoyment in the gentle and tranquil emotions arising from the progressive success of toil and study, from the limited industry which does everything itself. To create, to destroy—these are the two raptures of infancy. To create is a long, slow process; to destroy is quick and easy.

"It is a shocking and hideous thing to see a child partial to 'sport;' to see woman enjoying and admiring murder, and encouraging her child. That delicate and

* Cf. a recently published Essay, in the form of a letter to the present Premier, Mr. Gladstone, entitled *The Woman and the Age.* The author, one of the most refined thinkers of our times, has at once admirably exposed the utter sham as well as cruelty of a vivisecting science, and demonstrated the necessary and natural results to the human race from its shameless outrage upon, and cynical contempt for, the first principles of morality.

'sensitive' woman would not give him a knife, but she gives him a gun. Kill at a distance if it pleases you, for we do not see the suffering. And this mother will think it admirable that her son, kept confined to his room, will drive off *ennui* by plucking the wings from flies, by torturing a bird or a little dog.

"Far-seeing mother! She will know, when too late, the evil of having formed a bad heart. Aged and weak, rejected of the world, she will experience, in her turn, her son's brutality.

"Among too many children we are saddened by their almost incredible sterility. A few recover from it in the long circle of life, when they have become experienced and enlightened men. But the first freshness of the heart? It shall return no more.*

Although, as has already been indicated, Michelet evidently had not examined the *scientific* basis of akreophagy, yet all his aspirations and all his sympathies, it is also equally evident, were for the bloodless diet. With Locke and Rousseau, and many others before him, he presses upon mothers the vital import of not perverting the early preferences of their children for the foods prescribed by unsophisticated nature and their own truer instincts. In one of his books, the most often republished, in laying down rules for the education of young girls, he thus writes :—

"Purity, above everything, *in regimen and nourishment.* What are we to understand by this ?

"I understand by it that the young girl should have the proper nourishment of a child—that she should continue the mild, tranquilising, unexciting regimen of milk ; that, if she eats at your table, she will be accustomed not to touch the dishes upon it, which for her, at least, are poisons.

"A revolution has taken place. We have quitted the more sober French regimen, and have adopted more and more the coarse and bloody diet of our neighbours, appropriate to their climate much more than to ours. The worst of it all is that we inflict this manner of living upon our children. Strange spectacle! To see a mother giving her daughter, whom but yesterday she was suckling at her breast, this gross aliment of bloody meats, and the dangerous excitant wine ! She is astonished to see her violent, capricious, passionate ; but it is herself whom she ought to accuse as the cause. What she fails to perceive, and yet what is very grave, is that with the French race, so precocious, the arousing of the passions is so directly provoked by this food. Far from strengthening, it agitates, it weakens, it unnerves. The mother thinks it fine (*plaisant*) to have a child so preternaturally mature. All this comes from herself. Uuduly excitable, she wishes her child to be such another as she, and she is, without knowing it, the corruptress of her own daughter.

"All this [unnatural stimulation] is of no good to her, and is little better for you, Madame. You have not the heart, you say, to eat anything in which she has no share. Ah, well ! abstain yourself, or, at all events, moderate your indulgence in this food, good, possibly, for the hard-worked man, but fatal in its consequences to the woman of ease and leisure—regimen which *vulgarises* her, perturbs her, renders her irritable, or oppresses her with indigestion.

---

* *The Bird*, by Jules Michelet. English Translation. Nelson, London, 1870. See, too, his eloquent exposure of the scientific or popular error which, denying conscious reason and intelligence, in order to explain the mental constitution of the non-human races (as well that of the higher mammals as of the inferior species), has invented the vague and mystifying term "instinct."

R

For the woman and the child it is a grace—an amiable grace (*grace d'amour*)—to be, above all things, *frugivorous*—to avoid the coarseness and foulness (*fétidité*) of flesh-meats, and to live rather upon innocent foods, which bring death to no one (*qui ne coûtent la mort à personne*)—sweet nourishment which charms the sense of smell as much as it does the taste. The real reason why the beloved ones in nothing inspire in us repugnance but, in comparison with men, seem ethereal, is, in a special manner, their [presumed] preference for herbs and for fruits—for that purity of regimen which contributes not a little to that of the soul, and assimilates them to the innocency of the flowers of the field.*

XLV.

COWHERD. 1763—1816.

In any history of Vegetarianism it is impossible to omit record of the lives and labours of the institutors of a religious community who, in establishing humane dietetics as an essential condition of membership, may well claim the honourable title of religious reformers, and to whom belongs the singular merit of being the first and only founders of a Christian church who have inculcated a true religion of life as the *basis* of their teaching.

William Cowherd, the first founder of this new conception of the Christian religion, which assumed the name of the "Bible Christian Church," was born at Carnforth, near Lonsdale, in 1763. His first appearance in public was as teacher of philology in a theological college at Beverley. Afterwards, coming to Manchester, he acted as curate to the Rev. J. Clowes, who, while remaining a member of the Established Church, had adopted the theological system of Swedenborg. Cowherd attached himself to the same mystic creed, and he is said to be one of the few students of him who have ever read through all the Latin writings of the Swedish theologian. He soon resigned his curacy, and for a short time he preached in the Swedenborgian temple in Peter Street. There he seems not to have found the freedom of opinion and breadth in teaching he had expected, and he determined to propagate his own convictions, independently of other authority. In the year 1800 he built, at his own expense, Christ Church, in King Street, Salford—the first meeting-place of the reformed church.* His extraordinary eloquence and ability, as well as earnestness of purpose, quickly attracted a large audience, and may well have brought to recollection the style and matter of the great orator of Constantinople of the fourth century. One characteristic of his Church—perhaps unique at that time—was the non-appropriation of

---

* *La Femme*, vi. Onzième Edition. Paris, 1879.

sittings. Another unfashionable opinion held by him was the Pauline one of the obligation upon Christian preachers to maintain themselves by some "secular" labour, and he therefore kept a boarding school, which attained extensive proportions. In this college some zealous and able men, who afterwards were ordained by him to carry on a truly beneficent ministry, assisted in the work of teaching, of whom the names of Metcalfe, Clark, and Schofield are particularly noteworthy. Following out the principles of their Master, two of them took degrees in medicine, and gained their living by that profession. The Principal himself built an institute, connected with his church in Hulme, where, more recently, the late Mr. James Gaskill presided, who, at his death, left an endowment for its perpetuation as an educational establishment.

It was in the year 1809 that Cowherd formally promulgated, as cardinal doctrines of his system, the principle of abstinence from flesh-eating, which, in the first instance, he seems to have derived from "the medical arguments of Dr. Cheyne and the humanitarian sentiments of St. Pierre." He died not many years after this formal declaration of faith and practice, not without the satisfaction of knowing that able and earnest disciples would carry on the great work of renovating the religious sentiment for the humanisation of the world.

Of those followers not the least eminent was Joseph Brotherton, the first M.P. for Salford, than which borough none has been more truly honoured by the choice of its legislative representative. A printing press had been set up at the Institution, and, after the death of the Master, his *Facts Authentic in Science and Religion towards a New Foundation of the Bible*, under which title he had collected the most various matter illustrative of passages in the Bible, and in defence of his own interpretation of them, was there printed. It is, as his biographer has well described it, "a lasting memorial of his wide reading and research—travellers, lawyers, poets, physicians, all are pressed into his service—the whole work forming a large quarto common-place book filled with reading as delightful as it is discursive. Some of his minor writings have also been printed. He was, besides his theological erudition, a practical chemist and astronomer, and he caused the dome of the church in King Street to be fitted up for the joint purposes of an observatory and a laboratory. His microscope is still preserved in the Peel Park Museum. His valuable library, which at one time was accessible to the public on easy terms, is now deposited in the new Bible Christian Church in Cross

---

* This memorable building has been succeeded by the present well-known one in Cross Lane, where the Rev. James Clark, one of the most esteemed, as well as one of the oldest, members of the Vegetarian Society is the able and eloquent officiating minister.

Lane. The books collected exhibit the strong mind which brought them together for its own uses. This library is the workshop in which he wrought out a new mode of life and a new theory of doctrine—with these instruments he moulded minds like that of Brotherton, and so his influence has worked in many unseen channels." He died in 1816, and is buried in front of his chapel, in King Street, Salford.*

XLVI.

METCALFE. 1788—1862.

Amongst the immediate disciples of the founder of the new community, the most active apostle of the principles of Vegetarianism, William Metcalfe, to whom reference has been already made, claims particular notice. Born at Orton in Westmoreland, after instruction in a classical school kept by a philologist of some repute, he began life as an accountant at Keighley, in Yorkshire. His leisure hours were devoted to mental culture, both in reading and in poetic composition. Converted by Cowherd in 1809, in the twenty-first year of his age, he abandoned the flesh diet, and remained to the end a firm believer in the truths of "The Perfect Way." In the year following he married the daughter of the Rev. J. Wright who was at the head of the "New Church" at Keighley, and whom he assisted as curate. His wife, of highly-cultured mind, equally with himself was a persistent follower of the reformed mode of living. Sharing the experiences of many other dietary reformers, the young converts encountered much opposition from their family and friends, who attempted at one moment ridicule, at another dissuasion, by appealing to medical authority. Unmoved from their purpose, they continued unshaken in their convictions.

"They assured me," he writes at a later period, "that I was rapidly sinking into a consumption, and tried various other methods to induce me to return to the customary dietetic habits of society; but their efforts proved ineffectual. Some predicted my death in three or four months; and others, on hearing me attempt to defend my course, hesitated not to tell me I was certainly suffering from mental derangement, and, if I continued to live without flesh-food much longer, would unquestionably have to be shut up in some insane asylum. All was unavailing. Instead of sinking into consumption, I gained several pounds in weight during the first few weeks of my experiment. Instead of three or four months bringing me to the silent grave, they brought me to the matrimonial altar.

* These biographical facts we have transferred to our pages from an interesting notice by Mr W. E. A. Axon, F.R.S.L.

" She [his wife] fully coincided with me in my views on vegetable diet, and, indeed, on all other important points was always ready to defend them to the best of her ability—studied to show our acquaintances, whenever they paid us a visit, that we could live, in every rational enjoyment, without the use of flesh for food. As she was an excellent cook, we were never at a loss as to what we should eat. We commenced housekeeping in January, 1810, and, from that date to the present time, we have never had a pound of flesh-meat in our dwelling, have never patronised either slaughter-houses or spirit shops.

" When, again, in the course of time we were about to be blessed with an addition to our family, a renewed effort was made. We were assured it was impossible for my wife to get through her confinement without some *more strengthening food*. Friends and physicians were alike decided upon that point. We were, notwithstanding, unmoved and faithful to our principles. Next we were told by our kind advisers that the little stranger could not be sufficiently nourished unless the mother could eat a little 'meat' once a day ; or, if not that, drink a pint or half a pint of ale daily. To both proposals my wife turned a deaf ear ; and both she and the child did exceedingly well.* It may be proper to add here [remarks the biographer], that the 'little stranger' above referred to is the author of this *Memoir*,—that he is in the fifty-sixth year of his age, that he has never so much as *tasted* animal food, nor used intoxicating drinks of any kind, and that he is hale and hearty."

These experiences, it is scarcely necessary to remark, in the lives of followers of reformed dietetics, have been not seldom repeated.

In the Academy of Sciences, instituted by Dr. Cowherd, Metcalfe was invited to assume the direction of the " classical " department (1811). In the same year he took " Orders," and, at the solicitation of the secessionists from the Swedenborgian Communion (which, with some inconsistency, seems to have looked with indifference, or even dislike, upon the principles of akreophagy), he officiated at Adingham, in Yorkshire. By the voluntary aid of one of his admirers a church was built, to which was added a commodious school-room. He then resigned his position under Dr. Cowherd, and opened a grammar school in Adingham, where he was well supported by his friends.

The United States of America, however, was the field to which he had long been looking as the most promising for the mission work to which he had devoted himself ; and in this hope he had been sustained by his Master. In the spring of 1817 a company of forty-one persons, members of the Bible Christian community, embarked at Liverpool for Philadelphia. They comprised two clerics—W. Metcalfe and Jas. Clark—twenty other adults, and nineteen children. Of this band only a part were able to resist the numerous temptations to conformity with the prevalent social practices ; and the vast distances which separated the leaders from their

---

* *Memoir of the Rev. William Metcalfe, M.D.* By his son, Rev. Joseph Metcalfe, Philadelphia, 1866.

followers were almost an insuperable bar to sympathy and union. Settling in Philadelphia—for them at least a name of real significance—Metcalfe supported his family by teaching, while performing the duties of his position as head of the faithful few who formed his church. His day-school, which was attended by the sons of some of the leading people of the city, proved to be pecuniarily successful until the appearance of yellow fever in Philadelphia, which broke up his establishment and involved him in great difficulties; for upon his school he depended entirely for his living. He had many influential friends, who tempted him, at this crisis of his fortunes, with magnificent promises of support, if only he would desert the cause he had at heart—the propagandism of a religion based upon principles of true temperance and active goodness. Both moral and physical superiority pointed him out as one who could not fail to bring honour to any undertaking, and, had he sacrificed conviction to interest, he might have greatly advanced his material prospects. All such seductions he firmly resisted.

Meanwhile, through the pulpit, the schoolroom, and, more widely, through the newspapers, he scattered the seeds of the gospel of Humanity. But the spirit of intolerance and persecution, of self-seeking religionism, and of rancorous prejudice, was by no means extinct even in the great republic, and the (so-called) " religious " press united to denounce his humane teaching as well as his more liberal theology. Nor did some of his more unscrupulous opponents hesitate, in the last resort, to raise the war-cry of "infidel" and "sceptic." These assailants he treated with contemptuous silence ; but the principle of moral dietetics he defended in the newspapers with ability and vigour. In 1821 he published an essay on *Abstinence from the Flesh of Animals*, which was freely and extensively circulated. For several years his missionary labours appear to have been unproductive. In the year 1830 he made two notable converts— Dr. Sylvester Graham, who was at that time engaged as a "temperance" lecturer, and was deep in the study of human physiology; and Dr. W. Alcott. Five years later, the *Moral Reformer* was started as a monthly periodical, which afterwards appeared under the title of the *Library of Health*. In 1838-9 the *Graham Journal* was also published in Boston, and scientific societies were organised in many of the New England towns. The Bible was largely appealed to in the controversy, and a sermon of Metcalfe's had an extensive circulation through the United States. With all this controversy upon his hands, he was far from neglecting his private duties, and, in fact, his health was over-taxed in the close and constant work in the schoolrooms, overcrowded and ill-ventilated as they were. In the day and night school he was constantly employed, during one half of the year, from eight in the morning until ten at night; and Sunday brought him no remission of labour.

In the propagandism of his principles through the press he was not idle. The *Independent Democrat*, and, in 1838, the *Morning Star*, was printed and published at his own office—by which latter journal, in spite of the promise of support from political friends, he was a pecuniary loser to a large amount. *The Temperance Advocate*, also issued from his office, had no better success. Several years earlier, about 1820, it is interesting to note, he had published a tract on *The Duty of Abstinence from all Intoxicating Drinks;* and the founder of the Bible Christian Church in America can claim the merit of having been the first systematically to inculcate this social reform.

In the year 1847 the Vegetarian Society of Great Britain had been founded, of which Mr. James Simpson had been elected the first president. Metcalfe immediately proposed the formation of a like society in the United States. He corresponded with Drs. Graham, Alcott, and others; and finally an American Vegetarian Convention assembled in New York, May 15, 1850. Several promoters of the cause, previously unknown to each other (except through correspondence), here met. Metcalfe was elected president of the Convention; addresses were delivered, and the constitution of the society determined upon. The Society was organised by the election of Dr. William Alcott as president, Rev. W. Metcalfe as corresponding secretary, and Dr. Trall as recording secretary. An organ of the society was started in November, 1850, under the title of *The American Vegetarian and Health Journal*, and under the editorship of Metcalfe. Its regular monthly publication, however, did not begin until 1851. In that year he was selected as delegate to the English Vegetarian Society, as well as delegate from the Pennsylvania Peace Society to the "World's Peace Convention," which was fondly supposed to be about to be inaugurated by the *Universal Exhibition* of that year. The proceedings at the annual meeting of the Vegetarian Society of Great Britain, and the eloquent address, amongst others, of the American representative, are fully recorded in the *Vegetarian Messenger* for 1852. On this occasion Joseph Brotherton, M.P. presided.

Two years later he suffered the irreparable loss of the sympathising sharer in his hopes for the regeneration of the world. Mrs. Metcalfe died in the seventy-fourth year of her age, having been, during forty-four years, a strict abstinent. Her loss was mourned by the entire Vegetarian community. By far the larger part of the matter, as well as the expenses of publication, of the *American Vegetarian*, was supplied by the editor, and, being inadequately supported by the rest of the community, the managers were forced to abandon its further publication. The last volume appeared in 1854. It has been succeeded in later times, under happier circumstances, by the *Health Reformer* which is still in existence.

In 1855 Metcalfe received an invitation to undertake the duties attached to the mother church at Salford. Leaving his brother-in-law in charge of the church in Philadelphia, he embarked for England once more, and the most memorable event, during his stay in this country, was the deeply and sincerely lamented death of Joseph Brotherton, who for twenty years had represented Salford in the Legislature, and whose true benevolence had endeared him to the whole community. Metcalfe was chosen to preach the funeral eulogy, which was listened to by a large number of Members of Parliament and municipal officers, and by an immense concourse of private citizens. Returning to America soon afterwards, at the urgent request of his friends in Philadelphia, he was, in 1859, elected to fill the place of President vacated by Dr. Alcott, whose virtues and labours in the cause he commemorated in a just eulogy. His own death took place in the year 1862, in the seventy-fifth year of his age, caused by hemorrhage of the lungs, doubtless the effect of excessive work. His end, like his whole interior if not exterior life, was, in the best meaning of a too conventional expression, full of peace and of hope. His best panegyric is to be found in his life-work; and, as the first who systematically taught the truths of reformed dietetics in the "New World," he has deserved the unceasing gratitude of all sincere reformers in the United States, and, indeed, throughout the globe. By all who knew him personally he was as much loved as he was esteemed, and the newspapers of the day bore witness to the general lamentation for his loss.*

XLVII.

GRAHAM. 1794—1851.

As an exponent of the physiological basis of the Vegetarian theory of diet, in the most elaborate minuteness, the author of *Lectures on the Science of Human Life* has always had great repute amongst food reformers both in the United States and in this country. Collaterally connected with the ducal house of Montrose, his father, a graduate of Oxford, emigrated to Boston, U.S., in the year 1718. He must have

---

* See *Memoir of the Rev. William Metcalfe*. By his son, the Rev J. Metcalfe. Philadelphia; J. Capen. 1866.

attained an advanced age when his seventeenth child, Sylvester, was born at Suffield, in Connecticut. Yet he seems to have been of a naturally dyspeptic and somewhat feeble constitution, which was inherited by his son, whose life, in fact, was preserved only by the method recommended by Locke—free exposure in the open air. During several years he lived with an uncle, on whose farm he was made to work with the labourers. In his twelfth year he was sent to a school in New York, and at fourteen he was set for a short time to learn the trade of paper-making. " He is described as handsome, clever, and imaginative. ' I had heard,' he says, ' of noble deeds, and longed to follow in the field of fame.' Ill health soon obliged his return to the country, and at sixteen symptoms of consumption appeared. Various occupations were tried until the time, when about twenty years of age, he commenced as a teacher of youth, proving highly successful with his pupils. Again ill-health obliged the abandonment of this pursuit."*

At the age of thirty-two he married, and soon after became a preacher in the Presbyterian Church. Deeply interested in the question of "Temperance," he was invited to lecture for that cause by the Pennsylvania Society (1830). He now began the study of physiology and comparative anatomy, in which his interest was unremitting. These important sciences were used to good effect in his future dietetic crusade. At this time he came in contact with Metcalfe, by whom he was confirmed in, if not in the first instance converted to, the principles of radical dietary reform. " He was soon led to believe that no permanent cure for intemperance could be found, except in such change of personal and social customs as would relieve the human being from all desire for stimulants. This idea he soon applied to medicine, so that the prevention and cure of disease, as well as the remedy for intemperance, were seen to consist mainly in the adoption of correct habits of living, and the judicious adaptation of hygienic agencies. These ideas were elaborated in an *Essay on the Cholera* (1832), and a course of lectures which were delivered in various parts of the country, and subsequently published under the title of *Lectures on the Science of Human Life* (2 vols., Boston, 1839). This has been the leading text-book of all the dietetic and nearly all the health reformers since."†

---

* See Memoir in *Sylvester Graham's Lectures on the Science of Human Life.* Condensed by T. Baker, Esq., of the Inner Temple, Barrister-at-Law. Manchester: Heywood; London: Pitman.

† *The New American Cyclopædia.* Appleton, New York, 1861. It deserves remark in this place that, in no English cyclopædia or biographical dictionary, as far as our knowledge extends, is any sort of notice given of this great sanitary reformer. The same disappointment is experienced in regard to not a few other great names, whether in hygienic or humanitarian literature. The absence of the names of such true benefactors of the world in these books of reference is all the more surprising in view of the presence of an infinite number of persons—of all kinds—who have contributed little to the stock of true knowledge or to the welfare of the world.

The Science of Human Life is one of the most comprehensive as well as minute text books on scientific dietetics ever put forth. If it errs at all, it errs on the side of redundancy—a feature which it owes to the fact that it was published to the world as it was orally given. It therefore well bears condensation, and this has been judiciously done by Mr. Baker, whose useful edition is probably in the hands of most of our readers.   Graham was also the author of a treatise on *Bread and Bread-Making,* and "Graham bread" is now universally known as one of the most wholesome kinds of the "staff of life." Besides these more practical writings, for some time before his death he occupied his leisure in the production of a *Philosophy of Sacred History,* the characteristic idea of which seems to have been to harmonise the dogmas of the Jewish and Christian Scriptures with his published views on physiology and dietetics.   He lived to complete one volume only (12mo.), which appeared after his death.

Tracing the history of Medicine from the earlier times, and its more or less of empiricism in all its stages, Graham discovers the cause of a vast proportion of all the egregious failure of its professors in the blind prejudice which induces them to apply to the *temporary cure,* rather than to the *prevention,* of disease.   As it was in its first barbarous beginning, so it has continued, with little really essential change, to the present moment :—

"Everything is done with a view to *cure* the disease, without any regard to its cause, and the disease is considered as the infliction of some supernatural being. Therefore, in the progress of the healing art thus far, not a step is taken towards investigating the laws of health and the philosophy of disease.

"Nor, after Medicine had received a more systematic form, did it apply to those researches which were most essential to its success, but, like religion, it became blended with superstitions and absurdities.   Hence, the history of Medicine, with very limited exceptions, is a tissue of ignorance and error, and only serves to demonstrate the absence of that knowledge upon which alone an enlightened system of Medicine can be founded, and to show to what extent a noble art can be perverted from its capabilities of good to almost unmixed evil by the ignorance, superstition, and cupidity of men.   In modern times, anatomy and surgery have been carried nearly to perfection, and great advance has been made in physiology.   The science of human life has been studied with interest and success, but this has been confined to the few, while even in our day, and in the medical profession itself, the general tendency is adverse to the diffusion of scientific knowledge.

*   *   *   *   *   *   *

"The result is, that men prodigally waste the resources as if the energies of life were inexhaustible ; and when they have brought on disease which destroys their comforts, they fly to the physician, *not to learn by what violation of the laws of life* they have drawn the evil upon themselves, and by what means they can avoid the same ; but, considering themselves visited with afflictions which they have in no manner been concerned in causing, they require the physician's remedies, by which their sufferings may be alleviated. In doing this, the more the practice of the physician conforms to the *appetites* of the patient, the greater is his popularity and the more generously is he rewarded.

"Everything, therefore, in society tends to confine the practising physician to the department of therapeutics, and make him a mere curer of disease ; and the consequence is, that the medical fraternity have little inducement to apply themselves to the study of the *science of life*, while almost everything, by which men can be corrupted, is presented to induce them to become the mere panderers of human ignorance and folly ; and, if they do not sink into the merest empiricism, it is owing to their own moral sensibility rather than to the encouragement they receive to pursue an elevated scientific professional career.

"Thus the natural and acquired habits of man concur to divert his attention from the study of human life, and hence he is left to *feel* his way to, or gather from what he calls experience, all the conclusions which he embraces. It has been observed that men, in their (so-called) inductive reasonings deceive themselves continually, and think that they are reasoning from facts and experience, when they are only reasoning *from a mixture of truth and falsehood*. The only end answered by facts so incorrectly apprehended is that of making error more incorrigible. Nothing, indeed, is so hostile to the interests of Truth as facts incorrectly observed. On no subjects are men so liable to misapprehend facts, and *mistake the relation between cause and effect*, as on that of human life, health, and disease."

By the opponents of dietetic reform it has been pretended that climate, or individual constitution, must determine the food proper for nations or individuals :—

"We have been told that some enjoy health in warm, and others in cold climates some on one kind of diet, and under one set of circumstances, and some under another ; that, therefore, what is best for one is not for another ; that what agrees *well* with one disagrees with another ; that what is one man's meat is another man's poison ; that different constitutions require different treatment ; and that, consequently, no rules can be laid down adapted to all circumstances which can be made a basis of regimen to all.

"Without taking pains to examine circumstances, people consider the bare fact that some intemperate individuals reach old age evidence that such habits are not unfavourable to life. With the same loose reasoning, people arrive at conclusions equally erroneous in regard to nations. If a tribe, subsisting on vegetable food, is weak, sluggish, and destitute of courage and enterprise, it is concluded that vegetable food is the cause. Yet examination might have shown that causes fully adequate to these effects existed, which not only exonerated the diet, *but made it appear that the vegetable diet had a redeeming effect, and was the means by which the nation was saved from a worse condition.*

"The fact that individuals have attained a great age in certain habits of living is no evidence that those habits are favourable to longevity. The only use which we can make of cases of extraordinary old age, is to show how the human constitution is capable of sustaining the vital economy, *and resisting the causes which induce death.*

"If we ask *how* we must live to secure the best health and longest life, the answer must be drawn from physiological knowledge ; but if we ask *how long* the best mode of living will preserve life, the reply is, Physiology cannot teach you that. Probably each aged individual has a mixture of good and bad habits, and has lived in a mixture of favourable and unfavourable circumstances. Notwithstanding apparent diversity, there is a pretty equal amount of what is salutary in the habits and circumstances of each. Some have been 'correct' in one thing, some in another. All that is proved by instances of longevity in connexion with bad habits is, that such individuals are able to resist causes that have, in the same time, sent thousands of their fellow-beings to an untimely grave ; and, under a proper regimen, they would have sustained life, perhaps, a hundred and fifty years.

"Some have more constitutional [or inherited] powers to resist the causes of disease than others, and, therefore, what will destroy the life of one may be borne by another a long time without any manifestations of immediate injury. There are, also, constitutional peculiarities, but these are far more rare than is generally supposed. Indeed, such may, in almost every case, be overcome by a correct regimen. So far as the general laws of life and the application of general principles of regimen are considered, the human constitution is *one :* there are no constitutional differences which will not yield to a correct regimen, and thus improve the individual. Consequently, what is best for one is best for all. . . Some are born without any tendency to disease while others have the predisposition to particular diseases of some kind. But *differences result from causes which man has the power to control,* and it is certain that all can be removed by conformity to the laws of life for generations, and that the human species can be brought to as great uniformity, as to health and life, as the lower animals."

With Hufeland, Flourens, and other scientific authorities, he maintains that :—

"Physiological science affords no evidence that the human constitution is not capable of gradually returning to the primitive longevity of the species. The highest interests of our nature require that *youthfulness* should be prolonged. And it is as capable of being preserved as life itself, both depending on the same conditions. If there ever was a state of the human constitution which enabled it to sustain life [much beyond the present period], that state involved a harmony of relative conditions. The vital processes were less rapid and more complete than at present, development was slower, organisation more perfect, childhood protracted, and the change from youth to manhood took place at a greater remove from birth. Hence, if we now aim at long life, we can secure our object only by conformity to those laws by which youthfulness is prolonged."

As for the *omnivorousness* of the human animal :—

The ourang-outang, on being domesticated, readily learns to eat animal food. But if this proves that animal to be *omnivorous,* then the Horse, Cow, Sheep, and others are all omnivorous, for everyone of them is easily trained to eat animal food. Horses have frequently been trained to eat animal food,* and Sheep have been so accustomed to it as to refuse grass. All carnivorous animals can be trained to a vegetable diet, and brought to subsist upon it, with less inconvenience and deterioration than herbivorous or frugivorous animals can be brought to live on animal food. Comparative anatomy, therefore, proves that Man is naturally a frugivorous animal, formed to subsist upon fruits, seeds, and farinaceous vegetables.†

---

* The Greek story of the savage horses of the Thracian king who were fed upon human flesh, herefore, may very well be true.

† Graham here quotes various authorities—Linné, Cuvier, Lawrenc Bell, and others.

The *stimulating*, or alcoholic, property of flesh produces the delusion that it is, therefore, the most *nourishing* :—

"Yet by so much as the stimulation exceeds that which is necessary for the performance of the functions of the organs, the more does the expenditure of vital powers exceed the renovating economy ; and the exhaustion which succeeds is commensurate with the excess. Hence, though food which contains the greatest proportion of stimulating power causes a *feeling* of the greatest strength, it also produces the greatest exhaustion, which is commensurately importunate for relief; and, as the same food affords such by supplying the requisite stimulation, their *feelings* lead the consumers to believe that it is most strengthening. . . . Those substances, the stimulating power of which is barely sufficient to excite the digestive organs in the appropriation of nourishment, are most conducive to vital welfare, causing all the processes to be most perfectly performed, without any unnecessary expenditure, thus contributing to health and longevity.

"Flesh-meats average about *thirty-five per cent* of nutritious matter, while rice, wheat, and several kinds of pulse (such as lentils, peas, and beans), afford from *eighty to ninety-five per cent ;* potatoes afford twenty-five per cent of nutritious matter. So that one pound of rice contains more nutritious matter than two pounds and a half of flesh meat ; three pounds of whole meal bread contain more than six pounds of flesh, and three pounds of potatoes more than two pounds of flesh."

That the human species, *taken in its entirety*, is no more carnivorous *de facto* than it could be *de jure*, is apparent on the plain evidence of facts. In all countries of our Globe, with the exception of the most barbarous tribes, it is, in reality, only the ruling and rich classes who are kreophagist. The Poor have, almost everywhere, but the barest sufficiency even of vegetable foods :—

"The peasantry of Norway, Sweden, Denmark, Germany, Turkey, Greece, Italy, Switzerland, France, Spain, England, Scotland, Ireland, a considerable portion of Rrussia, and other parts of Europe subsist mainly on non-flesh foods. The peasantry of modern Greece [like those of the days of Perikles] subsist on coarse brown bread and fruits. The peasantry in many parts of Russia live on very coarse bread, with garlic and other vegetables, and, like the same class in Greece, Italy, &c., they are obliged to be extremely frugal even in this kind of food. Yet they are [for the most part] healthy, vigorous, and active. Many of the inhabitants of Germany live mainly on rye and barley, in the form of coarse bread. The potato is the principal food of the Irish peasantry, and few portions of the human family are more healthy, athletic, and active, when uncorrupted by intoxicating substances [and, it may be added, when under favourable political and social conditions]. But alcohol, opium, &c. [equally with bad laws] have extended their blighting influence over the greater portion of the world, and nowhere do these scourges so cruelly afflict the self-devoted race as in the cottages of the poor, and when, by these evils and neglect of sanitation, &c., diseases are generated, sometimes epidemics, we are told that these things arise from their poor, meagre, low, *vegetable* diet. Wherever the various sorts of intoxicating substances are absent, and a decent degree of cleanliness is observed, the vegetable diet is not thus calumniated.

"That portion of the peasantry of England and Scotland who subsist on their barley and oatmeal bread, porridge, potatoes, and other vegetables, with temperate, cleanly habits [and surroundings], are able to endure more fatigue and exposure than any other class of people in the same countries. *Three-fourths of the whole human family*, in all periods of time [excepting, perhaps, in the primitive wholly predatory ages] have subsisted on non-flesh foods, and when their supplies have been abundant, and their habits in other respects correct, they have been well nourished."

That the sanguinary diet and savagery go hand in hand, and that in proportion to the degree of carnivorousness is the barbarous or militant character of the people, all History, past and present, too clearly testifies. Nor are the carnivorous tribes conspicuous by their cruel habits only:—

"Taking all flesh-eating nations together, though some, whose other habits are favourable, are, comparatively, well-formed, as a general average they are small, ill-formed races ; and taking all vegetable-eating nations, though many, from excessive use of narcotics, and from other unfavourable circumstances, are comparatively small and ill-formed, as a general average they are much better formed races than the flesh-eaters.* It is only among those tribes whose habits are temperate, and who subsist on the non-flesh diet, that the more perfect specimens of symmetry are found.

"Not one human being in many thousands dies a *natural* death. If a man be shot or poisoned we say he dies a violent death, but if he is ill, attended by physicians, and dies, we say he dies a 'natural' death. This is an abuse of language—the death in the latter case being as truly violent as if he had been shot. Whether a man takes arsenic and kills himself, or by small doses or other means, however common, gradually destroys life, he equally dies a violent death. He only dies a natural death who so obeys the laws of his nature as by neither irritation nor intensity to waste his energies, but slowly passes through the changes of his system to old age, and falls asleep in the exhaustion of vitality.*

With Flourens he adduces a number of instances both of individuals and of communities who have attained to protracted ages by reason of a pure diet. He afterwards proceeds to prove from comparative physiology and anatomy, and, in particular, from the conformation of the human teeth and stomach (which, by an astounding perversion of fact, are sometimes alleged to be formed carnivorously, in spite of

---

* Professor Lawrence instances particularly "the Laplanders, Samoides, Ostiacs, Tungooses, Burats, and Kamtschatdales, in Northern Europe and Asia, as well as the Esquimaux in the northern, and the natives of Tierra del Fuego in the southern, extremity of America, who, although they live almost entirely on flesh, and that often raw, are the smallest, weakest, and least brave people of the globe."—*Lectures on Physiology.* Of all races the North American native tribes, who subsist almost entirely by the chase, are notoriously one of the most ferocious and cruel That the *omnivorous* classes in "civilised" Europe—in this country particularly—have attained their present position, political or intellectual, *in spite of their kreophagistic habits* is attributable to a complex set of conditions and circumstances (an extensive inquiry, upon which it is impossible to enter here) which have, *in some measure*, mitigated the evil results of a barbarous diet, will be sufficiently clear to every unprejudiced inquirer. If flesh-eating be the cause, or one of the principal causes, of the present dominance of the European, and especially English-speaking peoples, it may justly be asked—how is to be explained, *e.g.*, the dominance of the Saracenic power (in S. Europe) during seven centuries—a dominance in arms as well as in arts and sciences —when the semi-barbarous Christian nations (at least as regards the ruling classes) were *wholly* kreophagistic.

often-repeated scientific authority, as well as of common observation), the natural frugivorous character of the human species, and he quotes Linné, Cuvier, Lawrence, Bell, and many others in support of this truth.*

<hr>

XLVIII.

STRUVE. 1805—1870.

GERMANY, at the present day able to boast so many earnest apostles of humanitarianism, until the nineteenth century was some way advanced, had contributed little, definitely, to the literature of *Humane Dietetics*. A Haller or a Hufeland, indeed, had, with more or less boldness, raised the banner of partial revolt from orthodox medicine and orthodox living, but their heterodoxy was rather hygienic than humane. In the history of humanitarianism in Germany the honour of the first place, in order of time, belongs to the author of *Pflanzenkost, die Grundlage einer Neuen Weltanschauung*, and of *Mandaras' Wanderungen*, whose life, political as well as literary, was one continuous combat on behalf of justice, freedom, and true progress.

Gustav von Struve was born at München (Munich), October 11, 1805, from whence his father, who was residing there as Russian Minister, shortly afterwards removed to Stuttgart. The foundation of his education was laid in the gymnasium of that capital, where he remained until his twelfth year. From 1817 to 1822 he was a scholar in the Lyceum in Karlsruhe. Having finished his preparatory studies in those schools, he proceeded to the University of Göttingen, which, after a course of nearly two years, he exchanged for Heidelberg. Four years of arduous study enabled him to pass his first examination, and, as the result of his brilliant attainments and success, he received the appointment of *Attaché* to the Bundestag Embassy at Oldenberg.

With such an opening, a splendid career in the service of courts and kings seemed to be reserved for him. His family connexions, his great abilities, and his unusual acquirements at so early an age guaranteed to him quick promotion, with reward and worldly honour. But to figure in the service of the oppressors of the people—to waste in luxurious

* For one of the ablest and most exhaustive scientific arguments on the same side ever published we refer our readers to *The Perfect Way in Diet*, by Mrs. Algernon Kingsford, M.D. (Kegan Paul, London, 1881). Originally written and delivered as a Thesis for *le Doctorat en Médicine* at the Paris University, under the title of *L'Alimentation Végétale Chez L'Homme* (1880), it was almost immediately translated into German by Dr. A. Aderholdt under the same title of *Die Pflanzennah-rung bei dem Menschem*. It is, we believe, about to be translated into Russian. The humane and moral argument of this eloquent work is equally admirable and equally persuasive with the scientific proofs.

trifling the resources of a peasantry, supplied by them only at the cost of a life-time of painful destitution, to support the selfish greed and vain ostentation of the Jew—such was not the career which could stimulate the ambition of Struve.   The conviction that this was not his proper destiny grew stronger in him, and he soon abandoned his diplomatic position and Oldenberg at the same time.   Without wealth or friends, at variance with his relatives, who could not appreciate his higher aims, he settled himself in Göttingen (1831), and in the following year in Jena. His attempts to obtain fixed employment as professor or teacher, or as editor of a newspaper, long proved unsuccessful, for independent and honest thought, never anywhere greatly in esteem, at that time in Germany was in especial disfavour with all who, directly or indirectly, were under court influences.   Yet the three years which he lived in Göttingen and Jena supplied him with varied and useful experiences.

In 1833 he went to Karlsruhe.   After years of long patience and effort, he at length effected his object (to gain a position which should make it possible for him to carry out his schemes of usefulness for his fellow-beings), and, at the end of 1836, he obtained the office of Obergerichts-Advocat in Mannheim.   This position gave leisure and opportunity for the prosecution of his various scientific and philosophic pursuits, and to engage in literary undertakings.   He founded periodicals and delivered lectures, the constant aim of which was the improvement of the world around him.   At this period he wrote his philosophic romance, *Mandaras' Wanderungen* ("The Wanderings of Mandaras"), through which he conveys distasteful truths in accordance with the principles of Tasso.*

Struve's active political life began in 1845.   In that year were published *Briefwechsel zwischen einen ehemaligen und einen jetzigen Diplomaten* ("Correspondence between an Old and a Modern Diplomatist"), which was soon followed by his *Oeffentliches Recht des Deutschen Bundes* ("Public Rights of the German Federation") and his *Kritische Geschichte des Allgemeinen Staats-Rechts* ("Critical History of the Common Law of Nations").   In the same year he undertook the editorship of the *Mannheimer Journal*, in which he boldly fought the battles of political and social reform.   He was several times condemned to imprisonment, as well

---

* " Sai, che là corre il mondo ove più versi
  Di sue dolcesse il lusinghier Parnaso,
  E che'l Vero condito in molli versi
  I più schivi allettando ha persuaso.
  Così all' egro fanciul porgiamo aspersi
  Di soave licor gli orli del vaso :
  Succhi amari ingannato intanto ei beve,
  E dall' inganno sua vita riceve."
                              Gerusalemme Liberata, I.

as to payment of fines; but, undeterred by such persecution, the champion of the oppressed succeeded in worsting most of his powerful enemies.

In the beginning of 1847 he founded a weekly periodical, the *Deutscher Zuschauer* ("The German Spectator"), in which, without actually adopting the invidious names, he maintained in their fullest extent the principles of Freedom and Fraternity; and it was chiefly by the efforts of Struve that the great popular demonstration at Oldenberg of September 12, 1847, took place, which formulated what was afterwards known as the "Demands of the People." The public meeting, assembled at the same town March 9, 1848, which was attended by 25,000 persons, and which, without committing itself to the adoption of the term "republican," yet proclaimed the inherent Rights of the People, was also mainly the work of the indefatigable Struve. He took part, too, in the opening of the Parliament at Frankfurt. His principal production at this time was *Grundzüge der Staats-Wischenschaft* ("Outlines of Political Science"). This book, inspired by the movement for freedom which was then agitating, but, as it proved, for the most part ineffectually, a large part of Europe, is not without significance in the education of the community for higher political conceptions. Struve and F. Hecker took a leading part in the democratic movements in Baden. These attempts failing, after a short residence in Paris, he settled near Basel (Basle). There he published his *Grundrechte des Deutschen Volkes* ("Fundamental Rights of the German People"), and, in association with Heinzen, a *Plan für Revolutionizung und Republikanizung Deutschlands.* The earnest and noble convictions apparent in all the writings of the author, and the unmistakable purity of his aims, forced from the more candid of the opponents of his political creed recognition and high respect. Nevertheless, he narrowly escaped legal assassination and the *fusillades* of the Kriegsgericht or Military Tribunal.

Later the unsuccessful lover of his country sought refuge in England, and from thence proceeded to the United States (1850). Upon the breaking out of the desperate struggle between the North and South, he threw in his lot with the former, and took part in several battles. In America he wrote his historical work *Weltgeschichte* (12 vols.) and, amongst others, *Abeilard und Heloise.* In 1861 he returned to Europe, and, at different periods, wrote two of his most important books, *Pflanzenkost, die Grundlage einer Neuen Weltanschauung* ("Vegetable Diet, the Foundation of a New World-View"), and *Das Seelenleben, oder die Naturgeschichte des Menschen* ("The Spiritual Life, or the Natural History of Man'), in both of which he earnestly insists, not only upon

s

the vast and incalculable suffering inflicted, in the most barbarous manner, upon the victims of the *Table*, but, further, upon the demoralising influence of living by pain and slaughter :—

"The thoughts and feelings which the food we partake of provokes are not remarked in common life, but they, nevertheless, have their significance. A man who daily sees Cows and Calves slaughtered, or who kills them himself, Hogs 'stuck,' Hens plucked, or Geese roasted alive, &c., cannot possibly retain any true feeling for the sufferings of his own species. He becomes hardened to them by witnessing the struggles of other animals as they are being driven by the butcher, the groans of the dying Ox, or the screams of the bleeding Hog, with indifference. . . . Nay, he may come even to find a devilish pleasure in seeing beings tortured and killed, or in actually slaughtering them himself. . . .

"But even those who take no part in killing, nay, do not even see it, are conscious that the flesh-dishes upon their tables come from the Shambles, and that *their feasting and the suffering of others are in intimate connexion.* Doubtless, the majority of flesh-eaters do not reflect upon the manner in which this food comes to them, but this thoughtlessness, far from being a virtue, is the parent of many vices. . . . How very different are the thoughts and sentiments produced by the non-flesh diet!" *

The last period of his life was passed in Wien (Vienna), and in that city his beneficently-active career closed in August, 1870. His last broken words to his wife, some hours before his end, were, "I must leave the world . . this war . . this conflict!" With the life of Gustav Struve was extinguished that of one of the noblest soldiers of the Cross of Humanity. His memory will always be held in high honour wherever justice, philanthropy, and humane feeling are in esteem.

In *Mandaras' Wanderungen*, of a different inspiration from that of ordinary fiction, and which is full of refinement of thought and feeling, are vividly represented the repugnance of a cultivated Hindu when brought, for the first time, into contact with the barbarisms of European civilisation. To few of our English readers, it is presumable, is this charming story known; and an outline of its principal incidents will not be supererogatory here.

The hero, a young Hindu, whose home is in one of the secluded valleys of the Himalaya, urged by the solicitude of the father of his betrothed, who wishes to prove him by contact with so different a world, sets out on a course of travel in Europe. The story opens with the arrival of his ship at Leftheim (Livorno) on the Italian coast. Mandaras has no sooner landed than he is accosted by two clerics (*ordensgeistliche*), who wish to acquire the honour and glory of making a convert. But, unhappily for their success, like his predecessor Amabed, he had already

---

* See *Pflanzenkost: oder die Grundlage einer Neuen Weltanschauung*, Von Gustav Struve, Stuttgart, 1869. For the substance of the brief sketch of the life of Struve we are indebted to the courtesy of Herr Emil Weilshaeuser, the recently-elected President of the Vegetarian Society of Germany (Jan., 1882), himself the author of some valuable words on Reformed Dietetics.

on his voyage discovered that the religion of the people, among whom he was destined to reside, did not exclude certain horrible barbarisms hitherto unknown to him in his own unchristian land :—

"While still on board ship I had been startled when I saw the rest of the passengers feeding on the flesh of animals. 'By what right,' I asked them, 'do you kill other animals to feed upon their flesh?' They could not answer, but they continued to eat their salted flesh as much as ever. For my part, I would have rather died than have eaten a piece of it. But now it is far worse. I can pass through no street in which there are not poor slaughtered animals, hung up either entire or cut into pieces. Every moment I hear the cries of agony and of alarm of the victims whom they are driving to the slaughter-house,—see their struggles against the murderous knife of the butcher. Ever and again I ask of one or other of the men who surround me, *by what right* they kill them and devour their flesh ; but if I receive an answer, it is returned in phrases which mean nothing or in repulsive laughter."

In fact the Hindu traveller had been but a brief space of time in Christian lands when he finds himself, almost unconsciously, in the position of a *catechist* rather than of a *catechumen*. One day, for example, he finds himself in the midst of a vast crowd, of all classes, hurrying to some spectacle. Inquiring the cause of so vast an assemblage, he learns that some persons are to be put to death with all the frightful circumstances of public executions. After travelling through a great part of Germany, he fixes his residence, for the purpose of study, in the University of Lindenberg. In the society of that place he meets with a young girl, Leonora, the daughter of a Secretary of Legation, who engages his admiration by her exceptional culture and refinement of mind. On the occasion of an excursion of a party of her father's visitors, of some days, to an island on the neighbouring coast, the first discussion on humane dietetics takes place, when, being asked the reason of his *eccentricity*, he appeals to the ladies of the party, believing that he shall have at least *their* sympathy with the principles he lays down :—

"From you, ladies, doubtless I shall meet with approval. Tell me, could you, *with your own hands*, kill to-day a gentle Lamb, a soft Dove, with whom perhaps you yesterday were playing? You answer—No? You dare not say you could. If you were to say yes, you would, indeed, betray a hard heart. But why could you not? Why did it cause you anguish, when you saw a defenceless animal driven to slaughter? Because you felt, *in your inmost soul*, that it is wrong, that it is unjust to kill a defenceless and innocent being! With quite other feelings would you look on the death of a Tiger that attacks men, than on that of a Lamb who has done harm to no one. To the one action attaches, naturally, justice ; to the other, injustice. Follow the inner promptings of your heart,—no longer sanction the slaughter of innocent beings by feeding on their bodies *(befördern Sie nicht deren Tödtung dadurch dass Sie ihr Fleisch essen)*."

This exhortation, to his surprise, was received by all "the softer sex" with coldness, and even with signs of impatience, excepting Leonora, who acknowledged the force of his appeal and promised to the best of her power to follow his example. Pleased and encouraged by her approval, he proceeds :—

"Assuredly it will not repent you to have formed this resolution. The man who, with firmly-grounded habits, denies himself something which lies in his power, to spare pain and death to living and sentient beings, must become milder and more loving. The man who steels himself against the feeling of compassion for the lower animals, will be more or less hard towards his own species ; while he who shrinks from giving pain to other beings, will so much the more shrink from inflicting it upon his fellow-men."

Leonora, however, was a rare exception in his experience ; and the more he saw of Christian customs, the less did he feel disposed to change his religion, which, by the way, was of an unexceptionable kind. Some time before his leaving Lindenberg, the secretary's wife gave a dinner in his honour, which, in compliment to her guest, was without any flesh-dish. As a matter of course, the conversation soon turned upon Dietetics ; and one of the guests, a cleric, challenged the Hindu to defend his principles. Mandaras had scarcely laid down the cardinal article of his creed as a fundamental principle in Ethics—that it is unjust to inflict suffering upon a living and sensitive being, which (as he insists) cannot be called in question *without shaking the very foundations of Morality (welcher nicht die Sittenlehre in ihren Fundamenten erschüttern will)*—when opponents arise on all sides of him. A doctor of medicine led the opposition, confidently affirming that the human frame itself proved men to be intended for flesh-eating. Mandaras replied that :—

"It seemed to him, on the contrary, that it is the bodily frame of man that especially declares *against* flesh-eating. The Tiger, the Lion, in short, all flesh-eating animals seized their prey, running, swimming, or flying, and tore it in pieces with their teeth or talons, devouring it there and then upon the spot. Man cannot catch other animals in this way, or tear them in pieces, and devour them as they are. . . . . Besides he has higher, and not merely animal, impulses. The latter lead him to gluttony, intemperance, and many other vices. Providence has given him reason to prove what is right and what wrong, and power of will to avoid what he has discovered to be wrong. The doctor, however, in place of admitting this argument, grew all the warmer. 'In all Nature,' said he, 'one sees how the lower existence is serviceable to the higher. As man does, so do other animals seize upon the weaker, and the weakest upon plants, &c.' "

To this the Hindu philosopher in vain replies, *that* the sphere of man is *wider*, and ought therefore to be *higher* than that of other animals, for the larger the circle in which a being can freely move, the greater is the possible degree of his perfection ; *that*, if we are to place ourselves on the plane of the carnivora in one point, why not in all, and recognise also treachery, fierceness, and murder in general, as proper to man ?

that the different character of the Tiger, the Hyæna, the Wolf on the one side, and of the Elephant, the Camel, the Horse on the other, instruct us as to the mighty influence of food upon the disposition, and certainly not to the advantage of the flesh-eaters; *that* man is to strive not after the lower but the higher character, &c., &c.   To this the hostess replies: "This may be all very beautiful and good, but how is the housekeeper to be so skilful as to provide for all her guests, if she is to withhold from them flesh dishes?"   "Exactly as our housekeepers do in the Himalayan valley—exactly as our hostess does to-day," rejoins Mandaras.   He alleges many other arguments, and in particular the high degree of reasoning faculty, and even of moral feeling, exhibited by the miserable slaves of human tyranny.   Various are the objections raised, which, it is needless to say, are successfully overthrown by the champion of Innocence, and the company disperse after a prolonged discussion.

The second division of the story takes us to the Valley of Suty, the Himalayan home of Mandaras, and introduces us to his amiable family. A young German, travelling in that region, chances to meet with the father of Urwasi (Mandaras's betrothed), whom he finds bowed down with grief for the double loss of his daughter, who had pined away in the protracted absence of her lover and succumbed to the sickness of hope deferred, and of his destined son-in-law, who, upon his return to claim his mistress, had fallen (as it appeared) into a death-swoon at the shock of the terrible news awaiting him.   The old man conducts the stranger to the scene of mourning, where Damajanti, the sister of Mandaras, with her friend Sunanda, is engaged in weaving garlands of flowers to deck the bier of her beloved brother.   An interesting conversation follows between the European stranger and the Hindu ladies, who are worthy representatives of their countrywoman, Sakuntalà.*   Accidentally they discover that he is a flesh-eater.

Sunanda : Is it possible that you really belong to those men who think it lawful to kill other beings to feed upon their bleeding limbs?

Theobald : In my country it is the ordinary custom.   Do you not, in your country, use such food?

Damajanti : Can you ask? Have not other animals feeling? Do they not enjoy their existence?

Theobald : Certainly; but they are so much below us, that there can be no *reciprocity* of duties between us.

Damajanti : The higher we stand in relation to other animals, the more are we bound to disregard none of the eternal laws of Morality, and, in particular, that of Love.   Hateful is it, at all events, to inflict pain upon an innocent being capable of feeling pain.   Or do you consider it permissible to strike a dog, to witness the trembling of his limbs, and to hear his cries?

* See *Sakuntalà, or the Fatal Ring*, of the Hindu Shakspere Kalidâsa, the most interesting production of the Hindu Poetry.   It has been translated into almost every European language.

Theobald : By no means. I hold, also, that it is wrong to torture them, because we ought to feel no pleasure in the sufferings of other animals.

Damajanti : We ought to feel no *pleasure !* That is very cold reasoning. Detestation—disgust, rather, is the sensation we ought to have. Where this sentiment is real, there can be no desire to profit by the sufferings of others. Yet, where the feelings of disgust for what is bad are weaker than inclination to the self-indulgence which it promises, there is no possibility of their triumphing. For *gain* the butcher slaughters the victim ; for *horrible luxury* other men participate in this murder, while they devour the pieces of flesh, in which, a few moments before, the blood was still flowing, the nerves yet quivering, the life still breathing !

Theobald : I admit it: but all this is new to me. From childhood upwards I have been accustomed to see animals driven to the slaughter-house. It gave me no pleasure : rather it was a positively displeasing spectacle ; but I did not think about it— whether we have the right to slaughter for food, because I had never heard doubt expressed on the matter.

Sunanda : Ah ! Now I can well believe that the men in your country *must* be hard and cold. Every softer feeling *must* be hardened, every tenderer one be dulled in the daily scenes of murder which they have before their eyes, by the blood which they shed daily, which they taste daily. Happy am I that I live far from your world. A thousand times would I rather endure death than live in so horrible a land.

Damajanti : To me, too, residence in such a land would be torture. Yet, were I a man, had I the power of eloquence, I would go from village to village, from town to town, and vehemently denounce such horrors. I should think that I had achieved more than the founders of all religions, if I should succeed in inspiring men with sympathy for their fellow-beings. What is religious belief, if it tolerates this murder, or rather sanctions it ? What is all Belief without Love ? And what is a Love that *excludes from its embrace the infinitely larger part of living beings ?* Sweet and fair indeed is it to live in a valley which harbours only mild and loving people ; but it is greater, and worthier of the high destiny of human life, to battle amongst the Bad for Goodness, to contend for the Light amongst the prisoners of Darkness, What is Life without Doing ? We women, indeed, cannot, and dare not ourselves venture forth into the wild surge of rough and coarse men ; but it is our business at least to incite to all that is True, Beautiful, and Good ; to have regard for no man who is not ardent for what is noble, to accept none of them who does not come before us adorned with the ornament of worthy actions (der nicht mit dem Schmucke würdigen Thaten vor uns tritt).

This eloquent discourse takes place while the three friends are watching, during the night, at the bier of the supposed dead. At the moment when the last funeral rites are to be performed, equally with the spectators we are surprised and pleased at the unexpected resuscitation of Mandaras, who, it appeared, had been in a trance, from which at the critical moment he awoke. With what transports he is welcomed back from the confines of the shadow-land, may easily be divined. For some time they live together in uninterrupted happiness ; the young German, who had adopted their simple mode of living, remaining with them. In the intervals of pleasing labours in the field and the garden, they pass their hours of recreation in refined intellectual discourse and speculation, the younger ones deriving instruction from the experienced wisdom of the venerable sage. The conversation often turns upon the relations between the human and non-human races; ánd, in the course of one of his philosophical prelections, the old man, with profound insight, declares that " so long as other animals continue to be excluded from the circle of Moral Existence, in which Rights and Duties are recognised, so long is there no step forward in Morality to be expected. So long as men continue to support their lives upon bodies essentially like to their own, without misgiving and without remorse, so long will they be fast bound by blood-stained fetters (*mit blutgetränkten Fesseln*) to the lower planes of existence."

At length the sorrowful day of separation arrives. It is decided that Mandaras should return to Germany, a wider sphere of useful action than the Himalayan valleys presented; and an additional reason is found in the discovery that his mother herself had been German. With much painful reluctance in parting from beloved friends, he recognises the force of their arguments, and once more leaves his peaceful home for the turmoil of European cities. After suffering shipwreck, in which he rescues a mother and child—at the expense of what he had held as his most precious possession, a casket of relics of his beloved Urwasi—Mandaras lands once again at Livorno. He finds his old friends as eager as ever for proselytising "the heathen," and quite unconscious of the need of conversion for themselves. At the death of the aged father of Damajanti, she, with her friend Sunanda and Theobald, who still remains with them, and (as may have been divined) is the devoted lover of the charming Sunanda, determines to leave her ancestral abode and join her brother in his adopted German home. When they arrive at the appointed place of meeting they are overwhelmed with grief to find that he, for whose sake so long a pilgrimage had been undertaken, had been taken from them for ever. Having lost his passport he had been arrested on suspicion and imprisoned. In confinement he had shrunk from the European flesh-dishes, and, unsupplied with proper nourishment or a sufficiency of it, had died (in the true sense of the word) a *martyr*, to the last, to his moral principles. With great difficulty his final words in writing are discovered, and these, in the form of letters to his sister, declare his unshaken faith and hopes for the future of the World. There are, also, found short poems, which are published at the end of his Memoirs, and are fully worthy of the refined mind of the author of *Mandaras*. Thus ends a romance which, for beauty of idea and sentiment, may be classed with the *Aventures de Télémaque* of Fénélon and, still more fitly, with the *Paul et Virginie* of St. Pierre.*

The space we have been tempted to give to *Mandaras's Wanderings* precludes more than one or two further extracts from Struve's admirable writings. His *Pflanzenkost*, perhaps the best known, as it is his most complete, exposition of his views on Humane Dietetics, appeared in the year 1869. In it he examines Vegetarianism in all its varied aspects—in regard to Sociology, Education, Justice, Theology, Art and Science, Natural Economy, Health, War and Peace, the practical and real Materialism of the Age, Health, Refinement of Life, &c. From the section which considers the Vegetable Diet in its relations to National Economy we quote the following just reflections :—

"Every step from a lower condition to a higher is bound up with certain difficulties. This is especially the case when it is a question of shaking off habits strengthened by numbers and length of time. Had the human race, however, not the power to do so, then the step from Paganism to Christianity, from predatory life to tillage, in particular from savage barbarousness to a certain stage in civilisation, would have been impossible. All these steps brought many struggles in their train, which to many thousands pro-duced some hardships (*Schaden*) ; to untold millions, however, incalculable benefits. So, also, the steps onward from Flesh-Diet cannot be established without some disturbances. The great majority of men hold fast to old prejudices. They struggle, not seldom with senseless rage, against enlightenment and reason, and a century often passes away before a new idea has forced the way for the spread of new blessings.

" Therefore, we need not wonder if we, also, who protest and stand out against the evils of Flesh-Eating, and proclaim the advantages of the Vegetable Diet, find violent opponents. The gain which would accrue to the whole race of man by the acceptance of that diet is, however, so great and so evidently destined, that our final victory is certain. . . . .

" Doubtless the Political Economy of our days will be shaken to its foundations by the step from the flesh to the non-flesh diet ; but this was also the case when the nomads began to practise tillage, and the hunters found no more *game*. The relics of certain barbarisms must be shaken off. All barbarians, or semi-barbarians, will struggle desperately against this with their selfish coarseness (*eigenthümlichen Rohheit*). But the result will be that the soil which, under the influence of the Flesh-Régime supported one man only, will, with the unfettered advantages of the Vegetable Diet, support five human beings. Liebig, even, recognised so much as this—that the Flesh-Diet is twelve times more costly than the Non-Flesh."†

Struve's *Seelenleben*,‡ published in the same year with the *Pflanzenkost*, and his last important work, forms a sort of *résumé* of his opinions already given to the world, and is, therefore, a more comprehensive exposition of his opinions on Sociology and Ethics than is found in his

---

* *Mandaras' Wanderungen.* Zweite Ausgabe. Mannheim. Friedrich Götz. 1845. For a copy of this now scarce book we are indebted to the courtesy of Herr A. von Seefeld, of Hanover.

† *Pflanzenkost, die Grundlage einer neuen Weltauschauung.* Stuttgart, 1869. Cf. Liebig's *Chemische Briefe* ("Letters on Chemistry.")

‡ *Das Seelenleben ; oder die Naturgeschichte des Menschen.* Von Gustav Struve. Berlin : Theobald Grieben. 1869.

earlier writings. It is full of the truest philosophy on the Natural History of Man, inspired by the truest refinement of soul. In the section entitled *Moral* he well exposes the futility of hap-hazard speeches, meaning nothing, which, vaguely and in an indefinite manner addressed to the child, are allowed to do duty for *practical* moral teaching :—

" They tell children, perhaps, that they must not be cruel either to 'Animals' or to human beings weaker than themselves. But when the child goes into the kitchen, he sees Pigeons, Hens, and Geese slaughtered and plucked ; when he goes into the streets, he sees animals hung up with bodies besmeared with blood, feet cut off, and heads twisted back. If the child proceeds still further, he comes upon the slaughter-house. in which harmless and useful beings of all kinds are being slaughtered or strangled, We shall not here dwell upon all the barbarisms bound up in the butchery of animals ; but in the same degree in which men abuse their superior powers, in regard to other species, do they usually cause their tyranny to be felt by weaker human beings in their power.

" What avails all the fine talk about morality, in contrast with *acts of barbarism and immorality presented to them on all sides* ?

" It is no proof of an exalted morality when a man acts justly towards a person stronger than himself, who can injure him. *He alone acts justly who fulfils his obligatory duties (Verpflichtungen) in regard to the weaker.* . . He, who has no *human* persons under him, at least can strike his horse, barbarously drive his calf, and cudgel his dog. The relations of men to the inferior species are so full of significance, and exercise so mighty an influence upon the development of human character, that Morality wants a wider province that shall embrace those beings within it.

In the chapter devoted especially to Food and Drinks *(Speise una Trank)* Struve warns those whom it most concerns that :—

" The monstrous evils and abuses, which gradually and stealthily have invaded our daily foods and drinks, have now reached to such a pitch that they can no longer be winked at. He who desires to work for the improvement of the human species, for the elevation of the human soul, and for the invigoration of the human body, dares not leave uncontested the general dominant unnaturalness of living.

" With a people struggling for Freedom the Kitchen must be no murderous den *(Mördergrube)* ; the Larder no den of corruption ; the Meal no occasion for stupefac tion. In despotic states the oppressors of the People may intoxicate themselves with spirituous drink, and bring disease and feebleness upon themselves with unlawful and unwholesome meats. The sooner such men perish *(zu grunde gehen)* the better. But in free states (or in such as are striving for Freedom), Simplicity, Temperance, Sober- ness must be the first principles of citizen-life. No people can be free whose individual members are still slaves to their own passions.* Man must first free himself from these before he can, *with any success,* make war upon those of his fellow-men."

Weighty words coming from a student of Science and of Human Life. Still weightier coming from one who had devoted so large a part of his existence to assist, and had taken so active a part in, the struggles of the people for Justice and Freedom.

---

* " Weh' denen, die dem *Ewigblinden*
 Des Lichtes Himmelsfackel leihen ! "
 SCHILLER. *Das Lied von der Glocke.*

XLIX.

DAUMER. 1800—1875.

ONE of the earliest pioneers of the New Reformation in Germany, chiefly from what may be termed the religious-philosophical standpoint, and one whose useful learning was equalled only by his true conception of the significance of the religious sentiment, was born at Nürnberg, in the last year of the eighteenth century.

Of a naturally feeble constitution, unable to mix in the ordinary amusements of school-life, he found ample leisure for literature and for music, to which especially he was devoted. Much of his time, also, was given to theological, and, in particular, biblical reading, so that his mother unhesitatingly fixed upon the clerical profession as his future career. He attended the Gymnasium of his native town, at that time under the direction of Hegel, who exercised a permanent influence upon his mental development. In the eighteenth year of his age he proceeded to the University of Erlangen for the study of theology. Doubts, however, began to disturb his contentment with orthodoxy; and, more and more dissatisfied with its systems, the young student relinquished the course of life for which he had believed himself destined ; and, after attending the lectures of Schelling, he went to Leipsic to apply himself wholly to philology. Having completed the usual course of study, he was appointed teacher, and afterwards Professor of Latin in the Nürnberg Gymnasium (1827). Unpleasant relations with the Rector of the schools (whose orthodoxy seems to have been less questionable than his amiability), and also, in part, his feeble health, obliged him to resign this post, and from that time he gave himself up exclusively to literary occupations, which were, for the most part, in the domain of philosophic theology.

During his professoriate Daumer had written his *Urgeschichte des Menschengeistes* ("Primitive History of the Human Mind"), which was succeeded, at an interval of some years, by his *Andeutungen eines Systems Speculativer Philosophie* ("Intimations of a System of Speculative Philosophy"), in which he attempted to found and formulate a philosophic Theism. The unreality of the professions and trifling of those who had most reputation in the "religious" world, estranged him more and more from the prevalent interpretations of Christianity.

His *Philosophie, Religion, und Alterthum* appeared in 1833. Two years later his *Züge zu einer neuen Philosophie der Religion und Religionsgeschichte* ("Indications for a New Philosophy of Religion and History of Religion"). In 1842 was published *Der Feuer-und-Moloch-Dienst der Hebräer* ("The Fire and Moloch-Worship of the Hebrews"), and (1847) *Die Geheimnisse des Christlichen Alterthums* ("The Mysteries of Christian Antiquity"), in which he pointed out that human sacrifice, and even cannibalism, were connected with the old Baal-worship of the Jews, and maintained the newer religion to be, in one important respect, not so much a purification of Judaism, as an apparently retrograde movement to the still older religionism. Besides these and other philosophic writings, Daumer published a free translation of the Persian poet Hafiz. *Hafiz* was followed by *Mahomed und seine Werke: eine Sammlung Orientalischer Geschichte* ("Mahommed and his Actions: a Résumé of Oriental History") 1848 ; and in 1855 by *Polydora: ein' Weltpoetisches Liederbuch* ("Polydora : A Book of Lays from the World's Poetry").

In his *Anthropologismus und Kriticismus* ("Anthropology and Criticism"), 1844, are many assaults upon the orthodox dietetic practices ; and in *Enthullungen über Kaspar Hauser* ("Revelations in regard to Kaspar Hauser") he displays the noxious influences of flesh-eating upon a "wild boy of the woods," who had been deserted or lost by his parents in his childhood, and who had lived an entirely natural life in the forests, eating only wild fruits. When he had been reclaimed from the *savage* state, his guardians, it seems, thought that the most effectual method of "civilising" their charge was to force him to discard fruits for flesh. The result, as shown by Professor Daumer, who watched the case with the greatest interest, was not reassuring for the orthodox believers. The inveteracy of the practice of kreophagy, which blinds men to its essential barbarism, as well as its anti-ethical, anti-humanising influences, is eloquently insisted upon :—

"Among the reforms necessary for the triumph of true refinement and true morality, which ought to be our earnest aim, is the Dietetic one, which, if not the weightiest of all *(allerwichtigste)*, yet, undoubtedly, is one of the weightiest. Still is the 'civilised' world stained and defiled by the remains of a horrible barbarity ; while the old-world revolting practice of slaughter of animals and feeding on their corpses still is in so universal vogue, that men have not the faculty even of recognising it as such, as otherwise they would recognise it ; and aversion from this horror provokes censure of such eccentricity, and amazement at any manifestion of tendency to reform, as at something absurd and ridiculous—nay, arouses even bitterness and hate. To extirpate this barbarism is a task, the accomplishment of which lies in the closest relationship with the most important principles of humaneness, morality, æsthetics, and physiology. A foundation for real culture—a thorough civilising and refining of humanity—is clearly impossible so long as an organised system of murder and of corpse-eating *(organisirten Mord-und-Leichenfratz System)* prevails by recognised custom.

"That through a manner of living, of a character so fostering of corrupting and putrefying principles, is generated and nourished a whole host of diseases which,

otherwise, would not exist, is so easy to see, that only an extremely obstinate love of
flesh-meat can blind one to the fact. Before I renounced flesh-eating, which, unhappily,
I had not the courage to do before I had lived a half century, I suffered from time to
time from a frightful neuralgia, which tortured me many long days and nights. Since
I abstained from that diet I have rid myself of this evil entirely. Observations of
other individuals, in respect of the same and other maladies, have led me to the
same conclusion. Worms, for instance, from which it formerly suffered, have entirely
disappeared in a child, when it no longer was fed upon flesh.

"That through the *cadaverous* diet, also, very great disadvantages are derived to the
spiritual and moral nature of men, appears to me to be proved by my experience in the
case of my former foster-son, the celebrated Kaspar Hauser. This young man, maintained during his close confinement upon bread and water, for a long time after his
introduction to the world ate nothing else, and wished for nothing else, as food. While
he was accustomed, without ill-effect, to take bread-sops, oatmeal, and plain chocolate,
from flesh, which had for him an intolerable odour, he had conceived a violent aversion.
Living in this way he always looked sufficiently well-nourished, he developed a
remarkable intelligence, and exhibited an extraordinarily refined and tender feeling. He
was induced at last, but only by the most extraordinary caution and gradually, to
take the usual flesh-dishes, by being given at first only a few drops of flesh-soup in
his bread-sops, and, when he had grown in some measure accustomed to it, by infusing
stronger ingredients, and so on.

" There was now manifested the most disastrous change in his mind and disposition :
learning became for him strangely difficult—the nobility of his nature disappeared into
the background, and he turned out to be nothing more than a very ordinary individual.
They ascribed this, of course, to every other cause than to his habituation to the flesh-
diet. I myself was at that time very remote from the opinion of which I now am.
From my present standpoint, however, I certainly cannot doubt that dietetic
barbarism is for man of the most essential harm, not alone in a physical, but also in
an intellectual and moral, point of view, however much it may, at present, be taken
under the patronage of physiologists and physicians—upon no other ground, apparently,
than because they themselves, to a melancholy degree, are devotedly attached to this
inhuman diet. For, alas ! man is wont to make use of his reason to justify by
specious show of reasoning what he likes and delights in upon quite other grounds."[*]

Of the rest of the little band of the propagators of the truer
Philosophy in Germany no longer living—who resolutely bore aloft
the standard of the Humanitarian Creed, at a time when it was
yet more scouted and scorned by the infidels than even at the present
day—deserving as they are of everlasting gratitude and remembrance
at the hands of their more fortunate successors, the limits of this
book compel us to be content with recording here the witness of
one or two more only ; while for acquaintance with the numerous able
and eloquent expositions of their living representatives—of such earnest
humanitarian and social reformers as Ed. Baltzer, Emil Weilshäuser,
Theodor Hahn, Dr. Aderholdt, A. von Seefeld, R. Springer, and others—

---

* Quoted in *Die Naturgemässe Diät: die Diät der Zukunft*, von Theodor Hahn, Cöthen, 1859. For
the substance of biographical notice prefixed to this article we are again indebted to the kindness
of Herr Emil Weilshäuser, of Oppeln.

we must refer our readers, who wish to form an adequate idea of contemporary German *anti-kreophagistic* literature (as also in regard to the equally extensive contemporary English literature of the subject), to the original works themselves.

From *Der Weg zum Paradiese* ("The Way to Paradise") the following extract sufficiently represents the inspiration of the writer, Dr. W. Zimmermann :—

"Men are almost entirely everything that they are by the force of custom ; and this force, for the most part, resists every other power, and remains victorious over all. Reason itself, morality, and conscience are submissive to it. In the matter of Dietary Reform it displays itself as the enemy *par excellence (die Hauptmacht)*. People will fall back upon alleged *impossibilities*, although it is a question only of will and resolution. They will reject many of the dietetic propositions hitherto advanced as dangerous 'abstractions,' although they are founded in history, reason, and human destiny ; although a brief enquiry ought to suffice to convince one of the first importance of the Reform. For although one must suppose that all would prefer a long, healthy, and happy existence to a feeble, painful life upon the old regimen, yet will the majority of human beings think it easier to attempt to assuage their torments and pains by uncertain, and, by no means, unhazardous medicine, rather than to remove them by obedience to Nature's laws. As it is with most of the highest truths, so is it especially with Dietary Reform. People will reject it as an *abstraction,* and pronounce it an *impossibility*. In the future, however, by the greater number of the higher minds—for such a sacrifice of the lower and unnatural appetite we dare not expect from the ordinary run of men—will it be regarded in practice as a great blessing. For even now there are many exceptions in the social organism for whom Nature's laws are superior to unreasoning impulse ; for whom morality is superior to materialistic and mere sensual living ; for whom duty is superior to superfluity. Besides, we are advancing towards a humaner century ; and, as the present is a humaner time than the century before, so later will there be a milder *régime* than now. Just as, in our days, exposure of children, combats of gladiators, torture of prisoners, and other atrocities are held to be scandalous and shameful, while in earlier times they were thought quite justifiable and right, so in the future will the murder of animals, to feed upon their corpses, be pronounced to be immoral and indefensible. Already (1846) are associations being formed for the protection of these beings ; already now are there many who, like the nobler spirits of antiquity, apply to their diet the watchword of morality *(das Losungswort der Moral)* to do good and to abstain *from wrong is always, and above everything, possible,* and no longer give their sanction, by feeding on animals, to the torture and killing of innocent sentient beings.

"According to the *number* of proselytes will the importance of the evidence be adjudged. When thousands, practising natural diet, are observed in the midst of diseased flesh-eaters to be in the enjoyment of a prolonged, happy, old age, without disease and the sufferings of a vicious method of life, then will the way be laid down for *the many* to abandon the living upon the corpses of other animals."

Of a like inspiration is the indignant protest of another of the apostles of Humanitarianism in Germany :—

" What humiliation, what disgrace for us all, *that it should be necessary* for one man to exhort other men not to be inhuman and irrational towards their fellow-creatures !

Do they recognise, then, no mind, no soul in them—have they not feeling, pleasure in existence, do they not suffer pain? Do their voices of joy and sorrow indeed fail to speak to the human heart and conscience—so that they can murder the jubilant lark, in the first joy of his spring-time, who ought to warm their hearts with sympathy, from delight in bloodshed or for their 'sport,' or with a horrible insensibility and recklessness only to practise their aim in shooting! Is there no *soul* manifest in the eyes of the living or dying animal—no expression of suffering in the eye of a deer or stag hunted to death—nothing which accuses them of murder before the avenging Eternal Justice? . . . . Are the souls of all other animals but man mortal, or are they essential in their organisation? Does the world-idea *(Welt-Idee)* pertain to them also—the soul of nature—a particle of the Divine Spirit? I know not; but I feel, and every reasonable man feels like me, it is in miserable, intolerable contradiction with our human nature, with our conscience, with our reason, with all our talk of humanity, destiny, nobility; it is in frightful *(himmelschreinder)* contradiction with our poetry and philosophy, with our nature and with our (pretended) love of nature, with our religion, with our teachings about *benevolent design*—that we bring into existence merely to kill, to maintain our own life by the destruction of other life. . . . . It is a frightful wrong that other species are tortured, worried, flayed, and devoured by us, in spite of the fact that we are not obliged to this by necessity; while in sinning against the defenceless and helpless, just claimants as they are upon our reasonable conscience and upon our compassion, we succeed only in brutalising ourselves. This, besides, is quite certain, that man has no real pity and compassion for his own species, so long as he is pitiless towards other races of beings."*

◆•●•◆

L.

SCHOPENHAUER.   1788—1860.

THE chief interpreter of Buddhistic ideas in Europe, and whose bias in this direction is exercising so remarkable an influence upon contemporaneous thought, in Germany in particular, was born at Dantzig, the son of a wealthy merchant of that city. His mother, herself distinguished in literature, was often the centre of the most eminent persons of the day at Weimar. At a very early age devoted to the philosophies of Plato and of Kant, Arthur Schopenhauer studied at the Universities of Göttingen and Berlin. His course of studies, both scientific and literary, was, even for a German, unusually severe and searching; and his acquirements were encyclopædic in their range. Unlike most German students, it is worth noting, he was addicted neither to beer-drinking nor to duelling.

His most important writings are: *Die Welt als Wille und Vorstellung* ("The World as Will and Representation"), 2 vols; *Die Grund-*

---

* *Das Menschendasein in seinen Weltewigen Zügen und Zeichen.* Von Bogumil Goltz. Frankfurt.

probleme der Ethik ("The Ground-Problems of Ethics"); *Parerga und Paralipomena* ("Incidental and Neglected Subjects"), 2 vols; *Das Fundament der Moral* ("The Foundation of Morality"), 1840.

The peculiar characteristics of his philosophy are uncompromising opposition to the hollow doctrines of easy-going Optimism—an antagonism which, indeed, assumes the form of an exaggerated Pessimism—and (what especially distinguishes him from most systematisers and formularisers of morals) his making *Compassion* the principal, and, indeed, the exclusive source of moral action ; and it is his vindication of the *rights* of the subject species, in marked contrast with the silence, or even positive depreciation and contempt for them, on the part of ordinary moralists, which will always entitle him to take exceptionally high rank among reformers of Ethical systems, in spite of his exaggerations and short-comings in other respects. Dr. David Strauss *(Die Alte und die Neue Glaube)* thus writes of his claims on these grounds :—

" Criminal history shows us how many torturers of men, and murderers, have first been torturers of the lower animals. *The manner in which a nation, in the aggregate, treats the other species, is one chief measure of its real civilisation.* The Latin races, as we know, come forth badly from this examination ; we Germans not half well enough. Buddhism has done more, in this direction, than Christianity ; and Schopenhauer more than all ancient and modern philosophers together. The warm sympathy with sentient nature, which pervades all the writings of Schopenhauer, is one of the most pleasing aspects of his thoroughly intellectual, though often unhealthy and unprofitable, philosophy."

This, it is necessary to add, plainly is written in ignorance of the numerous writings of earlier and contemporaneous humanitarian dietists, to whom, of course, is due a higher, because more consistent and more logical, position than even Schopenhauer can claim, who, from ignorance of the physical and moral arguments of anti-kreophagy (it reasonably may be presumed), at the same time that he established the rights of the subject species on the firmest basis, and included them as an essential part of any moral code, yet, with a strange, but too common, inconsistency, did not perceive that to hand over the Cow, the Ox, or the Sheep, &c., to the butcher, is in most flagrant violation of his own ethical standard. While, then, the author of the *Foundation of Morality* cannot claim the highest place, absolutely; outside the ranks of anti-kreophagistic writers, a high rank may properly be conceded to him as one of the most eminent moralists who, short of entire emancipation, have done most to vindicate the position of the innocent non-human races.*

* Compare the remarks of Jean Paul Richter (1763-1825), in his treatise on Education, *Levana*, in which he, too, in scarcely less emphatic language, protests against the general neglect of this department of morals. Among other references to the subject, the celebrated novelist thus writes : "Love is the second hemisphere of the moral heaven. Yet is the sacred being of love

Especially has he denounced the horrible outrage upon the commonest principles of justice by the pseudo-scientific torturers of the physiological laboratory.† It is thus that he lays the foundations of morality :—

"A Pity, without limits, which unites us with all living beings—*in that* we have the most solid, the surest guarantee of morality. *With that* there is no need of casuistry. Whoso possesses it will be quite incapable of causing harm or loss to any one, of doing violence to any one, of doing ill in any way. But rather he will have for all long-suffering, he will aid the helpless with all his powers, and each one of his actions will be marked with the stamp of justice and of love. Try to affirm : 'this man is virtuous, only he knows no pity,' or rather : 'he is an unjust and wicked man : nevertheless, he is compassionate.' The contradiction is patent to everyone. Each one to his taste : but for myself, I know no more beautiful prayer than that which the Hindus of old used in closing their public spectacles (just as the English of to-day end with a prayer for their king). They said: 'May All that have life be delivered from suffering !'"

Enforcing his teaching that the principles and mainspring of all moral action must be justice and love, Schopenhauer maintains that the real influence of these first of virtues is tested, especially, by the conduct of men to other animals :—

"Another proof that the moral motive, here proposed, is, in fact, the true one, is, that in accordance with it the lower animals themselves are protected. The unpardonable forgetfulness in which they have been iniquitously left hitherto by all the [popular] moralists of Europe is well known. It is pretended that the [so-called] beasts have no rights. They persuade themselves that our conduct in regard to them has nothing to do with morals, or (to speak in the language of their morality) that we have no duties towards 'animals:' a doctrine revolting, gross, and barbarous, peculiar to the west, and which has its root in Judaism. In Philosophy, however, it is made to rest upon a hypothesis, admitted in the face of evidence itself, of an absolute difference between man and 'beast.' It is Descartes who has proclaimed it in the clearest and most decisive manner : and, in fact, it was a necessary consequence of his errors. The Cartesian-Leibnitzian-Wolfian philosophy, with the assistance of

---

little established. Love is an inborn but differently distributed force and blood-heat of the heart (*blütwärme des herzens*). There are cold and warm-blooded souls, as there are animals. As for the child, so for the lower animal, love is, in fact, an essential impulse ; and this central fire often, in the form of compassion, pierces its earth-crust, but not in every case. . . . . The child (under proper education) learns to regard all animal life as sacred—in brief, they impart to him the feeling of a Hindu in place of the heart of a Cartesian philosopher. There is here a question of something more even than compassion for other animals ; but this also is in question. Why is it that it has so long been observed that the cruelty of the child to the lower animals presages cruelty to men, just as the Old-Testament sacrifice of animals preshadowed that of the sacrifice of a man ? It is for *himself only* the undeveloped man can experience pains and sufferings, which speak to him with the native tones of his own experience. Consequently, the inarticulate cry of the tortured animal comes to him just as some strange, amusing sound of the air ; and yet he sees there life, conscious movement, both which distinguish them from the inanimate substances. Thus he sins against his own life, whilst he sunders it from the rest, as though it were a piece of machinery. Let life be to him [the child] sacred (*heilig*), even that which may be destitute of reason ; and, in fact, does the child know any other ? Or, because the heart beats under bristles, feathers, or wings, is it, *therefore*, to be of no account?"

† See a pamphlet upon this subject by Dr. V. Gützlaff—*Schopenhauer ueber die Thiere und den Thierschutz: Ein Beitrag zur ethischen Seite der Vivisectionsfrage.* Berlin, 1879.

entirely abstract notions, had built up the 'rational psychology,' and constructed an immortal *anima rationalis* : but, visibly, the world of 'beasts,' with its very natural claims, stood up against this exclusive monopoly—this *brevet* of immortality decreed to man alone—and, silently, Nature did what she always does in such cases—she protested. Our philosophers, feeling their scientific conscience quite disturbed, were forced to attempt to consolidate their 'rational psychology' by the aid of empiricism. They, therefore, set themselves to work to hollow out between man and 'beast' an enormous abyss, of an immeasurable width ; by this they would wish to prove to us, in contempt of evidence, an impassable difference. It was at all these efforts that Boileau already laughed :—

> ' Les animaux ont-ils des Universités ?
> Voit-on fleurir chez eux les Quatre Facultés ? '

In accordance with this theory, 'beasts' would have finished with no longer knowing how to distinguish themselves from the external world, with having no more consciousness of their own existence than of mine. Against these intolerable assertions one remedy only was needed. Cast a single glance at an animal, even the smallest, the lowest in intelligence. See the unbounded *egoism* of which it is possessed. It is enough to convince you that 'beasts' have thorough consciousness of their *ego*, and oppose it to the world—to the *non-ego*. If a Cartesian found himself in the claws of a Tiger, he would learn, and in the most evident way possible, whether the Tiger can distinguish between the *ego* and the *non-ego*. To these sophisms of the philosophers respond the sophisms of the people. Such are certain *idiotisms*, notably those of the German, who, for eating, drinking, conception, birth, death, corpse (when 'beasts' are in question), has special terms ; so much would he fear to employ the same words as for men. He thus succeeds in dissimulating, under this diversity of terms, the perfect identity of things.

"The ancient languages knew nothing of this sort of synonymy, and they simply called things which are the same by one and the same name. These artificial ideas, then, must needs have been an invention of the priesthood [*prêtraille*] of Europe, a lot of sacrilegious people who knew not by what means to debase, to vilipend the eternal essence which lives in the substance of every animated being. In this way they have succeeded in establishing in Europe those wicked habits of hardness and cruelty towards 'beasts,' which a native of High Asia could not behold without a just horror. In English we do not find this infamous invention ; that is owing, doubtless, to the fact that the Saxons, at the moment of the conquest of England, were not yet Christians. Nevertheless, the pendent of it is found in this particularity of the English language : all the names of animals there are of the *neuter gender* : and, as a consequence, when the name is to be represented by the pronoun, they use the neuter *it*, absolutely as for inanimate objects. Nothing is more shocking than this idiom, especially when the *primates* are spoken of—the Dog, for example, the Ape, and others. One cannot fail to recognise here a dishonest device (*fourberie*) of the priests to debase [other] animals to the rank of things. The ancient Egyptians, for whom Religion was the unique business of life, deposed in the same tombs human mummies and those of the Ibis, &c. ; but in Europe it would be an abomination, a crime, to inter the faithful Dog near the place where his master lies ; and yet it is upon this tomb sometimes that, more faithful and more devoted than man ever was, he has awaited death.

"If you wish to know how far the identity between 'beast' and man extends, nothing will conduct to such knowledge better than a little Zoology and Anatomy.

T

Yet what are we to say when an anatomical bigot is seen at this day (1839) to be labouring to establish an absolute, radical, distinction between man and other animals ; proceeding so far in enmity against true Zoologists—those who, without conspiracy with the priesthoods, without platitude, without *tartuferie,* permit themselves to be conducted by Nature and Truth—as to attack them, to calumniate them !

"Yet this superiority [of man over other mammals of the higher species] depends but upon a more ample development of the brain—upon a difference in one part of the body only; this difference, besides, being but one of *quantity.* Yes, man and other animals are, both as regards the moral and the physical, identical *in kind,* without speaking of other points of comparison. Thus one might well recall to them—these Judaising westerns, these menagerie-keepers, these adorers of 'reason'—that if *their* mother has given suck to them, Dogs also have *theirs* to suckle *them.* Kant fell into this error, which is that of his time and of his country : I have already brought the reproach against him. The morality of Christianity has no regard for 'beasts;' it is therein a vice, and it is better to avow it than to eternise it. We ought to be all the more astonished at it, because this morality is in striking accord with the moral codes of Brahmanism and of Buddhism.

"Between pity towards 'beasts' and goodness of soul there is a very close connexion. One might say without hesitation, when an individual is wicked in regard to them, that he cannot be a *good* man. One might, also, demonstrate that this pity and the social virtues have the same source. . . . That [better section of the] English nation, with its greater delicacy of feeling, we see it taking the initiative, and distinguishing itself by its unusual compassion towards other species, giving from time to time new proofs of it—this compassion, triumphing over that 'cold superstition' which, in other respects, degrades the nation, has had the strength to force it to fill up the chasm which Religion had left in morality. This Chasm is, in fact, the reason why in Europe and in N. America, we have need of societies for the protection of the lower animals. In Asia the Religions suffice to assure to 'beasts' aid and protection (?), and there no one thinks of Societies of that kind. Nevertheless in Europe, also, from day to day [rather by intervals of *decades*] is being awakened the feeling of the Rights of the lower animals, in proportion as, little by little, disappear, vanish, the strange ideas of man's domination over [other] animals, as if they had been placed in the world but for our service and enjoyment, for it is thanks to those ideas that they have been treated as *Things.*

"Such are, certainly, the causes of that gross conduct, of that absolute want of regard, of which Europeans are guilty towards the lower animals ; and I have shown the source of those ideas, which is in the *Old Testament,* in section 177 of the second volume of my *Parerga.*"*

Of the many eminent scientists who, in recent times, indirectly have affirmed the *wantonness* of slaughtering for human food, the most famous of European Chemists, Justus von Liebig, may seem to demand especial notice. THE founder of the science of Organic Chemistry and the method of Organic Analysis (1803-1873), educated at the Universities of Bonn and Erlangen, received his diploma of Doctor in Philosophy (physical

----

* *Le Fondement de La Morale,* par Arthur Scophenhauer, traduit de l'Allemand par A. Burdeau. Paris, Baillière et Cie, 1879.

and mathematical sciences) at the age of nineteen. Two years later, chiefly by the influence of Humboldt, he was named Professor Extraordinary of Chemistry at Giessen, whither a crowd of disciples flocked from all parts of Germany and from England. In 1832 he accepted a Chair at Munich. All the Scientific Societies of Europe were eager in offering him honorary distinctions.

It is his application of his Special Science to the advancement of Agriculture, and his more philosophic, though (it must be added) occasionally contradictory views upon the comparative values of Foods, which give him his best title to remembrance with posterity. We can enumerate only a few of his numerous works : *Ueber Theorie und Praxis der Landwirthschaft* ("Upon the Theory and Practice of Agricultural Economy),"Brunswick,1824,translated into English; *Anleitung zur Analyse Organische Körper* ("Introduction to the Organic Analysis of Bodies"), 1837 ; *Die Organische Chemie in ihren Anwendung auf Physiologie und Pathologie* ("Organic Chemistry in its Relationship to Physiology and Pathology"), 1839 ; "Researches upon Alimentary Chemistry," 1849 ; *Chemische Briefe* ("Letters upon Chemistry considered in Relation with Industry, Agriculture, and Physiology"), 1852.

Whatever opinions this eminent German Chemist may have published elsewhere inconsistent with the statements below, such inconsistency, no more than in the case of Buffon, can weaken the force of his more reasonable utterance. Upon the essential ultimate identity of the nutritive properties of animal and vegetable substance he thus clearly pronounces :—

"Vegetable fibrine and animal fibrine, vegetable albumen and animal albumen, differ at the most *(höchstens)* in form. If these principles in nourishment fail, the nourishment of the animal will be cut off ; if they obtain them, then the grass-feeding animal gets the same principles in his food as those upon which the flesh-eater entirely depends. Vegetables produce in their organism the blood of all beings. So that when the flesh-eaters consume the blood and flesh of the vegetable-eaters, they take to themselves exactly and simply the vegetable principles.

"Vegetable Foods, in particular Corn of all kinds, and through these Bread, contain as much iron as the flesh of Oxen or as other kinds of flesh.

"Certain it is, that of three men, of whom the one has fed upon ôx-flesh and bread, the other upon bread and cheese, the third upon potatoes, each considers it a peculiar hardship from quite different points of view; yet in fact the only difference between them is the action of the peculiar elements of each food upon the brain and nervous system. A Bear, who was kept in a zoological garden, displayed, so long as he had bread exclusively for nourishment, quite a mild disposition. Two days of feeding with flesh made him vicious, aggressive, and even dangerous to his attendant. It is well known that the *vis irritabilis* of the Hog becomes so excessive through flesh-eating that he will then attack a man.

"The flesh-eating man needs for his support an enormous extent of land, wider and more extensive even than the Lion and the Tiger. A nation of Hunters in a circumscribed territory is incapable of multiplying itself for that reason. The carbon necessary for maintaining life must be taken from animals, of whom in the limited area there can be only a limited number. These animals collect from the plants the elements of their blood and their organs, and supply them to the Indians living by the chase, who devour them unaccompanied by the substance *(stoffen)* which during the life of the animal maintained the life processes. While the Indian, by feeding upon a single animal, might contrive to *sustain* his life and health a certain number of days, he must, in order to gain for that time the requisite heat, devour *five* animals. His food contains a superfluity of nitrogenous substance. What is wanting to it during the greater portion of the year is the necessary quantity of carbon, and hence the inveterate inclination of flesh consumers for brandy.

"The practical illustration of agricultural superiority cannot be more clearly and profoundly given than in the speech of the North American Chief, which the Frenchman Crevecous has reported to us. The Chief, recommending to his tribe the practice of Agriculture, thus addressed it: 'Do you not observe that, while we live upon Flesh, the white men live [*in part*] upon Grain? That Flesh takes more than thirty months to grow to maturity, and besides is often scarce? That each of these miraculous grains of corn, which they bury in the earth, gives back to them more than a hundredfold? That Flesh has four legs upon which to run away, and we have only two to overtake them? That the Corn remains and grows where the white men sow it; that the winter, which for us is a time of toilsome hunting, is for them the time of rest? Therefore have they so many children, and live so much longer than we. I say, then, to each one who hears me: Before the trees over our wigwams have died from old age, and the maples have ceased to supply us with sugar, the race of the corn-planter will have exterminated the race of the flesh-eater, because the hunters determine not to sow.' "*

Liebig's views as to the mischievous effects of the propensity of farmers, and of so-called agriculturists, to convert arable into pasture land are sufficiently well known.†

---

* Quoted in *Die Naturgemässe Diät, die Diät der Zukunft*, von Theodor Hahn, 1859. We may note here that Moleschott, the eminent Dutch physiologist, and a younger contemporary of Liebig, alike with the distinguished German Chemist and with the French zoologist, Buffon, is chargeable with a strange inconsistency in choosing his place among the apologists of kreophagy, in spite of his conviction that "the legumes are superior to flesh-meat in abundance of solid constituents which they contain; and, while the amount of albuminous substances may surpass that in flesh-meat by one-half, the constituents of fat and the salts are also present in a greater abundance." (See *Die Naturgemässe Diät*, von Theodor Hahn, 1859). But, in fact, it is only too obvious *why* at present the large majority of Scientists, while often fully admitting the virtues, or even the superiority of the purer diet, yet after all enrol themselves on the orthodox side. Either they are altogether indifferent to humane teaching, or they want the courage of their convictions to proclaim the Truth.

† Among English philosophic writers, the arguments and warnings (published in the *Dietetic Reformer* during the past fifteen years) of the present head of the Society for the promotion of Dietary Reform in this country, Professor Newman, in regard to National Economy and to the enormous evils, present and prospective, arising from the prevalent insensibility to this aspect of National Reform are at once the most forcible and the most earnest. It would be well if our public men, and all who are in place and power, would give the most earnest heed to them. But this, unhappily, under the *present* prevailing political and social conditions, experience teaches to be almost a vain expectation.

APPENDIX.

---›•‹---

I.

HESIOD.

The original of the English version, given in the beginning of this work, is as follows :—

Νήπιοι, οὐδὲ ἴσασιν, ὅσῳ πλέον ἥμισυ Παντός,
Οὐδ' ὅσον ἐν Μαλάχῃ τε καὶ 'Ασφοδέλῳ μέγ' ὄνειαρ.

*   *   *   *   *

Χρύσεον μὲν πρώτιστα γένος μερόπων ἀνθρώπων
'Αθάνατοι ποίησαν 'Ολύμπια δώματ' ἔχοντες.
Ὥστε θεοὶ δ'ἔζωον ἀκηδέα θυμὸν ἔχοντες,
Νόσφιν ἄτερ τε πόνων καὶ ὀϊζύος· οὐδέ τι δειλὸν
Γῆρας ἐπῆν, αἰεὶ δε πόδας καὶ χεῖρας ὁμοῖοι
Τέρποντ' ἐν θαλίῃσι κακῶν ἔκτοσθεν ἀπάντων·
Θνῆσκον δ'ὡς ὕπνῳ δεδμημένοι· ἐσθλὰ δὲ πάντα
Τοῖσιν ἔην· καρπὸν δ'ἔφερε ζείδωρος ῎Αρουρα
Αὐτομάτη, πολλόν τε κἀὶ ἄφθονον· οἱ δ'ἐθελημοὶ
Ἥσυχοι ῎εργ' ἐνέμοντο σὺν ἐσθλοῖσιν πολέεσσιν,
['Αφνειοὶ μήλοισι, φίλοι μακάρεσσι θεοῖσι]*
Αὐτὰρ ἐπειδὴ τοῦτο γένος κατὰ γαῖα κάλυψεν,
Τοὶ μὲν δαίμονες εἰσι Διὸς μεγάλου διὰ βουλὰς
'Εσθλοί, ἐπιχθόνιοι, φύλακες θνητῶν ἀνθρώπων,†
Οἵ ῥα φυλάσσουσιν τε δίκας καὶ σχέτλια ἔργα,
'Ηέρα ἐσσάμενοι πάντῃ φοιτῶντες ἐπ' αἶαν,
Πλουτοδόται· καὶ τοῦτο γέρας βασιλήϊον ἔσχον.

*   *   *   *   *

Ζεὺς δὲ Πατὴρ τρίτον ἄλλο γένος μερόπων ἀνθρώπων
Χάλκειον ποίησε *   *
                              Οὐδέ τι σῖτον
῎Ησθιον, ἀλλ' ἀδάμαντος ἔχον κρατερόφρονα θυμόν,
'Απλητοι· μεγάλη δὲ βίη καὶ χεῖρες ἄαπτοι
'Εξ ὤμων ἐπέφυκον ἐπὶ στιβαροῖσι μέλεσσιν.
                    ῎Εργα καὶ ῎Ημεραι (Works and Days), passim.

---

* Μήλοισι Grævius, the famous German Scholar of the 17th century, maintains to mean here *Fruits*, not "Flocks," according to the vulgar interpretation, and the translation of Grævius, it will be allowed, is at least more consistent with the context than is the latter. It must be added that the whole verse bracketed is of doubtful genuineness.

† This remarkable passage, it is highly interesting to note, is the earliest indication of the idea of "guardian angels," which afterwards was developed in the Platonic philosophy ; and which, considerably modified by Jewish belief, derived from the Persian theology, finally took form in the Christian creed. Compare the beautiful idea of guardian angels, or spirits in the Prologue of the *Shipwreck* of Plautus.

II.

Extracts from "The Golden Verses" (Χρυσᾶ Ἔπη). *An Exposition of Pythagorean Doctrine, of the Third Century, B.C., in Hexameters.* (See pages 21, 22.)

κρατεῖν δ᾽ εἰθίζεο τῶνδε—
Γαστρὸς μὲν πρώτιστα, καὶ ὕπνου, λαγνείης τε,
Καὶ θυμοῦ· πρήξεις δ᾽ αἰσχρόν ποτε, μήτε μετ᾽ἀλλοῦ
Μήτ᾽ ἰδίῃ· πάντων δε μαλιστ᾽ αἰσχύνεο σαυτόν.
Εἶτα Δικαιοσύνην ἀσκεῖν ἐργῳ τε λόγῳ τε.
Μηδ᾽ ἀλογίστως σαυτὸν ἔχειν περὶ μηδὲν ἔθιζε·
Ἀλλα γνῶθε μὲν ὡς θανέειν πέπρωται ἅπασι.

*　　　　*　　　　*　　　　*　　　　*

Μηδεὶς μήτε λόγῳ σε παρείπῃ, μήτε τι ἐργῳ,
Πρήξαι μήτ᾽εἰπεῖν ὅ τι τοι μὴ βέλτερον ἐστι·
Εἰθίζου δε δίαιταν ἔχειν καθάρειον, ἄθρυπτον.

*　　　　*　　　　*　　　　*　　　　*

Μηδ᾽ὕπνον μαλακοῖσιν ἐπ᾽ ὄμμασι προσδέξασθαι
Πρὶν τῶν ἡμερινῶν ἔργων τρὶς ἔκαστον ἐπελθεῖν—
Πῆ παρέβην ; Τί δ᾽ἔρεξα ; Τί μοι δέον οὐκ ἐτελέσθη ;—
Ἀρξάμενος δ᾽ἀπὸ πρώτου ἐπέξιθι καὶ μετεπειτα
Δειλὰ μὲν ἐκπρήξας, ἐπιπλήσσεο· Χρηστὰ δε τέρπνου.
Ταῦτα πόνει, ταῦτ᾽ ἐκμελέτα· τούτων χρὴ ἐρᾶν.
Ταῦτα σε τῆς θείης Ἀρετῆς εἰς ἴχνια θήσει·
Ναὶ μὰ Τὸν ἀμετέρᾳ ψυχᾷ Παραδόντα Τετρακτύν,
Παγὰν ἀενάον Φύσεως　　*　　*　　*
　　　　　　Τούτων δε κρατήσας
Γνώσῃ ἀθανάτων τε Θεῶν, θνητῶν τ᾽ ἀνθρώπων
Σύστασιν, ἧτε ἔκαστα διέρχεται, ἧτε κρατεῖται.
Γνώσῃ δ᾽ῆ θέμις ἐστὶ, Φύσιν περὶ παντὸς ὁμοίην
Ὥστε σε μήτε ἄελπτ᾽ ἐλπίζειν, μήτε τι λήθειν.
Γνώσῃ δ᾽ἀνθρώπους αὐθαίρετα πήματ᾽ ἔχοντας
Τλήμονες, οἳ τ᾽ἀγαθῶν πέλας ὄντων οὐκ ἐσορῶσιν
Οὔτε κλύουσι· λύσιν δὲ Κακῶν παῦροι συνίσασι.
Ζεῦ Πάτερ, ῆ πολλῶν κε κακῶν λύσειας ἅπαντας,
Εἰ πᾶσιν δείξαις οἵῳ τῷ δαίμονι χρῶνται.
Ἀλλὰ σὺ θάρσει, ἐπεὶ θεῖον γένος ἐστὶ βροτοῖσιν,
Οἷς ἱερὰ προφέρουσα Φύσις δείκνυσιν ἔκαστα
Ὧν εἰ σοί μέτεστι, κρατήσεις ὧν σε κελεύω
Ἐξακέσας, ψυχὴν δὲ πόνων ἀπὸ τῶνδε σαώσεις.
Ἀλλ᾽εἴργου βρωτῶν ὧν εἴπομεν, ἔν τε καθάρμοις,
Ἐν τε λύσει ψυχῆς κρίνων, καὶ φράζευ ἔκαστα,
Ἡνίοχον γνώμην στήσας καθύπερθεν ἀρίστην·
Ἢν δ᾽ἀπολείψας σῶμα ἐς αἰθέρ᾽ ἐλεύθερον ἔλθῃς,
Ἔσσεαι ἀθάνατος, θεός, ἀμβρότος, οὐκ ἔτι θνητός.*

---

* See *Poetæ Minores Græci . . . Aliisque Accessionibus Aucta*. Edited by Thomas Gaisford. Vol. III. Lipsiæ, 1823.

III.

In *Texts from the Buddhist Canon*, Love or Compassion for all living beings is thus inculcated by Buddha, in a sermon addressed to a number of women (belonging to a class of hunters) whose husbands were then engaged on one of their predatory excursions :—

"He who is humane does not kill; he is ever able to preserve [his own ?] life. This principle is imperishable. Whosoever observes it, no calamity shall betide that man. Politeness, indifference to worldly things, hurting no one, without place for annoyance—this is the character of the Brahma Heaven. Ever exercising love towards the infirm; pure, according to the teaching of Buddha; knowing when sufficient has been had; knowing when to stop.

"There are eleven advantages which attend the man who practises compassion, and is tender to all that lives: his body is always in health (happy); he is blessed with peaceful sleep, and when engaged in study he is also composed; he has no evil dreams, he is protected by Heaven (Devas) and loved by men; he is unmolested by poisonous things, and escapes the violence of war; he is unharmed by fire or water; he is successful wherever he lives, and, when dead, goes to the Heaven of Brahma."

"When he had uttered these words, both men and women were admitted into the company of his disciples, and obtained rest.

"There was, in times gone by, a certain mighty King, called Ho-meh *(love-darkness)*, who ruled in a certain district where no tidings of Buddha or his merciful doctrine had yet been heard; but the religious practices were the usual ones of sacrifice and prayer to the gods for protection. Now it happened that the King's mother, being sick, the physicians having vainly tried their medicine, all the wise men were called to consult as to the best means of restoring her health. . . . On the King asking them [the Brahman priests] what should be done, they replied . . . sacrifices of a hundred beasts of different kinds should be offered on the four hills (or to the four quarters), with a young child, as a crowning oblation to Heaven. [Here follows a description of the King ordering a hundred head of Elephants, Horses, Oxen, and Sheep to be driven along the road from the Eastern Gate towards the place of sacrifice, and how their piteous cries rang through heaven and earth.— *Editor's Note.*] On this Buddha, moved with compassion, came to the spot, and preached a sermon on "Love to all that Live," and added these words :—

"If a man live a hundred years, and engage the whole of his time and attention in religious offerings to the gods, sacrificing Elephants and Horses, and other life, all this is not equal to *one act of pure love in saving life.*"

See *Texts from the Buddhist Canon, commonly known as Dhammapada— with accompanying Narratives—Translated from the Chinese,* by Samuel

Beal, Professor of Chinese, University College, London—Trübner, 1878: and the similar scene in *The Light of Asia,* where Buddha interposes at the moment of a religious sacrifice :—

" But Buddha softly said,
'Let him not strike, great King !' and therewith loosed
The victim's bonds, none staying him, so great
His presence was.   Then, craving leave, he spake
Of life which all can take but none can give,
Life, which all creatures love and strive to keep,
Wonderful, dear and pleasant unto each,
Even to the meanest ; yea, a boon to all
Where Pity is, for Pity makes the world
Soft to the Weak, and noble for the Strong.
Unto the dumb lips of his flock he lent
Sad pleading words, shewing how man, who prays
For mercy to the Gods, is merciless,
Being as God to those : albeit all Life
Is linked and kin, and what we slay have given
Meek tribute of the milk and wool, and set
Fast trust upon the hands that murder them.

.   .   .   .   .

"Nor, spake he, shall one wash his spirit clean
By blood ; nor gladden gods, being good, with blood ;*
Nor bribe them, being evil : nay, nor lay
Upon the brow of innocent bound beasts
One hair's weight of that answer all must give
For all things done amiss or wrongfully,
Alone—each for himself—reckoning with that
The fixed arithmic of the Universe,
Which meteth good for good and ill for ill,
Measure for measure, unto deeds, words, thoughts.

.   .   .   .   .

" While still our Lord went on, teaching how fair
This earth were, if all living things be linked
In friendliness, and common use of foods,
Bloodless and pure ; the golden grain, bright fruits,
Sweet herbs which grow for all, the waters wan,
Sufficient drinks and meats—which when these heard,
The might of gentleness so conquered them,
The priests themselves scattered their altar-flames
And flung away the steel of sacrifice :
And through the land next day passed a decree
Proclaimed by criers, and in this wise graved
On rock and column : 'Thus the King's will is :—
There hath been slaughter for the Sacrifice,
And slaying for the Meat, but henceforth none

---

* "Quum sis ipse nocens, moritur cur victima pro te?
Stultitia est, *morte alterius* sperare Salutem."

Shall spill the blood of life, nor taste of flesh,
Seeing that Knowledge grows, and Life is one,
And mercy cometh to the merciful.' "*

See also the annexed extracts from the Buddhist Sacred Scriptures, written probably about the third century B.C. :—

" The Short Paragraphs on Conduct."—The Kûla Sîlam.

1. "Now wherein, Vâsettha, is his [the true disciple's] Conduct good? Herein, O Vâsettha, that putting away the Murder of that which lives, he abstains from Destroying Life. The cudgel and the sword he lays aside; and, full of Modesty and Pity, he is compassionate and kind to all beings that have life.

" This is the kind of Goodness that he has.

[After strict prohibitions of Robbery and Unchastity, Gautama Buddha proceeds.]

4. "Putting away Lying, he abstains from speaking Falsehood. He speaks Truth. From the Truth he never swerves. Faithful and trustworthy, he injures not his fellow-men by deceit.

" This is the kind of Goodness that he has.

5. "Putting away Slander, he abstains from Calumny. What he learns here he repeats not elsewhere, to raise a quarrel against the people here. What he learns elsewhere, &c. Thus he lives as a binder together of those who are divided, an encourager of those who are friends, impassioned for Peace, a speaker of words that make for Peace.

" This, too, &c.

6. "Putting away Bitterness of Speech, he abstains from harsh language. Whatever word is humane, pleasant to the ear, lovely, reaching to the heart, urbane—such are the words he speaks.

7. " Putting away Foolish Talk, he abstains from Vain Conversation, &c.

8. "He abstains from Injuring any Herb [uselessly] or any Animal. He takes but one meal a day, abstaining from food at night-time, or at the wrong time, &c.

10. "He abstains from Bribery, Cheating, Fraud, and Crooked Ways.

" This, too, &c.

11. "He refrains from Maiming, Killing, Imprisoning, Highway-Robbery, Plundering Villages, or obtaining money by threats of Violence.

———

1. "And he lets his mind pervade one quarter of the World with thoughts of Love, and so the second, and so the third, and so the fourth. And thus the whole Wide World above, below, around, and everywhere, does he continue to pervade with heart of Love—far-reaching, grown great, and beyond measure.

---

* *The Light of Asia: or, The Great Renunciation (Mahâbhinishkramana).* Being the Life and Teaching of Gautama, Prince of India, and Founder of Buddhism (as told in verse by an Indian Buddhist). By Edwin Arnold. London: Trübner.—In the Hindu Epic, the *Mahâbhârata*, the same great principle is apparent, though less conspicuously :—

" The constant virtue of the Good is tenderness and love
To all that live in earth, air, sea—great, small—below, above:
Compassionate of heart, they keep a gentle will to each :
Who pities not, hath not the Faith. Full many a one so lives."
III.—Story of Savîtrî.

2. "Just, Vâsettha, as a mighty Trumpeter makes himself heard, and that without difficulty, in all the four directions, even so, of all Things that have Shape or Life, there is not one that he passes by or leaves aside ; but he regards them all with mind set free, and deep-felt love.

"Verily this, Vâsettha, is the way to a state of union with Brahmâ.

3. "And he lets his mind pervade all parts of the World with thoughts of Pity, Sympathy, and Equanimity.

.   .   .   .

9. "When he had thus spoken, the young Brâhmans, Vâsettha and Bhâradvâga, addressed the Blessed One, and said :—

'Most excellent, Lord, are the words of thy mouth, most excellent ! Just as if a man were to set up that which is thrown down, or were to reveal that which is hidden away, or were to point out the right road to him who has gone astray, or were to bring a Lamp into the Darkness, so that those who have eyes can see eternal forms—just even so, Lord, has the Truth been made known to us, in many a figure, by the Blessed One. And we, even we, betake ourselves, Lord, to the Blessed One, as our Refuge, to the Truth and to the Brotherhood. May the Blessed One accept us as disciples, as true believers from this time forth, so long as life endures ! "—*Buddhist Suttas*, Translated from Pâli, by T. W. Rhys Davids. *Sacred Books of the East*. Ed. by Max Müller, Clarendon Press, Oxford. 1881.

As for the older (sacerdotal) religionism of the Peninsula—that of Brahma—the force of Truth obliges us here to remark that, while the great mass of the Hindus continue to shrink with disgust and abhorrence from the Slaughter-house and from the sanguinary diet of their conquerors and rulers, Mohammedan and Christian, the richer classes, and even many of the Brahmins and priests have long conformed, in great measure at least, to Western dietetic practices ; and (the flesh of the Cow or Ox excepted), no more than other religionists do they scruple to violate the laws of their Sacred Books—the *Vedas*—which, however, are not so *humane* as the teaching of the great Founder of Buddhism, as preserved in the Buddhist Sacred Scriptures, the *Tripataka*, being more essentially ritual and ceremonial than its popular off-shoot. Yet there are traces in the sacred writings of Hinduism of a strong consciousness of the irreligionism of feeding upon slaughtered animals, as in the Laws of Manu, their Sacred Legislator, where it is laid down that :—

"The man who forsakes not the Laws, and eats not flesh-meat like a blood-thirsty demon, shall attain good-will in this world, and shall not be afflicted with Maladies."— (Quoted in the Works of Sir Wm. Jones, *vol. iii., 206.*)

"The man who perceives in his own soul the Supreme Good present in all beings acquires equanimity towards them all, and shall be absorbed, at last, in the highest Essence—even in that of the Almighty himself."—*Conclusion of the Laws of Manu.*

It is superfluous to insist upon the fact that inhabitants of the hotter and, in particular, of the tropical regions of the globe have, as a matter of course, even less valid pretexts for resorting to *butchering* than have the natives of colder climates ; and that proportionally, therefore, is the reprobation to which they are obnoxious. (See, among other recent testimony, that of Shib Chunder Bose in his interesting book—*The Hindus as they Are.* London: Ed. Stanford, 1881). The writer has usefully exposed the yearly-increasing evils to India from the example of English dietetic habits.

IV.

OVID.

THE original (the peculiar beauties of which cannot easily be represented in a modern idiom) of the English version already given in this work, with the concluding verses omitted in that translation, is here subjoined :—

Primusque animalia mensis
Arcuit imponi : primus quoque talibus ora
Docta quidem solvit, sed non et credita, verbis:—
" Parcite, mortales, dapibus temerare nefandis
Corpora. *Sunt Fruges ; sunt deducentia ramos*
Pondere Poma suo, tumidæque in vitibus Uvæ.
Sunt Herbæ Dulces ; sunt, quæ mitescere flammâ
Mollirique queant. Nec vobis lacteus Humor
Eripitur, nec Mella thymi redolentia florem.
Prodiga divitias alimentaque mitia Tellus
Suggerit : atque epulas sine Cœde et Sanguine præbet.
Carne Feræ sedant jejunia ; *nec tamen Omnes.*
Quippe Equus, et Pecudes, Armentaque gramine vivunt.
At quibus ingenium est immansuetumque ferumque—
Armeniæ Tigres, iracundique Leones,
Cumque Lupis Ursi—dapibus cum sanguine gaudent.
Heu quantum Scelus est—in viscera viscera condi,
Congestoque avidum pinguescere corpore corpus,
Alteriusque animantem animantis vivere leto !
Scilicet in tantis opibus, quas optima Matrum
Terra parit, *nil te nisi tristia mandere sævo*
Vulnera dente juvat, ritusque referre Cyclopum ?
Nec, nisi perdideris alium, placare voracis
Et male morati poteris jejunia ventris ?
At vetus illa Ætas, cui fecimus Aurea nomen,
Fœtibus arboreis et, quas humus educat, Herbis
Fortunata fuit : nec polluit ora Cruore.
Tunc et Aves tutas movere per aëra pennas,
Et Lepus impavidus mediis erravit in agris :
Nec sua credulitas piscem suspenderat hamo.
Cuncta sine insidiis, nullamque timentia Fraudem,
Plenaque Pacis erant. Postquam non utilis auctor

Victibus invidit (quisquis fuit ille virorum),
Corporeasque dapes avidam demersit in alvum.
Fecit iter sceleri ; primâque e cæde Ferarum
Incaluisse putem maculatum sanguine ferrum.
Idque satis fuerat ; nostrumque petentia letum
Corpora missa neci, salvâ pietate, fatemur:
Sed quàm danda neci, tàm non epulanda, fuerunt.

*   *   *

Quid meruistis, Oves, placidum pecus, inque tuendos
Natum homines, pleno quæ fertis in ubere nectar ?
Mollia quæ nobis vestras velamina Lanas
Præbetis, Vitâque magis quàm morte juvatis.
Quid meruêre Boves—animal sine fraude dolisque
Innocuum, simplex, natum tolerare labores ?
Immemor est demùm, nec Frugum munere dignus,
Qui potuit, curvi dempto modo pondere aratri,
Ruricolam mactare suum : qui trita labore
Illa, quibus toties durum renovaverat Arvum,
Tot dederat messes, percussit colla securi."
" Nec satis est quòd tale nefas committitur : *ipsos*
Inscripsêre Deos sceleri, numenque Supernum
Cæde Laboriferi credunt gaudere Juvenci !
Victima labe carens, et præstantissima formâ,
(Nam placuisse nocet), vittis præsignis et auro,
Sistitur ante aras, auditque ignara precantem :
Imponique suæ videt, inter cornua, fronti
Quas coluit fruges, percussaque sanguine cultros
Inficit in liquidâ prævisos forsitan undâ.
Protinus ereptas viventi pectore fibras
Inspiciunt : mentesque Deûm scrutantur in illis !*
" Unde fames Homini vetitorum tanta ciborum ?
Audetis vesci, *genus O Mortale !* Quod, oro,

---

* Compare the beautiful verses of Lucretius—who, almost alone amongst the poets, has indignantly denounced the vile and horrible practice of sacrifice—picturing the inconsolable grief of the Mother Cow bereft of her young, who has been ravished from her for the sacrificial altar :—

" Sæpe ante Deûm vitulus delubra decora
Thuricremas propter mactatus concidit aras
Sanguinis expirans calidum de pectore flumen.
At mater viridis saltus orbata peragrans
Noscit humi pedibus vestigia pressa bisulcis,
Omnia convisens oculis loca, si queat usquam
Conspicere amissum fœtum, completque querellis
Frondiferum nemus absistens, et crebra revisit
Ad stabulum desiderio perfixa Juvenci ;
Nec teneræ salices atque herbæ rore vigentes,
Fluminaque illa queunt summis labentia ripis
Oblectare animum, subitamque avertere curam,
Nec vitulorum aliæ species per pabula læta
Derivare queunt animum curâque levare."

(De Rerum Naturâ II.)

See also the memorable verses in which the rationalist poet stigmatises the vicarious sacrifice of Iphigeneia.—*Tantum Religio potuit suadere Malorum* (L).

Ne facite : et monitis animos advertite nostris.
Cumque Boûm dabitis cæsorum membra palato
Mandere vos vestros scite et sentite Colonos.

.   .   .   .   .   .

"Neve Thyestêis cumulemur viscera mensis.
Quàm male consuescit, quàm se parat ille cruori.
Impius humano, Vituli qui guttura cultro
Rumpit, et immotas præbet mugitibus aures ;
Aut qui vagitus similes puerilibus Hœdum
Edentem jugulare potest ; aut Alite vesci
Cui dedit ipse cibos—Quantum est, quod desit in istis
Ad plenum facinus !   Quò transitus inde paratur !
"Bos aret, aut mortem senioribus imputet annis :
Horriferum contra Boream Ovis arma ministret ;
Ubera dent saturæ manibus præstanda Capellæ.
Retia cum pedicis, laqueosque, artesque dolosas
Tollite : nec Volucrem viscatâ fallite virgâ,
Nec formidatis Cervos eludite pinnis,
Nec celate cibis uncos fallacibus hamos.
Perdite, si qua nocent : verùm hæc quòque perdite tantùm :
Ora vacent epulis, alimentaque congrua carpant."

Metamorphoseon, Lib. xv. 72-142, 462-478.

Nor is this the only passage in his writings in which the Pagan poet proves himself to have been not without that humaneness and feeling so rare alike in non-Christian and in Christian poetry. In the charming story of the visit of the disguised and incarnate Celestials to the cottage of the pious peasants, Philemon and Baucis, Ovid takes the opportunity to present an alluring picture of the innocent fruits which were placed before the divine guests—a picture which, probably, was present to Milton in recording the similar hospitality of Eve.

Among the fragrant dishes—" savoury fruits, of taste to please true appetite "—appear Figs, Nuts, Dates, Plums, Grapes, Apples, Olives, Radishes, Onions, and Endive, with Honey, Eggs, and Milk :—

" Ponitur hìc bicolor sinceræ bacca Minervæ,
Conditaque in liquidâ Corna autumnalia fæce :
Intubaque et Radix, et Lactis massa Coacti :
Ovaque, non acri leviter versata Favillâ.

.   .   .   .   .   .

Hìc Nux, hìc mista est rugosis Carica Palmis,
Prunaque, et in patulis redolentia Mala canistris,
Et de purpureis collectæ vitibus Uvæ.
Candidus in medio Favus est : super omnia vultus
Accessêre boni."   .   .   .   .

We are not surprised, however, that, notwithstanding all this variety of sufficient foods, ignorant peasants, imitating the vicious examples of

their rich neighbours, thought it due to " hospitality " to sacrifice life ;
and they were on the point of slaughtering the only *non-human* being
belonging to them—a Goose, the " guardian of the cottage "—when the
heavenly visitants intervene, and forbid the unnecessary barbarism :—

> "Unicus anser erat, minimæ custodia villæ,
> Quem Dîs hospitibus domini mactare parabant.
> Ille celer pennâ tardos ætate fatigat,
> Eluditque diu. Tandemque est visus ad ipsos
> Confugisse Deos. Superi vetuêre necari :
> ' Dîque sumus,' " &c.

When the rest of the inhabitants of Phrygia were, for their wickedness,
destroyed by indignant Heaven, the two old peasants, we may add, found
safety from the general *Deluge.* (*Metam.* viii. 664-688).*

It may be noted in this place that the great "Epicurean" poet, Horace
(Ovid's contemporary), *bon-vivant* though he was, and apparently un-
inspired by humanitarian feeling, yet now and again expresses his
conviction of the superiority of the Fruit to the Flesh banquet, and of
the greater compatibility of the former with the poetic genius. E.g.
Carmina I., 31. *Ad Apollinem:—*

> *Me pascunt Olivæ*
> *Me Cichorea levesque Malvæ.*
> (" Olives, Endives, and easily-digested Mallows are my fare.")

Satire II. 2. " Frugality. : "—

> " Quæ virtus et quanta, boni, sit vivere Parvo,
>
> .   .   .   .
>
> Discite non inter lances mensasque nitentes,
> Cum stupet insanis acies fulgoribus, et cum
> Acclinis falsis animus meliora recusat,
> Verum hic impransi mecum disquirite—
>   *Male Verum examinat omnis*
> *Corruptus judex.*
>
> .   .   .   .   .   .
>
> Cum labor extuderit fastidia, siccus, inanis
> Sperne cibum vilem : nisi Hymettia mella Falerno
> Ne biberis diluta. . . .
>   *Cum sale Panis*
> *Latrantem stomachum bene leniet.* . . .
>   *Non in caro nidore voluptas*
> *Summa, sed in te ipso. Tu pulmentaria quære*
> *Sudando :* pinguem vitiis albumque neque ostrea,
> Nec scarus aut poterit peregrina juvare lagois.
>
> .   .   .   .   .   .

---

* See, also, *Fasti*, already quoted above.
     "Pace Ceres læta est.   .   .   .   .   .
    A Bove succincti cultros removete Ministri, &c." IV. 407-416.

> *Num vesceris istâ*
> *Quam laudas, plumâ ?   Cocto num adest honor idem ?*
>
> .   .   .   .   .   .
>
> At vos
> Præsentes Austri, coquite horum obsonia.
>
> .   .   .   .   .   .
>
> Ergo
> Si quis nunc mergos suaves edixerit assos,
> Parebit pravi docilis Romana juventus.
>
> .   .   .   .   .   .
>
> Accipe nunc, victus tenuis quæ quantaque secum
> Afferat. Imprimis valeas bene. . . . . ."

His arraignment of the rich glutton, who obliges and allows the poor man to starve in the midst of plenty, is worthy of the morality of Seneca :—

> " Ergo,
> Quod superat, non est melius quo insumere possis ?
> *Cur eget indignus quisquam te divite ?* "

<hr>

V.

MUSONIUS (1st Century, A.D.),

A Stoic writer of great repute with his contemporaries, son of a Roman Eques, was born at Volsinii (Bolsena), in Etruria, at the end of the reign of Augustus. He was banished by Nero, who especially hated the professors of the *Porch ;* but by Vespasian he was held in extraordinary honour when the rest of the philosophers were expelled from Rome. The time of his death is uncertain. He was the author of various philosophical works which are characterised by Suïdas as " distinguished writings of a highly philosophic nature," who also attributes to him (but on uncertain evidence) letters to Apollonius of Tyana. We are indebted for knowledge of his opinions to a work (of unknown authorship) entitled *Memoirs of Musonius the Philosopher.* It is from this work that Stobæus (*Anthologion*), Aulus Gellius, Arrian, and others seem to have borrowed, in quoting the *dicta* of the great Stoic teacher. All the extant fragments of his writings are carefully collected by Peerlkamp (Haarlem, 1822). (See also Herr Ed. Baltzer's valuable monograph, *Musonius: Charakterbild aus Der Römischen Kaiserzeit.* Nordhausen, 1871) :—

" On diet he used to speak often and very earnestly, as of a matter important in itself and in its effects. For he thought that continence in meats and drinks is the beginning and groundwork of temperance.   Once, forsaking his usual line of argument, he spoke as follows:—

" ' As we should prefer cheap fare to costly, and that which is easy to that which is hard to procure, so also, that which is akin to man to that which is not so.   Akin to us is that from plants, grains, and such other vegetable products as nourish him well ; also what is derived from (other) animals—not slaughtered, but otherwise serviceable.   Of these foods the most suitable are such as we may use at once without fire, for such are readiest to hand.   Such are fruits in season, and some herbs, milk, cheese, and honeycombs.   Moreover such as need fire, and belong to the classes of grains or herbs, are also not unsuitable, but are all, without exception, akin to man.'

" Eating of flesh-meat he declared to be *brutal,* and adapted to savage animals.   It is heavier, he said, and hindering thought and intelligence ; the vapour arising from it is turbid and darkens the soul, so that they who partake of it abundantly are seen to be slower of apprehension.   As man is [at his best] most nearly related to the Gods of all beings on earth, so, also, his *food* should be most like to that of the Gods.   They, he said, are content with the steams that rise from earth and waters, and we shall take the food most like to theirs, if we take that which is *lightest and purest.*

" So our soul also will be pure and clear, and, being so, will be best and wisest, as Heracleitus judges when he says the clear soul is wisest and best.   As it is, said Musonius, we are fed far worse than the irrational beings ; for they, though they are driven fiercely by appetite as by a scourge, and pounce upon their food, still are devoid of cunning and contrivance in regard to their fare—being satisfied with what comes in their way, seeking only to be filled and nothing further.   But we invent manifold arts and devices the more to sweeten the pleasure of food and to deceive the gullet.   Nay, to such a pitch of daintiness and greediness have we come, that some have composed treatises, as of music and medicine, so also of cookery, which greatly increase the pleasure in the gullet, but ruin the health.   At any rate, you may see that those who are fastidious in the choice of foods are far more sickly in body—some even, like craving women, loathing customary foods, and having their stomachs ruined.   Hence, as good-for-nothing steel continually needs sharpening, so their stomachs at table need the continual whet of some strong tasting food. . . . .   Hence, too, it is our duty to eat for life, not for pleasure (only), at least if we are to follow the excellent saying of Socrates, that, while most men lived to eat, he ate to live.   For, surely, no one, who aspires to the character of a virtuous man, will deign to resemble the many, and live for eating's sake as they do, hunting from every quarter the pleasure which comes from food.

" Moreover, that God, who made mankind, provided them with meats and drinks for preservation, not for pleasure, will appear from this.   When food is most especially performing its proper function in digestion and assimilation, then it gives no pleasure to the man at all—yet we are then fed by it and strengthened.   *Then* we have no sensation of pleasure, and yet this time is longer than that in which we are eating.   But if it were for pleasure that God contrived our food, we ought to derive pleasure from it throughout this longer time, and not merely at the passing moment of consumption.   *Yet, nevertheless, for that brief moment of enjoyment we make provision of ten thousand dainties ;* we sail the sea to its furthest bounds ; *cooks are more sought after than husbandmen.*   Some lavish on dinners the price of estates, and that though their bodies derive no benefit from the costliness of the viands.

" Quite the contrary ; *it is those who use the cheapest food who are the strongest.*   For example, you may, for the most part, see slaves more sturdy than masters, country-folk than towns-folk, poor than rich—more able to labour, sinking less at their work, seldomer ailing, more easily enduring frost, heat, sleeplessness, and the like. Even if cheap food and dear strengthens the body alike, still we ought to choose the cheap ; for this is more sober and more suited to a virtuous man ; inasmuch as what is easy to procure is, for good men, more proper for food than what is hard—what is free from trouble than what gives trouble—what is ready than what is not ready. To sum up in a word the whole use of diet, I say that we ought to make its aim health and strength, for these are the only ends for which we should eat, and they require no large outlay."*

VI.

LESSIO. 1554—1623,

BORN at Brechten, a town in Brabant, of influential family, this noted Hygeist, at a very early age, exhibited so exceptional a disposition as to be known among his school-fellows as the "prophet." His ardour for learning was so intense as to cause him to forget the hours of meals, and to reduce his time for sleep to the shortest period possible. Having obtained a scholarship at the Arras College in Louvain, Lessio pursued the course of studies there with the greatest success, and by his fellow-students was proclaimed "prince of philologers." At the age of seventeen he entered the Society of Jesus. Two years later he was elected to the Chair of Philosophy at Douai. In 1585 he accepted the Professorship of Theology at Louvain.

So extraordinary were the respect and veneration which he had attracted in his Order and from all who had access to him, that not only did his death cause the greatest regret, but (as we are assured) his friends contended among themselves for possession of every possible relic and memento " of one who had composed so admirable works." He was interred before the high altar of the church of his college in Louvain. Held in high honour during life, after his death so rare an ornament of his Church was signally eulogised by the Pope, Urbano VIII. ; and he was even believed to have worked miracles. His praises are especially recorded in a book entitled *De Vitâ et Moribus R. P. Leonardi Lessii*— reprinted at Paris, 1644.

Principal Writings : *De Justitiâ et de Jure Actionum Humanarum, &c.* (reprinted seven times). Many of the propositions, it seems, eventually came under the censure of the Theological Faculty, the Bishops, and the Pontiffs.

* *Florilegium* of Stobæus—(17-43 and 18-38), quoted by Professor Mayor in *Dietetic Reformer*, July, 1881. In the erudite and exhaustive edition of Juvenal, by Professor Mayor (Macmillan, Cambridge), will be found a large number of quotations from Greek and Latin writers, and a great deal of interesting matter upon frugal living.

U

Quæ Fides et Religio sit Capessenda, Consultatio. Anvers, 1610. In the estimation of S. François de Sales, a work "not so much that of Lessio as of an Angel of the Judgment (Ange du Grand Conseil)."

Hygiasticon (Anvers, 1613-14, 8vo); it is superfluous to remark, his really valuable work. It was translated from the Latin into French by Sebastian Hardy, with the title of *Le Vrai Régime de Vivre pour la Conservation du Corps et de l'Ame.* Paris, 1646. Another editor, *La Bonnodière*, added notes, republishing it under the title of *De la Sobriété et de Ses Avantages.* Paris, 1701.

"Lessio," writes the author of the article in the *Biographie Universelle*, "having been condemned by the physicians to have no more than two years longer to live, himself studied the principles of *Hygiene*, was struck by the example of Cornaro, resolved to imitate him, and found himself so well from such imitation that he translated his book (*Della Vita Sobria*), joining to it the results of his own experience, to which he owed the prolongation of his life by forty years." For the rest, he was a man of extensive erudition; and Justus Lipsius celebrates, in some fine verse, the variety of his talents. (See *Biog. Universelle Ancienne et Moderne.* À Paris, chez Michaud, 1819.)

The *Hygiasticon* is prefaced by testimonials from three eminent physicians, setting forth their concurrence in the principles of the author. The English translation (1634) has prefixed to it addresses, in verse, to him; one of which is by Crashaw, the friend of Cowley, and a *Dialogue between Glutton and Echo*, also in verse. Affixed to this edition are an English version of Cornaro, by George Herbert, and a translation of an anonymous treatise by another Italian writer—*That a Spare Diet is better than a Splendid and Sumptuous One: A Paradox.*

In his chap. v. "Of the Advantages which a Sober Diet brings to the Body, and first, That it freeth almost from all Diseases"—Lessio promises the adherents of it, that in the first place :—

"It doth free a man and preserve him from almost all manner of diseases. For it rids him of catarrhs, coughs, wheezings, dizziness, and pain in the head and stomach. It drives away apoplexies, lethargies, falling-sickness, and other ill-affections of the brain. It cures the gout in the feet and in the hands; the sciatica and diseases in the joints. It also prevents crudity (indigestion), the parent of all diseases. In a word, it so tempers the humours, and maintains them in an equal proportion, that they hurt not any way, either in quantity or quality. And this both reason and experience do confirm. For we see that those who keep themselves to a sober course of diet are very seldom, or rather never, molested with diseases; and if at any time they happen to be oppressed with sickness, *they do bear it much better, and sooner recover than those others whose bodies are full fraught with ill-humours.*

"I know very many who, though they be weak by natural constitution, and well grown in years, and continually busied in employments of the mind, nevertheless by the help of this temperance, live in health, and have passed the greater part of their lives, which have been many years long, without any notable sickness. . . . . .

" The self-same comes to pass in wounds, bruises, puttings out of joint, and breaking of bones ; in regard that there is either no flux at all of ill-humours, or, at least, very little of that part affected. . . . . Furthermore an abstinent diet doth arm and fortify against the plague ; for the venom thereof is much better resisted if the body be clear and free—wherefore Sokrates brought to pass that he himself was never sick of the plague, which ofttimes greatly wasted the city of Athens, where he lived, as Laertius writeth. The third commodity of the diet is that, although it doth not cure such diseases as are incurable in their own nature, yet it doth *so much mitigate and allay them as that they are easily borne,* and do not much hinder the functions of the mind. This is seen by daily experience."

Lessio proceeds to descant upon the other benefits of the reformed regimen—such as that it prolongs life (other things being equal) to extreme old age, produces cheerfulness, activity, memory, and the like.*

MOFFET, another hygienic writer of the sixteenth century, demands indignantly :—

"Till God (*i.e.*, Superstition or Fraud) would have it so [the slaying of other animals for food], who dared to touch with his lips the remnant of a dead carcase ? or to set the prey of a wolf, or the meat of a falcon, upon his table ? Who, I say, durst feed upon those members which, lately, did see, go, bleat, low, feel, and move ? †

" Nay, tell me, can civil and human eyes yet abide the slaughter of an innocent 'beast,' the cutting of his throat, the smashing him on the head, the flaying of his skin, the quartering and dismembering of his limbs, the sprinkling of his blood, the ripping up of his veins, the enduring of ill-savours, the heaving of heavy sighs, sobs, and groans, the. passionate struggling and panting for life, which only hard-hearted butchers can endure to see ?

" Is not the earth sufficient to give us meat, but that we must also rend up the bowels of ' beasts,' birds, and fishes ? Yes, truly, there is enough in the earth to give us meat ; yea, verily, and choice of meats, needing either none or no great preparation, which we may take without fear, and cut down without trembling ; which, also, we may mingle a hundred ways to delight our taste, and feed on safely to fill our bellies." —*Health's Improvement,* by Dr. W. Moffet (ed. 1746), as quoted by Ritson. The author died in 1604.

THE author of the *Anatomy of Abuses,* a writer of the same period, denouncing the unnatural and luxurious living of his time, compares the two diets with equal force and truth :—

" I cannot persuade myself otherwise, but that our *niceness* and *cautiousness* in diet hath altered our nature, distempered our bodies, and made us subject to hundreds of

---

* " *Hygiasticon : On the Right Course of Preserving Life and Health unto Extreme Old Age ; together with Soundness and Integrity of the Senses, Judgment, and Memory.* Written in Latin by Leonard Lessius, and now done into English. The second edition. Printed by the printers to the Universitie of Cambridge, 1634." Lessio, like his master Cornaro, Haller, and many other advocates of a reformed diet, was influenced not at all by humanitarian, but by health reasons only.

† Cf. Plutarch—*Essay on Flesh-Eating.*

,diseases and *discrasies* (indigestions) more than ever our forefathers were subject unto, and consequently of shorter life than they. . . . Who are sicklier than they who fare deliciously every day? Who is corrupter? Who belcheth more? Who looketh worse? Who is weaker and feebler than they? Who hath more filthy phlegm and putrefaction (replete with gross humours) than they? And, to be brief, who dieth sooner than they?

"Do we not see the poor man who eateth brown bread (whereof some is made of rye, barley, *peason*, beans, oats, and such other gross grains), and drinketh small drink, yea, sometimes water, and feedeth upon milk, butter, and cheese—I say do we not see such a one healthfuller, stronger, fairer complexioned, and longer-living than the other that fares daintily every day ; and how should it be otherwise ?"—*Stubbes's Anatomy of Abuses*, 1583. Quoted by Ritson *(Abstinence from Flesh : A Moral Duty.)*

VII.

COWLEY. 1620—1667.

AMONG the poets of the age second only to Milton and to Dryden. *The Garden*, from which we extract the following just sentiments, is prefixed by way of dedication to the *Kalendarium Hortense* of John Evelyn, his personal and political friend. *The Gardener's Almanac*, it is worthy of note, is one of the earliest prototypes of the numerous more modern treatises of the kind. It had reached a tenth edition in 1706.

" When Epicurus to the world had taught
  That pleasure is the chiefest good,
  (And was, perhaps, i'th' right, if rightly understood),
 His life he to his doctrine brought,
 And in a garden's shade that Sovereign pleasure sought :
  Whoever a true *Epicure* would be.
  May there find cheap and virtuous luxury.
Vitellius his table which did hold
As many creatures as the ark of old—
  That fiscal table to which every day
  All countries did a constant tribute pay—
Could nothing more delectable afford
  Than Nature's Liberality—
  Helped with a little Art and Industry—
Allows the meanest gardener's board.
  The wanton Taste no Flesh nor Fowl can choose,
  For which the Grape or Melon it would lose,
  Though all th' inhabitants of Earth and Air
  Be listed in the Glutton's bill of fare.

 . . . . . .

 Scarce any Plant is growing here.
 Which against Death some weapon does not bear.

Let Cities boast that they provide
For life the ornaments of Pride;
But 'tis the Country and the Field
That furnish it with Staff and Shield.

The Garden.   Chertsey, 1666.

VIII.

TRYON.   1634—1703,

ONE of the best known of the seventeenth century humane Hygeists, was born at Bibury, a village in Gloucestershire. His father was a tiler and plasterer, who by stress of poverty was forced to remove his son, when no more than six years of age, from the village school, and to set him at the work of spinning and carding, (the woollen manufacture being then extensively carried on in Gloucestershire). At eight years of age he became so expert, he tells us, as to be able to spin four pounds a day, earning two shillings a week. At the age of twelve he was made to work at his father's employment. At this period he first learned to read. He next took to keeping sheep. With the sum of three pounds, realised by the sale of his four sheep, he went to London to seek his fortune, when seventeen years old, and bound himself apprentice to a "castor-maker," in Fleet Street. His master was an Anabaptist—"an honest and sober man;" and, after two years' apprenticeship, Tryon adopted the same religious creed. All his spare time was now devoted entirely to study; and, with the usual ardour of scholars who depend upon their own talents and exertions, he scarcely gave any time to food or sleep. The holiday period, too, spent by his fellow-apprentices in eating and drinking, and gross amusements, was utilised in the same way. Science, and Physiology in particular, attracted his attention.

At the age of twenty-three he first adopted the reformed diet, "my drink being only water, and food only bread and some fruit, and that but once a day for some time; but afterwards I had more liberty given me by my guide, Wisdom, to eat butter and cheese; my clothing being mean and thin; for, in all things, self-denial was now become my real business." This strict life he maintained for more than a year, when he relapsed, at intervals, during the next two years. At the end of this period he had become confirmed in his reform, and he remained to the end strictly akreophagist, and, indeed, strictly frugal, " contenting myself with herbs, fruits, grains, eggs, butter and cheese for food, and pure water for drink." About two years after his marriage he made voyages to Barbadoes and to Holland in the way of trade—"making beavers." He finally settled himself in England, and at the age of forty-eight he published his first book on *Dietetics*.

His brief autobiography, from which the above facts are drawn, ends at this period. His editor adds, as to his appearance and character: "his aspect easily discovered something extraordinary; his air was cheerful, lively, and brisk; but grave with something of authority, though he was of the easiest access. Notwithstanding he was of no strong make, yet, through his great temperance, regularity, and by the strength of his spirits and vigour of his mind, he was capable of any fatigue, even to his last illness, equally with any of the best constitutions of men half his years. Through all his lifetime he had been a man of unwearied application, and so indefatigable that it may be as truly said of him as it can be of any man that he was never idle; but of such despatch that, though fortune had allotted him as great multiplicity of business as, perhaps, to any one of his contemporaries, yet, without any neglect thereof, he found leisure to make such a search into Nature, that perhaps few of this age equalled him therein: and not only into Nature, but also into almost all arts and sciences, of some whereof he was an improver, and of all innocent and useful ones **an** encourager and promoter."[*]

In spite of that penetration of mind and justness of thought which influenced him to abandon the cruelty and coarseness of the orthodox diet, the author of *The Way to Health* could not free himself from certain of the credulous fancies of his age; and, it must be admitted, his writings are by no means exempt from such prejudices. It is as a moral reformer that he has deserved our respect, and of his numerous books the following are noteworthy :—

A Treatise on Cleanliness in Meats and Drinks. London, 1682.

The Way to Health, Long Life, &c. 1683, 1694, 1697. 3 vols., 8vo.

Friendly Advice to the Gentlemen-Planters of the East and West Indies. London, 1684.

The Way to Make All People Rich: or, Wisdom's Call to Temperance and Frugality. 1685.

Wisdom's Doctrine: or, Aphorisms and Rules for Preserving the Health of the Body and the Peace of the Mind. 1696.

England's Grandeur and the Way to Get Wealth: or, Promotion of Trade Made Easy and Lands Advanced. 1699. 4to.

---

[*] *Some Memoirs of the Life of Mr. Thomas Tryon, late of London, Merchant. Written by Himself.* London, 1705.

Nothing can be more just or forcible than these expostulations :—

"Most men will, in words, confess that there is no blessing this world affords comparable to health. Yet rarely do any of them value it as they ought to do till they feel the want of it. To him that hath obtained this goodly gift the meanest food—even bread and water—is most pleasant, and all sorts of exercise and labour delightful. But the contrary makes all things nauseous and distasteful. What are full-spread Tables, Riches, or Honours, to him that is tormented with distempers ? In such a condition men do desire nothing so much as *Health*. But no sooner is that obtained, but their thoughts are changed, forgetting those solemn promises and resolutions they made to God and their own souls, going on in the old road of *Gluttony*, taking little or no care to continue that which they so much desired when they were deprived of it.

"Happy it were if men did but use the tenth part of that care and diligence to preserve their minds and bodies in Health, as they do to procure those dainties and superfluities which do generate Diseases, and are the cause of committing many other evils, there being but few men that do know how to use riches as they ought. For there are not many of our wealthy men that ever consider that as little and mean food and drink will suffice to maintain a *lord* in perfect health as it will a *peasant*, and render him more capable of enjoying the benefits of the Mind and pleasures of the Body, far beyond all ' dainties and superfluities.' But, alas ! the momentary pleasures of the *Throat-Custom*, vanity, &c., do ensnare and entice most people to exceed the bounds of necessity or convenience ; and many fail through a false opinion or misunderstanding of Nature—childishly imagining that the richer the food is, and the more they can cram into their bellies, the more they shall be strengthened thereby. But experience shews to the contrary ; for are not such people as accustom themselves to the richest foods, and most *cordial* drinks, generally the most infirm and diseased ?

"Now the sorts of foods and drinks that breed the best blood and finest spirits, are Herbs, Fruits, and various kinds of Grains ; also Bread, and sundry sorts of excellent food made by different preparations of Milk, and all dry food out of which the sun hath exhaled the gross humidity, by which all sorts of Pulses and Grains become of a firmer substance. So, likewise, Oil is an excellent thing, [in nature more sublime and pure than Butter. . . . .

As to the unsuspected cause of the various diseases so abundant :—

"Many of the richest sort of people in this nation might know by woful experience, especially in London, who do yearly spend many hundreds, I think I may say thousands, of pounds on their *ungodly paunches*. Many of whom may save themselves that charge and trouble they are usually at in learning of *Monsieur Nimble-heels*, the Dancing-Master, how to go upright ; for their bellies are swollen up to their chins, which forces them ' to behold the sky,'* but not for contemplation sake you may be sure, but out of pure necessity, and without any more impressions of reverence towards the Almighty Creator than their fellow-brutes ; for their brains are sunk into their bellies ; *injection and ejection* is the business of their life, and all their precious hours are spent between the platter and the glass and the close-stool. Are not these fine fellows to call themselves *Christians* and *Right-Worshipfuls*." †

In his xiv chapter, " Of Flesh and its Operation on the Body and Mind," Tryon employs all his eloquence in proving that the practice of

---

* Os homini sublime dedit, cœlumque tueri.—Ovid, *Met.* I.

† Compare Seneca and Chrysostom, above.

slaughtering for food is not only cruel and barbarous in itself, but originates, or, at all events, intensifies the worst passions of men.

Eulogising the milder manners of the followers of Pythagoras, and of the Hindus generally, he tells his countrymen that :—

"The very same, and far greater, advantages would come to pass amongst Christians, if they would cease from contention, oppression, and (what tends and disposes them thereunto) the killing of other animals, and eating their flesh and blood ; and, in a short time, human murders and devilish feuds and cruelties amongst each other would abate, and, perhaps, scarce have a being amongst them. For *separation* has greater power than most imagine, whether it be from evil or from good ; for whatever any man separates himself from, that property in him presently is weakened. Likewise, *separation* from cruelty does wonderfully dispel the dark clouds of ignorance, and makes the understanding able to distinguish between the good and evil principles— first in himself, and then in all other things proportionably. But so long as men live under the power of all kinds of uncleanness, violence, and oppression, they cannot see any evil therein. For this cause, those who do not separate themselves from these evils, but are contented to follow the multitude in the left-hand-way, and resolve to continue the religion of their fore-fathers—though thereby they do but continue mere *Custom*, the greatest of tyrants—'tis, I say, impossible for such people ever to under- stand or know anything *truly*, either of divine or of human things. . . . .

"It is a grand mistake of people in this age to say or suppose : That Flesh affords not only a stronger nourishment, but also more and better than Herbs, Grains, &c. ; for the truth is, it does yield more stimulation, *but not of so firm a substance, nor so good as that which proceeds from the other food ;* for flesh has more matter for corruption, and nothing so soon turns to putrefaction. Now, 'tis certain, such sorts of food as are subject to putrify *before* they are eaten, are also liable to the same afterwards. Besides, Flesh is of soft, moist, gross, phlegmy quality, and generates a nourishment of a like nature ; thirdly, Flesh heats the body, and causeth a drought; fourthly, Flesh does breed great store of noxious humours ; fifthly, it must be considered that 'beasts' and other living creatures are subject to diseases * and many other inconveniences, and uncleannesses, surfeits, over-driving, abuses of cruel butchers, &c., which renders their flesh still more unwholesome. But on the contrary, all sorts of dry foods, as Bread, Cheese, Herbs, and many preparations of Milk, Pulses, Grains, and Fruits ; as their original is more clean, so, being of a sound firm nature, they afford a more excellent nourishment, and more easy of concoction ; so that if a man should exceed in quantity, the Health will not, thereby, be brought into such danger as by the superfluous eating of flesh. . . . .

"What an ill and ungrateful sight is it to behold dead carcasses, and pieces of bloody, raw, flesh ! It would undoubtedly appear dreadful, and no man but would abhor to think of putting it in his mouth, had not Use and Custom from generation to generation familiarised it to us, which is so prevalent, that we read in some countries the mode is to eat the bodies of their dead parents and friends, thinking they can no way afford them a more noble sepulchre than their own bowells. And because it is *usual*, they do it with as little regret or nauseousness as others have when they devour the leg of a Rabbit or the wing of a Lark. Suppose a person were bred up in a place where it were not a *custom* to kill and eat flesh, and should come into our Leadenhall Market, or

* If Tryon could point to diseases among the victims of the shambles in the 17th century, what use might he not make of the epidemics or endemics of the present day ?

view our Slaughter Houses, and see the communication we have with dead bodies, and how blythe and merry we are at their funerals, and what honourable sepulchres we bury the dead carcasses of beasts in—nay, their very guts and entrails—would he not be filled with astonishment and horror ? Would he not count us cruel monsters, and say we were *brutified*, and performed the part of beasts of prey, to live thus on the spoils of our fellow-creatures ?

" Thus, Custom has awakened the inhuman, fierce nature, which makes killing, handling, and feeding upon flesh and blood, without distinction, so easy and familiar unto mankind. And the same is to be understood of men killing and oppressing those of their own kind ; for do we not see that a soldier, who is trained up in the wars of bloody-minded princes, shall kill a hundred men without any trouble or regret of spirit, and such as have given him no more offence than a sheep has given the butcher that cuts her throat. If men have but Power and Custom on their side, they think all is well.

Whatever may be thought of the zealous attempt of the pious author to meet the assertions of the (practical) materialists, who draw their arguments from the Jewish Sacred Scriptures, or elsewhere, his replies to the common subterfuges or prejudices of the orthodox dietists are able and conclusive. His *humane* arguments, indeed, are worthy of the most advanced thinkers of the present day; and those who are versed in the anti-kreophagist literature of the last thirty years—in the controversy in the press, and on the platform—will, perhaps, be surprised to find that the ordinary prejudices or subterfuges of this year " of Grace " are identical with those current in the year 1683. We wish that we could transcribe some of these replies. We cannot forbear, however, to quote his representation of the changed condition of things under the imagined humanitarian *régime* :—

" Here all contention ceaseth, no hideous cries nor mournful groans are heard, neither of man nor of ' beast.' No channels running with the blood of slaughtered animals, no stinking shambles, nor bloody butchers. No roaring of cannons, nor firing of towns. No loathsome stinking prisons, nor iron grates to keep men from enjoying their wife, children, and the pleasant air ; nor no crying for want of food and clothes. No rioting, nor wanton inventions to destroy as much in one day as a thousand can get by their hard labour and travel. No dreadful execrations and coarse language. No galloping horses up hills, without any consideration or fellow-feeling of the victim's pains and burdens. No deflowering of virgins, *and then exposing them and their own young to all the miseries imaginable.* No letting lands and farms so dear that the farmer must be forced to oppress himself, servants, and cattle almost to death, and all too little to pay his rent. No oppressions of inferiors by superiors ; neither is there any want, because there is no superfluity nor gluttony. No noise nor cries of wounded men. No need of chirurgeons to cut bullets out of their flesh ; nor no cutting off hands, broken legs, and arms. No roaring nor crying out with the torturing pains of the gout, nor other painful diseases (as leprous and consumptive distempers), except through age, and the relics of some strain they got whilst they lived intemperately. Neither are their children afflicted with such a great number of diseases ; but are as free from distempers as lambs, calves, or the young ones of any of the 'beasts' who are preserved sound and healthful, because they have not outraged God's law in Nature, the breaking of which is the foundation of most, or all, cruel diseases that afflict mankind; there being nothing that makes the difference between Man and 'Beasts' in health, but only superfluity and intemperance, both in quality and in quantity."

His chapter, in which he deals with the relations between the sexes and the married state, shews him to have been as much in advance of his time, in a sound knowledge and apprehension of Physiology, and of the laws of Health, in that important part of hygienic science, as he was in the special branch of Diet.*

Affixed to this work is a very remarkable Essay, in the shape of *A Dialogue between an East-Indian Brachman and a French Gentleman, concerning the Present Affairs of Europe.* In this admirable piece, the author ably exposes the folly no less than the horrors of war—and, in particular, *religious* war—all which he ultimately traces to the first source—the iniquities and barbarism of the Shambles. The Dialogue is worthy of the most trenchant of the humanitarian writers of the next century. It was by meeting with *The Way to Health* that Benjamin Franklin, in his youth, was induced to abandon the flesh-diet, to which revolutionary measure he ascribes his success, as well as health in after life.

IX.

HECQUET. 1661—1737.

THIS meritorious medical reformer, at first intended for the Church, happily (in the event) adopted the profession which he has so truly adorned, by his virtues, as well as by his enlightened labours. After a long and severe course of Anatomy and Physiology, in 1684 he was admitted as " Doctor " at Reims, and as Fellow (*Agrégé*) in the College of Physicians in his native town. He then returned to Paris to perfect himself in physiological science. Disgusted with the *tricasseries* which were excited against him by the members of his profession, he withdrew (in 1688) to Port-Royal-des-Champs, where he succeeded Hamon, who had just died, as physician. Here he practised the reforms he taught, while he devoted himself to the most laborious works of charity, giving all his time and attention to the poor for several leagues round, and travelling the distances, great as they were, on foot.

---

* *The Way to Health, Long Life, and Happiness: or a Discourse of Temperance, and the Particular Nature of all things Requisite for the Life of Man. . . . . . The Like never before Published. Communicated to the World, for the General Good, by Philotheos Physiologus* [Tryon's nom de plume.] *London, 1683.* It is (in its best parts) the worthy precursor of *The Herald of Health,* and of the valuable hygienic philosophy of its able editor—Dr. T. L. Nichols.

His health enfeebled by excessive labour in this way, he was induced to retire from his post at Port-Royal, and he went back to the capital where, having gone through the necessary formalities, he was regularly enrolled as Doctor of the Paris University, receiving the official hat after an examination of "rare success" (1697).

Soon afterwards the Faculty named him *Docteur-Régent,* and appointed him to the post of Professor of *Materia Medica.* "Hecquet had soon numerous and illustrious patients, and his services were eagerly sought for, particularly in religious communities and in hospitals. He attached himself to that of Charity." In 1712 he was named Dean of the Faculty. In the midst of so much work, he found time to publish several medical books.

"He exercised his art with a noble disinterestedness. The poor were his favourite patients. He presented himself at the houses of the rich only when absolutely obliged, or when courtesy required it. He had much studied his art, and contributed with all his power, to advance it, as well by his writings as by his guidance and encouragement of young physicians. . . . He was in correspondence with the most famous savants and physicians of his age. His style in Latin is correct, and does not want eloquence; in French he is more negligent, and a little unpolished. He was animated (*vif*) in debate, and strongly attached to his opinions; but he sought Truth in good faith."

Amongst his numerous works are :—

De l'Indécence aux Hommes d'Accoucher les Femmes, et de l'Obligation de Celles-ci de nourrir leurs enfants. (On the Indecency of Male Physicians Attending Women in Child-Birth) 1708. *Traité des Dispenses du Carême,* 1709—his most celebrated book. *De la Digestion et des Maladies de l'Estomac,* 1712. *Novus Medicinæ Conspectus cum Appendice De Peste,* 1722. "He there combats the various systems upon the origin of diseases, which he attributes to the disorders which supervene, in accordance with the laws which direct the movement of the blood :" the Plague, upon which he writes, was desolating the south of France at that time. Also, at this period, various *brochures* upon the Small-Pox.

La Médecine, la Chirurgie, et la Pharmacie des Pauvres (1740-2), his most popular book—*La Brigandage de la Médecine* (1755), which he supplemented with *Brigandage de la Chirurgie, et de la Pharmacie*—will sufficiently mark his attitude towards the orthodox Schools of Medicine of his day. *Le Naturalisme des Convulsions dans les Maladies* (1755), with several other books upon the same subject. The history of the *Convulsionnaires* occupies a curious episode in the religious history of the period, as it has occupied, and, in some measure still, in fact, occupies the attention of physiologists and psychologists of our own age. Hecquet, with the physiologists of the present time, attributes the phenomena to physical and natural causes. *La Médecine Naturelle:* " in this work the author alleges that it is not in the blood only that is to be sought the causes of maladies, but also in the nervous fluid."*

The books in which he treats of reform in Dietetics are the *Traité des Dispenses* and *La Médecine des Pauvres.*

However *dietetically* heterodox and heretical, the author of *The Treatise on Dispensations* was of unsuspected ecclesiastical as well as theological orthodoxy ; yet he takes occasion, at the outset of his book, to reproach his Church with its indifferentism towards so essentially important a matter as Dietetics—scientific or moral :—

" It will, perhaps, be found that much theology enters into this undertaking. We acknowledge it. One might even expect that some zealous ecclesiastic or other would have done himself the credit of sustaining so beautiful a cause (que quelque ecclesias-tique zélé se seroit fait gloire de soutenir une si belle cause). It might be hoped, especially in an age like ours, when physical science is in honour and for the benefit of everyone, and in which Medicine has become the property of every condition . . . It ought then to have been the duty of so many Abbés, Monks and Religious Orders, who invest themselves with the titles of physicians—who receive their pay, who fill their employments—to advocate this part of ecclesiastical discipline [abstinence]. But, instead of doing so, though they undertake the care of the body, they, in fact, apply themselves solely to the *healing* of maladies . . . . . One can see enough of it, nevertheless, to be convinced that the public has gained less from their *secrets* than they themselves, while their patients die more than ever under their hands. . . . .

In Chap. VI., *Que les Fruits, les Grains, les Legumes sont les Alimens les plus Naturels à l'Homme,* after appealing to *Gen.* i. and " the Garden of Eden," Hecquet proceeds to insist that our foods should be analogous and consistent with the juices which maintain our life ; and these are Fruits, Grains, Seeds, and Roots. But prejudice, of long standing, opposes itself to this truth. The false ideas attached to certain traditional terms have warped the minds of the majority of the world, and they have succeeded in persuading themselves that it is upon stimu-lating foods that depend the strength and health of men. From thence has come the love of wine, of spirituous liquors, and of gross meats. The ambiguity (équivoque) comes from confounding the idea of Remedy with that of Food.

" Here the greater part of the world take alarm. 'How,' say they, ' can we be supported on Grains, which furnish but dry meal, fitter to cloy than to nourish ; on Fruits, which are but condensed water; with vegetables, which are fit but for manure (fumier) ? ' But this meal, well prepared, forms Bread, the strongest of all aliments, this condensed water is the same that has caused the Trees to attain so great bulk,

---

*See *Biog. Universelle,* Art. *Philippe Hecquet*

this *fumier* becomes such only because they prepare vegetables badly, and eat of them to excess. Besides, how can men affect to fear failure in strength, in eating what nourishes even the most robust animals, who would become even formidable to us, if only they knew their own strength."

In Chap. VII., *Que l'Usage de la Viande n'est pas le plus naturel à l'Homme, ni absolument Nécessaire*, he remarks :—

"It is incredible how much Prejudice has been allowed to operate in favour of [flesh] meat, while so many facts are opposed to the pretended necessity of its use."

Having entered into the physiological argument, now so well-worn, among other reasons he adduces the fact that "the soundest part of the world, or the most enlightened, have believed in the obligation to abstain from flesh," and "the very nature of flesh, which is digested with difficulty, and which furnishes the worst juices."

Nature being uniform in her method of procedure, is anything else necessary to determine whether Man is intended to live upon flesh-meats than to compare the organs which have to prepare them for his nourishment, with those of animals whom Nature manifestly has destined for carnage? And herein it may be clearly recognised, since men have neither fangs nor talons to tear flesh, that it is very far from being the food most natural to them.

He quotes numerous examples of eminent persons, as well as of nations in all times, and adds, as an argument not easy to be answered, that :— "It is proved it would not be difficult to nourish animals who live on flesh with non-flesh substances, while it is almost impossible to nourish with flesh those who live ordinarily upon vegetable substances."

Hecquet devotes several chapters to a description of various Fruits and Herbs, and also of various kinds of Fish, which he holds to be much less objectionable and more innocent food than flesh. Comparing the two diets, we must acknowledge :—

"It causes our nature to revolt, and excites horror to eat raw flesh, and as it is presented to us naturally ; and it becomes supportable for us to the taste and to the sight only after long preparation of cooking, which deprives it of what is inhuman and disgusting in its original state; and, often, it is only after *many* various preparations and strange seasonings that it can become agreeable or sanitarily good. It is not so with other meats : the majority, as they come from the hand of Nature, without cookery and without art, are found proper to nourish, and are pleasant to the taste —plain proof that they are intended by Nature to maintain our health. Fruits are of such property that, when well-chosen and quite ripe, they excite the appetite by *their own virtue*, and might become, without preparation, sufficing. . . . If Vegetables or Fish have need of fire to accommodate them to our nature, the fire appears to be used less to *correct* these sorts of foods than to penetrate them, to make them soft and tender, and to develope what in them is most proper and suitable for health. . . . In fine, it is clear that vegetables and fish have need of less, and less strange and récherché, condiments—all sensible marks that these aliments are the most natural and suited to man."*

Hecquet's *Traité des Dispenses* received the formal approval and com-mendation of several "doctors regent" of the Faculty of Medicine of the Paris University, which testimonies are prefixed to the second edition of 1710. With his English contemporary, Dr. Cheyne, and other medical reformers, however, he experienced much insult and ridicule from anonymous professional critics.

X.

POPE.   1688—1744.

Primâque e cœde ferarum
Incaluisse putem maculatum sanguine ferrum.
(Ovid *Metam.* XV. 106).

"I cannot think it extravagant to imagine that mankind are no less, in proportion, accountable for the ill use of their dominion over the lower ranks of Beings, than for the exercise of tyranny over their own species. The more entirely the inferior creation is submitted to our power, the more answerable we should seem for the mis-management of it ; and the rather, as the very condition of Nature renders these beings incapable of receiving any recompense in another life, for their ill-treatment in this.

"It is observable of those noxious animals, who have qualities most powerful to injure us, that they naturally avoid mankind, and never hurt us unless provoked, or necessitated by hunger. Man, on the other hand, *seeks* out and pursues even the most inoffensive animals on purpose to persecute and destroy them. Montaigne thinks it some reflection on human nature itself, that few people take delight in seeing 'beasts' caress or play together, but almost every one is pleased to see them lacerate and worry one another.

"I am sorry this temper is become almost a distinguishing character of our own nation, from the observation which is made by foreigners of our beloved *Pastimes*— Bear-baiting, Cock-fighting, and the like. We should find it hard to vindicate the destroying of anything that has Life, merely out of wantonness. Yet in this principle our children are bred, and one of the first pleasures we allow them is the licence of inflicting Pain upon poor animals. Almost as soon as we are sensible what Life is ourselves, we make it our Sport to take it from other beings. I cannot but believe a very good use might be made of the fancy which children have for Birds and Insects. Mr. Locke takes notice of a mother who permitted them to her children ; but rewarded or punished them as they treated well or ill. This was no other than entering them betimes into a daily exercise of Humanity, and improving their very diversion to a Virtue.

"I fancy, too, some advantage might be taken of the common notion, that 'tis ominous or unlucky to destroy some sorts of Birds, as Swallows or Martins. This opinion might possibly arise from the confidence these Birds seem to put in us, by building under our roofs, so that it is a kind of violation of the laws of Hospitality to murder them. As for Robin-red-breasts, in particular, 'tis not improbable they owe

---

* *Traité des Dispenses, &c.* Par Philippe Hecquet, M.D., Paris. Ed. 1709.

their security to the old ballad of the *Children in the Wood*. However it be, I don't know, I say, why this prejudice, well-improved and carried as far as it would go, might not be made to conduce to the preservation of many innocent beings, who are now exposed to all the wantonness of an ignorant barbarity. . . . . .

" When we grow up to be men we have another succession of sanguinary Sports—in particular, *Hunting*. I dare not attack a diversion which has such Authority and Custom to support it ; but must have leave to be of opinion, that the agitation of that exercise, with the example and number of the chasers, not a little contribute to resist those checks which Compassion would naturally suggest in behalf of the Animal pursued. Nor shall I say, with M. Fleury, that this sport is a remain of the Gothic Barbarity ; but I must animadvert upon a certain custom yet in use with us, barbarous enough to be derived from the Goths or even the Scythians—I mean that savage compliment our Huntsmen pass upon ladies of quality who are present at the death of a Stag, when they put the knife into their hands to cut the throat of a helpless, trembling, and weeping creature.

> " *Questuque cruentus,*
> *Atque imploranti similis**

" But if our ' Sports ' are destructive, our *Gluttony* is more so, and in a more inhuman manner. Lobsters roasted alive, Pigs whipt to death, Fowls sewed up,† are testimonies of our outrageous Luxury. Those who (as Seneca expresses it) divide their lives betwixt an anxious Conscience and a Nauseated Stomach, have a just reward of their gluttony in the diseases it brings with it. For human savages, like other wild beasts, find snares and poison in the provisions of life, and are allured by their appetite to their destruction. I know nothing more shocking or horrid than the prospect of one of their kitchens covered with blood, and filled with the cries of Beings expiring in tortures. It gives one an image of a giant's den in a romance, bestrewed with the scattered heads and mangled limbs of those who were slain by his cruelty.

" The excellent Plutarch (who has more strokes of good nature in his writings than I remember in any author) cites a saying of Cato to this effect :—*That 'tis no easy task to preach to the Belly which has no ears.* Yet if (says he) we are ashamed to be so out of fashion as not to offend, let us at least offend with *some* discretion and measure. If we kill an animal for our provision, let us do it with the meltings of compassion, and without tormenting it. Let us consider that it is, in its own nature, cruelty to put a living being to death—we, at least destroy a soul that has sense and perception.‡

" History tells us of a wise and polite nation that rejected a person of the first quality, who stood for a justiciary office, only because he had been observed, in his youth, to take pleasure in teasing and murdering of Birds. And of another that expelled a man out of the Senate for dashing a bird against the ground who had taken refuge in his bosom. Every one knows how remarkable the Turks are for their Humanity in this kind. I remember an Arabian author, who has written a Treatise to

---

* "That lies beneath the knife,
Looks up, and from her butcher begs her life.'
Æn. VII. (Pope's translation.) Quoted first by Montaigne. *Essais*
† And, Pope might have added, a more diabolical torture still—calves bled to death by a slow and lingering process—hung up (as they often are) head downwards. Although not universal as it was some ten years ago, this, among other Christian practices, yet flourishes in many parts of the country, unchecked by legal intervention.
‡ See Article, Plutarch, above.

show how far a man, supposed to have subsisted in a desert island, without any instruction, or so much as the sight of any other man, may, by the pure light of Nature, attain the knowledge of Philosophy and Virtue. One of the first things he makes him observe is the benevolence of Nature, in the protection and preservation of her creatures.* In imitation of which, the first act of virtue he thinks his self-taught philosopher would, of course, fall into, is to relieve and assist all the animals about them in their wants and distresses. . . . .

"Perhaps that voice or cry, so nearly resembling the human, with which Nature has endowed so many different animals, might purposely be given them to move our Pity, and prevent those cruelties we are to apt to inflict upon our Fellow Creatures."

Pope quotes, in part, the admirable verses of Ovid, Metam. XV., with Dryden's translation—and an apposite *fable* of the Persian Pilpai, which illustrates the base ingratitude of men who torture and slaughter their fellow labourers.—"I know it" (this common ingratitude) said the Cow, " by woful experience ; for I have served a man this long time with milk, butter, and cheese, and brought him, besides, a Calf every year—but now I am old, he turns me into this pasture with design to sell me to a butcher, who, shortly, will make an end of me."—*The Guardian*, LXI, May 21, 1713.

With Pilpai or Bidpai's fable, compare that of La Fontaine on the same subject—*L'Homme et la Couleuvre*.

—◆◆◆—

XI.

CHESTERFIELD. 1694—1773.

To the expression of the opinion or feeling of Lord Chesterfield on butchering, given, in its place, in the body of this work (page 140), is here subjoined the remainder of his paper in *The World*. The value of such testimony may be deemed proportionate to the extreme rarity of any protests of this sort from those who, by their influential position, are the most *bound* to make them :—

"Although this reflection [the fact of the preying of the stronger upon the weaker throughout Nature] had force enough to *dispythagorise* me *before my companions* [in his college at the University of Oxford] *had time to make observations upon my behaviour, which could by no means have turned to my advantage in the world,* I for a great while retained so tender a regard for all my fellow-creatures, that I have several times brought myself into imminent peril by putting butcher-boys in mind, that their Sheep were going to die, and that they walked full as fast as could reasonably be expected, without the cruel blows they were so liberal in bestowing upon them. As I commonly came

---

* So far, at least, as the *natural and necessary wants* of each species are concerned.—That "Nature" is regardless of suffering, is but too apparent in all parts of our globe. It is the opprobrium and shame of the human species that, placed at the head of the various races of beings, it has hitherto been the *Tyrant*, and not the *Pacificator*.

off the worst in these disputes, and as I could not but observe that I often aggravated, never diminished, the ill-treatment of these innocent sufferers, I soon found it necessary to consult my own ease, as well as security, by turning down another street, whenever I met with an adventure of this kind, rather than be compelled to be a spectator of what would shock me, or be provoked to run myself into danger, without the least advantage to those whom I would assist.

" I have kept strictly, ever since, to this method of fleeing from the sight of cruelty, wherever I could find ground-room for it ; and I make no manner of doubt, that I have more than once escaped the horns of a Mad Ox, as all of that species are called, that do not choose to be tortured as well as killed. But, on the other hand, these escapes of mine have very frequently run me into great inconveniences. I have sometimes been led into such a series of blind alleys, that it has been matter of great difficulty to me to find my way out of them. I have been betrayed by my hurry into the middle of a market—*the proper residence of Inhumanity.* I have paid many a six-and-eightpence for non-appearance at the hour my lawyer had appointed for business ; and, what would hurt some people worse than all the rest, I have frequently arrived too late for the dinners I have been invited to at the houses of my friends.

" All these difficulties and distresses, I began to flatter myself, were going to be removed, and that I should be left at liberty to pursue my walks through the straightest and broadest streets, when Mr. Hogarth first published his Prints upon the subject of Cruelty.* But whatever success so much ingenuity, founded upon so much humanity, might deserve, all the hopes I had built of seeing a Reformation, proved vain and fruitless. I am sorry to say it, but there still remain in the *streets* of this metropolis, more scenes of Barbarity than, perhaps, are to be met with in all Europe besides. Asia (at least in the larger population of it—the Hindus) is well known for compassion to ' brutes ' ; and nobody who has read Busbequius, will wonder at me for most heartily wishing that our common people were no crueller than Turks.

" I should have apprehensions of being laughed at, were I to complain of want of compassion in our Laws [!] ; the very word seeming contradictory to any idea of it. But I will venture to own that to me it appears strange, that the men against whom I should be enabled to bring an action for laying a little dirt at my door, may, with *impunity,* drive by it half-a-dozen Calves, *with their tails lopped close to their bodies and their hinder parts covered with blood.* . . . . .

" To conclude this subject—as I cannot but join in opinion with Mr. Hogarth, that the frequency of murders among us is greatly owing to those scenes of Cruelty, which the lower ranks of people are so much accustomed to ; *instead of multiplying such scenes,* I should rather hope that some proper method might be fixed upon either *for preventing them,* or removing them out of sight ; so that our infants might not grow up into the world in a familiarity with blood.

" If we may believe the Naturalists, that a Lion is a gentle animal until his tongue has been dipped in blood, *what precaution ought we to use to prevent MAN from being inured to it, who has such superiority of power to do mischief.—The World,* No. LXI., Aug. 19, 1756.

---

* *The Four Stages of Cruelty,* in which, beginning with the torture of other animals, the legitimate sequence is fulfilled in the murder of the torturer's mistress or wife.

XII.

JENYNS.   1704—1787.

A SUPPORTER of the Walpole Administration, he represented the county of Cambridge, and during twenty-five years held the office of Commissioner of the Board of Trade.   He wrote papers in *The World* and other periodicals, and published two volumes of Poems. His principal book is the *Free Enquiry into the Origin of Evil*, in which he seeks to reconcile the obvious evils in the constitution of things with his optimistic creed.   Johnson, who, with all his orthodoxy, was pessimistic, severely criticised this apology for Theism.   In striking contrast with the indifferentism of the vast majority of his class, his just and humane feeling is sufficiently remarkable.   The line of reasoning, in his comprehensive arraignment of the various atrocities perpetrated, sanctioned, or condoned by English Society or English Law in the last century, and which, for the most part, still continue (it is scarcely necessary to add), *logically* leads to the abolition of the Slaughter-House— the fountain and origin of the evil :—

"How will Man, that sanguinary Tyrant, be able to excuse himself from the charge of those innumerable cruelties inflicted on his unoffending subjects, committed to his care, and placed under his authority, by their common father ? To what horrid deviations from these benevolent intentions are we daily witnesses ! No small part of Mankind derive their chief amusement from the deaths and sufferings of inferior Animals. A much greater part still, consider them only as engines of wood or iron, useful in their several occupations. The Carman drives his Horse as the Carpenter his nail by repeated blows ; and so long as these produce the desired effect, and they both go, they neither reflect nor care whether either of them have any sense of feeling.

"The Butcher knocks down the stately Ox with no more compassion than the Blacksmith hammers a horse-shoe, and plunges his knife into the throat of the innocent Lamb with as little reluctance as the Tailor sticks his needle into the collar of a coat.* If there are some few who, formed in a softer mould, view with pity the sufferings of these defenceless beings, *there is scarce one who entertains the least idea that Justice or Gratitude can be due to their Merits or their Services.*

"The social and friendly Dog, if by barking, in defence of his master's person and property, he happens unknowingly to disturb his rest—the generous Horse, who has carried his ungrateful master for many years, with ease and safety, worn out with age and infirmities contracted in his service, is by him condemned to end his miserable days in a dust-cart, where the more he exerts his little remains of spirit, the more he is whipped to save his stupid driver the trouble of whipping some other less obedient to the lash. Sometimes, having been taught the practice of many unnatural and useless feats in a

---

* Which is the accomplice *really guilty ?* The ignorant, untaught, wretch who has to gain his living some way or other, or those who have been entrusted with, or who have assumed, the control of the public conscience—the statesman, the clergy, and the schoolmaster? Undoubtedly it is upon these that almost all the guilt lies, and always will lie.

Riding-House, he is, at last, turned out and consigned to the dominion of a hackney-coachman, by whom he is every day corrected for performing those tricks which he has learned under so long and severe a discipline. [Add the final horrors of the *Knackers' Yard*, to which sort of hell the worn-out Horse is usually consigned.]

"The Sluggish Bear, in contradiction to his nature, is taught to dance, for the diversion of an ignorant mob, by placing red-hot irons under his feet. The majestic Bull is tortured by every mode that malice can invent, for no offence but that he is unwilling to assail his diabolical tormentors.* These and innumerable other acts of Cruelty, Injustice, and Ingratitude are every day committed—not only with impunity, but *without censure, and even without observation.* . . .

"The law of self-defence, undoubtedly, justifies us in destroying those animals that would destroy us, that injure our properties, or annoy our persons; but not even these, whenever their situation incapacitates them from hurting us. . . .

" If there are any [there are vast numbers even now], whose tastes are so vitiated, and whose hearts are so hardened, as to delight in such inhuman sacrifices [the tortures of the Slaughter-House and of the Kitchen], and to partake of them without remorse, they should be looked upon as demons in human shape, and expect a retaliation of those tortures *which they have inflicted on the Innocent for the gratification of their own depraved and unnatural appetites.*

" So violent are the passions of anger and revenge in the human breast, that it is not wonderful that men should persecute their real or imaginary enemies with cruelty and malevolence. But that there should exist in Nature a being who can receive pleasure from giving pain would be totally incredible, if we were not convinced by melancholy experience that there are not only many—but that this unaccountable disposition is in some manner inherent in the nature of men.† For as he cannot be taught by example, nor led to it by temptation, nor prompted to it by interest, it must be derived from his native constitution. ‡

" We see children laughing at the miseries which they inflict on every unfortunate animal who comes within their power. All Savages are ingenious in contriving and executing the most exquisite tortures, and [not alone] the common people of all countries are delighted with nothing so much as with Bull-Baitings, Prize-Fightings, 'Executions,' and all spectacles of cruelty and horror. . . . They arm Cocks with artificial weapons which Nature had kindly denied to their malevolence, and with shouts of applause and triumph see them plunge them into each other's hearts. They view with delight the trembling Deer and defenceless Hare flying for hours in the utmost agonies of terror and despair, and, at last, sinking under fatigue, devoured by their merciless pursuers. They see with joy the beautiful Pheasant and harmless Partridge drop from their flight, weltering in their blood, or, perhaps, perishing with wounds and hunger under the cover of some friendly thicket, to which they have in vain retreated for safety. . . . And to add to all this, they spare neither labour· nor expense to preserve and propagate these innocent animals for no other end than to multiply the objects of their persecution.

---

* Bull-baiting, in this country, has been for some years illegal; but that moralists, and other writers of the present day, while boasting the abolition of that popular *pastime*, are silent upon the equally barbarous, if more fashionable *sports* of Deer-hunting, &c., is one of those inconsistencies in logic which are as unaccountable as they are common.

† " That is," remarks Ritson, " in a state of Society influenced by Superstition, Pride, and a variety of prejudices equally unnatural and absurd."

‡ " The converse of all this is true. He is certainly taught by example, and by temptation, and prompted by (what he thinks is) interest."—Note by Ritson in *Abstinence from Flesh a Moral Duty.*

" What name should we bestow upon a Supreme Being whose whole endeavours were employed, and whose whole pleasure consisted, in terrifying, ensnaring, tormenting, and destroying mankind ; whose superior faculties were exerted in fomenting animosities amongst them, in contriving engines of destruction, inciting them to use them in maiming and murdering each other ; whose power over them was employed in assisting the rapacious, deceiving the simple, and oppressing the innocent ? Who, without provocation or advantage, should continue, from day to day, void of all pity and remorse, thus to torment mankind for diversion ; and, at the same time, endeavouring, with the utmost care, to preserve their lives and propagate their species, in order to increase the number of victims devoted to his malevolence ? I say, what name detestable enough could we find for such a being. Yet if we impartially consider the case, and our intermediate situation, with respect to inferior animals, just such a being is a ' Sportsman,' [and let us add, by way of corollary, *à fortiori* one who consciously sanctions the daily and hourly cruelties of the Slaughter-House and the Butcher."]—*Disquisition II.* "On Cruelty to Animals," by Soame Jenyns.

XIII.

PRESSAVIN. 1750.

An eminent Surgeon of Lyon, in the Medical and Surgical College of which city he held a professorship, and where he collected an extensive Anatomical Museum. At the Revolution of 1789 he embraced its principles with ardour, and filled the posts of Municipal Officer and of Procureur de la Commune. On the day of the Lyon executions, under the direction of the revolutionary tribunals, Sept. 9, 1792, Pressavin intervened, and attempted to save several of the condemned. In the Convention Nationale, to which he had been elected deputy, he voted for the execution of the King ; in other respects he was opposed to the extreme measures of the violent revolutionists, and in Sept., 1793, he was expelled from the Society of the Jacobins. In 1798 he was named Member of the Council of Five Hundred, for two years, by the department of the Rhone. The date of his death seems to be uncertain.

His chief writings are :—

Traité des Maladies des Nerfs, 1769. *Traité des Maladies Vénériennes, où l'on indique un Nouveau Remède*, 8vo., 1773. Last, and most important, *L'Art de Prolonger la Vie et de Conserver la Santé*, 8vo. Paris, 1786. It was translated into Spanish, Madrid, 8vo., 1799.

Pressavin thus expresses his convictions as to the fatal effects of Kreophagy :—

" We cannot doubt that, if Man had always limited himself to the use of the nourishment destined for his organs, he would not be seen, to-day, to have become the victim of this multitude of maladies which, by a premature death, mows down (moissonne) the greatest number of individuals, before Age or Nature has put bounds to the career of his life. Other Animals, on the contrary, almost all arrive at that term without having experienced any infirmity. I speak of those who live free in the fields ; for those whom we have subjected to our needs (real or pretended), and whom we call *domestic*, share in the penalty of our abuses, experience nearly the same alteration in their temperament, and become subject to an infinity of maladies from which Wild Animals are exempt.

"Men, then, coming from the hands of Nature, lived a long time without thinking of immolating living beings to gratify (s'assouvir) their appetite. They are, without doubt, those happy times which our ancient poets have represented to us under the agreeable allegory of the *Golden Age*. In fact Man, *by natural organisation* mild, nourishing himself only on vegetable-foods, must have been originally of pacific disposition, quite fitted (bien propre) to maintain among his fellows that happy Peace which makes the delights of Society. Ferocity, I repeat it, is peculiar to carnivorous animals ; the blood which they imbibe maintains that character in them. . . . .

" But if this faculty (reflection), which is called Reason, has furnished Man with so great resources for extending his enjoyments and increasing his well-being, how many evils have not the multiplied abuses, which he has made of them, drawn upon him ? That which regards his Food is not the one of them which has *least* contributed to his degradation, as well physical as moral. . . . . .

" Among other evidences of this, country-people, who subsist upon the non-flesh diet, are exempt from the multitude of maladies which engender corruption of the juices of the blood, such as *humoral*, putrid, and malign fevers, from Apoplexy, from *Cachexy*, from Gout, and from an infinity of miserable disorders—their offspring ; they arrive at a very advanced Age, free from the infirmities which early affect our old *Sybarites*. On the contrary, the inhabitants of towns, who make flesh their principal food, pass their lives miserably, a prey to all these maladies which one may regard, for that reason, endemic among them.

"Another very evident proof that Flesh is not a food natural to man is that, whoever has abstained, during a certain time, when he goes back to it—it is rare that this new regimen does not soon become in him the germ of a disease, the graver in proportion to the abstinence from that food. We have opportunities of observing this after the Fasts of the Catholics—in the majority of those who have faithfully practised abstinence from flesh."

He admits that there may be some constitutions, whose organs of digestion have been so corrupted by the long use of flesh, that a *sudden* change may be unadvisable ; but a gradual reform cannot but be always beneficial :—

"I do not doubt that Apoplexy, that fatal Malady so common among the rich people of the towns, might be escaped by those who are threatened with it, by entire abstinence from flesh. A Sanguine or humoral *plethora* is always the predisposing cause of this disease. A sudden rarefaction of the blood or of the humours in the vessels is the proximate cause of it ; this rarefaction takes place only by the predisposition of the juices of the body to corruption."

Pressavin devotes a considerable proportion of his Treatise to the arguments from Comparative Physiology.—While firmly persuaded both of the unnaturalness, and of the fatal mischiefs, of the diet of blood,* he expresses his despair of an early triumph of Reason and Humanity by means of a general dietetic reformation.†

XIV.

SCHILLER. 1759—1805.

AFTER Goethe the greatest of German Poets, began life as a surgeon in the army. In his twenty-second year he produced his first drama, *Die Räuber* ("The Robbers"). Some passages in it betrayed the "cloven hoof" of revolutionary, or at least democratic, bias, and he brought upon himself the displeasure of the sovereign Duke of Würtemberg, in consequence of which he was forced to leave Stuttgart. His principal dramas are *Wallenstein, Wilhelm Tell, Die Jungfrau von Orleans, Maria Stuart,* and *Don Carlos,* of which *Wallenstein* is, usually, placed first in merit. Even greater than the dramatic power of Schiller is the genius of his ballad poetry, and in lyrical inspiration he is the equal of Goethe. *Das Lied von der Glocke* ("The Lay of the Bell"), one of his most widely-known ballads, is also one of the most beautiful in its kind.

In prose literature, his *Briefe Philosophische* ("Philosophical Letters"), and his correspondence with his great poetical rival, are the most interesting of his writings.

In *Das Eleusische Fest* ("The Eleusinian Feast") and *Der Alpenjäger* ("The Hunter of the Alps") are to be found the humanitarian sentiments as follow :—

> Schwelgend bei dem Siegesmahle
> Findet sie die rohe Schaar,
> Und die blutgefüllte Schaale
> Bringt man ihr zum Opfer dar
> Aber schauernd, mit Entsetzen,
> Wendet sie sich weg und spricht :
> ' *Blut'ge Tigermahle* netzen
> Eines Gottes Lippen nicht.
> Reine Opfer will er haben
> Früchte, die der Herbst bescheert—

---

* Among living enlightened medical authorities of the present day, Dr. B. W. Richardson, F.R.S., perhaps the most eminent hygeist and sanitary reformer in the country now living, has delivered his testimony in no doubtful terms to the superiority of the purer diet. In his recent publication *Salutisland* he has banished the slaughter-house, with all its abominations, from that model State. See also his *Hygieia.*

† *L'Art de Prolonger la Vie et de Conserver la Santé : ou, Traité d'Hygiène.* Par M. Pressavin, Gradué de l'Université de Paris ; Membre du Collège Royal de Chirurgie de Lyon, et Ancien Demonstrateur en Matière Medicale-Chirurgicale. A Lyon, 1786.

Mit des Feldes frommen gaben
Wird der Heilige verehrt.

Und sie nimmt die Wucht des Speeres
Aus des Jäger's rauher hand ;
Mit dem Schaft des Mordgewehres
Furchet sie den leichten Sand,
Nimmt von ihres Kranzes Spitze
Einen Kern mit Kraft gefüllt,
Senkt ihn in die zarte Ritze,
Und der Trieb des Keimes schwillt.*

———

Mit des Jammers Stummen Blicken
Fleht sie zu dem harten Mann,
Fleht umsonst, denn, loszudrücken,
Legt er schon den Bogen an ;
Plötzlich aus der Felsenspalte
Tritt der Geist, der Bergesalte

Und mit seinen Götterhänden
Schützt er das gequälte Thier :
" *Musst du Tod und Jammer Senden* "
Ruft er " bis herauf zu mir ?
Raum fur alle hat die Erde
Was verfolgst du meine Heerde ?" †

———————•———————

XV.

BENTHAM.   1749—1832.

THIS great legal reformer was educated at Westminster, and at the
age of thirteen proceeded to Queen's College, Oxford.   At the age of six
teen he took his first degree in Arts.   The mental uneasiness with
which he signed the obligatory test of the " Thirty-nine Articles "
he vividly recorded in after years.   At the Bar, which he soon afterwards
entered, his prospects were unusually promising ; but unable to reconcile
his standard of ethics with the recognised morality of the Profession, he
soon withdrew from it.   His first publication,—*A Fragment on Government*, 1776—which appeared without his name, was assigned to some of
the most distinguished men of the day.   His next, and principal work, was
his *Introduction to the Principles of Morals and Legislation* (1780), not
published until 1789.   At this period he travelled extensively in the East
of Europe.   *Panopticon: or the Inspection-House* (on prison discipline),
appeared in 1791.   The *Book of Fallacies* (reviewed by Sidney Smith, in

———

* *Die Eleusische Fest.*
† *Der Alpenjäger.*   See also Göthe—*Italienische Reise*, XXIII. 42 ; *Aus Meinem Leben*, XXIV. 23 ;
Werther's Leiden ; Brief 12.

the *Edinburgh*), in which the "wisdom of our ancestors" delusion was mercilessly exposed (1824), is the best known, and is the most lively of all his writings. *Rationale of Judicial Procedure*, and the *Constitutional Code*, are those which have had most influence in effecting legislative and judicial reform.

Bentham stands in the front rank of legal reformers; and as a fearless and consistent opponent of the iniquities of the English Criminal Law, in particular, he has deserved the gratitude and respect of all thoughtful minds. Yet, during some sixty years, he was constantly held up to obloquy and ridicule by the enemies of Reform, in the Press and on the Platform; and his name was a sort of synonym for *utopianism* and revolutionary doctrine. In his own country his writings were long in little esteem; but elsewhere, and in France especially, by the interpretation of Dumont, his opinions had a wider dissemination. In *Morals*, the foundation of his teaching is the principle of the greatest Happiness of the Greatest Number; that other things are good or evil in proportion as they advance or oppose the general Happiness, which ought to be the end of all morals and legislation.

Not the least of his merits as a moralist is his assertion of the rights of other animals than man to the protection of Law, and his protest against the culpable selfishness of the lawmakers in wholly abandoning them to the capricious cruelty of their human tyrants. The most eminent of the disciples of Bentham, John Stuart Mill (who found himself forced to defend the teaching of his master, in this respect, against the sneers of Whately, Archbishop of Dublin, and others), repeats this protest, and declares that—

"The reasons for legal intervention in favour of children apply not less strongly to the case of those unfortunate slaves and victims of the most brutal part of mankind, the lower animals. It is by the grossest misunderstanding of the principles of Liberty, that the infliction of exemplary punishment on ruffianism practised towards these defenceless beings has been treated as a meddling by Government with things beyond its province—an interference with domestic life. The domestic life of domestic tyrants is one of the things which it *is the most imperative on the Law to interfere with.* And it is to be regretted that metaphysical scruples, respecting the nature and source of the authority of governments, should induce many warm supporters of laws against cruelty to the lower animals to seek for justification of such laws in the incidental consequences of the indulgence of ferocious habits to the interest of human beings, *rather than in the intrinsic merits of the thing itself.* What it would be the duty of a human being, possessed of the requisite physical strength, to prevent by force, if attempted in his presence, it cannot be less incumbent on society generally to repress. The existing laws of England are chiefly defective in the trifling—often almost nominal—maximum to which the penalty, even in the worst cases, is limited." (*Principles of Political Economy*, ed. 1873.)

The observations both of Bentham and of Mill upon this subject, slighted though they are, are pregnant with consequences. It is thus that the former authority expresses his opinion :—

"What other agents are those who, at the same time that they are under the influence of man's direction, are susceptible of Happiness? They are of two sorts : (1) Other Human beings, who are styled *Persons.* (2) Other Animals who, on account of their interests having been neglected by the insensibility of the ancient Jurists, stand degraded into the class of *Things.* Under the Gentoo and Mahometan religions, the interests of the rest of the animal kingdom seem to have met with *some* attention. Why have they not, universally, with as much as those of human beings, allowance made for the differences in point of sensibility? *Because the Laws that are have been the work of mutual fear*—a sentiment which the less rational animals have not had the same means, as men have, of turning to account. Why *ought* they not [to have the same allowance made]? No reason can be given . . .

"The day has been (and it is not yet past) in which the greater part of the Species, under the denomination of *Slaves,* have been treated by the Laws exactly upon the same footing—as in England, for example, the inferior races of beings are still. The day *may* come, when other Animals may obtain those rights *which never could have been withholden from them but by the hand of Tyranny.* The French have already (1790) recognised that the blackness of the skin is no reason why a human being should be abandoned, without redress, to the caprice of a tormentor.

"It may come one day to be recognised that the number of the legs, the villosity of the skin, or the termination of the *os sacrum,* are reasons equally insufficient for abandoning a sensitive being to the same fate. What else is it should fix the insuperable line? Is it the faculty of reason, or, perhaps, the faculty of discourse? But a full-grown Horse or Dog is, beyond comparison, a more rational, as well as more conversable animal, than an infant of a day, or a week, or even of a month old. But suppose the case were otherwise, what would it avail? The question is not, can they reason? Nor is it, can they talk? But, *can they suffer?*" *

XVI.

SINCLAIR. 1754—1835.

THIS celebrated Agricultural Reformer and active promoter of various beneficent enterprises was a most voluminous writer. During sixty years he was almost constantly employed in producing more or less useful books. He was born at Thurso Castle, in Caithness, and received

---

* *Introduction to the Principles of Morals and Legislation* (page 311). By Jeremy Bentham, M.A., Bencher of Lincoln's Inn, &c.; Oxford: Clarendon Press, 1876. It must be added that the assumption (on the same page on which this cogent reasoning is found), that man has the right to *kill* his fellow-beings, for the purpose of feeding upon their flesh, is one more illustration of the strange inconsistencies into which even so generally just and independent a thinker as the author of the *Book of Fallacies* may be forced by the "logic of circumstances." Among recent notable Essays upon the Rights of the Lower Animals (the *right to live* excepted) may here be mentioned— *Animals and their Masters,* by Sir Arthur Helps (1873), and *The Rights of an Animal,* by Mr. E. B. Nicholson, librarian of the Bodleian, Oxford (1877).

his education at the Edinburgh High School, and at the Universities of Glasgow and Oxford. In 1775 he was admitted a member of the Faculty of Advocates, and afterwards was called to the English Bar. Five years later he was elected to represent his county in the Legislature ; and for more than half a century Sir John Sinclair occupied a prominent position in the world of politics, as well as of science and literature. His reputation as an Agriculturist extended far and wide throughout Europe and America; and statesmen and political economists, if they did not aid them as they ought to have done, professed for his labours the highest esteem.

His principal writings are: (1) *A History of the Revenue of Great Britain*, 3 vols. ; (2) *A Statistical Account of Scotland*, a most laborious work ; (3) *Considerations on Militias and Standing Armies ;* (4) *Essays on Agriculture ;* (5) Not the least important, *The Code of Health and Longevity*, in which the sagacious and indefatigable author has collected a large number of interesting particulars in regard to the diet of various peoples. Comparing the two diets, he asserts :—

" The Tartars, who live wholly on animal food, possess a degree of ferocity of mind and fierceness of character which form the leading feature of all carnivorous animals. On the other hand, an entire diet of vegetable matter, as appears in the Brahmin and Gentoo, gives to the disposition a softness, gentleness, and mildness of feeling directly the reverse of the former character. It also has a particular influence on *the powers of the mind*, producing liveliness of imagination and acuteness of judgment in an eminent degree."

Sir John Sinclair elsewhere quotes the following sufficiently condemnatory remarks from the *Encyclopédie Methodique*, vol. vii., part 1:—

" The man who sheds the blood of an Ox or a Sheep will be habituated more easily than another to witness the effusion of that of his fellow-creatures. Inhumanity takes possession of his soul, and the trades, whose occupation is to sacrifice animals for the purpose of supplying the [pretended] necessities of men, impart to those who exercise them a ferocity which their relative connections with Society but imperfectly serve to mitigate."—*Code of Health and Longevity*, vol. i., 423, 429, and vol. iii., 283.*

---

* Compare the *Voyages* of Volney, one of the most philosophical of the thinkers of the eighteenth century, who himself for some time seems to have lived on the non-flesh diet. Attributing the ferocious character of the American savage, " hunter and butcher, who, in every animal sees but an object of prey, and who is become an animal of the species of wolves and of tigers," to such custom, this celebrated traveller adds the reflection that " the habit of shedding blood, or simply of seeing it shed, corrupts all sentiments of humanity." (See *Voyage en Syrie et en Egypte*.) See, too, Thevenot (the younger), an earlier French traveller, who describes a Banian hospital, in which he saw a number of sick Camels, Horses, and Oxen, and many invalids of the feathered race. Many of the lower Animals, he informs us, were maintained there for life, those who recovered being sold to Hindus exclusively.

XVII.

BYRON. 1788—1824.

" As we had none of us been apprised of his peculiarities with respect to food, the embarrassment of our host [Samuel Rogers] was not little, on discovering that there was nothing upon the table which his noble guest could eat or drink. Neither [flesh] meat, fish, nor wine would Lord Byron touch ; and of biscuits and soda water, which he asked for, there had been, unluckily, no provision. He professed, however, to be equally well pleased with potatoes and vinegar ; and of these meagre materials contrived to make rather a hearty meal. . . .

" We frequently, during the first months of our acquaintance dined together alone. . . . . Though at times he would drink freely enough of claret, he still adhered to his system of abstinence in food. *He appeared, indeed, to have conceived a notion that animal food has some peculiar influence on the character* ; * and I remember one day, as I sat opposite to him, employed, I suppose, rather earnestly over a 'beef-steak,' after watching me for a few seconds, he said in a grave tone of inquiry,—' Moore, don't you find eating *beef-steak* makes you ferocious ?' "—*Life, Letters, and Journals of Lord Byron,* by Thomas Moore. New Edition. Murray, 1860.

In these Memorials of Byron, reference to his aversion from all "butcher's meat" is frequent ; and for the greater part of his life, he seems to have observed, in fact, an extreme abstinence as regards eating ; although he had by no means the same repugnance for fish as for flesh-eating. That this abstinence from flesh-meats was founded upon physical or mental, rather than upon moral, reasons, has already been pointed out. Nor, unhappily, was he as abstinent in drinking as in eating ; to which fact, in great measure, must be attributed the failure of his purer eating to effect all the good which, otherwise, it would have produced.

The observations of the author of a book entitled *Philozoa,* published in 1839, and noticed with approval by Schopenhauer, are sufficiently worthy of note, and may fitly conclude this work :—

" Many very intelligent men have, at different times of their lives, abstained wholly from flesh ; and this, too, with very considerable advantage to their health. Mr. Lawrence, whose eminence as a surgeon is well known, lived for many years on a vegetable diet. Byron, the poet, did the same, as did P. B. Shelley, and many other distinguished *literati* whom I could name. Dr. Lambe and Mr. F. Newton have published very able works in defence of a diet of herbs, and have condemned the use of flesh as tending to undermine the constitution by a sort of slow poisoning. Sir R. Phillips has published *Sixteen Reasons for Abstaining from the Flesh of Animals,* and a large society exists in England of persons who eat nothing which has had life.

" The most attentive researches, which I have been able to make into the health of

----

* This feeling occasionally appears in his poems, as, for instance, when describing a " banquet " and its flesh-eating guests, he wonders how " Such bodies could have souls, or souls such bodies."

all these persons, induce me to believe that vegetable food is the natural diet of man. I tried it once with very considerable advantage. My strength became greater, my intellect clearer, my power of continued exertion protracted, and my spirits much higher than they were when I lived on a mixed diet. I am inclined to think that the 'inconvenience' which some persons profess to experience from vegetable food is only *temporary*. A few repeated trials would soon render it not only safe but agreeable, and a disgust for the taste of flesh, *under any disguise*, would be the result of the experiment. The Carmelites, and other religious orders, who subsist only on the productions of the vegetable world, live to a greater age than those who feed on flesh ; and, in general, frugivorous persons are milder in their disposition than other people. The same quantity of ground has been proved to be capable of sustaining a *larger* and stronger population* on a vegetable than on a flesh-meat diet ; and experience has shown *that the juices of the body are more pure, and the viscera much more free from disease, in those who live in this simple way.*

"All these facts, taken collectively, point to a period in the history of civilisation when men will cease to slay their fellow-mortals for food, and will tend to realise the fictions of Antiquity, and of the Sybilline oracles respecting a 'Golden Age.'' †

---

* Note on this point the words of the late W. R. Greg, to the effect that "the amount of human life sustained on a given area may be almost indefinitely increased by the substitution of vegetable for animal food ;" and his further statement—"A given acreage of wheat will feed at least ten times as many men as the same acreage employed in growing ' mutton.' It is usually calculated that the consumption of wheat by an adult is about one quarter per annum, and we know that good land produces four quarters. But let us assume that a man living on grain would require two quarters a year ; still one acre would support two men. But a man living on [flesh] meat would need 3lbs. a day, and it is considered a liberal calculation if an acre spent in grazing sheep and cattle will yield in ' beef' and ' mutton' more than 50lb. on an average—the best farmer in Norfolk having averaged 90lb., but a great majority of farms in Great Britain only reach 20lb. On these data it would require 22 acres of pasture land to sustain one adult person living on [flesh] meat. It is obvious that in view of the adoption of a vegetable diet lies the indication of a vast increase in the population sustainable on a given area."—*Social and Political Problems (Trübner).*

† " Of the Cruelty connected with the Culinary Arts " in *Philozoa ; or, Moral Reflections on the Actual Condition of the Animal Kingdom, and on the Means of Improving the Same ;* with numerous Anecdotes and Illustrative Notes, addressed to Lewis Gompertz, Esq., President of the Animals' Friend Society : By T. Forster, M.B., F.R.A.S., F.L.S., &c. Brussels, 1839. The writer well insists that, however remote may be a *universal* Reformation, every individual person, pretending to any culture or refinement of mind, is morally bound to abstain from sanctioning, by his dietetic habits, the revolting atrocities " connected with the culinary arts, of which Mr. Young, in his Book on Cruelty, has given a long catalogue."

APPENDIX TO THE
ILLINOIS EDITION

———

[IN 1896, Howard Williams issued his revised and expanded edition of *The Ethics of Diet*. To ensure that this 2003 edition represents Williams's amplified efforts and reformulated thoughts, the following appendix includes the new material with which he augmented the original edition. This appendix contains Williams's introduction to the 1896 edition, entries on historical figures that he added to that edition, and new appendixes that were interpolated into the 1883 appendix. Because the 1896 edition was typeset in a font different from that of the first edition, the material here has been reset. For help with amassing the material added to the 1896 edition, I thank Andrew Rowan and the Humane Society of the United States as well as Beth Beathard, the typist.—Carol J. Adams]

1.

INTRODUCTION TO THE 1896 EDITION.

The most marked characteristics of the prevailing thought of the present age, obviously, are Pessimism, Indifferentism, and Unbelief. In various, Protean, shapes they permeate Philosophy (social and political), Legislation, Religion, the fashionable Literature, and even Art; and they are reflected plainly in the general tone and intercourse of society: the persuasion, that is to say, that real and permanent progress and happiness for the world in general are, in the very nature and constitution of things, a mere utopian dream, with the consequent logical conviction of the uselessness of serious efforts to *eradicate* the ultimate causes of the innumerable frightful wrongs and miseries which always, more or less, have infested our infinitesimal globe. Not that these latent paralysing and fatal influences are peculiar to our time. The Creed of Despair predominated, indeed, far more disastrously under the rule of the old theologies, most of all in what are termed the Mediæval Ages—*oppressa gravi sub religione*. But such creed was alike natural and inevitable under the domination of barbarous scientific and moral ignorance, and of its natural result, barbarous superstition.

What, on a first and superficial examination, must seem to be inexplicable in modern society, is the remarkable fact that, while the advance of Reason, and the marvellous discoveries of Natural Science, with the consequent huge development of material civilization have—for the ruling and richer classes, at least—so greatly advanced all material conditions of human existence, yet this pessimism in thought (sometimes affected, but generally too sincere) and this unbelief in practice still flourish—if not with so frightful results in the regions of theology or religion, none the less in almost everything that concerns terrestrial and actual public interest. That there abound the most various "charitable" associations, all of them good in intention, doubtless, and many of them excellent in practice—a fact which may be objected as in contradiction to this assertion—paradox, as it seems, in reality supports it. They aim, one and all, not at the eradication of the underlying causes—in the principles of the orthodox systems, an impossible and even impious attempt—of Wrong and Wretchedness, but only at some sort of mitigation of them; at their best, at what are mere temporary remedies for, in the orthodox persuasion, an *essentially incurable* disease.

Of the inveteracy of this Philosophy of Scepticism and Indifferentism, perhaps the most conspicuous outward and visible sign is the general constant attitude in regard to War, and in regard to the Destitution of the masses of populations—next to fanatical Superstition, the two greatest evils that have oppressed and oppress the human world. Recently, not only the present but even the *permanent* necessity of the state of war and of huge military establishments has been emphatically reaffirmed, on a public occasion, in this country, by the newly-appointed Commander-in-Chief of the British Army, who derided as mere utopianism, and even as mischievous fanaticism, all protest against the fostering of the war-spirit with all its actual horrors and all its directly and indirectly demoralizing influences. And this, as far as appeared, without dissent from either political party in the State, from the fashionable organs of public opinion, or from the representative of the various religious sects. * As for Pauperism and Destitution (in spite of occasional spasmodic appeals on the part of a more zealous philanthropist here and there, which die out almost as speedily as they blaze up) the creed of the most part of Society may be epitomized in the constantly-repeated formula, "the [destitute] poor you always shall have with you."†

---

* So far are they from protesting against the perpetuation of war, and of its necessary accompaniment, vast military establishments, the war-spirit is openly glorified by, at least, the more orthodox as a sort of divine institution.

† *Ptochous* (not *Penêtas*) in the original of the well-known quotation—meaning, literally, "crouching and suppliant poor." One of the most significant signs of the Pessimistic in-

To indicate here all the various phases of this prevailing and fatal Unbelief, in western civilization, is altogether beyond the limits of an Introduction. What is more necessary, and more to the purpose of this work, is to inquire into the *unrecognized* causes of the prevalence of so desolating a creed in face, and in spite of, a material, external civilization, present and prospective, such as never before has been known to the (more fortunate part of the) human world.

Modern Science—Comparative Physiology and Comparative Anatomy—has established, beyond all possibility of doubt and dispute, the oneness of the higher non-human races, essentially, with the human in mental no less than in physical organization. Reluctantly, in spite of all the inveterate prejudices of human arrogance, it has inevitably been forced to admit the only absolutely differentiating character to be the defect of *articulate* speech, and therefore of processes of sustained ratiocination. Only by the most inveterate adherents of the old orthodoxies (secular or religious) will the subterfuge of the non-possession of "a soul"—the *non anima* of the Italian peasant—nowadays seriously be resorted to. No more for the non-human than for the human is the matter of natural, or moral, rights any longer determinable by theological and metaphysical theories. Countless numbers of these so highly-organized and, therefore, so highly-sentient, fellow-beings and fellow-labourers of man daily are done to death to supply the foods pretended to be necessary for the support of his existence, or rather for the support of the life of the otiose part of his species; the vastly larger proportion of the toiling masses all over the globe necessarily begin debarred from the unnatural luxury. They are done to death, for the most part—it is notorious—with every circumstance of cruelty and barbarity. The treatment of these victims of the sanguinary diet—it is equally notorious—antecedent to the slaughter-house, in various ways, is, in an incalculable number of instances, atrocious and callously unfeeling in the extreme. This awful fact of thus living by the indescribably agonizing suffering of infinite numbers of their innocent fellow-beings—fellow-beings in all the essential properties of life as well as in all the accidents of mortality—however much the majority of kreophagist society may vainly attempt to disguise or to suppress it, none the less is there, obtruding itself in every street and in every (unreformed) kitchen: latent in the consciousness of all who have

---

fluence is the great number—a number tending constantly to increase—of self-destroyers, in all classes of the population. But, obviously, the most disgraceful circumstances of Christian civilization are the scenes of starvation (slow or rapid, as the case may be) in the very midst of the extremest superfluousness of luxuries; the frightful suffering resulting from what affectedly is styled the "Social Evil"; and the hideous brutality of all kinds (caused in large measure by the sale of alcohol, but indirectly and originally by destitution) with which the journals are crowded, and might yet more be crowded.

enough of intelligence or enough of sensibility for the least serious reflection, and who glance at all below the (comparatively) smooth and pleasant surface, under which the seething mass of agony is so industriously hidden away from the view of fashionable society—refined only in name. And this reflection, and these glances beneath the surface, revealing all that is meant by Butchery, of necessity insensibly induce the conviction that, as the human animal (as represented, at least, by the richer classes) must live at the expense of enormous and unutterable suffering, all efforts at radical improvement of the world of being are vain and are doomed to failure. For the religionist sections of Christian society this Creed of Hopelessness is strengthened by the (theoretical) belief that the terrestrial life is of small significance as compared with the post-terrestrial, and, besides, that the non-human, allowed no share in the latter, can have no substantial claim to more than at most what is vaguely termed "kind treatment," which by no means excludes daily butchery and other horrible tortures for tens of thousands of innocent and beautiful beings.

Such prevailing prejudice, conviction, or pretence (as the case may be), generally latent and unrecognized as it is, is the ultimate or principal source of the religion of Despair—sometimes openly affirmed, but, for the most part, secretly entertained. Many other causes obviously, more or less, contribute to it, but they, also, may be traced, in the last resort, to the diet of blood and slaughter, of which they are the subsidiary, if indirect, effects. The fatal influence of Butchery is no new theory. It has been affirmed more or less strongly by all the higher thinkers who have attempted the re-generation of the race, although their admonitions have been invariably ignored or relegated into the distant background by the accredited guides of the public conscience. Thus it is that radical Reformation in diet assumes a position of the very highest importance for the vital interests of human society, no less than it rests upon the highest principles of Ethics—universal Right and universal Compassion. This great Reformation is based upon the widest no less than upon the deepest foundations, upon arguments which (in the order of their significance) briefly may be summarized as follow:—

Firstly, upon the *original physical organization* of the human animal—upon the arguments of Comparative Physiology. Without such basis it could have no sure support.

Secondly, upon Justice. Abstractedly regarded, in any impartially constructed code of morality, zoophagy, so far, at least, as the higher mammals are concerned—the cow, or the deer, for example—is little removed, in the degrees of immorality, from anthropophagy commonly called cannibalism. Abstractedly considered, indeed, it has less justification that the old ortho-

dox practice of anthropophagy, practised, as that has been, by wholly barbarous peoples, and, for the most part, from necessity.

Thirdly, upon Humaneness, in the twofold sense of gentleness or "humanity," and of refinement of life—its original meaning. This is so obvious a truth
as to need, it might be presumed, no insistance. To allege the failure of (wholly
or partially) non-carnivorous people in these higher virtues does not invali-
date the argument, for the sufficient reason that never yet has the humaner
living been founded on the only true principles of Justice and Compassion.
In those countries where the masses have been, or are, non-carnivorous—the
privileged class (the Brahmins of Hindustan excepted) and the middle classes,
for the most part, have never been other than carnivorous—either merely
climatic or merely superstitious reason has been the cause of the purer living.
Yet even so, the manners of the more frugivorous peoples, under less unfavourable sociological, political, or religious environment, contrast, in many respects,
certainly not to the advantage of the more barbarously fed nations.

Fourthly, upon the widespread Degradation and Demoralization which
are the direct consequence of wholesale butchery. Apart from all other consideration, the simple fact that tens of thousands of human beings are dedicated to, and daily and hourly occupied in, the vicarious work of slaughter,
and that under the most revolting conceivable conditions—continually and
literally wading in blood—in every land of Christendom, is enough, in itself,
to condemn Kreophagy in the court of final appeal, the court of conscience.
In that latest development and boast of Western civilization—the city of
Chicago, where thousands of men and thousands of acres are continually
saturated with blood—this diabolical work has been reduced to what truly
may be termed the art of murder in its most revolting shape! But, on smaller scale, it is being daily perpetrated in every city, town, and village of Christendom—most extensively in this land of ours, which prides itself on being
in the van of civilization, of morals, and of religion. Let it further be considered, that these agents for Society, in this sanguinary occupation, are drawn,
necessarily, from the dregs of the population—that, by the very necessity of
their trade, their originally callous nature is made yet more callous by these
daily scenes of agonizing suffering—that they naturally and very excusably
seek to drown and deaden what remains of better feeling in the oblivion of
intoxicating drinks—that a large proportion of these pariahs of Christianity
physically are unfitted for the work, and that many of them are mere youths—
that the very nature of their horrible trade aggravates natural ferocity and
savagery—and the contrast between professions of morality or of religion
on the part of Society, the affectations of superiority of Christian ethics to

those of wholly barbarous or of less highly "civilized" nations, and the actual facts becomes yet more marked.

Fifthly, upon National Economy. The enormous, the incalculable waste of national resources, induced by the prevalence of Pastoralism, and the tremendous evils directly resulting from it, have been exposed repeatedly by every profounder writer on Political Economy.* Perhaps its most permanent evil is the enormous increase of Pauperism, the immediate consequence of the continuous forced immigration of the rural population into the now long overgrown cities, with the necessary result of incalculable addition to the sum of human and non-human suffering.

Sixthly, upon the Public Health. The variety, virulence, and extensiveness of human diseases, originating in the enormous and always-increasing consumption of "butchers' meat," repeatedly has been affirmed by Royal Commissions as well as by individual scientific witnesses. The constantly multiplying demands on the butchers induces unnatural breeding and feeding, often, also, under the most insanitary conditions. Hence multiplicity of disease for the various victims of human gluttony, which by just retribution is passed on, in the form of phthisis and other endemic maladies, to the human species. The less obvious, but not the less certain, aggravation of many human diseases, by large consumption of flesh (whether diseased or not) is a fact which, although neglected by the majority of the medical profession, has yet been indubitably ascertained and affirmed by the more thoughtful part of it.† The remark of Seneca, made eighteen centuries ago, is yet more justifiable now—"count the number of cooks [butchers]: you will no longer wonder at the innumerable number of diseases" (*Ep.* xcv.).

Seventhly, upon Health and Domestic Economy in their relations to individual members of the Social State. This argument has been, and is, so abundantly insisted upon in the constantly-increasing vast and varied literature of Vegetarianism, to which easy access is obtainable, that it is superfluous and supererogatory to emphasize it in this place.

Objections to the humaner diet fall under two heads, and may be distinguished as the *genuine* and the *non-genuine,* the sophistical, puerile, and popular, and the quasi-philosophical. The former, which may briefly be dismissed,

---

* The incalculably disastrous results of the shameful neglect of agriculture by Governments have been most forcibly exposed, of late, in one of the severest indictments of existing social institutions or pretentions—*Conventional Lies of Our Civilization,* of Max Nordau. Seventh Edition, 1895. See Section entitled, *The Economic Lie.*

† The latest Royal Commission for inquiry into the effects of the eating of flesh of *tuberculosis*-affected animals, which published its Report last year (1895), confirms, by the strongest evidence (not only the possibility but) the actual fact of the unsuspected transmission of virulent disease by this common means.

assume the following shapes:—(1) "The animals" were *created* wholly and
solely for human use or pleasure—an arrogant assumption so forcibly ridiculed
in the well-known verse of the *Essay on Man*—and to man is given by Deity
or Nature absolute, arbitrary, power over them all; the right to kill wholesale
for food being the first and most undoubtedly legitimate exercise of this despotic dominion, all the various suffering and horrors involved in it notwithstanding. The best and briefest answer to this arbitrary assumption is given in
the reply of Montaigne—"Shew to us this *sealed charter* of so enormous pretensions." *Who* has conferred on the human animal so tremendous power and
privilege, at the expense of the rights of all other of his fellow-beings, however harmless, intelligent, and beautiful—however near to himself in physical
and mental organization? Far from originally having been the head and lord
of "creation," during countless ages he was forced to struggle merely to maintain existence on the globe against the more powerful and almost equally intelligent forms of life. After those long ages of precarious and most wretch-
edly-barbarous existence and struggle, what Power but that of Might secured
for him, by the slow development of mental capacities, the final dominance?
The *device* of the human, as against the non-human, equally as against his
fellow-man, might well be openly, as it is actually, adopted, "Might is Right."
(2) Which may be styled the superlatively puerile reason—what is to become
of them if they are not butchered and eaten? They would soon eat us up, the
earth would be over-run, etc. To answer this puerility seriously is not easy. How
do the domesticated victims of the (orthodox) table come into existence if not
by human intervention, by which they are bred for the slaughter-house? Cease
to breed beyond the necessary and legitimate requirements of human interest, and the question is settled at once.

(3) And, perhaps, the most popular objection—which may be distinguished
as the "John Bull" sophistry. It is asserted (a) that the conquering and dominant races always have been flesh-eaters, that the non-flesh eating peoples,
on the contrary, always have been conquered and ruled by the kreophagist
nations; (b) that the British dominion, in particular, has been gained and maintained by beef-eating. To this the answer is, that the assertion is founded on
pure ignorance or pure prejudice. It is not here possible to examine the fallacy at length. But it may confidently be affirmed (a) that—some savage tribes,
such as the North Americans, excepted—there never has been a nation the
masses of which have consisted of flesh-eaters. In every case of the conquering people, the flesh-eaters have been limited to the dominant classes—to the
chiefs and rulers; while almost the whole body of their armies have been composed, in reality, of vegetable-eaters, or almost exclusively vegetable-eaters.
This is true of the old Italian or Roman armies (in the Western world), and

yet more certainly true of the Eastern empires—the Egyptian; the Assyrian, the Saracenic; and is almost equally true (in the modern Western world) of the Teutonic, the Russian, and other predominating races; (b) that the asserted inferiority in military courage and endurance of the non-kreophagist nations or races is an assumption of prejudice. It is enough to refer to the most often cited instance—the alleged Hindu inferiority.

That the Hindus have always been vanquished in conflict with the Muhammedan Tartars is commonly taken to be an historical fact. On the contrary, it has been pointed out by one of the highest authorities on Hindu history during the eleven centuries of intermittent Muhammedan invasion of Hindustan, these invaders—with all the advantages of greater homogeneousness and military organization, and with the choice of times or points of attack—have yet frequently been repulsed by the invaded, wanting though these were in all the advantages of national cohesion, and even (by comparison, at least) of fanaticism—so powerful a stimulant to mere brutal courage and strength.* It is indubitably true that the half-starvation diet, such as that of the mass of the Hindu populations, and that of populations nearer home, is not the ideal living, whether for War or for Peace; and that, as especially respects the Hindu, to that semi-starvation (joined to climatic influences) is owing want of robust vigour. But what is significant, for the present argument, is the incontestable fact of resistance on the part of the *least carnivorous* peoples of the globe, during so many ages, to the onslaughts of one of the *most carnivorous* races, so often with success. Of recent instances of almost non-carnivorous barbarian tribes in conflict with European armies, that of the Zulus—unsurpassed in daring and vigour in war—or of the hardly less carnivorous Soudanese Arabs, will occur to every reader. But proof of the entire fallacy of the *beef-eating* argument lies nearer to hand. The great mass of the Russian and Turkish armies, *e.g.,* undoubtedly consist of, or the most part, non-carnivorous combatants—a fact which was stated, with expression of wonder, by the newspaper correspondents in the war of 1879; who, also remarked on the extraordinarily greater healing qualities of the vegetable foods, exhibited in the cases of either combatant, but especially in that of the Turkish soldier.

The actual causes of the dominance of this country is to be sought elsewhere than in *beef-eating*—the inveterate prejudice of the "true-born Englishman" notwithstanding. They are various and complex. Intermixture of northern and

---

* See *"The Indian Empire,* and *A Brief History of the Indian Peoples,"* ix., by Sir W. Hunter (1892). Southern India, after so many centuries of fierce conflict, was not subjugated until the middle of the sixteenth century, a subjugation which continued not more than a hundred years.

more enterprising races, geographical position, a fortunate soil productive of various natural and mineral wealth, freer institutions, in spite of not a few great reverses of military fortune (which, by the way, patriotic historians have been too careful to conceal or to extenuate) have all contributed to the gradual growth of British ascendency. Only prejudice, ignorance, or sophistry will deny that had the diet of the Rulers, as is that of the larger part of the Ruled, been generally non-carnivorous, *on reasoned moral principles,* ascendency in all that makes a people truly great would have been yet more firmly established than is its undisputed naval and commercial dominance, at least it cannot be matter for sincere doubt that, in such case, the history of our island would not be deformed, as it is in almost every page of its annals, by hideous brutalities social as well as political.

The less sophistical and less unphilosophical class of objections, which demand more serious consideration, may thus be stated:—

(1) Appeal is made to the general order and constitution of Nature. In the general constitution of life on our globe, cruel suffering and slaughter, it is objected, are the constant and normal condition of things—the Strong preying on the Weak in endless succession; and, it is asked, why should the human species form an exception to the general rule, and hopelessly fight against the laws, as it seems, of the universe. To this statement it is to be replied first, that, although too certainly a cruel and unceasing internecine war has been waged upon this atomic globe of ours from the first origin of life until now, yet, apparently, there has been going on a slow, but not uncertain, progress, through the ages, towards the ultimate elimination of the more cruel phenomena of existence—that, if the carnivora form a very large proportion of living beings, yet the non-carnivora are in the majority; and, lastly (what is still more to the purpose), that man most evidently, by his origin and physical organization, belongs not to the former but to the latter. Besides and beyond which incontrovertible fact, it is to be added that, in proportion as he boasts himself—and, as he is at his best (or, rather, at his better), and only so far, he boasts himself with justness—to be the highest of all the gradually ascending and co-ordinated series of living beings, so is he, in that proportion, bound to prove his right to the supreme place and power, and his asserted claims to moral as well as to mental superiority, by just conduct. In brief, in so far only as he proves himself to be the beneficent ruler and pacificator—and not the selfish tyrant—of the world, can he have any just title to (moral) pre-eminence.

(2) If Butchery be generally abandoned, what is to be done with the *surplus* life—*e.g.* of the male young of the cow, and of the domesticated barnfowl—and with the domesticated animals, now bred for slaughter, when past the age of breeding or of service to human needs or convenience? This ob-

jection, at the first statement, of no little (apparent) force, has often been met—not with all the satisfactoriness possible under less barbarous or semi-barbarous social conditions than prevail now, but—with the practically logical and reasonable reply that, as, at birth, the surplus young of other non-humans—*e.g.* dogs and cats—constantly are being put out of existence, so with equal reason the lives of the surplus young of the victims of the butcher would be ended *at birth*. Not, indeed, with the accustomed brutal callousness and barbarity with which the canine or feline—and to the special shame of the human species, it must be added, even the human—young are constantly put out of existence, but by a public, and, as far as possible, painless mode of death. And as the canine and feline bodies are not cooked and eaten—at all events, by authorised custom—so, in the nature of things, there is no absolute reason for the devouring of the bovine or other thus sacrificed young. Under more natural national custom, it is to be added, the bovine male young would be reared for the plough. As for the disposal of the non-human slaves grown old and past service, in a humaner and juster civilization it will be deemed a breach of the sacred laws of justice and humaneness callously to send to the butcher and to the knacker's yard those who so long have supplied (legitimate) wants in giving milk, wool, etc., and rendered other hard but indispensable service. They will be pensioned off (so to say) and remunerated—as far as that is possible—by free life and by kind treatment and care. Under the present prevailing barbarous code of Morality, this justice, enjoined by Plutarch, is not likely to be practised but by very few humaner persons. But the most elementary sense of morality points out the weighty obligation, at the least, to end the existence of worn-out dependants with the very minimum of suffering; and this private and public duty, it is equally obligatory, should be guaranteed by the most careful official inspection.

(3) If Butchery be generally abandoned, it is further demanded with some show of reason, how or where will the substitutes for leather and other substances necessary in manufactures be procurable? Such difficulty, however, is, in reality, founded upon a contracted apprehension of facts and phenomena. For it is a reasonable and sufficient answer that the whole history of civilization, as it has been a history of the slow, but, upon the whole, continuous advance of the human race in the arts and sciences, contributing to the debarbarising of the human life, so, also, has it proved that *demand creates supply*—that it is the absence of *demand* alone which permits the various substances, no less than the various forces, yet latent in Nature to remain uninvestigated and unused. Nor can any thoughtful person, who knows anything of the history of Science or Discovery, doubt that the resources of Nature and the mechanical ingenuity of man are all but boundless. Already, in spite of the absence of any

demand for them outside the ranks of anti-kreophagists, various non-animal substances have been proposed, in some cases used, as substitutes for the prepared skins of the victims of the slaughter-house. That, in the event of general demand for such substitutes, an active competition would spring up there can be no reasonable doubt. Besides, it must be taken into account that the process of conversion of the flesh-eating—that is to say, of the richer sections of—populations, to the bloodless diet only too certainly will be very tardy and very gradual.

(4) How is the large population of a country so populous—or, rather, of cities and towns so populous—as England or Belgium to be fed without butchery? Supplies of corn and other vegetable products, it is contended, are insufficient and, owing to variableness of climate as well as to artificial social causes, uncertain in quantity. To this the reasonable reply is, the *variety* of nutritious non-flesh foods, native and foreign, possible to be supplied, is practically inexhaustible. If the *farinacea,* under present adverse political and social conditions—far more adverse than any natural cause—cannot be grown in sufficient quantity in this country, there are regions where they are grown and from which they are imported in limitless abundance; while there remain every sort of root-crops and many sorts of the hardier fruits, for the cultivation of the English farmers and fruit-growers, and for the support of the population of the country. Added to the native products are imported fruits and other various vegetable productions of every sort, grown in rich abundance in other parts of the globe, so that the just observation of Evelyn, the disciple of Ray, in his advocacy of the purer diet, that we have "variety [of non-flesh foods] in more abundance than any of the former ages could show," is infinitely more true of the close of the nineteenth than of the seventeenth century. By neglect of the proper functions of Government, by the dominance of a wholly selfish economic system or no-system, and, in particular, owing to irrational or privileged methods of distribution (which are designed for the enriching of only certain classes in the social state, not for the good of the masses) the country generally derives little benefit from all this variety of wholesome and excellent foods, and from the enormous extension of commercial activity.

Beyond the obvious and ordinary sources of food supplies to the masses, one method there is, neglected by the responsible authorities, at once scientific and commonsense, of supplying deficiency (real or imagined) of the innocent diet. It has long been pointed out by one of the leading scientific and sanitary authorities of this day that, not only is it possible for chemical science to combine and *concentrate* vegetable substances in such forms as entirely to obviate at once the difficulty or inconvenience (such as it may be) of *bulk*; but that it is the shame of Science that it has not long ago achieved this incalcula-

bly important triumph of rational dietetics.* Herein lies, obviously, the poten-
tiality of at once immeasurable advance in human living, and in the economy
of wearisome and monotonous culinary labour.

Such are the answers, briefly summarized, both to apparent (but, in fact, only
imagined) difficulties, and to the objections of sophistry, subterfuge, or puerility. But what it is of the first importance to affirm—and it is incumbent upon
all earnest Food Reformers to emphasise it upon all possible occasions—at the
very outset of the consideration of the whole matter of Reformation in diet, is
the indisputably vital truth of the *supreme obligations* of Right and of Humanity. In earlier human history the sacred mission of the first protesters against
Kreophagy, undoubtedly, was chiefly limited to the *unmaterialising*, in its most
materialistic aspect, of human life, and to found the purer living on a spiritual
basis—the principles of Justice, Humaneness, and Compassion being a subordinate consideration. In later times, when these more secular virtues had
assumed a higher position in moral codes, the yet more sacred mission of the
little band of the Higher Thinkers, in successive ages, has been first and foremost to *unbarbarize* human life in its essentially most *inhuman* aspect, and to
establish the great Reformation upon the supreme and surest foundation of
the Higher Morality—upon the foundation of the inherent natural, or moral,
rights of all innocent, highly-organized beings. In consonance with such higher
interpretation of the obligations of the Moral Law, these pioneers of the truer
Civilization have constantly denounced the gross immorality, the enormous
wrong of living by the daily agonizing suffering of millions of our fellow-beings;
affirming it to be the principal source of the miseries and injustices of human
existence—the just, reflected, retribution for so enormous abuse of human
power. Far from neglecting, indeed, the only less important considerations—
of criminal waste of national wealth, of the direct demoralization of large masses
of the people, and the indirect corrupting influences on society in general, of
health, and the virtues of temperance and simplicity—they yet have justly
placed in the forefront the gross violation of the first principles of morality. A
creed which inculcates Humanity, Justice, and Compassion as the first and most
obligatory of virtues, that—as they have rightly pronounced—in the moral
nature of things, is the only religion, under whatever name, that ultimately can
succeed, be universal, and world-compelling.

Thus immeasurably most important, most significant for the welfare of the
world (using "world" in its full and proper sense), it is the highest office and the
highest wisdom of all altruistic promoters of the great social Revolution (subor-

* See *Foods for Man, Animal and Vegetable: A Comparison.* By Sir B. W. Richardson
(Heywood, Manchester, 1891.) An important contribution to the philosophical and scientific literature of Dietary Reform.

dinating, while reasonably, also, propagating, the less comprehensive and less cosmopolitan ideas of dietetic reform), to press to the front the greatest of all obligations—*Universal Justice.* In strict consistence with the Higher Morality and the Higher Philosophy, they will denounce *first and foremost* the most shocking of the brutalities and the atrocities connected with the Slaughter-House. They will, at all times and in all places, hold up to the most indignant reprobation (reprobation which the complete indifference of the accredited guides of the public conscience makes all the more an urgent duty and an unceasing toil), first, the indescribable, supreme, social iniquity of the continued sanction by the English State of *free, unsupervised, unrestricted* dens of butchery—of all the many shames and shams of Christian morality, the greatest; the continued employment of the unskilled, untrained, and often most brutalized part of the population as the vicarious agents, in the work of slaughter; the continued legalization of the "cattle-ship," and all its frightful horrors; such especial atrocities as the *bleeding* of young cows, the horribly-revolting way of slaughtering swine, the forced, mechanical, cramming of poultry, the inexpressible atrocity of butchering in the presence of the rest of the victims; the foul state of the majority of those secret dens of a realized *Inferno;* the various multiplied barbarous mutilations and tortures for the mere gratification of the vitiated palate of the *gourmets.* * Such are some of the atrocious cruelties of the orthodox diet, the unceasing stigmatizing of which infinitely transcends, in obligation and in significance, the most interesting of subordinate dietetic questions.

For the present age of all but invincible selfish interests, if but even a few steps towards the final stopping of the principal source of so various brutalities and selfishnesses of the human race be made—in the language of one of the most eloquent and (in the true sense of a greatly-abused word) most enthusiastic of the prophets of Humanity, J. A. Gleïzès—it would be of infinite significance for the world. Cessation from the butchery of the higher mammals—equal in physical organization and, in an impartial estimate, far superior in real worth to the brutal, cruel, or vicious human animal—and, in particular, from that of the most beneficent of all the victims of the slaughter-house, the Cow, is that step which, indisputably, would be the most significant.

---

* As these sheets are passing through the press, there appears in a periodical devoted to the cause of the higher morality and of the higher civilization, (*Humanity,* Sept., 1896), an illustration of one of the many horrible phases of the Slaughter-house, of the most instructive kind. It is contained in the narrative of "A Visit to Deptford," by Mr. Ernest Bell. The scenes of savagery and ferocity described could not be surpassed in the most uncivilised regions of the globe. To the attention of the guides of the public conscience, by whom these *daily* horrors of a true *Inferno,* (at contemplation of which even a Virro might feel some compunction), are, for the most part, regarded with so much lightness of heart, the narrative in question is, especially, to be commended.

Fifteen years have passed since the publication of the first edition of the *Ethics of Diet,* which originally appeared in the pages of the *Dietetic Reformer.* * The present edition has been much enlarged, and usefulness of the book for reference has been increased by a full Index. As in the former edition, living writers—numerous and important as they are—have been excluded by the plan of the work, which has been designed to be a history of the past literature of the Ethics of Diet. In selection of matter, the leading principle of the writer has been that of impartial, unsectarian eclecticism. To render the biographical sections more interesting for the general reader, some account of the productions of the various writers, beyond the strictly dietetic limits, has, in most instances, been given. An attempt has been made at a more rational spelling of foreign personal and geographical names; but the attempt has been restricted by the force of English orthographical custom.

2.

SAKYA MUNI.†—590–510 (?) B.C. (CHAPTER III, 1896)

In the history of the development of thought—in the revelation whether of moral or of physical truth—few facts are more remarkable than the coincidence of simultaneous announcement by independent and sometimes far-separated thinkers. Whether the philosopher of Samos, or the great religious revolutionist of the East, have the priority of claim to the assertion of the sublime moral truths of Anti-kreophagy may be matter of doubt. But all probability seems to be in favour of the Eastern; since from the (remoter) East—from Persia and Hindustan—in the earlier periods of history, the most influential religious, or semi-religious, ideas always have emanated. In respect to flesh-eating, it is certain that to some extent, and in some degree, before the age of the Buddha, abstinence from animal food formed one of the sacred dogmas of Brahmanism and Vedas. But the principle rested wholly, or almost wholly, on religious or ascetic dogma with that sacerdotal caste-religion.‡ It was the great Hindu

---

* It has had the exceptional honour of appearing in a Russian version by Count Tolstoi (1893), himself the author of a highly-important Essay directed against Butchery—"The First Step."

† Siddhartha is the personal; Sakya-Muni the tribal (conjoined with the distinguishing epithet, "the wise"); Gotama or Gautama the family name; Buddha, the *religious* or prophetic title, meaning "the Enlightened"—the noblest of distinguishing epithets that can be applied to the religious revolutionist.

‡ In the earlier period of Brahminism (if any deduction may safely be drawn from the utterance of perhaps an exceptionally moral writer, as to the prevailing sentiment or practice) humane and juster ideas seem to have been more conspicuous than in later periods. But it is not always easy to determine what may be genuine and what interpolated in sa-

prophet who first proclaimed it as a great moral truth, and based it upon the sublime doctrine of universal justice and compassion.

Siddhartha, or Sakya Muni, according to the constant traditions of his life, was the only son of the Raja of Kapilavastu, a region of the peninsula lying on the southern slopes of the Himálayas. Educated in all the luxury of an eastern court, the young prince was led to renounce the grandeur and privileges of his order, profoundly moved by the frightful sights of various suffering and misery, on all sides, as he drove or wandered through the streets of the capital. Abandoning, at the age of twenty, his father's palace and his wife—the story of his silent farewell to her, silent from the fear that he might be deterred from his purpose by her entreaties adds much pathos to this great act of renunciation, a frequent theme of the Buddhist sacred Scriptures—he set out as a wanderer, in the miserable dress of a mendicant. He seems to have arrived, on his first pilgrimage, in the district of Magadha on the Ganges, whose king, or raja, he is said to have converted to the New Way. Soon afterwards he retired into meditative and perfect solitude in the jungles of Gayá, where he practised the extremes of austerity and abstinence—imaginary virtues which he had learned from two Brahmin hermits. Here he remained six years with five disciples.

"Instead of gaining peace of mind by self torture he sank into a religious depair, during which the Buddhist Scriptures affirm that the enemy of man-

---

cred literature. In the *Code of Manu,* which assumed final shape not before the first century, B.C., the following text is almost worthy of Buddhism itself. "He who injures animals that are not injurious, from a wish to give himself pleasure, adds nothing to his own happiness living or dead—while he who gives no creature willingly the pain of confinement or death, but seeks the good of all sentient animals, enjoys bliss without end." (Quoted in *Sacred Anthology* by M. D. Conway, Fifth edition.) In the great Hindu epic—descriptive of social and religious life some twelve hundred years B.C., when evidently the idea of the sacredness of life was unknown—the *Mahabhârata,* occurs a sentiment of the highest morality—but it scarcely can be other than an interpolation of a later hand:

> "The constant virtue of the Good is tenderness and love
> To all that live in earth, air, sea, great, small, below, above.
> Compassionate of heart, they keep a gentle will to each:
> *Who pities not hath not the faith.* Full many a one so lives."
> —III. *Story of Savitri.*

The inspiration of the subjoined precept in the *Hitopadesa,* as quoted by Sir E. Arnold, could not be excelled:

> "True religion—'tis not blindly prating what the Gurus prate,
> But to love, as God has loved them, all things, be they great or small,
> And true bliss is when a sane mind doth a healthy body fill—
> And true knowledge is the knowing what is good and what is ill."

See Appendix ["Humanity of the Higher Hindu Scriptures," item 11, pp. 384–86, in the 2003 edition] for further quotation from the Hindu Sacred Scriptures.

kind, Mará, wrestled with him in bodily shape. Torn with doubts whether all his penance availed anything, the haggard hermit fell senseless to the earth. When he recovered the mental agony had passed. He felt that the path to salvation lay not in self-torture in mountain, jungles, or caves, but in reaching a higher life to his fellowmen. He gave up penance. His five disciples, shocked by this, forsook him; and he was left alone in the forest. The Buddhist Scriptures depict him as sitting serene under a fig-tree, while demons whirled round him with flaming weapons. From this temptation in the wilderness he came forth with his doubts for ever laid at rest, seeing his way clear, and henceforth to be known as Buddha, the Enlightened."*

He now proceeded to the Holy City, and began as a religious reformer his self-imposed mission, destined to influence a third part of the population of our globe. The deer forest, near to Benares, witnessed his first public preaching to the people—for, unlike that of the Brahmin priesthood, his religion was so far from being narrowly exclusive, that he intended it to embrace the world and the poorest and the most despised without respect of class or caste. He soon attached to himself sixty disciples, whom he commissioned to preach in the neighbouring countries. Among his earnest followers were women—as significant an innovation upon the established sacerdotalism as the breaking down of the barriers of caste. "Princes, merchants, artisans, Brahmins, husbandmen, and serfs, noble ladies, and repentant women were added to those who believed." The field of his labours extended throughout the larger part of Northern Hindustan. Having added largely to the number of his followers of the "Excellent Way," he revisited his father's palace as a preaching mendicant, in dingy yellow robes, and with a shaven head, the well-known characteristic of the Buddhist monks of the present day. His family embraced the faith, and his wife became the first *réligieuse,* and the most devoted of his adherents.

He began his public preaching at the age of thirty-six and continued it during forty-four years. Foretelling his death, he addressed his disciples in the following solemn words: "Be earnest, be thoughtful, be holy. Keep steadfast watch over your own hearts. He who holds fast to the law and discipline, and faints not, he shall cross the ocean of life and make an end of sorrow. The world is fast bound in fetters, I now give a deliverance, as a physician who brings heavenly medicine. Keep your minds on my teaching; all other things change, this changes not, no more shall I speak to you. I desire to depart. I desire eternal rest, Nirvána." His last recorded words were "Work out your salvation with diligence." His divinely-beneficent life was protracted to an advanced age, and probably ended about the year 500.

---

* *History of the Indian Peoples,* v., by Sir W. W. Hunter, 1892. Compare the *Indian Empire,* v., and Prof. Rhys Davids' *Buddhism.*

The original, most characteristic, most important principles of the Excellent Way, or Excellent Law, as Siddhartha entitled his truly revolutionary moral teaching, are (1) the abolition of caste (2) the sacredness of all life, and the obligation of observance of justice and compassion to all beings. (3) The doctrine of Nirvána, or final deliverance and *cessation* from the sufferings of existence (as generally interpreted) by the merging of the individual vital principle into the universal spirit. In brief, the final rest of the human soul. The dogma of the *Metempsychosis* he derived and developed from Brahminism. Closely connected with this tenet of the transmigratory soul is that of *Karma*, as it is called in Hindu language. It teaches the doctrine of Free Will in its practical shape—that each human being must work out his own salvation which, in any particular stage of existence, depends wholly upon the character of his actions in his receding form of life. Necessarily such a belief obviates at once all sacerdotal pretensions. "The secret of Buddha's success was, that he brought spiritual deliverance to the people. He preached that salvation was equally open to all men, that it must be earned, not by propitiating imaginary deities, or by our own conduct. He thus did away with sacrifices, and with the priestly claims of the Brahmins as mediators between God and man. What a man sows that he must reap. As no evil remains without punishment, and no good deed without reward, it follows that neither priest nor god can prevent each act from bringing about its own consequences. Misery or happiness in this life is the unformidable result of our conduct in a past life, and our actions here will determine our happiness or misery in a life to come. When any creature dies, he is born again in some higher or lower state of existence, according to his merit or demerit. His merit or demerit consists of the *sum total of his actions in all previous lives.* A system like that in which our whole well-being, past, present, and to come depends on ourselves, leaves little room for a personal God."*

The philosophical defect in the creed, morally considered, obviously is that it seems to introduce the principle of *vicariousness* in a new shape. In no possible sense can a man be said to be responsible for the deeds of an occupant of a wholly foreign body; for to affirm the identity of "soul" of a series of various animal forms (human and non-human), or that they would not, of necessity, be influenced by their environment, to the apprehension of the Western mind must appear entirely out of the question. To the subtler and more imaginative intellect of the Hindu or Chinese there would be no such inherent logical difficulty. To the surpassing attractiveness of the personal

---

* *History of the Indian Peoples:* v., by Sir W. W. Hunter.
See also, Prof. Rhys Davids' *Buddhism,* Oldenberg's *Buddha Sein Leben,* and Senart's *Essai sur la Légende du Bouadha.*

character, as well as to the high morality of the public teaching of Siddhartha, every competent biographer or inquirer has paid tribute of admiration, which is fairly represented in the following eloquent words of a distinguished authority on Hindu literature and philosophy. "A generation ago, little or nothing was known in Europe of the great faith of Asia, which, nevertheless, had existed during twenty-four centuries, and at this day surpasses in the number of its followers, and the area of its prevalence, any other form of Creed. Four hundred and seventy millions of our race live and die in the tenets of Gautama, and the spiritual dominions of this ancient teacher extend at the present time from Nepaul and Ceylon, over the whole Eastern Peninsula, to China, Japan, Thibet, Central Asia, Siberia and even Swedish Lapland. India itself might fairly be included in this magnificent empire of belief; for though the profession of Buddhism has, for the most part, passed away from the land of its birth, the mark of Gautama's sublime teaching is stamped ineffaceably upon modern Brahminism, and the most characteristic habits and convictions of the Hindus are clearly due to the benign influence of Buddha's precepts. More than a third of mankind, therefore, owe their moral and religious ideas to the illustrious prince whose personality, though imperfectly revealed in the existing sources of information, cannot but appear the highest, gentlest, holiest and most beneficent, with one exception, in the history of thought. Discordant in frequent particulars, much over-laid by corruptions, inventions, and misconceptions the Buddhistic books yet agree in the one point of recording nothing, no single act or word—which mars the perfect purity and tenderness of this Indian teacher, who united the truest princely qualities with the intellect of a sage and the passionate devotion of a martyr. In point of age most other creeds are youthful, compared with this venerable religion, which has in it the eternity of a universal hope, the immortality of a boundless love, an indestructible element of faith in final good, and the proudest assertion ever made of human freedom."*

The characteristic of the Buddhist Gospel, which differentiates its promulgator from all other founders of religions, and which undoubtedly, forms its surpassing and compelling charm—compelling even for those who are scarcely conscious of the secret actual influence—is the divine *compassion* which lay at the foundation of the truly *protestant* creed of its founder. This unique religious, or rather moral, superiority cannot better be illustrated than in the subjoined passage from the *Light of Asia*. Siddhartha, in the course of his benefi-

---

* Preface to *The Light of Asia,* by Sir E. Arnold, 1878—an elegant versification of the story of the life and doctrine of the Buddha, which has been received with applause in Buddhist countries, no less than in Europe and North America, where numerous editions witness to its popularity.

cent mission, comes upon a number of Brahmin priests, with the king of the
country, on the point of offering one of their sanguinary, vicarious, sacrifices:

> "But Buddha softly said,
> 'Let him not strike, great king!' and therewith loosed
> The victim's bonds, none staying him, so great
> His presence was. Then, craving leave, he spake
> Of life which all can take but none can give,
> Life which all creatures love and strive to keep,
> Wonderful, dear, and pleasant unto each,
> Even to the meanest: yea, a boon to all
> Where pity is; for pity makes the world
> Soft to the weak and noble for the strong.
> Unto the dumb lips of his flock he lent
> Sad, pleading, words, shewing how man, who prays
> For mercy to the Gods, is merciless,
> Being as God to those: albeit all Life
> Is linked and kin; and what we slay have given
> Meek tribute of the milk and wool, and set
> Fast trust upon the hands that murder them.
>
>   ✿   ✿   ✿   ✿
>
> Nor, spake he, shall one wash his spirit clean
> *By blood:* nor gladden Gods, being good, with blood.*
> Nor bribe them, being evil; nay, nor lay
> Upon the brow of innocent bound beasts
> One hair's weight of that answer all must give
> For all things done amiss or wrongfully,
> Alone—each for himself—reckoning with that
> The fixed arithmic of the universe,
> Which meteth good for good, and ill for ill—
> Measure for measure unto deeds, thoughts, words.
>
>   ✿   ✿   ✿   ✿
>
> While still our Lord went on, teaching how fair
> This earth were, if all living things be linked
> In friendliness, and common use of foods,
> Bloodless and pure—the golden grain, bright fruits,
> Sweet herbs, which grow for all, the waters wan,

---

* Compare the similar utterance of the higher morality and feeling of Ovidius, in the celebrated and beautiful passage in the *Metamorphoses,* in which he presents the Pythagorean creed:

> Nec satis est quōd tale nefas committitur: *ipsos*
> *Inscripsêre deos sceleri,* numenque supernum
> Cæede laboriferi credunt gaudere juvenci!

And that other memorable text of another non-Christian poet:

> "Quùm sis ipse nocens, moritur cur victima pro te?
> Stultitia est *morte alterius* sperare salutem."

Sufficient drinks and meats—which when these heard,
The might of gentleness so conquered them,
The priests themselves scattered their altar-flames
And flung away the steel of sacrifice;
And through the land next day passed a decree
Proclaimed by criers, and in this wise graved
On rock and column: thus the king's will is—
'There hath been slaughter for the sacrifice,
And slaying for the meat, but henceforth none
Shall spill the blood of life, nor taste of flesh;
Seeing that knowledge grows and life is one,
And mercy cometh to the merciful.'"

We have space only for a few typical precepts of the Great Teacher, which are taken from a meritorious manual (published within recent years) the *Imitation of Buddha*. The passages quoted are to be found in the various Buddhist sacred writings:—

"All beings desire happiness, therefore to all extend you benevolence—Because he has pity upon every living being *therefore* is a man to be called Holy—Hurt not others with that which pains yourself—Whether any man kill with his own hand, or command any other to kill; or whether he only see with pleasure the act of killing: all is equally forbidden by this Law—He came to remove the sorrow of all living things. I will ask you, if a man in worshipping, sacrifices a sheep *and so does well,* wherefore not his child . . . and *so do better?* Surely . . . there is no merit in killing a sheep!" [addressed, apparently to the Sacerdotal order]—Our Scripture saith: 'Be kind and benevolent to every being, and spread peace in the world!'—The practice of Religion involves as a first principle, a loving, compassionate heart for all beings—"Hear ye all this maxim, and, having heard it, keep well: whatsoever is displeasing to yourselves, never do to another—In this mode of Salvation there are no distinctions of rich and poor, male and female, priests and people. All are equally able to arrive at the blissful state."*

---

* Quoted from *The Imitation of Buddha,* compiled by E. M. Bowden (Methuen and Co., London, 1891). An admirable little manual of various Buddhistic Scriptures, Chinese as well as Hindu, which should be in the hands of all who wish to learn how much of high worth is to be found in these sacred books. In all cases, to the quotations are subjoined the authorities from which they are extracted. That the higher morality of all sacred books has always been infinitely less in esteem with religionists than their ceremonial teaching, is a melancholy truth, to which Buddhism is no exception. The characteristic principle of the Founder—that of justice and compassion to the non-human races—it is gravely to be suspected, has long been far less honoured than the ritualistic developments of later times. That contact with European civilisation during the past three centuries, has tended to affect, for the worse, the Hindu tenderness for, and treatment of, the subject species is too certain.

3.

EMPEDOKLÊS. *CIRCA* 450, B.C. (CHAPTER IV, 1896)

The most remarkable of the poet-philosophers of Antiquity—the highly eulo-gised of the greatest of Latin poets—Empedoklês, of Agrigentum (the modern Girgenti) in Sicily, may fitly be regarded as Pythagorean in his ethical principles; although commonly classed as of the (so-called) Eleatic School. It is possible that he may have heard the celebrated instructor of Euripides and of Periklês, Anax-agoras, with whom his physics seem to have been partly in agreement. It is un-necessary to discuss, in detail, his abstruse physical theories, which, like those of the old Hellenic savants, in general, can have little interest for the ordinary read-er. It will be enough to state the distinguishing feature of his philosophy—a sort of Zoroastrianism in physics—to have consisted in the theory that the princi-ples of Benevolence (Φιλία) and Malevolence are in constant antagonism in the physical universe—that Benevolence or Love is the motive power which alone tends to hold together the discordant cosmic elements. As for the human ani-mal himself, "the Agrigentine" taught Love, conjoined with pure Intellect, to be the only true means of his attaining the higher knowledge. He was idealist in rejecting with Plato, and the School of Berkeley, and Kant in modern times, the reality of the impressions of the senses. Of the four elements, with Hera-kleitus, "the weeping philosopher," he taught Fire to be the chief of the *primor-dia rerum,* and the central soul of the universe.

So great was his fame for the higher *Gnosis* and in particular, for medical science, that even during his life-time he was regarded as semi-divine, and he himself was commonly asserted to have aspired to divine honours. In fact, it must be owned that in spite of, or perhaps, by reason of his marvellous attain-ments, he seems, with other transcendental intellects, not to have been able wholly to resist that "last infirmity of noble minds," the temptation of Fame. But there can scarcely be a doubt that much of the extravagant assumptions, as well as opinions, attributed to him by gossiping writers hostile to Pythagore-anism, is pure fable; otherwise it is improbable that Lucretius would have made him the subject of special eulogy in his rationalistic poem.[*]

Only fragments of his philosophical poems *On Nature*, the *Discourse on Medicine* (some 470 verses), and the *Lustral Precepts* remain.[†] It is thus that

---

[*] In the eloquent verses in which, after having sung the praises of Sicily as "for many reasons the object of admiration of the world," he proceeds to glorify "the Acragantine," as "its most illustrious, saint-like, a marvellous, and precious offspring," so that "he seems to be scarcely of human birth" *De Rerum Naturá* I, 715–735.

[†] In this last (Καθαρμοί) the poet insists (as far as can be collected from the fragment) on moral conduct as the only true and efficacious method of averting disease and other evils.

he sings of the "Golden Age":—"Then every animal was tame, and familiar with men—both mammals and birds; and mutual love prevailed. The trees flourished with perpetual leaves and fruits, and ample crops adorned their boughs throughout the year. Nor had these happy people any War-God, nor had they any mad violence for their divinity. Nor was their monarch Zeus or Kronos or Poseidon, but Queen Kypris [the divinity of Love]. Her favour they besought with fragrant essences, and censers of pure myrrh, and frankincense, and with golden honey. The altars did not reek with the blood of oxen."

Elsewhere he affirms:—"Blessed is the man who has obtained the riches of the wisdom of God. Wretched is he who has a false opinion about things divine. He [the Supreme] may not be approached. He is all pure mind, holy, and infinite,* darting with swift thought through the Universe." Exhorting the world to abandon the foul diet of blood, he exclaims:—"Will you not put an end to this accursed slaughter? Will you not see that you are destroying yourselves in blind ignorance of soul?"† By birth and connections Empedoklês belonged to the aristocratic or wealthy classes, who oppressed the masses of the people; yet, like the Roman Gracchi, he threw in his lot with the party of revolution, and assisted in the expulsion of the reigning tyrant. "His zeal, in the establishment of political equality, is said to have been manifested by his disinterested support of the poor, by his severity in denouncing the overbearing conduct of the aristocrats, and in his declining the sovereignty which was offered to him."‡ A distinguished writer of the present day has compared him to the great English prophet-poet. "Empedoklês resembles Shelley in the quality of his imagination, and in many of his utterances. The belief in a beneficent, universal, soul of Nature, the hatred of animal food, the love of all innocent things moving or growing on the face of the Earth, the sense of ancient misery and present evil are all—allowing for the difference of centuries and race and education—points in which the Greek and the English poets meet in community of nature."§ Aristoteles thus refers to his Anti-kreophagist teaching: "As Empedoklês affirms on the subject of not butchering that which has life: 'for this principle is not right here and wrong there; but is a principle of law to all'" (*Rhetorike* I. 13: *Of Acts of Injustice and Matters of Equity.*)

---

* Cp., the remarkable, similar, expressions of the elder Plinius: Quisquis est Deus, si modo est alius, quâcunque in parte, totus en sensus, totus visus, totus auditus, totus animus, *His. Nat.* I. I.

† See *Studies of the Greek Poets,* by J. A. Symonds, 1873, from which the above fragments are quoted. One of the most considerable of the poems of Matthew Arnold has *Empedoklês on Etna* as its subject.

‡ *Dictionary of Greek and Roman Biography,* Ed. By Dr. W. Smith, 1859.

§ *Studies of the Greek Poets,* by J. A. Symonds, 1873. *The Fragments,* with commentary, edited by F. W. Sturz, were published at Leipzig, 1805.

4.

ASOKA 250 B.C. (CHAPTER VI, 1896)

The first great Council of the New Religion, consisting of five hundred of the disciples, was held immediately upon the death of the Master, at Patna, in the year 543. A second great Council assembled, a century later, to draw up the canon of Buddhist Scripture and to settle disputed points of faith or ritual. But it was not until the conversion of Asoka, King of Behar, the (better) Constantine of Buddhism, that it entered upon that career of peaceful and beneficent conquest, which eventually brought under its influence all Asia east of the Ganges.

Asoka, grandson of Chandra Gupta (famous in having been one of the Hindu princes who appeared in the camp of the Greek Alexander, in the Punjab, and who afterwards established himself at Patna as the most powerful sovereign of the whole of N. Hindustan,) embraced the faith of Buddha about the year 257. Of a type very superior to that of kings in general, whether Asiatic or European, he was peace-loving, humane, just, and wise. One of his first acts was to check growing corruptions. At Patna the third and most important Council met at his summons, when one thousand doctors fixed the sacred canon. Royal edicts, confirming the decisions of the Council were published throughout his empire, and some of them are still found engraved on columns and on rocks throughout the peninsula. A minister of Justice and Religion was appointed by him to promote the true faith.

In striking contrast to Brahminism—which has been almost more exclusive and jealous of its caste privileges than even Judaism—"one of the first duties of Buddhism being to proselytize, an officer was especially charged with the welfare of the aborigines to whom its missionaries were sent. Asoka did not think it enough to convert the inferior races, without looking after their material interests. Wells were to be dug, and trees planted along the roads. A system of medical aid was established throughout his kingdom, and the conquered provinces as far as Ceylon, for man *and beast*. Officers were appointed to watch over public morality [not in the limited European sense of the much abused term,] and to promote instruction among the women as well as the youth. Asoka recognised proselytism, by *peaceful means*, as a State duty. The Rock Inscriptions record how he sent forth missionaries to the utmost limits of the barbarian countries, to intermingle with all unbelievers, for the spread of religion—'They shall mix equally,' enjoins the edict, 'with soldiers, Brahmins, and beggars—with the outcast and despised, both within the kingdom and in foreign countries, teaching better things.' Conversion is to be effected by persuasion not by the sword [in striking contrast with the two younger rivals, especially with that of Islam]. Buddhism was at once the most intensely missionary religion in the world, and the most tolerant. This character of a proselytising Faith, which wins its victories by peaceful means, so strongly impressed upon it by Asoka, has remained a prominent feature of Buddhism to the present day."

The same high authority, whom we are here quoting, thus remarks on the influence of the doctrine of the Metempsychosis, as interpreted by the wisest of the followers of Sakya-Muni. "Buddhism carried *transmigration* to its utmost spiritual use, and proclaimed our own actions to be the sole ruling influence on our past, present, and future states. It was thus led into denial of any external being or God, who could interfere with the immutable law of cause and effect as applied to the soul. But, on the other hand, it linked together mankind as parts of one universal whole, and denounced the isolated self-seeking of the human heart as the *heresy of individuality*. Its mission was to make men moral, kinder to others, and happier themselves—not to propitiate imaginary deities. Accordingly it founded its teaching on man's duties to his neighbours instead of any obligation to [Gods or] God, and constructed its ritual on a basis of relic worship, or the commemoration of good men, instead of on sacrifice." (*History of the Indian Peoples,* by Sir W. W. Hunter, Clarendon Press, Oxford, 1862).

Of its inherent vitality, as well as of its essentially missionary character, Sir W. Hunter remarks:—"During the last thousand years Buddhism has been a banished religion from its native home. But it has won greater triumphs in its exile than it could ever have achieved in the land of its birth.* It has created a Literature and Religion for nearly half the human race, and it is supposed by its influence on early Christianity to have affected the beliefs of a large part of the other half. Five hundred millions of men, or forty *per cent* of the inhabitants of the world, still follow the teaching of Buddha. . . . During twenty-four centuries, Buddhism has encountered and outlived a series of powerful rivals. At this day, it forms, with Christianity and Islam, one of the three great religions of the world, and (by far) the most numerously followed of the three. The noblest survivals of Buddhism in India are to be found not among any peculiar body,† but in the religion of the people—in that principle of the *brotherhood* of Man,

---

* Another point of resemblance to historical Christianity. For the very close resemblance—no less curious than it is close—*ritually* of the Papal (and especially its monastic) system to the Buddhist (in its later development), see *The Chinese Empire* by the Abbé Huc.

† Buddhism has left its influence, in the peninsula of Hindustan, in the most interesting and most estimable of its present religions—that of the Jains. They number about 500,000. "Like the Buddhists they deny the authority of the *Veda,* except in so far as it agrees with their own doctrines. They disregard sacrifice, practise a strict morality, believe that their

with the re-assertion of which each new revival of Hinduism starts; in the asylum which the great sect of Vaishnavs affords to women, who have fallen victims to caste rules—to the widows and the outcast, in that gentleness and charity to all men, which take the place of the *Poor Law* in India, and give a high significance to the half-satirical epithet of the 'mild' Hindu."

For some twelve centuries alternately triumphant and oppressed by the Old Religion, it exercised an extensive and permanent influence, yet perceptible even in the orthodox Hinduism. Contemporary with the ascendancy of Buddhism was that of the (better) science of Medicine. Between the age of Asoka and Siláditya (the Asoka of the seventh century, A.D.,) famous schools of Medicine seem to have been established in many parts of Northern India—and the name of Charaka, an eminent doctor of the science, became as distinguished in Hindustan as Hippokrates in Hellas. At the final expulsion of what may be termed (in its better sense) the Protestant Faith from its original home, at the beginning of the tenth century, Medicine no longer flourished, and soon relapsed into the old stereotyped routine of caste-practice. But in the age of the second restorer of the popular Creed in the seventh century, it was at its height of prosperity. Its principle seat was the great monastery or rather coenobite establishment of Nalenda, near Gavà in Bengal, the University (as it may be styled) of Hindustan and which, probably, indirectly influenced, in some sort, the later medicinal Universities of Europe through Arab science. In this vast collection of buildings resided ten thousand students who devoted themselves to the sciences, as well as to the practice of their religious ritual; supported wholly by the revenues granted to them by their royal patrons. Their diet we may reasonably presume to have been of that bloodless sort, which has been imitated in the "monastic" establishments of Christendom only in two or three exceptional instances.

All that is known of the history of Buddhism, in the middle period, is obtained from the records of two enterprising Chinese travellers of that faith. At the beginning of the fifth century of our era, Fah Hian began his remarkably extensive journey through Hindustan, where he found his religion everywhere flourishing. He carried home revised copies of the Buddhist Scriptures, which he had studied at Patali-Poutra (Patna) the centre of the Faith at that epoch. He notes the large number of its public hospitals, where the sick and diseased were re-

---

past and future states depend upon their own actions rather than upon any external deity, and scrupulously reverence the vital principle in man and 'beast.' . . . Their charity is boundless, and they form the chief supporters of the 'beast' hospitals, which the old Buddhistic tenderness for [other] animals has left in many of the cities of India." *Indian Empire*, v. Many of the older travellers, and writers on Hindustan, mention those noble humanitarian institutions.

ceived and treated free of expense, and supplied with food as well as medicine. Two centuries later (about 630) a second Chinese monk, a yet more distinguished Buddhist traveller, named Hiouen Tsiang, traversed the same ground, and extended his travels still further. His records of the social life of the peoples whom he encountered are of high interest, and he gives a very favourable report of the influences of Buddhism upon the national manners. There seem to have been no death-sentences—most offences being punishable by fines. His account of the beneficence of the principal Hindu monarch of that period, Siláditya, already mentioned, and of his distribution, every five years, of all his treasures to the poor at the city, now called Allahábad (the Muhammedan name), gives a picture of royal benevolence, not often imitated by kings and emperors. The prince, so we are assured, divested himself of all his royal dress and insignia, and put on the rags of a beggar; thus commemorating the great Renunciation of the Founder of Buddhism.* Hiouen Tsiang reports of the general condition of the people, and of the encouragement given to Science—and in particular, of the higher practice of Medicine—equally favourably with his predecessor; and, altogether, the Hindu peninsula seems to have been in a happier state then, perhaps, ever before or since.

5.

OLIVER GOLDSMITH. 1728–1774. (CHAPTER XXXIII, 1896)

In the lighter provinces of English Literature, of the XVIII Century, indisputably the most deservedly eminent name is that of the author of *The Deserted Village* and *The Vicar of Wakefield*. As is the case with other distinguished humanitarian writers of the age, as has already been remarked, the principles of Goldsmith, in regard to humane dietetics, were better than his practice, his sensibility stronger than his self-control. But, not the less, he supplies, in notable degree, example of that *latent* aversion from the cruel diet, which is developed in every feeling mind in proportion as it obeys the promptings of the enlightened conscience and reason.

The facts in the adventurous career of Goldsmith so often have been repeated, that it is unnecessary to reproduce here more than the leading events. Of Anglo-Irish birth, his youth was passed in the obscure village of Lissoy—the supposed original of Auburn—in the county of West Meath, of which his father held the incumbency; and at the age of seventeen he entered Trinity College, Dublin, where his experiences seem to have been no happier than those of Swift some sixty years earlier. After an interval of desultory occupa-

---

* See *History of India,* by Talboys Wheeler. Cp. Cunningham's *Corpus Inscriptionum.*

tions, he began the study of medicine in Edinburgh, continued at Leyden, at that time a general resort of the disciples of Asklepius. He then wandered, on foot, over a large part of Europe—Holland, Belgium, France, Switzerland, Italy,—paying his way mainly (as it appears) by his skill on the flute, and acquiring in the surest, though not in the most luxurious, fashion various experience of men and manners. He returned to England in 1756, after two years' absence. Encountering various rather obscure fortunes, Goldsmith made his first literary appearance in the *Monthly Review,* the property of the bookseller and publisher Griffiths. Among his contributions to that periodical was a review of the *Essai sur les Mœurs et L'Esprit des Nations* of Voltaire, for whom he always had high esteem and admiration, and of Smollett's continuation of Hume's *History of England,* and Gray's *Odes.* The *Critical Review* (edited by Smollett), the *British Magazine,* the *Ladies' Magazine,* the *Bee,* and some other of the even then swarming crowd of ephemeral literature were enlivened by his always more or less instructive no less than entertaining papers. More famous than any of these periodicals, however, by his contributions became the Public Ledger, in which appeared, first in 1760, the *Letters of the Citizen of the World*—the most entertaining satire, of the sort, which had yet been provided for the reading public; its prototype, the *Letires Persanes* of Montesquieu, hardly excepted. Of the hundred and twenty-five *Letters,* ninety-eight appeared in the *Public Ledger.* Two of his more considerable productions, at this time, were his *Enquiry into Polite Literature** and a *Life of Voltaire* (in the *Lady's Magazine*). Translations, compilations, abridgements innumerable for the publishers brought, if not the large sums of money now frequently gained by periodical or other purveyors for the idler reading public, at least some relief from pressing pecuniary anxieties.

His masterpieces, appeared in the following order of time: *The Traveller: or a Prospect of Society,* 1724; *Vicar of Wakefield* 1766; the *Deserted Village,* 1770; *She Stoops to Conquer,* 1773. Of his numerous compilations, *A History of the Earth and of Animated Nature* (1772), in eight volumes (for which he

---

* In this most considerable of his Essays, which contains many just observations and which is written in a much more perspicuous and pleasing though perhaps, less "profound," style than the majority of similar productions, the essayist satirising the useless pedantry of the Academies and Universities, proceeds—"There are, it is true, several societies in this country [Germany] instituted to promote knowledge. His late Majesty [George II], as elector of Hannover, has established one at Göttingen, at an expense of not less than a hundred thousand pounds. This University has already pickled monsters [monstrosities] and *dissected live puppies without number.* Their transactions have been published in the learned world at proper intervals, since their institution, and will, it is hoped, one day give them just reputation. But had the fourth part of the immense sum above mentioned been given in proper rewards to genius, in some neighbouring countries, it would have rendered the name of the donor immortal, and added to the real interest of Society." (*The Present State, etc.* IV).

was to receive eight hundred guineas), is the most meritorious, though not the most accurate. By the publication of the *Traveller* he first made for himself the reputation of an English classic; and he was deemed worthy of admission to the "Literary Club," where Johnson figured as the most prominent and, Gibbon excepted, the most distinguished member. His common sense, his independence, his sensibility, his perspicuousness, his pleasing style, mark him out among writers of "Magazine" literature as easily the first.

In an early number of the *Citizen of the World,* the cosmopolitan Chinese traveller records his astonishment at the prevailing contradiction between the pretentions and the practice (dietetically) of the less barbarous classes of mind of European Society; and, in an apologue, displays the proper retribution of selfish gourmandism:—

"The better sort here pretend to the utmost compassion for animals of every kind: to hear them speak, a stranger would be apt to imagine they could hardly hurt the gnat that stung them. They seem so tender, and so full of pity, that one would take them for the harmless friends of the whole creation, the protectors of the meanest insect or reptile that was privileged with existence. And yet (would you believe it?) I have seen the very men who have thus boasted of their tenderness, at the same time devour the flesh of six different animals tossed up in a fricassee. Strange contrariety of conduct! *they pity and they eat the objects of their compassion!* The lion roars with terror over his captive; the tiger sends forth his hideous shriek to intimidate his prey; no creature shows any fondness for its short-lived prisoner, except a man and a cat.

"Man was born to live with innocence and simplicity, but he has deviated from Nature; he was born to share the bounties of heaven, but he has monopolized them; he was born to govern the 'brute creation,' but he has become their tyrant. If an epicure now shall happen to surfeit on his last night's feast, twenty animals the next day are to undergo the most exquisite tortures, in order to provoke his appetite to another guilty meal. Hail, O ye simple, honest brahmins of the East! Ye inoffensive friends of all that were born to happiness as well as you! You never sought a short-lived pleasure from the miseries of other creatures! You never studied the tormenting arts of ingenious refinement; you never surfeited upon a guilty meal! How much more purified and refined are all your sensations than ours! You distinguish every element with the utmost precision: a stream untasted before is a new luxury, a change of air is a new banquet, too refined for Western imaginations to conceive.

"Though the Europeans do not hold the transmigration of souls, yet one of their doctors has, with great force of argument and great plausibility of reasoning, endeavoured to prove that the bodies of animals are the habitations of demons and wicked spirits, which are obliged to reside in these prisons till the resurrection pronounces their everlasting punishment; but are previously condemned to suffer all the pains and hardships inflicted upon them by man, or by each other, here. If this be the case, it may frequently happen that, while we whip pigs to death, or boil live lobsters, we are putting some old acquaintance, some near relation, to excruciating tortures, and are serving him up to the very same table where he was once the most welcome companion.

"Kabul," says the Zendavesta, "was born on the rushy banks of the river Mawra; his possessions were great, and his luxuries kept pace with the affluence of his fortune; he hated the harmless brahmins, and despised their holy religion; every day his table was decked out with the flesh of a hundred different animals, and his cooks had a hundred different ways of dressing it, to solicit even satiety. Not withstanding all his eating, he did not arrive at old age; he died of a surfeit caused by intemperance: upon this, his soul was carried off, in order to take its trial before a select Assembly of the souls of those animals whom his gluttony had caused to be slain, who were now appointed his judges.

"He trembled before a tribunal, to every member of which he had formerly acted as an unmerciful tyrant: he sought for pity but found none disposed to grant it. 'Does he not remember,' cries the angry Boar, 'to what agonies I was put, not to satisfy his hunger, but his vanity? I was first hunted to death, and my flesh scarce thought worthy of coming once to his table. Were my advice followed, he should do penance in the shape of a hog, whom in life he most resembled.'

"'I am rather,' cries a Sheep upon the bench, 'for having him suffer under the appearance of a lamb; we may then send him through four or five transmigrations in the space of a month.'—'Were my voice of any weight in the assembly,' cries a Calf, 'he should rather assume such a form as mine; I was bled every day, in order to make my flesh white, and at last killed without mercy.'—'Would it not be wiser,' cries a Hen, 'to cram him in the shape of a fowl, and then smother him in his own blood, as I was served?' The majority of the Assembly were pleased with this punishment, and were going to condemn him without further delay, when the Ox rose up to give his opinion,—'I am informed,' says this counsellor, 'that the prisoner at the bar has left a wife with child behind him. By my knowledge in divination, I foresee that this child will be a son, decrepit, feeble, sickly, a plague to himself and all about him. What say you, then, my companions, if we condemn the father to animate the body of his own son; and by this means make him feel in himself those miseries his intemperance must otherwise have entailed upon his posterity?' The whole Court applauded the ingenuity of his torture: they thanked him for his advice. Kabul was driven once more to revisit the earth; and his soul, in the body of his own son, passed a period of thirty years, loaded with misery, anxiety, and disease."

In a later number the Chinese philosopher sends home a highly entertaining and instructive report of "A Visitation Dinner"—the customary annual feast provided for his clerical subordinates by their ecclesiastical superior, which might serve as a fitting pendant to the immortal picture of Hogarth, "A Civic Feast" (LVIII). Among the most instructive and philosophic of the reflections of the inquisitive traveller from the Central Kingdom are his remarks upon European (and, especially English) funeral ceremonies (XII); the usual causes and pretexts of Wars (XVII); on Voltaire (XLIII); the morals of English Electioneering—another (literary) pendant to Hogarth's satires in his *Humours of an Election*. "No festival in the world," writes the disciple of Confutse, "can compare with it for eating. Their gluttony, indeed, amazes me. Had I five hundred heads, and were each head furnished with brains, yet they would all be insufficient to compute the number of cows, pigs, turkeys, and geese that, upon this occasion, die for the good of their country. To say the truth, eating seems to make a grand ingredient in all English parties of religion, business, or amusement. When a church is to be built, or a hospital endowed, the directors assemble and, instead of consulting upon it, they *eat* upon it—by which means the business goes on with success. When the poor are to be relieved, the officers appointed to dole out public charity assemble to *eat* upon it. Nor has it ever been known that they filled the bellies of the poor till they had previously satisfied their own. But, in the election of magistrates, this people seem to exceed all bounds: the merits of a candidate are often measured by the number of his *treats.* His constituents assemble, *eat* upon him, and lend their applause—not to his integrity or sense, but to the quantities of his beef and brandy." (CXII).

At another time, the philosopher reflects upon the overweening claims of the human species—: "Mankind have ever been prone to expatiate in the praise of human nature. The dignity of man is a subject that has always been their favourite theme. They have declaimed with that ostentation, which usually accompanies such as are sure of having a partial audience; they have obtained victories, because there are none to oppose. Yet, from all I have ever read or seen, men appear more apt to err by having too high than by having too despicable an opinion of their nature; and, by attempting to exalt their original place in Nature, they depress their real value in it" (CXV). One of the best of the scenes is "A City Night-Piece" (as it is entitled in the *Bee,* in which it first appeared) in which certain of the ordinary scenes of the London streets at night are faithfully pictured. History repeats itself. "Who are those," exclaims the horror-struck inquirer, "who make the streets their couch, and find a short repose from wretchedness at the doors of the opulent? They are strangers, wanderers, and orphans, whose circumstances are too humble to expect redress, and their miseries are too great even for pity. Their wretchedness excites even horror. Some are without the covering even of rags, and others emaciated with disease. The world has disclaimed them. Society turns its back upon their distress, and has given them up to hunger and nakedness. These poor shivering women have once seen happier days, and been flattered for their beauty. They have been prostituted to the heartless, luxurious, villain, and are now turned out to meet the severity of winter. Perhaps, now lying at the doors of their betrayers, they sue to wretches whose hearts are insensible, or to debauchees who may curse, but will not relieve them. The slightest misfortunes of the great"—he reflects—"the most imaginary uneasiness of the rich, are aggravated with all the power of eloquence, and held up to engage our attention and sympathetic sorrow. The poor weep unheeded, persecuted by every subordinate species of tyranny; and every law, which gives to others security, becomes an enemy to them!" To the *Bee,* a weekly short-lived periodical, Goldsmith contributed some interesting pieces, besides the one already quoted. The best are those "*On Education*" and "*On Political Frugality.*" In the latter the force of his remarks upon Ale houses—that "it would be our highest interest to have the greatest part of them suppressed"—is not weakened by the lapse of a hundred and fifty years. Of the *Essays*—1785–1765— that on national prejudice, in which is exposed that sort of patriotism which consists in absurd contempt for all nationalities but one's own, (one of the chief hindrances to international comity and chief provocations of war) is by no means without significance for the present time. "Should it be alleged in defence of national prejudice, that it is the natural and necessary growth of love to our country, and that, therefore, the former cannot be destroyed without hurting the latter, I answer that this is gross fallacy and delusion."

In his two descriptive poems—the most charming of the kind—Goldsmith advances certain principles of political or national economy which, however opposed to the dogmas of the orthodox schools, recommend themselves to every unprejudiced thinker as equally just and rational. Especially he will not be deluded by the prevailing prejudice that the prosperity and happiness of a nation is to be measured by the amount of money riches in the hands of capitalists, while the mass of the people are abandoned to a state of chronic destitution. The picture of the frightful results of the selfish policy of sacrificing the interests of Agriculture, as well as the rights of masses of the people to the ambitions of Landlordism is well known to all readers; and, in particular, the often quoted verses in the *Deserted Village:*—

> "Ill fares the land, to hastening ills a prey,
> Where wealth accumulates and men decay."

Absence of these contrasts of the extreme of luxury and the extreme of destitution—in highly "civilised" countries so scandalously prominent—in other lands serve, in the judgement of the "traveller" (who is the poet himself), to compensate the hardy peasantry of the mountain-land even for all the severities of Nature:—

> "Yet still e'en here content can spread a charm,
> Redress the clime, and all its rage disarm.
> Though poor the peasant's hut, his feasts though small,
> He sees his little lot the *lot of all.*
> Sees no contiguous palace rear its head
> To shame the meanness of his humble shed,
> No costly lord the sumptuous banquet deal
> To make him loathe his vegetable meal."

It is in that most widely-read of all prose fictions, his *Vicar of Wakefield,* in the ballad of the "Hermit" (or "Edwin and Angelina," as he first entitled it) that Goldsmith exhibits the loftier poetic inspiration when he represents his recluse as having abandoned the diet of blood.:—

> "No flocks, that roam the valleys free,
> To slaughter I condemn;
> Taught by the Power that pities me,
> I learn to pity them."

6.

WILHELM ZIMMERMANN, 1819–1885. (CHAPTER LIII, 1896)

The name of the author of the *Way to Paradise* will always be joined, with that of Struve, as one of the two earliest pioneers of the Naturgemässe Diät in Germany.

Born at Heinichen near Jena, the son of a tenant-farmer of several manorial properties, Zimmermann early displayed his mental superiority. Obtaining his degree of doctor in Philosophy, he held for some time the position of head of a Polytechnic School at Halle. But his aversion from the reaction of politics of the Prussian Government induced him to fix his residence at Leipzig, where he devoted himself chiefly to the study of National Economy. About the year 1867 he founded a gymnasium or school, for the formation of that much-abused as well as much-neglected science, which he directed during the next seven years. In 1873 he withdrew from public teaching to the quiet of home life. The origin of his revolt from orthodox dietetics is interesting and instructive. While yet at school, punished on one occasion for refusing some especially gross dish set before him, (by an enforced abstinence from flesh meat altogether, for several weeks), he was astonished at finding that health and vigour of mind and body, far from failing had been greatly improved. The young student was thus first led to inquire into the history of the ethics of the subject. Rousseau's *Emile* seems to have been a chief means of establishing his new faith. *

But it was a chance experience, at a later period, which induced him definitely to put it into practice for the rest of his life. Travelling with pupils under his charge, in England, in the year 1840, he had fixed his temporary residence at Alcott House, at Richmond, a vegetarian establishment. So favourably-impressed was he by the good results of the manner of life, as respected both the directorate and the pupils, that, as he records, this humane establishment "in the midst of the self-seeking, pretences, and deceptions of morally and

---

* See *Studies of the France of Voltaire and Rousseau,* by Frederica Macdonald, lately published, in which the great and beneficial influences of Rousseau upon modern education are ably vindicated.

physically corrupt world-waste (wastenwelt) appeared to me as a paradisiac oasis." In this oasis he long lingered, and there he wrote his *Way to Paradise.*

"*Der Weg Zum Paradies,*" writes Robert Springer, "next to Struve's *Mandaras,* is the most significant work which prepared the way for Vegetarianism in Germany. Indebtedness to the writings of Shelley and to the *Thalysie* of Gleïzès is apparent. The ethical influence of the Natural Diet, and its capability of arresting the ever-increasing corruption of morals, is affirmed in this book in the inspired, moving, language of a gospel. . . . The results of the publication of the book were surprising. It brought over new adherents to the Reformation and originated the first union for its promotion."*

From *Der Weg zum Paradies* ("The Way to Paradise") the following extracts sufficiently represent the inspiration of the writer.

"Men are almost entirely everything that they are by the force of custom; and this force, for the most part, resists every other power, and remains victorious over all. Reason itself, morality, and conscience are submissive to it. In the matter of Dietary Reform it displays itself as the enemy *par excellence* (*die Hauptmacht*). People will fall back upon alleged difficulties, although it is a question only of will and resolution. They will reject many of the dietetic propositions hitherto advanced as dangerous 'abstractions,' although they are founded in history, reason, and human destiny; although a brief enquiry ought to suffice to convince one of the first importance of the Reform. For although one must suppose that all would prefer a long, healthy, and happy existence to a feeble, painful life upon the old regimen, yet will the majority of human beings think it easier to attempt to assuage their torments and pains by uncertain, and, by no means, unhazardous medicine, rather than to remove them by obedience to Nature's laws. As it is with most of the highest truths, so it is especially with Dietary Reform. People will reject it as an *abstraction* and pronounce it an *impossibility.* In the future, however, by the greater number of the higher minds—for such a sacrifice of the lower and unnatural appetite we dare not expect from the ordinary run of men—will it be regarded as a great blessing. For even now there are many exceptions in the social organism, for whom Nature's laws are superior to unreasoning impulse; for whom morality is superior to materialistic and mere sensual living; for whom duty is superior to superfluity. Besides, we are advancing towards a humaner century: and, as the present is a humaner time than the century before, so later will there be a milder *régime* than now. Just as, in our days, exposure of children, combats of gladiators, torture of prisoners, and other atrocities are held to be scandalous and shameful, while in earlier times they were thought quite justifiable and right, so in the future will the murder of animals, to feed upon their corpses, be pronounced to be immoral and indefensible. Already (1846) are associations being formed for the protection of these beings; already

---

* See *Enkarpa.* The full title of Dr. Zimmermann's book is "*Der Weg Zum Paradies:* Eine Beleuchtung der Hauptursachen des physichmoralischen Verfalls der Culturvölker, sowie naturgemässe Vorschläge diesen Verfall zu Sühnen. Ein zeitgemässer Aufruf an Alle denen eigenes Glück und Menschenwohl am Herzen liegt." The first edition appeared in 1843, a second in 1846. A new edition was published by Springer in 1884.

now are there many who, like the nobler spirits of antiquity, apply to their diet the watchword of morality (*das Losungswort der Moral*).

"According to the number of proselytes will the importance of the evidence be judged. When thousands, practising natural diet, are observed in the midst of the diseased flesh-eaters to be in the enjoyment of a prolonged, happy, old age, without disease and the sufferings of a vicious method of life, then will the way be laid down for the many to abandon the living upon the corpses of other animals."

Of a like inspiration is the indignant protest of another of the apostles of Humanitarianism in Germany:—

"What humiliation, what disgrace for us all, that it should be necessary for one man to exhort other men not to be inhuman and not irrational towards their fellow creatures! Do they recognise, then, no soul in them—have they not feeling, pleasure in existence, do they not suffer pain? Do their voices of joy and sorrow indeed fail to speak to the human heart and conscience—so that they can murder the jubilant lark in the first joy of his spring time, who ought to warm their hearts, with sympathy, from delight in bloodshed, or for their 'sport,' or with a horrible insensibility and recklessness only to practise their aim in shooting! Is there no soul manifest in the eyes of the living or dying animal—no expression of suffering in the eye of a deer or stag hunted to death—nothing which accuses them of murder before the avenging Eternal Justice? . . . . . Are the souls of all other animals but man mortal, or are they essential in their organisation? Does the world-idea (*Welt-idee*) pertain to them also—the soul of nature—a particle of the Divine Spirit? I know not; but I feel, and every reasonable man feels like me, it is in miserable, intolerable contradiction with our human nature, with our conscience, with our reason, with all our talk of humanity, destiny, nobility! It is in frightful (*himmelschande*) contradiction with our poetry and philosophy, with our nature and with our (pretended) love of nature, with our religion, with our teaching about *benevolent design*—that we bring into existence merely to kill, to maintain our own life by the destruction of other life. . . . . It is a frightful wrong that other species are tortured, worried, flayed and devoured by us, in spite of the fact that we are not obliged to this by necessity; while in sinning against the defenceless and helpless, just claimants as they are upon our reasonable conscience and upon our compassion, we succeed only in brutalising ourselves. This, besides, is quite certain, that man has no real pity and compassion for his own species, so long as he is pitiless towards other races of beings."*

7.

EDUARD BALTZER, 1814–1886. (CHAPTER LIV, 1896)

The founder of the first German Vegetarian Society and one of the most prolific and distinguished of vegetarian writers, was the son of an Evangelical clergyman. Born in the village of Hohenleine, near Delitsch, in Prussia, he was

* *Das Menschendasein in seinen Weltewigen Zügen und Zeichen.* Von Bogunmil Goltz, Frankfurt.

educated at the Universities of Leipsic and Halle, where he seems chiefly to have studied Theology. In his twenty-second year he received the post of Hospital-Preacher in Delitsch, and in 1847 he accepted that of preacher at St. Nikolas' in Nordhausen (in Prussian Saxony) offered to him by the recently constituted Committee of the Free Religious Community in that town. This position he held until 1882, when, being in failing health—he was of a naturally somewhat feeble physical constitution—he withdrew altogether into private life. In his thirty-fourth year Baltzer had been chosen by the Church directorate at Delitsch as representative in the National Conference (National-Versammlung) at Berlin, and he had been a member of its publishing committee. In Nordhausen not only did he acquire the highest esteem of his fellow-thinkers by his admirable Addresses and by the amiability of his character, but, also, the high respect of his fellow-citizens universally.

It was in the memorable year 1867 that at Nordhausen the Vegetarier-Verein came into existence by the inspiration and under the direction of Baltzer; and to its *propaganda* he dedicated his talents, his learning, and his enthusiastic labours. A complete catalogue of his writings in furtherance of the great Cause fills large spaces in the valuable *Guide to Vegetarian Literature* of Robert Springer.° Especially did he devote his time and learning to the lives and writings of the Greek Fathers of dietetic philosophy; and his monographs on Pythagoras, Empedokles, and Porphyry, with that on the Latin Stoic Musonius, are highly interesting contributions to its literature. Among his other writings are:—*Die Naturliche Lebensweise,* in four parts; *Die Sittliche Seit de Vegetarianismus,* Berlin; *Vegetarianismus und Kultur, Vegetärianismus und Æsthetik; Funf Bücher vom Wahren Menschenthum.* Leipsic, 1880. Of the *Natural Way of Living* the fourth part is devoted to the examination of the ideas of the Jews (as they appear in their sacred scriptures), of the Essenians, Ebionites, of Paul of Tarsus of the Zend-Avesta and Zoroastrianism. In the concluding section, *Was ist Vernunftige Lebensweise?* ("What is Reasonable Living?") Baltzer justly states the actual aims and purpose of the life according to Nature:—"*Naturalness* as a principle is not so to be understood, as if we wished, blindly imitating her, to bring back certain states of pure Nature. That is the assertion only of jesters, or of mistaken followers of our creed. Human nature is perfectible: *but it is so in its proper fashion.* As that is called 'reasonable,' which is really according to the laws of Reason, so that is here called 'nature' which springs from the laws of human nature, which are fundamental. The art of the rational life or of

---

° *Der Wegweiser für die Vegetarische Literatur,* von Robert Springer: Zeite Auflage Vehrmerhte. Nordhausen, 1880.

vegetarianism, therefore, is the *conscious fulfilling of the conditions of life imposed on us by Nature* (Lebensbedingungen).

"This conscious practical exhibition of our life, according to its own natural laws, has reference, consequently, to the whole man—not, by any means, alone to Eating and Drinking. For whatever conception of human nature one may have; that the body, the mind, and soul stand one to the other in mutual very essential relationship is, in fact, admitted on all sides. Even he who does not share the conviction of the oneness of man's nature, who distinguishes between the life-rule of the body, of the mind and of the soul, is obliged to recognise none the less in them mutual interaction and influence. Vegetarianism, or the art of the rational life, is this threefold dietetics reduced to actual practice. At the present time the reasonable science of life (Lebenskunst) remains opposed to the prevailing conceptions and customs, especially as concerns the nourishment, maintenance, and building up of the corporeal existence. . . . .

"Our opponents think to embarrass us by saying, 'Have you really considered what you propose? It would be a complete overturning of the existing relations!' We answer very composedly, Yes, certainly, and, exactly because we have considered it, and have proved it in our literature, does it seem to be worth all the labour to resort to this fundamental principle in place of always twisting about unprofitably in a circle of sham reforms. Our system introduced generally into human life, will, without doubt, give a different face to the earth. Agriculture and horticulture will flourish as never before. But we shall not, in addition, profane the earth's fruits by driving the world of lower animals to the Slaughter-House, and by converting corn and mead into alcohol only to involve the race in an *Odyssey*° of diseases of mind, soul, and body.

"Our system is no mere ménu (speisezettel), nor is it a *Sectarian* creed. It is the practical philosophy of the rational individual and social life, which makes us to be at peace with ourselves and with Nature. Our system is no narrow *asceticism,* nor is it a paltry calculation of niggardliness or of avarice. It is the truest and noblest poetry of life, that divine breath which merits the name of the perfect religion. Our system is no [political] revolution. But, in proportion as its recognition becomes the common property, will there be new and better ordering of human things; and the greater social evils become, the mightier will the force of the counter-action of our principles become! Our system is not our own, as though it had been invented by us. It has been implanted in our common nature, and the better part of the world has always had a more or less clear anticipation of it, but it is the task of the Future to elevate it to a *recognised Gospel.* The individual already can take his happy part in it, but the full harvest can only be possible in the human Society at large."

8.

HENRY DAVID THOREAU, 1817–1862. (CHAPTER LVI, 1896)

In preceding sections a brief history of the origin of the Reformation in the United States of North America, begun by the energetic labours of the Manches-

---

° The expression of Porphyrius in his Treatise on Abstinence. See above.

ter missionaries has been given. Among the most eminent of the whole, or in part, converts, Sylvester Graham, author of the elaborate physiological argument for the Natural Living, the *Science of Life;* Thoreau, (author of *Walden*); and Horace Greeley (the well known politician and journalist) are, perhaps, the best known.° In the Republic (so-called) of the West, no less than under the effete political institutions of Europe, however, the only just basis of the *Public Interest* has been ignored by those who should be the first to accept it.

Of all teachers of men, who have practically sought to return to Nature or, at least, to a less unnatural manner of life, Thoreau is the most remarkable and most distinguished by the force and originality of his character and writings. He has variously been described as the "poet naturalist," the "bachelor of Nature," and the "American Rousseau." His birthplace was Concord, Massachusetts. After graduating at Harvard University, he passed almost the whole of his life in his native village, where he enjoyed the friendship of Emerson— the most uncompromising modern eulogist of Plato—and other leaders of the Transcendental movement; and he himself became famous by his two years' experiment of "plain living and high thinking" in the woods of Walden. Supporting himself by occasional work as a surveyor, or pencil-maker, or gardener (for he was master of many crafts) he devoted his leisure to an enthusiastic study of wild nature, and to the writing of books in which, in many respects, is displayed a high, large, idealism tempered with a singularly shrewd and practical knowledge of life. He died at the village-town of Concord of inherited phthisis, in 1862.

The Transcendental philosophy, to which his distinguished friends adhered, by no means seems necessarily always to imply the transcending the materialistic diet. Thoreau himself, although he did not carry it out so far as altogether to renounce that unnatural nourishment, was an ardent follower of Natural Life, in general; and his efforts to bring his fellow men to live less conventionally and artificially, and to have a better appreciation of *living* Nature, with his reasoned frugality (in the accustomed sense of the term) entitle him to be regarded, in so small degree, as a representative of the more practical view of Idealism. Yet, with characteristic and, sometimes, too indiscriminating aversion from all formulas and systems, he carefully disclaims any distinct philosophy of Compassion, although with some apparent inconsistency he was, essentially, one of the humanest of modern writers. In fact, his contempt for all

---

° In his autobiographical reminiscences Horace Greeley affirms his faith in the superiority of the diet prescribed by Nature. For a useful abstract of the elaborately-minute scientific work of Graham, the reader is referred to the compendious edition of T. Baker, Barrister at Law, published by the Vegetarian Society.

the mischievous, or futile, systems of philosophy or morals, in esteem, carried him too far in the opposite direction, and seems to have not allowed him to appreciate the enormous value of an *organisation* or system, which should be dedicated to just aims, and to the regeneration of the World. That, practically, he was at variance with his avowed horror of systematised "philanthropy," his vehement opposition to human slavery, and the fact of his being the first to denounce the judicial murder of John Brown (the Abolitionist-martyr) sufficiently establish.

One of his most remarkable faculties was the magnetic attraction—derived from his sympathy and from his association with them in his hermit-life— exercised by him over the non-human races, which gives him something of a pythagorean character, and a unique position among "naturalists." And he expresses the pythagorean faith, or approaches very near to it, in declaring: "No humane person will wantonly murder any being who holds his life by the same tenure as himself. The hare, in her extremity, cries like a child. I warn you, Mothers," he apostrophises the principal educators of the race, "that my sympathies do not always make the usual philanthropic distinctions." Upon the interpretation of the qualifying particle *wantonly,* obviously, depends a great deal of possible divergence of opinion, and, it may be added, it is a very indeterminate and variable quantity.

The author of *Walden* seems, during the greater part of his career, to have been practically a food reformer. It was, however, in the Latin sense of 'humanity,' yet more than in the better modern meaning, that he advocated the return to the bloodless food; but he clearly discerned the necessity of applying the much-talked of "humanities," of the Schools to human diet in any moral creed which regards consistency and reality rather than mere sophistry and word-knowledge. In spite of some appearance, in one or two of his earlier essays, of a scarcely sufficient appreciation of all the significance of the revolution, his final verdict in favour of abstinence from flesh is given in no uncertain language:—

"It may be vain to ask *why* the imagination will not be reconciled to flesh eating. I am satisfied that it is not. Is it not a reproach that man is a carnivorous animal? True, he can and does live, in a great measure, by preying upon other animals; but this is a miserable way, as any one who will go snaring rabbits, or slaughtering lambs, may learn and he will be regarded as a benefactor to his race, who shall teach man to confine himself to a more innocent and wholesome diet. Whatever my own practice may be, I have no doubt that it is a part of the destiny of the human race, in its gradual improvement, to leave off eating animals, as surely as the savage tribes have left off eating each other, when they came in contact with the more civilised."

From which confession it may be conferred, with much confidence, that, had this eminent opponent of conventionality and artificiality been thoroughly instructed in the philosophy of Dietetic Reform, he would have made it a question not only of the *future* destinies of the race but of *present* and universal obligation.

His principal writings, next to his masterpiece *Walden,* are:—*A Week on the Concord River* (1849); *Excursions* (1863); *Letters* (1864); *Anti-Slavery and Reform Papers* (1866).*

9.

RICHARD WAGNER, 1813–1883. (CHAPTER LVII, 1896)

Of the distinguished champions of the Humaner Life, of recent times, none is more enthusiastic or more eloquent in affirming its high obligations than the eminent prophet of the Art of the Future, and the most intellectual of all the great masters of Music. In the whole history of Art no reformer has excited at once so much admiration (if, also, so vehement hostility) as the great Poet-Composer of the Nibelungen Trilogy. But what is of still higher interest and significance, for all for whom the debarbarising of human life is a supreme aspiration, is his eloquent insistence on the despised but none the less certain fact of the incompatibility of higher, genuine, Art and Æsthetics with a blood-stained diet.

Leipzig, the famous literary emporium of Germany and of the world, has the honour of having given birth to Richard Wagner. At the age of twenty-one he wrote his first opera *Die Feen* (*"The Fairies."*) At the same period he was engaged in writing articles for the liberal press. His next musical opera was *Rienzi,* in the style of Meyerbeer, for which, as for all his operas, he wrote the libretto. *Der Fliegender Holländer,* the first work to make him famous as the founder of the (newer) Romantic school of Music, appeared in his twenty-eighth year; but, like very many other works of genius and of future wide-world fame, it was not to owe its success, in the first instance, to those who should have been the first to be able to recognise its merits. From the directors or managers of the theatres it had for some time a cold reception. In the following year, however, it was produced at Dresden with great applause, with his *Rienzi.* In 1845, he published *Der Tannhauser,* when Schumann predicted, if his melody should equal his intellectual genius, reputation for him as the first musician of the century.

When the political-reform movements, which agitated Western Europe during the year 1848 began, Wagner, throwing in his lot with the patriots, had

---

* For the best and most complete account of the life and writings of Thoreau the reader is referred to the excellent biography by H. S. Salt (in *Great Writers Series,* Walter Scott, 1896).

to flee from Dresden, where he held the post of Court Capel-Meister. Betaking himself to Paris, he proceeded to Zürich, where he published his first prose-works, *Kunst und Revolution,* and *Kunstwerk der Zukunft* ("The Art of the Future") 1849; which was soon followed by his *Oper und Drama* (in three volumes). *Art and Revolution* displays his high conceptions, and hopes for the Future. He affirms Art hitherto to have been prostituted (or, at least, misused) for the most part, for the mere entertainment of frivolous and capricious Society, and expresses his ambition to raise it again to the lofty place held by it in Hellenic antiquity. With Schiller he hoped that, possibly, out of the Opera might be developed to a nobler form and meaning the tragic drama, as the latter had been developed from the Choral Ode by the Hellenes; but he almost despairs of the capabilities of the human animal, in general, for perception of the true aims of æsthetical education. He had not yet adopted the sure and steadfast faith as it is in Humanitarianism. In the "Art of the Future" he develops, however, the leading conviction that the low prevailing estimate and state of Art are inextricably connected with social conditions. In Franz Liszt Wagner had, from the first, an ardent admirer and supporter. By the good offices of the distinguished Pianist his next great opera, "Lohengrin" was brought out successfully at Weimar about this time: and, at the instigation of Liszt, Wagner formed the plan and began the composition of his great Trilogy of the "Nibelungen Lied," not completed until long afterwards. In 1857 was composed the *Tristan und Isolde;* and in 1868 the *Meistersinger* was first performed at Munchen. The great successes of the Franco-German war had stirred national enterprise in various directions, and "Wagner Unions" were formed in some of the chief cities, with the purpose of raising the sum of 90,000 marks for the production of the Ring des Nibelungen. At the invitation of his constant friend and ardent admirer, the King of Bavaria (Ludwig), he took up his residence at Bayreuth, where the famous Wagner Theatre was begun. From all parts of the world—an unprecedented event in the history of musical art—an immense concourse of people crowded the building and precincts at the presentation of the "Nibelungen." His last work in dramatic music, *Parsifal,* was performed at Bayreuth, in 1882.

In that year he was awaited with eager expectation in this country, for superintendence of the production of his principal works; but it is more interesting to record that the English Vegetarian Societies sent an invitation to him to a public banquet, to be given in his honour on the occasion of his visit. His first and last visit to England had been made in 1855. The proposed second visit was never undertaken; and in the next year the illustrious poet-composer, attacked by lung disease, died at Venice.

To Schopenhauer Wagner, on one occasion, expresses his full assent to that philosopher's principle—that compassion is the first law of Morals. In accordance with this conviction is his most important essay, "Art and Religion," written. Formerly a sharer in the thoroughly Pessimistic philosophy, of which the famous author of the "World as Will and Representation," is at once the most elaborate and the most humanitarian exponent, Wagner was converted from the philosophy of despair by the *New Existence* of Gleïzès. From the moment of his acquaintance with that noble assertion of the proper place and destiny of the human species in the plan of Nature, Wagner became a firm believer in the Gospel of Hope. Eulogising the better appreciation of the *connexion* of living existence by the Buddhist, and even by the Brahman he observes:—

"This teaching (of the sinfulness of murdering and living upon our fellow beings) was the result of a deep metaphysical recognition of a truth; and, if the Brahman has brought to us the consciousness of the most manifold phenomena of the living world, with it is awakened the consciousness that the sacrifice of one of our near kin is, in a manner, the slaughter of one of ourselves: that the non-human animal is separated from man only by the *degree* of its mental endowment and, what is of more significance than mental endowment, that it has the faculties of pleasure and of pain, has the same desire for life as the most reason-endowed portion of mankind. . . . We learn that about the middle of the last century English speculators bought up the rice-harvest of India, and thereby induced a famine in the land, which carried off three millions of the natives. Not one of these famishing people could bring himself to butcher his domesticated animals, and, after their masters had perished, they, too, perished of hunger. . . .

"If the physiologists are still in disagreement whether man is designed by Nature exclusively for the fruit diet, or for flesh eating, History, at all events, shows him as far advanced in the character of an animal of prey. He overruns the world, subjugates the fruit-nourished nations; through the subjugation of other preceeding conquerers forms huge empires, and governments; and orders his civilisation with the purpose of enjoying his prey in security. . . . Falling ever deeper, blood and murdered corpses seem to him the only worthy food for these world-conquerers. With the Hindu peoples the Thyestean feast would have been impossible. With such shocking pictures, however, could the human imagination delight itself—for human and non human murder was now in full swing. And has the imagination of the modern civilised man any right to turn away in horror from such Thyestean pictures, when he is accustomed day by day to the spectacle (e.g.) of a Parisian slaughter-house, in its early morning work; perhaps, even spectator of a famous field of butchery on the evening of some glorious victory? In truth we have only got so far beyond the Thyestean banquet, that a heartless hypocrisy and deception has become possible to us, in a matter which, with our earlier ancestors, was boldly exposed to view, and avowed in all its natural horribleness.

From the first, amidst the rage for predatory pursuits and for bloodshed, it has ever been familiar to the consciousness of the true philosopher that the human races suffer from a disease which maintains them in a state of demoralisation. . . .

He affirms that had the horrid practice of sacrificial and vicarious slaughter on the blood drenched altars of Antiquity any true moral teaching for the modern ages, the slaughter houses would no longer flow with daily streams and baths of blood (*ein taglicher Blutbad*) but would be thoroughly and for ever cleansed (von sauberen wasser durcbspülten Schlachthäusern) that no longer would the tables be spread with mutilated and disfigured limbs of murdered dependants (Leichentheile ermordeter Hausthiere).

Wagner insists on the immense importance of *union* between the Associations for protection against cruelty—and especially between those directed against butchery, and against experimental torture. Those who seek to gain the assistance of the mass of the people, he justly reminds them, will obtain effectual results only if, with compassion they unite a thoroughly reasonable and vigourous presentation (einen verstandnissvollen Durchdringung) of the deeper meaning of Vegetarianism; and that such unions would present so effective a front to the enemies of humanity as could not be despised. The union of the working classes and a reasonable socialistic propaganda with abolition or strict regulation of the drink traffic, he regards as the next most important agencies for the regeneration as well of Art as of man himself.[*]

In his open letter to Earnest Von Weber, author of "the Torture Chambers of Science" (*Die Folten-kammern des Wissenchaft*), Wagner affirms the heavy responsibilities alike of Governments, Society and Religion, in face of inhuman crimes against Humanity and Justice.[†]

Of no contemporary personage have more biographies been written. Among the most recent eulogists are Dr. Parry, *"The Art of Music"* (1893) *"Wagner and his Works"* and *"The Story of Wagner's Life"* by H. Finck, (1893), London. At this moment (Jan. 1896) Part three of the fourth volume of his Prose Works has appeared in an English translation.

10.

ANNA KINGSFORD, M.D., 1846–1888. (CHAPTER LVIII, 1896)

For even the most ordinary or superficial student of the history of Morals it must be matter for some wonder that, in the whole literature of Reformed Dietetics, among women-writers (who fill, during the past hundred years, so large a space in the annals of book making) no name occurs before the end of the third quarter of the present century.[‡] Nor, however the fact is to be ex-

---

[*] *Kunst und Religion* III.

[†] See *Offener Briet Ueber die Vivisection* von Richard Wagner, Berlin and Leipzig 1880. Among his other works are *Das Indenthum er Musik* and *Programmatische Erlauteruugen*.

[‡] The first book in the annals of English Literature written by a woman (it is a curious fact) is on hunting "sports." The authoress, Juliana Berncrs, was Prioress of St. Albans in the first half of the xv. century.

plained, up to this period does there appear to have been any successor at all—in the propaganda of the higher morality—to the Theanos or Hypatias of the non-Christian ages.*

Distinguished by rare refinement of intellect and by the yet nobler endowment of a true sensibility—*nostri pars optima sensûs*—the first woman teacher of the Humaner Life, in modern time, is the authoress of the *Perfect Way in Diet*. Anna Bonus descended, on the paternal side from an Italian family of distinction. Her father was a ship-owner in London, and her birthplace was Stratford in Essex. "The childhood shews the man, as morning shews the day," is one of those instances of proverbial philosophy which experience proves to be not universally true. Of the future eloquent prophetess of the Humaner Life, however, the well known verse of Milton seems to be, exceptionally, true. At the age of thirteen she already discovered considerable fancy and feeling in some verses, which she entitled *River Reeds*.† With the poet of the *Eloisa to Abelard* "she lisped in numbers"; but what manifested more unmistakably the young girl's imaginative faculties—developed, in later years, in so marvelous a degree and in so abnormal a manner—were her vivid child-fancies of Fairy-land and her dream-visions. The volume of *Dreams, and Dream Stories,* published by her in after years, is inspired by this extremely imaginative faculty no less than by high and true sensibility. At the age of twenty she became the wife of Mr. Algernon Kingsford, incumbent of Atcham in Shropshire, a devoted admirer of her moral and intellectual excellence. She had previously contributed some charming tales, characterised by all the peculiar graces of her style, to the magazines and periodicals.

It was in the year 1873 that Anna Kingsford embraced the cause of the Dietary Reformation; and, in the following year, she delivered a series of Addresses before the Vegetarian Society in Manchester, inspired with the highest and best enthusiasm. "The keen and searching fire of criticism," she declares in the opening Address, "which burns around us may well be likened to the famous cauldron of the enchantress Medea. Into it is put the old worn-out body of the world's past creeds and theories, inert, decrepid, powerless to touch any longer the minds and hearts of the people. But out of the purifying furnace springs the aspiration of the new Age, vigorous and strong, full of life and youth and purpose. So it comes about that the popular movements of our

---

* In this country two of the more prominent promoters of the cause, among literary women, at this moment, are Mrs. Leigh Hunt Wallace (Editress of the *Herald of Health*) and Mrs. Sibthorpe (Editress of *Shafts*). Recently, a "Vegetarian Union of Women" has been formed, which includes already many enthusiastic and energetic members.

† See *Anna Kingsford: Her Life, Letters, Diary, and Work, by her Collaborator Edward Maitland:* 2 vols. Redway, 1896.

time are the result, as a rule, of criticism applied to past ideas. Of late, people
have dared to ask why, in old times, wives and daughters were subjected to
their male relatives, and practically denied the dignity of humanity. As a result of this inquiry we have the agitation for women's rights. Other people,
again, have questioned the morality of flesh-eating habits which have prevailed
so generally in European countries, and, by consequence, the Vegetarian societies rise into being. Reform is the cry of our day. With us the inquiry to be
made is not 'What did our fathers think?' or 'What has been the belief and
practice of the Past?,' but (more reasonably), 'What should *we* think?' 'What
should be the belief and practice of the future?' Intellect is ever on the march.
The spirit of man is never [ought never to be] contented with a possession of
a by-gone age: his nature and the law of his being compel him to a continual
striving after the highest and the best—that is, the Divine."* Somewhat later
a tale of exceptional charm and beauty of idea "The Turquoise Ring," from
her hand, appeared in the *Dietetic Reformer.* At this time she is described by
one who knew her—herself an active worker in the cause of Humanity—(Mrs.
Fenwick Miller) as "the most faultlessly beautiful woman I ever beheld."

She was at the age of twenty-seven when she first met her future fellow-worker
and intimate friend, the author of the much admired philosophical fiction *The
Pilgrim and the Shrine;* the rationalistic principles of which, however, were to
be abandoned for those of the mystic-transcendental creed, or what, perhaps,
may be termed a highly developed neo-Platonism. Remarkable correspondence
of ideas in that direction of thought or fancy, and strong sympathy of feeling,
attracted, as by a natural affinity, these two distinguished prophets of a Humanitarianism which was thus based upon, or closely connected with, such mystic
transcendentalism. But, probably, the strongest mutual attraction was hatred and
horror of a false science, which, as Mr. Edward Maitland well expresses it, meant,
"the demonisation of the race; the reconstitution of human society on the Ethics of hell, the peopling of the earth with fiends instead of with beings really
human. It was the character of the race of the future that was at stake."†

In the year 1877 Anna Kingsford undertook the editorship of the *Ladies
Own Paper,* "a journal of progress, taste, and art, which she conducted with a
heroic disregard of financial considerations, and in which she honorably distinguished herself as an advocate of the emancipation of her sex at a time when
such advocacy was rarer than it is today." Still more honorably did she distin-

---

* These admirable Addresses have been published by the Vegetarian Society, under the
title of "Some of the Aspects of the Vegetarian Question."

† See the tract on *Vivisection* the (joint production of Edward Carpenter and Edward
Maitland,) published by the Humanitarian League, 1894. One of the most meritorious ex-
posures of the true character of this pseudo-science is Jules Charles Scholl's *Ayez-Pitie* 1881.

guish herself by her assaults, in that fashionable periodical, upon experimental torture, which then, it seems, first attracted her attention, and of which she was soon to be the most eloquent, and it must be added, an (exceptionally) consistent assailant. She then, too, consequently upon this conscientious and courageous braving of the selfish prejudices of Society, made the acquaintance of Frances Power Cobbe, now, for a long time, protagonist in the desperate conflict with the scientific Inquisition and its innumerable and all powerful supporters.

Forced to exile herself from her home in Shropshire, as a sufferer from acute pulmonary disease and from asthma, she formed the resolution of studying medicine at Paris (at that period women were still debarred from medical degrees in this country) with the principal purpose of qualifying herself scientifically for the tremendous contest against the Inquisitors of a science falsely so-called as well as to support more effectually the cause of Diet-reform. At the urgent wish of her husband she was accompanied by her sympathetic friend, and they took up their residence near the Medical Schools of Paris until Mrs. Kinsgford had taken her degree. She passed all the severe examinations of the Medical Boards with brilliant success, in spite of the determined hostility of some of the professors, who looked with suspicion upon her undisguised horror of their atrocious vivisectional and 'pathological' experiments. Her confrère (the word is here used in its most real meaning) and biographer records, with fulness of highly interesting detail, her terrible experiences during this eventful period of her life—the intense anguish caused to her by the knowledge of the State and Society-sanctioned hideous crimes against humanity, and by the consciousness of her powerlessness to stay them—and they are narrated with the greatest force and vividness. From the lecture-room she was driven away altogether by the horrifying cries which resounded from the dens of the tortured victims—*sospiri, pianti, e alti guai*, of the true infernos°—as she passed, in her first ignorance, through the wards of the hospitals: horrors devised by the perverted ingenuity of inquisitors masquerading in the character of scientific philanthropists. It is thus that she describes the first practical experience of the character of these torture-chambers:—"Very shortly after my entry as a student at the Paris Faculty, I was one morning, while studying alone at the Natural History Museum, suddenly disturbed by a burst of screams

---

°   "sights of woe,
Regions of anguish, doleful shades, where peace
And rest can never dwell, hope never comes,
That comes to all: but torture without end
Still urges."
            Par. Lost I.

of a character more distressing than words can convey, proceeding from some chamber on another side of the building. I called the porter in charge at the Museum, and asked him what it meant. He replied with a grin 'it is only the dogs being vivisected in M. Béclard's laboratory'! There swept over me a wave of such extreme mental anguish that my heart stood still under it. It was not sorrow, nor was it indignation merely that I felt, it was nearer despair than these. It seemed suddenly as if all the laboratories of torture throughout Christendom stood open before me, with their manifold, unutterable, agonies exposed, and the awful future which [a practically] atheistic° science was everywhere making for the world, rose up and stared me in the face; and then and there, burying my face in my hands, with tears of agony I prayed for strength and courage to labour effectually for the abolition of so vile a wrong, and to do, at least, what one heart and one voice might to root out this curse of torture from the lands [of Christendom]." Seldom has a solemn vow been so faithfully fulfilled. Upon taking the degree of Doctor in Medicine, (in July, 1880,) Dr. Anna Kingsford presented, as her *thèse pour le doctorat* at the Faculty of Medicine at Paris, a vindication of the non-flesh diet under the title of *De l'Alimentation Végétale Chez l'Homme.* It is to be added that, to the disgrace of the "Faculté," before they authorised the formal reading on it, they insisted upon the elimination of the *moral* arguments of the Thesis. Within the year it was translated into German by Dr. A. Aderholdt, an eminent scientific advocate of its principles in Germany. In the following year it appeared in its English version, under the well-known title of *The Perfect Way in Diet,* and was welcomed by the Vegetarian Societies of England and Germany with great applause.

Beginning with a clear demonstration of the physiological proofs of the physical organisation of man as that of a frugivorous animal, the Thesis proceeds to a compendious survey, by way of comparison, of the national eating habits of various peoples of the Globe. The chemical proofs are next brought to view, and the disease-producing effects of flesh meats both as originally forbidden to men by Nature, and as rendered yet more deleterious as derived, in ever increasing proportion, from the unnaturally-reared and nourished victims of the table. These patent facts are abundantly illustrated and fortified by the authoritative witness of various scientific and public authorities. The barbarities of butchery itself, and of cattle-transport, come next under con-

---

° Inasmuch as a large proportion of the practicers or advocates or apologists of the pseudo-scientific Inquisition are religionists (many of them of unimpeachable orthodoxy), the force of truth obliges us to remark that (theoretical) "Atheism" can be charged, by no means, with the exclusive infamy of that peculiar institution. A preferable epithet is *Hypocritical:* for pretence, in every direction, is one of its principal characteristics.

sideration. Appeals to arguments, derived from refinement of life, and, finally, from National Economy, with reflections on the cruel national "sports" (the
natural accompaniment of barbarous living), the cruelties of Commerce, of
Superfluous Luxury, and of Avarice, conclude one of the most valuable contributions made to the literature of Food-Reform, most especially to the scientific part of it.

In the division of the Thesis treating of Anatomy and Physiology, the newly
made doctor—a title justly deserved in the present instance—thus combats
certain of the common scientific as well as popular fallacies. "If we have consecrated," she admonishes the assembled savants, "to the sketch of comparative anatomy and physiology a paragraph which may seem a little wearisome
in detail, it is because it appears necessary to combat certain erroneous opinions affecting the structure of man, which obtains credence not only in the
vulgar world but even among otherwise instructed persons. How many times,
for instance, have we not heard people speak with all the assurance of conviction about the canine teeth and "simple stomach" of man as certain evidence
of his natural adaptation for a flesh-diet! At least we have demonstrated one
fact—that if such arguments are valid, they apply with even greater force to
the anthropoid apes, whose "canine teeth" are much longer and more powerful than those of man; and the scientist must make haste, therefore, to an-
nounce a rectification of their present division of the animal kingdom in order to class with the carnivora, and their proximate species, all those animals
who now make up the order of *Primates*. And yet, with the solitary exception
of man, there is not one of these last that does not, in a natural condition,
absolutely refuse to feed on flesh! M. Pouchet observes that all the details of
the digestive apparatus in man, as well as his dentition, constitute 'so many
proofs of his frugivorous origin; an opinion shared by Owen, who remarks that
the anthropoids and all the *quadrumana* derive all their alimentation from
fruits, grains, and other succulent and nutritive vegetable substances; and that
the strict analogy which exists between the structure of the animals and that
of man clearly demonstrates his frugivorous nature. This is, also, the view taken
by Gassendi, Cuvier, Linné, Professor Lawrence, Charles Bell, Flourens, and
a great number of other eminent writers."*

It may be recorded, in this place, that the latest version of the *thesis* is that
which has been made into the Dutch language by Dr. F. Van Eeden, a distinguished physician who, in a very valuable introduction, affirms the supremacy of compassion among the various obligations to abstinence. "Flesh eating

---

* *The Perfect Way in Diet, A Treatise advocating a Return to the Natural and Ancient
Food of our Race* (Kegan Paul, Trench & Co., 1881.)

and vivisection are simply immoral, and for that reason alone must be abolished. *We* know our duty, and we act up to it, giving free way to the sentiment of compassion."

Soon after having taken her degree Dr. Anna Kingsford returned to England, and began practice as a physician, passing part of the year in London (in Park Lane, and elsewhere) and the less inclement period of the year in her Shropshire home. But a nobler, and an imperative, sense of duty would not allow her to remain content with the ordinary routine of the medical profession. She felt herself called by the inner voice of conscience to engage in a yet holier Cause than that of the virgin-deliverer of France—to fight on behalf of the Oppressed who have no helpers. With untiring energy of soul and devotion to this supreme duty she visited various parts of France, Italy, and Switzerland to urge war *à outrance* with the Claude Bernards, Paul Berts, Schiffs, and the chief false prophets of physiological science. A nominal member of the Papal Church, she had been drawn to it, in her extreme youth, by the imposing ritual and æsthetic magnificence, probably, much more than by any calm inquiry into the confident claims of that marvellously-organised ecclesiastical institution. She had sought, in the fond enthusiasm of her as yet disillusioned hopes, interviews with the Cardinal-Archbishop of Paris, and afterwards with the supreme Pontiff himself, with the purpose of engaging those powerful hierarchs in her humane *propaganda* so far, at least, as the Scientific Inquisition is concerned. That her hopes were disappointed it is scarcely necessary to add. The natural affinities and deepest sympathies of Anna Kingsford, it may here be added—though she seems never to have formally separated from the Creed of her early choice or fancy— were yet mainly or, rather, wholly, with the humane and gentle teaching of the founder of the religion which includes two-thirds of the human race. The especially characteristic (speculative) doctrine of Buddhism—the metempsychosis— she embraced with all the ardour of her temperament and made it, indeed, a cardinal feature of her faith.

In the year 1884, in conjunction with her coadjutor, she founded the Hermetic Society, of which she was president, and whose meetings were held in the rooms of the Asiatic Society in Albemarle Street. She previously had presided, for some time, over the lately instituted English Theosophical Society. We are here concerned only with the humanitarian, and practical, *propaganda* of this highly gifted and (on the religious, spiritual, or metaphysical side of her manifold character), it must be added, also, highly-transcendental leader of the mystic thought of the day. Her energy of soul, and activity in the practical humanitarian work (so vast and, apparently, so hopeless) were incessant, whether in addressing large audiences, in writing, or in the more private assemblies of the drawing-room. Her striking personality, her commanding figure, and the charm of her presence, no less than her calm and persuasive eloquence, at once secured the rapt attention of every audience, however, at first, indifferent or even hostile it might be, and, often, even its applause.

One of the audacious eulogies of experiment by torture—usually character-ised by cynical contempt for all idea of pity or compassion—had appeared in the *Révue des Deux Mondes*. Dr. Anna Kingsford took up the challenge, and published a reply under the title of *Roi ou Tyran,* which she dedicated to Victor Hugo and to Victor Schölcher. She cites, among many other typical instances of atrocity, an experiment of Bert, who, having *curarised* a dog—curari paralys-ing, as is well known, the power of movement while preserving the sensibility entire—"laid bare the nerves, irritating them, during ten hours, by means of electric currents. After which this dog, still living, owing to the artificial respira-tion which had been maintained to prevent him from succumbing, was aban-doned a whole night in the Laboratory, mutilated and agonising, without the power of articulating a single cry of pain." She overthrows the unfounded as-sertions of the writer of the apology (Charles Richet), and clearly demonstrates that the physiological "Laboratory," to which they annually sacrifice nearly 43,000 dogs (exclusive of equally highly-organised victims, horses, cows, deer, monkeys and others) is but a *traditional routine,* a mass of prejudices and errors, of which true Science will hereafter rid itself, as it already has got rid, after long strug-gles, of countless superstitious barbarities in the past.

Anna Kingsford, her delicate constitution worn out by incessant labours, and suffering from a pulmonary disease, the fatal effects of which she had so long warded off by means of the purer diet, in the winter of 1887 sought mitiga-tion of her sufferings in the South of France and, later, in Rome. In that city she seems to have contracted malarial fever; and, with her constant and faith-ful friend, who tended her during her protracted painful illness with the ut-most assiduousness and solicitude, she returned to England in the January of 1888. She breathed her last in the month of February of that year—an irrep-arable loss to the Cause of Humaneness and Compassion. But, for herself, the final release from the heavy load not alone of physical suffering, but of men-tal anguish from the consciousness of the continuously-inflicted tortures of the helpless and innocent, amid the revoltingly selfish indifference of the sur-rounding world,* was not to be lamented by her friends. Of her it may be affirmed, with especial truth, "all was conscience and tender heart." Great as were her intellectual gifts, it is that ardent indignation against cruelty and, especially, against the most iniquitous forms of it—and the sacrifice of self in

---

* The *indifferent, che visser senza infamia e senza lodo,* from one point of view, are ob-noxious to severer censure, than even the actual apologists.

the highest and holiest of Causes—which constitute her most enduring title to honour with all who know how to distinguish true merit.

The representatives alike of Humanity, of Theosophy, and Spiritualism, in the Press, gave expression to their high admiration of her virtues and of the consciousness of their irreparable loss. From one of these tributes to her memory we extract the following just remarks: "The key-note to her teaching was the word Purity. She held that man, like everything else, is only at his best when pure, and her insistancy upon a vegetable diet—which she justified upon grounds at once physiological, chemical, hygienic, economical, moral, and spiritual—was based upon the necessity to his perfection of purity of blood and tissue obtainable only upon a regimen drawn direct from the fruits of the earth, and excluding the products of the slaughter of innocent beings. In thus teaching, she had the strongest of all personal motives. She ascribed her own delicacy of constitution to the violation of the law of purity by her ancestors; and her knowledge of the cruelties perpetrated in the world—especially those enacted in the name of Science—robbed life of all joyousness for her, and made the earth a hell, from which she was eager to escape. Her scorn and contempt for a Society which, by tolerating Vivisection, consented to accept for itself (pretended) benefits obtained at such terrible cost of suffering to others, were beyond all expression. . . Tall, slender, and graceful of form; of striking beauty of face and delicacy of complexion, intelligence of expression, and vivacity of manner, with a noble brow, grey, deep-set eyes, a profusion of golden-auburn hair, a full mouth, a rich musical voice, admirable elocution—alike artist, poet, orator, and philosopher—Anna Kingsford was as a diamond with many facets; and the admiration and affection with which she inspired her friends, masculine and feminine alike, was of the most fervent kind."* In one of the better inspired French periodicals appeared a highly

---

* *Anna Kingsford: Her Life, Letters, Diary, and Work. By her Collaborator Edward Maitland. Illustrated with Portraits, Views, and Facsimiles—In Two volumes.* Volume II. London, George Redway, 1896. For a very eloquent address on the Humaner Diet the reader is referred to volume II. xxx. One of the many reasons for the Reformation the modern Hypatia thus well signalises: "one of the greatest hindrances to the advancement and the enfranchisement of the sex is due to the luxury of the age, which demands so much time, study, money, and thought to be devoted to what is called the 'pleasures of the table.' A large class of men seems to believe that women were created chiefly to be 'housekeepers,' a term which they apply almost exclusively to ordering dinners and superintending their preparation. Were this office connected only with the garden, the field, and the orchard, the occupation might truly be said to be *refined, refining,* and worthy of the best and most gentle lady in the land. But connected, as it actually is with slaughter-houses, butchers' shops, and dead carcases, it is an occupation at once unwomanly, inhuman, and barbarous in the extreme. Mr. Ruskin has said that the criterion of a beautiful action or of a noble thought is to be found in song, and that an action about which we cannot make a poem is not fit for humanity. Did he ever apply this test to *flesh-eating*?"

interesting notice of her literary and humanitarian career, by Edouard Mill. "C'est Mde. Anna Kingsford," concludes the writer, "qui la première, en Angleterre, par de savantes conférences et de nombreux écrits a combattu Pasteur depuis qu'il se vante de prevenir *le rage* en le faisant nâître et renâître, à perpetuitè, dans une ou plusieures espèces animales, au moyen d'inoculations multiples." As for her personal qualities, "il se dégage de toute sa personne" adds the eulogist-critic, "une sorte de rayonnement magnétique qui attire, captive, et charme, et qui inspire à tous, amis et ennemis, un respect mêlé d'affection. Elle possède, au plus haut dégré, cette suprême vertu des forts et des miséricordieux qui est comme le diamant de l'âme humaine—*la bonté.* Et il est impossible de voir Mdle Kingsford, de l'approcher, de l'entendre, ou de lire ce qu'elle a écrit sans rendre hommage à son abnégation, à son courage, à son intelligence, et sans être pénétré d'un sentiment profond de sympathie et d'admiration."[*]

Important treatises, in various departments of the literature of Reformed Dietetics, by living writers, have appeared within the last decade. Of these, the following are among the most considerable or most representative:—

Essays on Diet by Emeritus Professor Newman: London and Manchester— the best and most complete manual on the subject considered from the National Economic point of view; the *Horrors of Butchery* by Josiah Oldfield, M.A., B.C.L., London,—illustrated: the most complete exposure of the various barbarities connected with Kreophagy; *Foods for Man, Animal and Vegetable: A Comparison* by Sir. B. W. Richardson M.D., London and Manchester, a scientific vindication of the Humaner Diet by one of the highest medical and sanitary authorities; *Die Erste Stufe* ("The First Step") by Count Leo Tolstoi, Leipzig—a social, religious, and humanitarian, eloquent appeal to the conscience of Society; *"Why am I Vegetarian?* (in English and German) by Professor Mayor, President of the Vegetarian Society, a vindication of the Reformed Diet principally from the point of view of simplicity of living; *The Case Against Butcher's Meat,* by Charles Forward, with Preface by the Hon. Dudley Campbell, (illustrated): a compendious scientific and humanitarian treatise; *Hygieine Kai Ethike* ("Hygiene and Ethics") by Platon Drakouli, Athens—scientific and humanitarian; *Essays on Vegetarianism* by A. F. Hills (President of the London Vegetarian Society)—eloquent appeals, chiefly on the religious and on the purity of living phases; *Introduction to the Perfect Way in Diet* by Dr. Van Eeden, Scheveringen, Holland—humanitarian; *Le Vegetarisme* by Dr.

---

[*] *L'Encyclopédie Contemporaine,* Sept. 18, 1887. Of the many eulogies of Anna Kingsford, which appeared after her decease, that by Mrs. Fenwick Miller is especially noteworthy. See *Anna Kingsford* vol. II. xx.

Bonnejoy, Paris, chiefly hygienic; *Manchester Vegetarian Lectures*—a comprehensive review of the various phases of dietary reform by various writers; *A Plea for Vegetarianism* by H. S. Salt—an able vindication; *A Plea for a Broken Law* by H. J. Williams; chiefly from the theological and religious point of view; *Report of the International Congress of Vegetarians* at Chicago (in special issue of the *Hygienic Review*) in which various phases of the Dietary Reformation are treated by various authorities. To which may be added *Behind the Scenes in Slaughter Houses* by H. F. Lester, (published by the Humanitarian League, 1893); and *Cattle-Ships* by Samuel Plimsoll, (illustrated). The two last publications, not directly written from the anti-kreophagist position, are yet, indirectly, powerfully persuasive to it.

Of histories of the general literature connected with the Dietary Reformation, the most comprehensive, since the appearance of *La Nouvelle Existence* of Gleïzès, is *Enkarpa: Cuiturgeschichte der Menschheit im Lichte der Pythagoraischen Lehre,* by Robert Springer, Berlin, 1883;* of periodicals, established in the interest of a true religion, the most recent in this country, is *The Herald of the Golden Age,* founded and edited by Sidney Beard, author of the Essay—*Is Flesh eating by Christians Morally Defensible?*

11.

HUMANITY OF THE HIGHER HINDU SCRIPTURES.
(APPENDIX III, 1896)

The following dialogue is taken (from the MS. of H.H. Wilson), from the *Padma Puràna: Bhùmi Khandu* (as quoted in *The Sacred Anthology,* of Moncure Conway, 1876). To a king, or raja, surrounded by priests, suddenly appears "a certain person of splendid form," and addresses him thus:—

"O king, in vain art thou governing thy kingdom with [so-called] justice. Know thou me as one adoring virtue, before whom even gods must bow. I speak truth, and never falsehood." King Vena said, "But what is thy virtue and thy religion, and what works dost thou perform?"

The stranger thus spoke, "That object, virtue, worshipped by the gods, is the source of all honours. *Mercifulness* is above all those virtues which are performed to obtain *salvation*. Attend, O king! I perform no ceremonies, nor study the Vedas, nor practise austerities nor incantations. What are offerings to the gods? Our highest work is to reverence the justest men." King Vena said, "What is the nature of this, thy virtue of mercifulness?" The stranger answered, "O king, when the life of a human being ends, the body no longer has a sepa-

---

* This extensive and meritorious work appeared after the publication of the first edition of "The Ethics of Diet."

rate existence, but re-unites with the elements of which it was composed. The friends of that person are affected with grief, and they perform a sacrifice, and continue to offer sacrifices on the day of his death. This is *delusion*. Where do those decayed ones dwell, and on what do they subsist? . . . .

"And concerning the ceremonies and austere devotion in honour of your gods, hear what I shall say. In all their rites innocent animals are sacrificed, and in one even a man is slain on the altar. This is called the bestowal of gifts. But he who destroys an innocent being, even in the most solemn ceremony, has effected *only evil*. What virtue can there be in a ceremony where even innocence is no barrier against (deified) vanity? What benefit does the performer obtain from it? He eats the dust of the ground.

"Know this well, that whatever ceremonies, prescribed in the Vedas, bring needless *pain or death* contain no virtue and conduct to no beatitude. A Veda void of *mercy* is a holy scripture only in name.

"O king, not even a god could possess virtue, did he not also possess *mercy*. And he alone is the true worshipper of God—be he Brahman or Pariah—who cherishes all beings with goodness and with compassion."

See, also, quotation from the Buddhist *Somadeva* for similar expression of faith. (*Sacred Anthology*, page 217, ed. 1876).

In a highly interesting and important volume published recently, by the Maharaja of Bhavnagar, "A Collection of Prakrit and Sanscrit Inscriptions," the chief of all these monuments of early Hindustan—the Rock Inscriptions of Asoka—(about 250 B.C.), are thus described by Professor Dowson:—

"Asoka was a convert to Buddhism; but his edicts have few distinctive marks of that, or any formal, religion. They are entirely free from vaunts of power and dignity. They inculcate a life of morality and compassion—a practical religion, not one of rites and ceremonies. They proscribe the slaughter of animals, they enjoin obedience to parents, affection for children, friends and dependants, universal benevolence and unreserved toleration. They would seem to have been set up at a time when there were few differences between Buddhists and Brahmins, and their apparent object was to unite the people in the bond of peace by a religion of morality and charity free from dogma and ritual.

"Writing on the India of to-day, it is impossible not to desire to call attention to the wise and earnest words in which Asoka taught the doctrine of universal toleration, not as a matter of state policy, but as a duty man owes to man. 'Proper treatment of servants and subordinates, sincere self-restraint towards all that has life and breath, sincere regard for good men, whatever they be, Brahmins or Buddhists—these things consecrate religion.'" (*A Collection of Prakrit and Sanscrit Inscriptions*. Published by the Bhavnagar Archæological Department, etc. Bhavnagar: The State Printing Press, 1895).

One of the *Edicts* (viii.) of Asoka is especially directed against the hunting of deer; and it is commended to the serious attention of Christian and English deer-hunters.

12.

VAN RHOER'S INTRODUCTION TO HIS VERSION OF PORPHYRY'S *APOCHE*. (APPENDIX V, 1896)

Vergönnt mir die wahrheit offen zu sagen:
   Wilderes als uns selbst giebt es auf Erden nicht mehr.
Nichts ist sicher annoch vor uns und unsern Verbrechen,
   Alles flüchtet vor uns fürchtet nur Raub und Betrug.
Weh uns! Wo soll hinaus des Leibes gierige Wollust
   Und des Gaumens Begehr unser Geschlecht denn noch ziehn?
Doch Dir genüget es nicht zu tödten das mächtige Hornthier:
Ach, in des Kälbchens geblüt tauchst Du den grausamen Stahl!
Ja von dem Enter der Mutter sogar mit mörderhänden
Reissest Du, grausamer mensch, weg das unschuldige Lamm!

        ❀   ❀   ❀   ❀   ❀

Und der zur Norm ist gesetst an die Spitze der irdischen Schöpfung,
   Er nun allein, der mensch, ist es, der Alles verschlingt.
Könnte das Vorbild doch der Thierwelt ihn nun bekehren,
   Viele belehren ihn ja milder und frömmer zu sein.

        ❀   ❀   ❀   ❀   ❀

Halt nicht frugales mahl für gemeinere Leute nur passend—
Aber für Helden und Herren zieme sich blutige kost!

        ❀   ❀   ❀   ❀   ❀

Thor Du! der Du beklagst die sich mehrenden krankheitsformen:
Trägst bei dem prunkenden mahl sie Dir ja selber doch auf.
"Leb wie Du willst," sprichtst Du, "wir müssen die Erde beherrschen,
   Alles in Leben muss sich richten nach unserm gesetz!"
Also schmeicheln wir uns in ungemessenen stolze,
   Dünken uns wunder wie hoch über der alten Geschlecht!

        ❀   ❀   ❀   ❀   ❀

13.

SOME SCIENTIFIC EVIDENCES FOR THE UNNATURALNESS AND FOR THE UNHEALTHINESS OF FLESH-EATING. (APPENDIX VIII, 1896)

"The natural food of man, to judge from his structure, appears to consist principally of the fruits, roots, and other succulent parts of vegetables."—CUVIER, *Le Régne Animal.*

"The apes, whom man nearly resembles in his dentition, derive their staple food from fruits, grain, the kernels of nuts, and other forms [of vegetable products] in which the most sapid and nutritious tissues of the vegetable kingdom are elaborated; and the close resemblance between the quadrumanous and the human dentition shows that man was, from the beginning, adapted to eat the fruit of the trees of the garden."—PROFESSOR OWEN, *Odontography*, 471.

"It is, I think, not going too far to say that *every fact connected with the human organization* goes to prove that man was originally formed a frugivorous animal. This opinion is principally derived from the formation of his teeth and of his digestive organs, as well as from the character of his skin, and the general structure of his limbs."—THOMAS BELL, *Anatomy, Physiology, and Diseases of the Teeth*.

"The teeth of men have not the slightest resemblance to those of the carnivorous animals, excepting that their enamel is confined to the external surface. He possesses, indeed, teeth called 'canine'; but they do not exceed the level of others, and are obviously unsuited to the purposes which the corresponding teeth execute in carnivorous animals. Thus we find, whether we consider the teeth and jaws, or the immediate instruments of digestion, that the human structure closely resembles that of the simiæ [apes], all of whom, in their natural state, are completely herbivorous [frugivorous]."—PROFESSOR LAWRENCE, *Lectures on Physiology*, 189, 191.

"It is a vulgar error to regard [flesh] meat, in any form, as necessary to life."—SIR H. THOMPSON, F.R.C.S., *Diet in Relation to Age and Activity*.

"In respect to *the propagation of disease*, it seems to me just to declare that the danger is much less, and much more easily preventible, on the vegetarian than on the animal system of diet. I think, too, I ought to add that some constitutional diseases, such as scrofula, gout, rheumatism, obesity, and certain forms of troublesome dyspepsia or indigestion, are more favoured by an animal than by a vegetable diet. As to strength of body [or of mind], when the vegetarian diet is conducted on a sensible scale, and is supplemented judiciously by additions of milk, butter, cheese and eggs, I can have no doubt that the whole of the animal strength and power of work, *physical and mental,* belonging to any man or woman, can be got out of it."—SIR B. W. RICHARDSON, M.D., F.R.S., in *Foods for Man* (1891).

"Chemistry is no more antagonistic to vegetarianism than is biology. Flesh food is certainly not necessary to supply the nitrogenous products required for the repairs of tissue. Therefore a *well-selected* diet from the vegetable kingdom is perfectly fitted, from a chemical point of view, for the nutrition of man. Further than this, I will add that I entirely disapprove of the modern rage for concentrated essences of flesh-meat. The habitual use of such things as *Bovril* and *Bouillon* (often made worse by the addition of brandy and gin) is, to my mind, so terrible an evil, that I should feel it to be a dereliction of my duty, as a public officer of health, if I did not take every opportunity of condemning the practice."—F. J. SYKES, M.D., B.Sc.

"*One-fifth* of the total amount of [flesh] meat consumed is derived from animals killed in a state of malignant or chronic disease."—PROFESSOR GAMGEE (*Fifth Report to the Privy Council*).

"We have obtained ample evidence that food derived from tuberculous animals can produce *tuberculosis* in healthy animals. In the absence of direct experiments on human subjects, we infer that man, also, can acquire *tuberculosis* by feeding upon materials derived from tuberculous food-animals [*sic.* viz., the victims of butchery]. The actual amount of tuberculous disease, among certain classes of food-animals, is so large as to afford to man frequent occasions for contracting tuberculous disease through his food."—*Report of the Royal Commission on Tuberculosis*, May, 1895.

INDEX

Howard Williams (1837–1931) was a humanitarian
and vegetarian whose book *The Ethics of Diet*
is considered a foundational document in the history
of vegetarianism.

Carol J. Adams is an independent scholar and
the author of a number of books on topics ranging
from vegetarianism to animal rights to violence
against women. Her path-breaking work
*The Sexual Politics of Meat: A Feminist-Vegetarian
Critical Theory* was published in a
tenth-anniversary edition in 2000.

The University of Illinois Press
is a founding member of the
Association of American University Presses.

University of Illinois Press
1325 South Oak Street
Champaign, IL 61820-6903
www.press.uillinois.edu